Numerical Methods for Bifurcations of Dynamical Equilibria

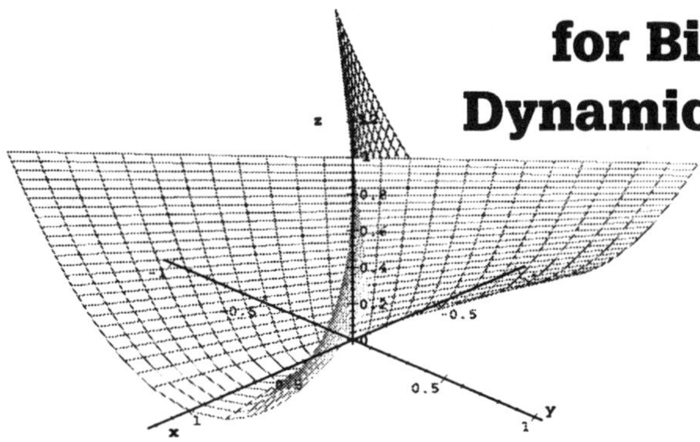

Numerical Methods for Bifurcations of Dynamical Equilibria

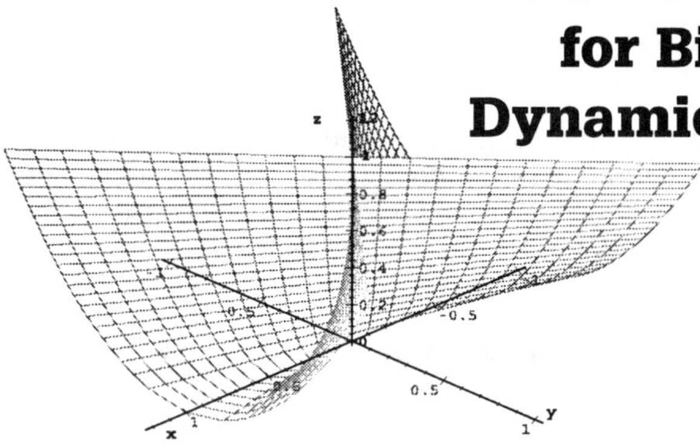

Willy J. F. Govaerts
University of Gent
Gent, Belgium

Society for Industrial and Applied Mathematics
Philadelphia

10 9 8 7 6 5 4 3 2 1

Library of Congress Cataloging-in-Publication Data

Govaerts, Willy J.F.
Numerical methods for bifurcations of dynamical equilibria / Willy J. F. Govaerts.
p. cm.
Includes bibliographical references and index.
ISBN 0-89871-442-7 (pbk.)
1. Differentiable dynamical systems. 2. Differential equations—Numerical solutions. 3. Bifurcation theory. I. Title.

QA614.8.G68 2000
515'.352 21—dc21

99-044796

To my wife Nicole, with love.

Contents

Preface

This book describes numerical methods for the detection, computation, and continuation (following paths) of equilibria and bifurcation points of equilibria of dynamical systems. In the enormous field of differential equations this is a modest subfield with the particular attraction of having many links with other research fields, in particular the geometric theory of differential equations, numerical analysis, and linear algebra. The problems that are studied usually arise in other sciences (physics, chemistry, biology, engineering, economy) and so have an interest of their own.

On the other hand, the numerical problems in dynamical systems theory have often influenced developments in numerical linear algebra and in numerical analysis; cf. the books [75], [76], [210], [182], [228].

Another link is with software development. The only realistic way to make numerical methods widely available is to include them in software. Much work in this direction has been done, and implementations of many of the algorithms discussed in this book are freely available. A Web site with information on software for dynamical systems is maintained by H. Osinga at http://www.cds.caltech.edu/~hinke/dss/.

This book is meant for those who want to apply numerical methods to bifurcation and dynamical systems problems. It is assumed that the reader is familiar with the basic techniques in analysis, numerical analysis, and linear algebra as they are usually taught in undergraduate courses in science and engineering; some reference to standard textbooks on these topics may be necessary. People with little understanding of dynamical systems theory should be able to read the book, starting with the Introduction and Chapter 1. Nevertheless, it is unlikely that they can fully appreciate what is being done and why it is interesting. Fortunately, there are many good books on dynamical systems, e.g., [55], [127], [240], [193], [164].

The book as a whole is not written as a textbook for teaching. However, it is structured in such a way that certain topics can be isolated relatively easily and studied separately or used for a course. For example, Chapter 2 might be used for a short course on numerical continuation (it contains a large number of exercises). Chapters 3 and 4 could be used in a somewhat advanced course on numerical linear algebra; they contain results that are not usually found in standard texts. Chapters 6 and 7 (singularity theory) might be used for an advanced numerical analysis course (provided that one skips the more theoretical results).

Many people have influenced this book and contributed to it in various ways. Vladimir Janovsky (Prague, Czech Republic) suggested writing it and provided many ideas, both on the structure of the book and on details. In particular, the approach for dealing with Bogdanov–Takens points in §5.1 is his.

I learned a lot from Yuri A. Kuznetsov (Pushchino, Russia, and Utrecht, Netherlands). His hand is particularly visible in Chapters 9 and 10. Many of the pictures were drawn using the software package CONTENT, which is being developed by him and V. V. Levitin, originally at CWI, Amsterdam. I thank them for making their software available and for the excellent collaboration and many illuminating discussions.

In particular Yuri encouraged Bart Sijnave (Gent, Belgium) to implement further methods for equilibrium bifurcations and other numerical algorithms in CONTENT. Working on this was very helpful by forcing me to formulate the numerical methods in a precise way. I thank Yuri and Bart for their patience in dealing with my suggestions. Their contribution to this book is fundamental.

Many people contributed by making suggestions, correcting errors, or asking questions. I mention some of them, taking the risk of forgetting others. John Pryce (RMCS, Shrivenham, UK) taught me numerical analysis and the importance of software. The example on combustion in §1.3 and Exercise 3.8.5 are his suggestions. Alastair Spence (Bath, UK) introduced me to numerical work on equilibrium bifurcations; the introduction to symmetry-breaking bifurcation in §1.4 is his suggestion. Klaus Boehmer (Marburg, Germany) provided a working environment during several visits to Marburg and we had many useful discussions (usually with Vladimir Janovsky) on partial differential equations, symmetry breaking, singularities, and linear algebra. John Guckenheimer (Cornell University, Ithaca, NY, USA) gave me a taste for dynamical systems, the importance of neural models, and the difficulties in writing interactive software.

I further thank Wolf-Juergen Beyn (Bielefeld, Germany), Eusebius Doedel (Concordia University, Montreal, Canada), Alexander Khibnik (United Technologies, East Hartford, CT, USA) and Bodo Werner (Hamburg, Germany) for usually critical but always helpful discussions. Andre Vanderbauwhede (Gent, Belgium), Dirk Roose (Leuven, Belgium), and Kurt Lust (Cornell University) were equally helpful in several ways. In particular, Kurt Lust suggested a major improvement of the proof of Proposition 4.4.24 and several other corrections.

I further thank my employer, the Fund for Scientific Research FWO–Vlaanderen, for additional support through grant S 2/5–AV.E.3. The Department of Applied Mathematics and Computer Science of the University of Gent was a stimulating environment. Gilbert Crombez suggested many detailed corrections. During coffee breaks, Marnix Van Daele suggested Exercise 3.8.2, Hans De Meyer brought the corruption model in §9.4.3 to my attention, and Joris Van der Jeugt provided the reference to the proof of Proposition 4.4.23.

Working with the publications staff at SIAM was a real pleasure. I am grateful for their fast and efficient handling of my manuscript and their care in the production of the book.

Notation

TP	Turning point
LP	Limit point
H	Hopf point
N	Neutral saddle point
BP	Branching point
BT	Bogdanov–Takens point
ZH	Zero Hopf point
DH	Double Hopf point
CP	Cusp point
GH	Generalized Hopf point
HN	Hopf neutral saddle point
DN	Double neutral saddle point
RT	Rectangular eigenvalues point
BTH	Bogdanov–Takens–Hopf point
BTN	Bogdanov–Takens neutral saddle point
RDH	Resonant double Hopf point
RDN	Resonant double neutral saddle point
ST	Swallowtail bifurcation
ZA	Triple equilibrium bifurcation
ZB	Double equilibrium bifurcation

Introduction

For the purposes of this book a dynamical system is a system of ordinary differential equations (ODEs) of the form

$$\frac{dx}{dt} \equiv \dot{x} = G(x, \alpha), \tag{0.1}$$

where $x \in \mathbb{R}^N, \alpha \in \mathbb{R}^m, G(x, \alpha) \in \mathbb{R}^N$; x is called the state variable and G is a (usually nonlinear) function of the state variable and the parameter α. The space in which x lives is called phase space. For each fixed value of α, $G(x, \alpha)$ is called a vector field in the phase space.

The number N may be small; even the case $N = 2$ can lead to difficult mathematical and numerical problems. On the other hand, a lot of current interest is in the case where (0.1) represents a discretized partial differential equation (PDE) so that N can be very large as well.

In classical texts on ODEs (e.g., [62]) the stress is on individual solution curves of (0.1) and their properties, i.e., on the behavior of the solutions to (0.1) for fixed values of α. The fundamental local existence and uniqueness theorem (e.g., [62], [136]) tells us that if an initial value condition $x(0) = x_0$ is given and G is a continuously differentiable ($\mathcal{C}^1$-)function, then system (0.1) has a unique solution in a neigborhood U of $t = 0$. (In fact, a Lipschitz condition instead of continuous differentiability is sufficient.) Such solution is called an orbit or trajectory. Thus the vector field G defines a function $\phi(x, t, \alpha)$ with (x, t) in a neighborhood of $(x_0, 0)$ such that

$$\frac{\partial}{\partial t}(\phi(x, t, \alpha)) = G(\phi(x, t, \alpha), \alpha), \phi(x_0, 0, \alpha) = x_0. \tag{0.2}$$

If we define ϕ_t by $\phi_t(x) = \phi(x, t, \alpha)$, then ϕ_t is a transformation of (part of) the phase space and we have the semigroup property $\phi_{t+s} = \phi_t \phi_s$. The family of transformations ϕ_t is called the flow of the dynamical system, and a sketch of it in phase space is called a phase portrait.

The focus of dynamical systems theory is the behavior of phase portraits under changes in the parameter α in (0.1). Values of α for which this behavior changes qualitatively are called bifurcation values. This is made precise by introducing the notion of topological equivalence of vector fields; the requirement is that there exists a homeomorphism that preserves orbits and the orientation of the flows (but not necessarily their parameterization by time). A vector field is called structurally stable if all sufficiently small $\mathcal{C}^1$-perturbations of it are topologically equivalent to the unperturbed vector field;

parameter values of α for which (0.1) is not structurally stable are called bifurcation values.

The simplest solutions of (0.1) are the equilibrium solutions; i.e., those for which x is constant and

$$G(x, \alpha) = 0. \tag{0.3}$$

In the neighborhood of an equilibrium solution a vector field is structurally stable if and only if it is hyperbolic, i.e., the Jacobian matrix G_x has no eigenvalues on the imaginary axis.

Periodic orbits are another type of solution to (0.1). They can be either stable or unstable depending on the values of the Floquet multipliers, i.e., the eigenvalues of the linearization of the return map, the so-called monodromy matrix.

Another well-studied solution type are the homoclinic and heteroclinic connections, i.e., solutions which both for $t \to \infty$ and $t \to -\infty$ tend to finite equilibrium points x_∞ and $x_{-\infty}$; they are called homoclinic if $x_\infty = x_{-\infty}$ and heteroclinic otherwise.

Periodic orbits can further bifurcate into invariant tori, i.e., homeomorphic images of the ordinary torus $\mathbb{R}^3$ in the phase space that are invariant under the flow of (0.1). Chaotic behavior and strange attractors are also quite common, even in apparently simple examples.

There are several well-known routes to increasingly complex behavior. For example, for fixed values of α the system (0.1) generically has equilibrium solutions. If one component of α is freed, one may follow a curve of equilibria. Such a path generically contains Hopf bifurcation points, i.e., points where the Jacobian matrix of the vector field has a pair of pure imaginary eigenvalues and (under some nondegeneracy conditions) a curve of periodic orbits branches off. If the path of periodic orbits is followed, one may expect further bifurcations. One possibility is period doubling, another is torus bifurcation. It may also happen that the period itself tends to infinity, i.e., the periodic orbit degenerates into a homoclinic orbit.

The numerical study of dynamical systems of course involves the computation of orbits. This is a classical problem for small systems and normally does not present difficulties on small time scale. However, the long-term behavior of the computed orbit and its relation to the true orbit is still an active area of research, in particular in the case of strange attractors. The computation of orbits in the case of discretized PDEs is also an active area of research. The computation of periodic orbits and their stability and the computation of homoclinic and heteroclinic connections are now well understood for small systems; cf. [29], [30], [47], [81], [94], [95].

If a system has several parameters, one would like to get a global understanding of the behavior. Ideally, one might subdivide the parameter space into a number of regions so that in each region the system (0.1) has a qualitatively known behavior. Except in some simple cases this is a hopeless task. However, there is at least something that one can do in this direction. It is based on the idea of an organizing center. Let us explain this in some detail, since it is fundamental to the approach that we follow.

The basic idea is as follows. Consider an equilibrium bifurcation found in the system (0.1) for a particular value of α. If we classify the behavior of all its possible small perturbations (this turns out to be possible in many nontrivial cases), then we have some a priori information about what we can expect to find in the particular natural perturbation that is found by varying the parameter α in (0.1).

It turns out that some equilibrium bifurcation points are more complex than others in the sense that there is a richer variety of perturbed phase portraits. In fact, there is a hierarchy of complexities in equilibrium bifurcation points, and this hierarchy can to some extent be classified independently of the particular vector field. In the most common cases the classification is determined by the eigenvalues of the Jacobian matrix of the vector field that lie on the imaginary axis and by the coefficients of the Taylor expansion of the vector field in the equilibrium point (we always assume as much smoothness of G as we need), i.e., by local information in the equilibrium bifurcation point itself.

Typically the type of an equilibrium bifurcation point is characterized by a codimension and a normal form. The codimension is a nonnegative integer that tells us how many parameters we need for the generic occurrence of the type. The normal form is a representation of the bifurcation in its simplest possible form. For a complete understanding we also need a universal unfolding of the bifurcation, i.e., a perturbed form of the normal form with as many parameters as the codimension indicates and such that all possible other perturbations factor through this particular universal perturbation. This is hard analytical work that has been done for most common types of equilibrium bifurcations.

In a problem of the form (0.1) it is of interest to know what equilibrium bifurcation points it contains; this gives information about the dynamic behavior of the system for a set of parameter values with nonzero Lebesgue measure.

We note that periodic orbits, homoclinic connections, two-dimensional invariant tori, Shil'nikov homoclinic connections, and chaotic behavior all appear in generic unfoldings of equilibrium bifurcation points with codimension less than or equal to 2. In fact, this is often the only method to prove the occurrence of such phenomena in a given model. So the study of equilibrium bifurcations has consequences far beyond the geometry of the solution set to (0.3). In Chapter 9 we give two examples in the case of the generalized Hopf (GH) bifurcation.

The recognition problem for equilibrium bifurcations is the problem of giving necessary and sufficient conditions for an equilibrium bifurcation to belong to a certain class. For the numerical implementation we want them to be expressed in terms of the values of the Taylor coefficients of G at the bifurcation points and quantities derivable from these (e.g., the singular vector of a singular Jacobian matrix).

We will distinguish between linearly and nonlinearly determined bifurcations. The former are those that are determined by the Jacobian matrix G_x, although some inequality conditions may still be imposed on the higher-order derivatives of G. The latter are those for which equality relations are imposed on the higher derivatives of G. This distinction is not too relevant for the behavior of the dynamical system but makes numerical sense. The methods for linearly determined bifurcations use pure linear algebra techniques and can at least for low codimension be organized in a systematic way. The nonlinearly determined bifurcations can use the analytic classifications after a reduction by (a numerical form of) the Lyapunov–Schmidt or center manifold reductions. If only the geometric properties of the equilibrium surface are involved, the Lyapunov–Schmidt reduction leads to singularity theory. To study all dynamic bifurcations a center manifold reduction needs to be done; however, it is usually possible to translate the obtained classification results back into the unreduced problem; cf. [167]. All this is reflected in the way the book is organized.

Chapter 1 is introductory and mainly intended for people with little background in dynamical systems or numerical analysis. It provides some standard examples from population dynamics and combustion theory, which allows us to discuss limit points, Hopf bifurcation, stability, symmetry breaking, and the need for numerical continuation techniques.

Chapter 2 gives an elementary introduction to manifold theory (embedded in finite-dimensional Euclidean spaces) and to branches and limit points. It then discusses numerical continuation methods in more detail, since this is fundamental to most algorithms in the book. The chapter contains a large number of numerical exercises dealing with the use of continuation methods to study nonlinear systems of equations.

In Chapter 3 we study bordered matrices. The intended main application is to obtain numerically computable local defining systems for manifolds of matrices with a prescribed rank defect or Jordan type. Furthermore, backward stable numerical methods are discussed to solve linear systems using black box solvers for a nearly singular matrix whose bordered extension is well conditioned. From a more theoretical side (but important for further numerical work) we illustrate the flexibility of the approach by discussing invariant subspaces of parameter-dependent matrices.

In Chapter 4 we discuss several methods with which to compute the two codimension-1 equilibrium bifurcations, i.e., limit points and Hopf points. In the case of limit points the classical Moore–Spence system uses the tangent vector as an additional unknown. This system is often used in software and we prove its regularity in the case of a quadratic turning point. An alternative with direct bordering of the Jacobian matrix is also given. As an application we compute turning points in a one-dimensional Brusselator model, i.e., a discretized PDE. In the Hopf case several methods are discussed with $3n+2, 2n+2$, and $n+2$, respectively, unknowns. The relation with Bogdanov–Takens points and zero-sum pairs of eigenvalues is established.

We further discuss Hopf bifurcations with the use of the bialternate product of matrices. This is a recently revived method that avoids the use of the imaginary part of the Hopf eigenvalue in the characterization of the Hopf point. We spend some effort to give a good introduction to the theory behind this matrix construction since this is not usually found in textbooks on linear algebra or numerical linear algebra.

In Chapter 5 linearly determined bifurcations are discussed. The chapter is organized around the three codimension-2 bifurcations of this type, namely, Bogdanov–Takens (BT), zero Hopf (ZH), and double Hopf (DH). In the first case, G_x has two zero eigenvalues (geometric multiplicity 1), in the second case a zero eigenvalue and a Hopf pair $\pm i\omega, \omega > 0$, in the third case two Hopf pairs $\pm i\omega_1, \pm i\omega_2$, $\omega_1 > 0, \omega_2 > 0$. The numerical methods use bordered matrices and are based on the properties of the corresponding manifolds of matrices. In particular, the link is established between the regularity of the defining systems, transversality properties of matrix manifolds, and genericity of the natural unfolding.

The methods are generalized to higher-order cases, including all codimension-3 cases. In particular, triple zero, zero DH, and triple Hopf can be handled in this way, as well as resonant DH ($\pm i\omega_1, \pm i\omega_2$ with $\omega_1 = \omega_2$). The examples include computations in the case of a realistic model of a neuron, developed at Cornell University, that is extremely rich in complicated dynamic behavior.

In Chapter 6 we discuss singularity theory with no distinguished bifurcation parameter, i.e., the (numerical) study of the local geometric structure of the solutions to (0.3) near limit points. The main computational tool in this study is the generalized Lyapunov–Schmidt reduction (cf. [146], [148], [149]). It allows us to reduce a problem of the form (0.3) to a problem in smaller dimensions $g(y,t) = 0$, where $y, g(y,t) \in \mathbb{R}^k$, k is the rank defect of G_x, $g_y = 0$ at the bifurcation point and t is the shift of α. We discuss this reduction and present a classification list of singularities up to codimension 4 based on the analysis in [104]. We note that all cases with rank defect 2 have at least codimension 4, so they are fairly exotic in the generic situation (although common in situations with symmetry).

In Chapter 7 singularities with a distinguished bifurcation parameter are considered; these are often called λ-singularities since λ is the traditional name for the distinguished parameter. In this setting it is not the dynamic behavior in phase space with constant parameters that is classified but rather the behavior of the equilibrium solutions in the product of phase space and a given (distinguished) unfolding parameter. This leads to a different classification somewhat outside mainstream dynamical systems theory but quite valuable in applications with a natural distinguished bifurcation parameter. This approach is implicitly present in classical numerical work as [150], [151], [68], [152] where a continuation parameter acts as a distinguished parameter; it became very popular when its analytic base was explained in great detail in [109] and [110].

We note, however, that in a numerical study with continuation methods there is not necessarily a relation between the distinguished bifurcation parameter of the classification and a continuation variable. For example, the notion of a pitchfork bifurcation is often understood in the context of a continuation variable, which is then implicitly identified with a distinguished parameter. But one might also follow a path of pitchfork bifurcations by freeing a parameter unrelated to the pitchfork singularity (we give numerical examples to illustrate this). Similarly, one can find cusp curves in the state-distinguished parameter space and follow them if sufficient parameters are available (we do this also in an example). On the other hand, a cusp curve also arises generically in parameter space in a two-parameter problem for the so-called cusp bifurcation, a codimension-2 equilibrium bifurcation with no distinguished bifurcation parameter.

The computational examples in Chapter 7 are based on the generalized Lyapunov–Schmidt reduction and explicitly given lists of defining conditions and nondegeneracy conditions for the λ-singularities; these were taken from [109] with some minor reformulations to make a direct implementation easier.

In Chapter 8 we deal with the case of equivariant dynamical systems. The numerical methods are first discussed in the simplest (but fundamental) case of $\mathbb{Z}_2$-symmetry. We emphasize the use of a symmetry-adapted bordering in the Lyapunov–Schmidt reduction for the detection, computation, and continuation of $\mathbb{Z}_2$-equivariant germs. This method allows us to use the full analytical classification of these germs in [109] routinely. We discuss both the cases of corank 1 (symmetry breaking) and corank 2 (mode interaction). To show the power of the methods we compute a corank-2 point on a curve of cusp points in the case of a discretized one-dimensional PDE, using the classification results in [65] to locate this point in the hierarchy of singularities.

The discussion of other symmetry groups is more sketchy since the real problems here are group theoretic rather than numerical (we give the appropriate references to the literature). However, we discuss the use of symmetry-adapted bases and of symmetry-adapted borderings for a given isotropy subgroup of a given group. This allows us to compute the branching phenomena in the fundamental cases covered by the equivariant branching lemma; the computation of other branching phenomena is still an active area of research.

In Chapter 9 we discuss nonlinearly defined bifurcations other than the singularities in Chapter 6. This includes in particular GH (i.e., vanishing of the first Lyapunov coefficient), which is a codimension-2 phenomenon, and double and triple equilibrium bifurcation points, which are codimension-3 phenomena that are generically found on curves of BT points. Our discussion is based on [167], where explicit normal form coefficients for all codimension-2 bifurcations of equilibria are given. We discuss numerically two examples where the presence of a GH point has a strong influence on the dynamic behavior of the system. The first is the classical Hodgkin–Huxley model for the giant axon of the squid, the second is a more recent model for corruption in democratic societies.

In Chapter 10 we make some remarks on the case of large dynamical systems. Such systems typically arise from discretizations of PDEs. A serious discussion of this topic would require a separate book, since many new issues arise, in particular, the relation between the continuous and the discretized problem and the efficient handling of the data. We restrict ourselves to a few topics. First, we discuss the basic issues (computation of orbits and continuation of steady states) in the case of one-dimensional PDEs. Second, we indicate how methods for small dynamical systems can in principle be extended to large ones, in particular the use of subspace reduction methods. For example, though the use of bialternate products of matrices (with dimension $N(N-1)/2$ by $N(N-1)/2$) seems unsuitable for large systems, the picture changes if one assumes that a subspace reduction has been done. Present research, including developments in numerical linear algebra, points in that direction. Third, we call the attention of the reader to research that we consider particularly promising.

Chapter 1

Examples and Motivation

In this chapter we introduce the basic ideas of dynamical systems theory, i.e., equilibria, stability, turning point bifurcations, Hopf bifurcations, periodic orbits, and symmetry-breaking branching. We describe these phenomena in standard simple examples in population dynamics and combustion. We further discuss the need for numerical methods—in particular, discretization and numerical continuation.

1.1 Nonlinear Equations and Dynamical Systems

We will deal with nonlinear equations, routinely denoted

$$G(x) = 0, \tag{1.1}$$

where $x, G(x) \in R^N$. The first derivative or Jacobian matrix G_x is the $N \times N$ matrix whose (i,j)th entry is the partial derivative

$$\frac{\partial G_i(x)}{\partial x_j},$$

where G_i is the ith component of G and x_j is the jth component of x.

The second derivative of G is the bilinear mapping $G_{xx} : \mathbb{R}^N \times \mathbb{R}^N \to \mathbb{R}^N$ for which

$$(G_{xx}[p,q])_i = \sum_{j,k=1}^{N} \frac{\partial^2 G_i}{\partial x_j \partial x_k} p_j q_k,$$

where $p, q \in \mathbb{R}^N$. For simplicity we will usually write $G_{xx}pq$ instead of $G_{xx}[p,q]$.

Third- and higher-order derivatives of G are defined similarly, and we use notation like $G_{xxx}pqr$ in the same way.

Here N may be small; even $N = 1$ leads to a nontrivial theory. But N can also be huge if (1.1) represents a discretized boundary value problem or integral equation. When (1.1) contains parameters, we write it as

$$G(x, \alpha) = 0 \tag{1.2}$$

with $\alpha \in \mathbb{R}^m$. In many examples m is small, although certain realistic models require 20 or more parameters. Then typically only a few parameters are free in any particular application.

The solution x to (1.2) in general depends on α. If G_x is nonsingular at a solution point, then this dependence is smooth by the implicit function theorem. If it is singular, then α is called a bifurcation value of (1.2). Then the local behavior of the solutions to (1.2) may be quite simple or quite complicated. A simple example is the case $G(x, \alpha) \equiv x^2 - \alpha$ ($N = m = 1$) at the point $(x, \alpha) = (0, 0)$. The study of the general situation is the subject of singularity theory. We deal with this in Chapters 6 and 7.

The classical eigenvalue problem

$$Au - \lambda u = 0 \tag{1.3}$$

with $A \in \mathbb{R}^{n \times n}$, $u \in \mathbb{R}^n$, $\lambda \in \mathbb{R}$ is a particular case of (1.2). For each value of λ, $u = 0$ is a solution to (1.3). The bifurcation values are the eigenvalues of A; for them (1.3) admits more than one solution.

Sometimes it is useful to introduce parameters artificially for numerical reasons. If the solution to (1.1) is hard to find with standard methods, then one may embed (1.1) in a parameterized problem; i.e., one chooses a function $G(x, \beta)$, $\beta \in \mathbb{R}$, such that

$$\begin{aligned} G(x, 1) &\equiv G(x), \\ G(x, 0) &\equiv G_1(x) \end{aligned}$$

with $G_1(x)$ a function such that one can solve the equation

$$G_1(x) = 0.$$

In some cases it is then possible to continue numerically the solution to

$$G(x, \beta) = 0$$

from $\beta = 0$ to $\beta = 1$. Such methods are called homotopy methods (see Chapter 2).

We will usually deal with parameterized dynamical systems, i.e., ordinary differential equations (ODEs) routinely denoted as

$$\frac{dx}{dt} = G(x, \alpha) \tag{1.4}$$

with initial value

$$x(0) = x_0. \tag{1.5}$$

A solution curve to (1.4)–(1.5) for a given α is called an *orbit* or *trajectory*. If there exists a real number $T > 0$ such that $x(T) = x(0)$, then the orbit is periodic with period T. To study (1.4)–(1.5) one may start with the simplest solutions, i.e., the constant solutions $x(t) \equiv x_0$. They are called steady states or equilibrium solutions. For these (1.2) must hold. From the study of (1.2) it is further possible to obtain information on the behavior of the solutions to (1.4) in a neighborhood of the equilibrium solutions. This is an important object of study in dynamical systems theory. We will first consider a few examples.

1.2 Examples from Population Dynamics

Mathematical biology is a rich source of problems and examples in nonlinear dynamics. We will discuss several of them. The reader with a deeper interest in the field is recommended to consult [190]. In this section we restrict ourselves to some examples from population dynamics. These are popular because they can be formulated easily, seem to have some biological relevance (at least in a qualitative sense), and present nice features that can be worked out in detail. Recently, they have been used as building blocks in more complicated systems that combine fast (environmental) and slow (genetical) dynamics [158]. Our examples are based on [153].

1.2.1 Stable and Unstable Equilibria

Let $x(t)$ be the population density of a living organism at time t. Since its reproduction is presumably proportional to the already existing density the simplest model for the time evolution of $x(t)$ is

$$\frac{dx}{dt} = G(x,\beta) \equiv \beta x \qquad (\beta > 0), \qquad x(t_0) = x_0, \tag{1.6}$$

where β is the (constant) reproduction rate. The solution is

$$x(t) = x_0 e^{\beta(t-t_0)}. \tag{1.7}$$

So $x(t)$ tends to ∞ for $t \to \infty$.

To obtain a somewhat more plausible model, one might assume that the reproduction rate decreases if the population density approaches a critical threshold. In a simple form this means replacing in (1.6) the constant β by a function

$$\beta = \beta_1\left(1 - \frac{x}{x_1}\right) \qquad (\beta_1 > 0, x_1 > 0). \tag{1.8}$$

Now β_1 is the highest possible reproduction rate. If $x < x_1$, then $\beta > 0$ and the population density increases; if $x > x_1$ then $\beta < 0$ and the density decreases.

The solution to

$$\frac{dx}{dt} = G(x,\beta_1,x_1) \equiv \beta_1\left(1 - \frac{x}{x_1}\right)x, \qquad x(t_0) = x_0 \tag{1.9}$$

is given by

$$x(t) = \frac{x_0 x_1 e^{\beta_1(t-t_0)}}{(x_1 - x_0) + x_0 e^{\beta_1(t-t_0)}}. \tag{1.10}$$

A solution to (1.9) is given in Figure 1.1. If we omit the initial value condition, then (1.9) admits two equilibrium solutions, namely, $x \equiv 0$ and $x \equiv x_1$. In this case $G_x(0) = \beta_1 > 0$ and $G_x(x_1) = -\beta_1 < 0$. Clearly $x(t)$, as defined in (1.10), converges to x_1 if x_0 is sufficiently close to it. On the other hand, for no initial value $x_0 > 0$ it converges to 0. The equilibrium $x \equiv 0$ is therefore unstable in an intuitive sense while $x \equiv x_1$ is stable. We discuss this in a more general setting in §1.5.

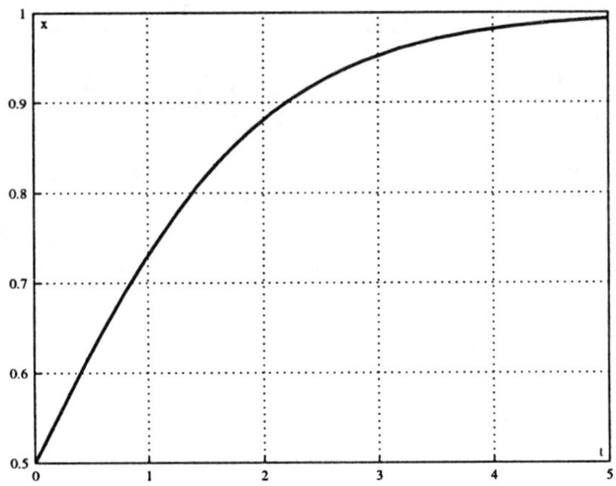

Figure 1.1: Solution to (1.9) for $x_0 = 0.5, x_1 = 1.0, \beta_1 = 1.0, t_0 = 0$.

1.2.2 A Set of Bifurcation Points

A more general model allows immigration and emigration. If the population density in an adjacent region is u_e, then a possible model is

$$\frac{dx}{dt} = \beta_1 \left(1 - \frac{x}{x_1}\right) x + \alpha_0 (x_e - x) \tag{1.11}$$

with α_0 a new parameter. For $\alpha_0(x_e - x) > 0$ immigration dominates; for $\alpha_0(x_e - x) < 0$ emigration dominates.

We can simplify (1.11) by introducing

$$\alpha_1 = \frac{\beta_1 - \alpha_0}{\beta_1} x_1, \tag{1.12}$$

$$\alpha_2 = \frac{\alpha_0}{\beta_1} x_1 x_e, \tag{1.13}$$

$$t = \tau \frac{x_1}{\beta_1}. \tag{1.14}$$

This leads to

$$\frac{dx}{d\tau} = G(x, \alpha) \equiv -x^2 + \alpha_1 x + \alpha_2, \tag{1.15}$$

where $\alpha = (\alpha_1, \alpha_2)$ is the new parametervector.

The equilibrium solutions of (1.15) satisfy

$$G(x, \alpha) \equiv -x^2 + \alpha_1 x + \alpha_2 = 0 \tag{1.16}$$

and are given by

$$x_\pm = \frac{\alpha_1}{2} \pm \sqrt{\left(\frac{\alpha_1}{2}\right)^2 + \alpha_2}.$$

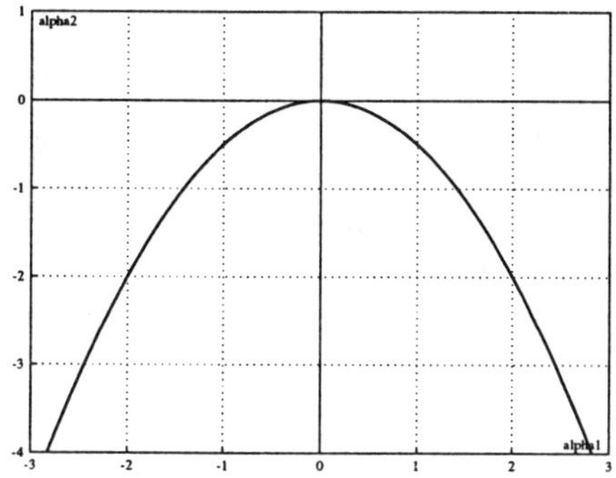

Figure 1.2: Bifurcation set of (1.16).

The equilibria of (1.15) coincide if and only if

$$\alpha_2 + \left(\frac{\alpha_1}{2}\right)^2 = 0. \tag{1.17}$$

The curve in the parameter plane determined by (1.17) is called the bifurcation set. See Figure 1.2. To get a picture of the surface determined by (1.16) we consider several sections.

1. $\alpha_2 = 0$. Then (1.16) reduces to

$$G(x, \alpha) \equiv -x^2 + \alpha_1 x \equiv x(-x + \alpha_1) = 0.$$

   In the (α_1, x) plane the solution consists of the lines $x = 0$ and $x = \alpha_1$ that intersect in $(0, 0)$.

2. $\alpha_2 \neq 0$. In the (α_1, x) plane the equation

$$G(x, \alpha) \equiv x(-x + \alpha_1) + \alpha_2 = 0$$

   with α_2 fixed represents a hyperbola with asymptotes $x = 0$ and $x = \alpha_1$. (See Figure 1.3.)

In the parameter region where two different equilibria exist $((\frac{\alpha_1}{2})^2 + \alpha_2 > 0)$ we can rewrite (1.15) as

$$\frac{dx}{d\tau} = -(x - x_+)(x - x_-), \tag{1.18}$$

where x_-, x_+ denote the smallest and largest root of (1.16), respectively. Then x is decreasing for $x < x_-$ and for $x > x_+$; it is increasing for $x_- < x < x_+$. Hence x_+ is a stable equilibrium and x_- an unstable equilibrium in the same (informal) sense as in §1.2.1.

In Figure 1.4 we show the stability diagrams in the (α_1, x) plane for $\alpha_2 = -0.5$. We note that there are two disconnected branches and each branch has a turning point where

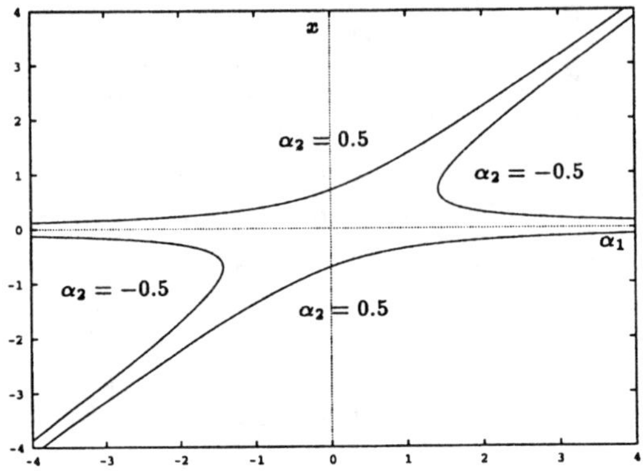

Figure 1.3: Sections of (1.16) for $\alpha_2 = .5$ and $\alpha_2 = -.5$.

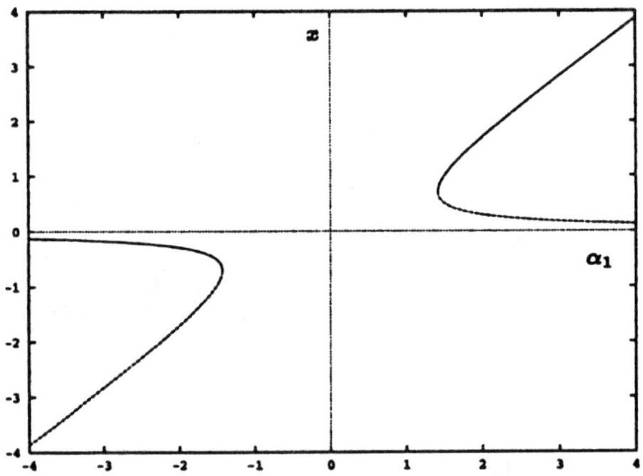

Figure 1.4: Stability of the equilibria of (1.15) for $\alpha_2 = -0.5$.

the stability changes. The solid (upper) lines correspond with stable equilibria and the dashed (lower) lines correspond with unstable equilibria.

We remark that in the bifurcation diagram for $\alpha_2 = 0$ also an exchange of stability occurs in the point $(0, 0)$. For $\alpha_2 > 0$ there are again two disconnected branches (see Figure 1.3); the upper one has only stable equilibria and the lower one has only unstable equilibria (not dashed in Figure 1.3).

The Jacobian of $G(x, \alpha_1, \alpha_2)$ is $[G_x, G_{\alpha_1}, G_{\alpha_2}] \equiv [-2x + \alpha_1 \quad x \quad 1]$ and so has always full rank 1. The solution surface therefore is a smooth manifold with dimension $3 - 1 = 2$ (manifolds will be discussed in Chapter 2) without singular points. For each fixed α_2 the Jacobian $[G_x, G_{\alpha_1}] = [-2x + \alpha_1 \quad x]$ is rank deficient in $(x = 0, \alpha_1 = 0)$. This point satisfies $G(x, \alpha) = 0$ only if $\alpha_2 = 0$.

1.2.3 A Cusp Catastrophe

Another choice for the reproduction rate is

$$\beta = \beta_1 \left[1 - \left(\frac{x}{x_1} \right)^2 \right]. \tag{1.19}$$

The corresponding equation for the population density is

$$\frac{dx}{dt} = \beta_1 \left[1 - \left(\frac{x}{x_1} \right)^2 \right] x + \alpha_0 (x_e - x). \tag{1.20}$$

If we set

$$\begin{cases} \alpha_1 &= \left(1 - \dfrac{\alpha_0}{\beta_1}\right) x_1^2, \\ \alpha_2 &= \dfrac{\alpha_0 x_e x_1^2}{\beta_1}, \\ \tau &= t \dfrac{\beta_1}{x_1^2}, \end{cases}$$

then (1.20) reduces to

$$\frac{dx}{d\tau} = -x^3 + \alpha_1 x + \alpha_2. \tag{1.21}$$

Now define

$$G(x, \alpha) \equiv -x^3 + \alpha_1 x + \alpha_2. \tag{1.22}$$

For every fixed α_1, $G(x, \alpha) = 0$ determines a cubic curve

$$\alpha_2 = x^3 - \alpha_1 x \tag{1.23}$$

in the (x, α_2) plane. For $\alpha_1 \leq 0$, α_2 is a nondecreasing function of x. For $\alpha_1 > 0$ (Figure 1.5) α_2 attains a minimum $-\frac{2}{3\sqrt{3}}\alpha_1^{3/2}$ in a point L where $x = \sqrt{\frac{\alpha_1}{3}}$ and a maximum $\frac{2}{3\sqrt{3}}\alpha_1^{3/2}$ in a point U where $x = -\sqrt{\frac{\alpha_1}{3}}$.

Furthermore, α_2 is increasing in the intervals $]-\infty, -\sqrt{\frac{\alpha_1}{3}}[$ and $]\sqrt{\frac{\alpha_1}{3}}, \infty[$; α_2 is decreasing in $]-\sqrt{\frac{\alpha_1}{3}}, +\sqrt{\frac{\alpha_1}{3}}[$.

Of course we would prefer to have x as a function of α_2. If the solution curve to (1.23) is parameterized by α_2, then we are in trouble at the points U and L because x is not a smooth function of α_2 near these points. Again, U and L are turning points of the curve.

The surface determined in (x, α_1, α_2)-space by (1.23) is represented in Figure 1.6. The points (U,L) of all curves $\alpha_1 = \textit{constant}$ together form a curve of turning points. If we project the solution surface of $G(x, \alpha) = 0$ orthogonally onto the parameter plane (α_1, α_2), then the projection of the turning point curve satisfies the equation obtained by eliminating x from

$$\begin{cases} G(x, \alpha) &\equiv -x^3 + \alpha_1 x + \alpha_2 = 0, \\ \dfrac{d\alpha_2}{dx} &\equiv 3x^2 - \alpha_1 = 0, \end{cases} \tag{1.24}$$

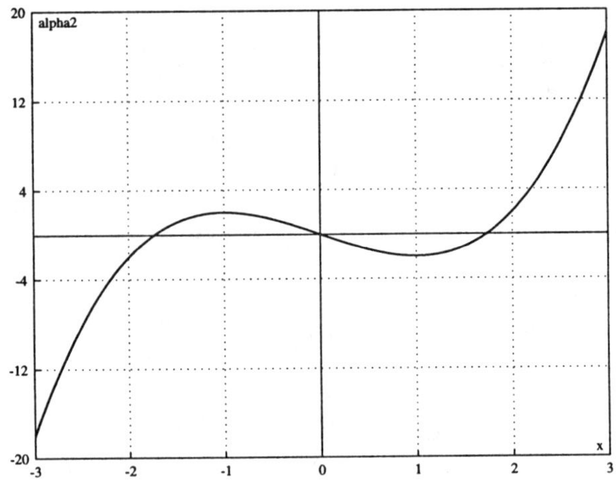

Figure 1.5: The function (1.23) for $\alpha_1 = 3$.

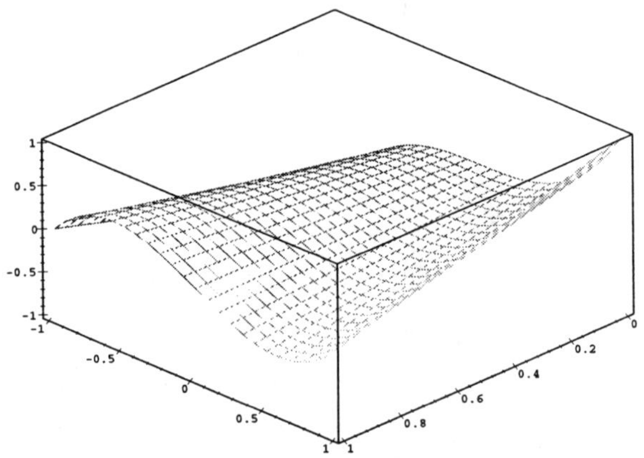

Figure 1.6: Representation of (1.23) for $0 \le \alpha_1 \le 1, -1 \le x \le 1$.

i.e.,

$$\alpha_2^2 = \frac{4}{27}\alpha_1^3. \tag{1.25}$$

This is called a *cusp curve.* We represent it in Figure 1.7. On the curve (1.25) the equation has one simple and one double solution except for the point $(0,0)$ where it has one triple solution. In the points above the curve the equation $G(x, \alpha_1, \alpha_2) = 0$ has one simple solution in x; below the curve it has three simple solutions.

Remarks.

1. The point $(0,0)$ of the above cusp curve is in the (α_1, α_2)-projection of the section $\alpha_1 = 0$. This section satisfies $G(x, 0, \alpha_2) \equiv -x^3 + \alpha_2 = 0$. This is called a *hysteresis curve.* See Figure 1.8.

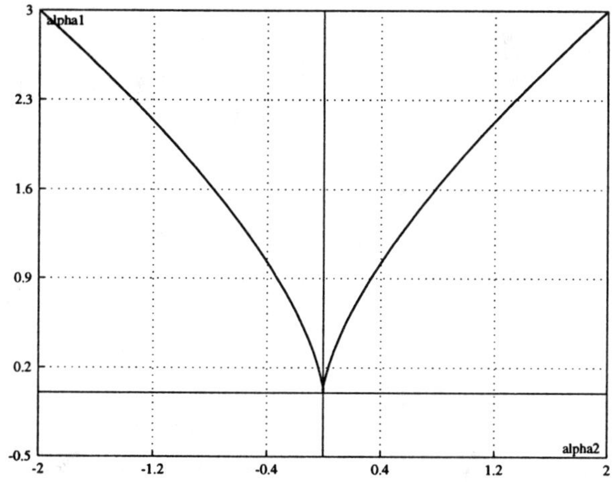

Figure 1.7: Representation of (1.25).

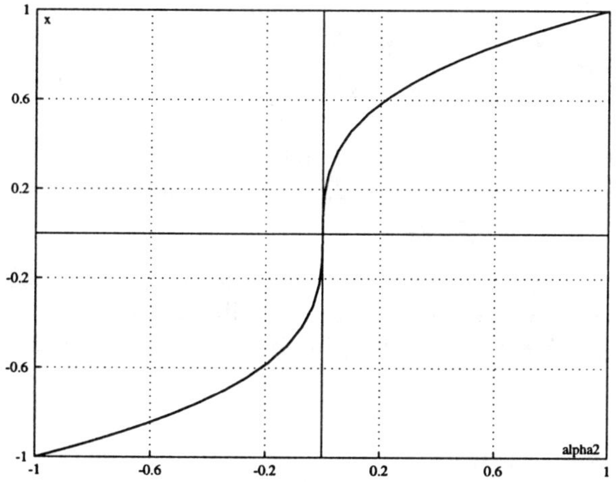

Figure 1.8: Section $\alpha_1 = 0$ of (1.23).

2. In the plane $\alpha_2 = 0$ the equation is

$$G(x, \alpha_1, 0) \equiv -x^3 + \alpha_1 x = 0.$$

The solution in (x, α_1)-space consists of two branches:

$$\begin{cases} x &= 0, \\ x^2 &= \alpha_1. \end{cases}$$

The point $(0, 0)$ is called a pitchfork bifurcation (Figure 1.9).

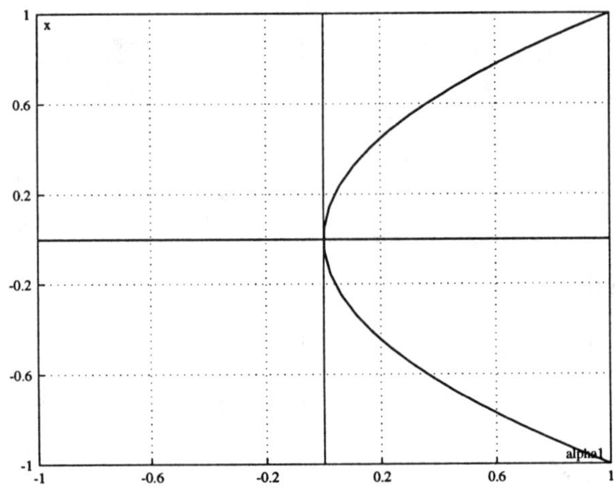

Figure 1.9: Pitchfork bifurcation in $(0,0)$.

3. The Jacobian $[G_x \quad G_{\alpha_1} \quad G_{\alpha_2}]$ of $G(x, \alpha_1, \alpha_2) \equiv -x^3 + \alpha_1 x + \alpha_2$ is

$$[-3x^2 + \alpha_1 \quad x \quad 1]$$

and has always full rank 1. Also for each fixed α_1 the remaining Jacobian has full rank. For fixed α_2 the remaining Jacobian is rank deficient in $(x = 0, \alpha_1 = 0)$. This corresponds to a point on the surface only if $\alpha_2 = 0$.

4. The system (1.24) is equivalent to the system $G(x, \alpha) = G_x(x, \alpha) = 0$. A point that satisfies these requirements is called a limit point. We note that limit points (unlike turning points) are defined in dynamical systems without reference to a parameter.

1.2.4 A Hopf Bifurcation

We now consider a model of a predator–prey interaction that allows a periodic time behavior of the two population densities. Intuitively this can be understood from different reproduction rates.

In the mathematical model let $\tilde{x}, \tilde{y}$ be the population densities of prey and predator, respectively. Consider the coupled system

$$\tilde{x}_t = \beta_x \left[1 - \left(\frac{\tilde{x}}{\tilde{x}_1}\right)^2 - \left(\frac{\tilde{y}}{\tilde{y}_1}\right)^2\right] \tilde{x} + \alpha_x(\tilde{x}_0 - \tilde{x}) - \tilde{\gamma}_x \tilde{y}, \tag{1.26}$$

$$\tilde{y}_t = \beta_y \left[1 - \left(\frac{\tilde{x}}{\tilde{x}_1}\right)^2 - \left(\frac{\tilde{y}}{\tilde{y}_1}\right)^2\right] \tilde{y} + \alpha_y(\tilde{y}_0 - \tilde{y}) + \tilde{\gamma}_y \tilde{x} \tag{1.27}$$

with

- β_x, β_y the maximal reproduction rates of prey and predator, respectively;

- $[1-(\frac{\tilde{x}}{\tilde{x}_1})^2-(\frac{\tilde{y}}{\tilde{y}_1})^2]$ a measure for the environmental pressure by the two species;
- $\tilde{x}_1, \tilde{y}_1$ the maximal densities of prey and predator that the environment can support;
- $\tilde{x}_0, \tilde{y}_0$ the (constant) densities in an adjacent region;
- α_x, α_y the immigration–emigration coefficients;
- $\tilde{\gamma}_x, \tilde{\gamma}_y$ parameters that describe the predator–prey interaction.

We now rescale and introduce other parameters by

$$x = \frac{\tilde{x}}{\tilde{x}_1}, \quad y = \frac{\tilde{y}}{\tilde{y}_1}, \quad x_0 = \frac{\tilde{x}_0}{\tilde{x}_1}, \quad y_0 = \frac{\tilde{y}_0}{\tilde{y}_1},$$

$$\gamma_x = \tilde{\gamma}_x \frac{\tilde{y}_1}{\tilde{x}_1}, \qquad \gamma_y = \tilde{\gamma}_y \frac{\tilde{x}_1}{\tilde{y}_1}.$$

Then (1.26), (1.27) are transformed into

$$x_t = \beta_x[1 - x^2 - y^2]x + \alpha_x(x_0 - x) - \gamma_x y, \tag{1.28}$$

$$y_t = \beta_y[1 - x^2 - y^2]y + \alpha_y(y_0 - y) + \gamma_y x. \tag{1.29}$$

We also introduce the following simplifying assumptions:

$$\beta_x = \beta_y = \gamma_x = \gamma_y = 1, \quad x_0 = y_0 = 0, \quad \lambda = 1 - \alpha_x = 1 - \alpha_y. \tag{1.30}$$

So we get

$$x_t = [1 - x^2 - y^2]x - (1 - \lambda)x - y, \tag{1.31}$$

$$y_t = [1 - x^2 - y^2]y - (1 - \lambda)y + x. \tag{1.32}$$

Or, in matrix notation,

$$\begin{bmatrix} x \\ y \end{bmatrix}_t = \begin{bmatrix} \lambda & -1 \\ 1 & \lambda \end{bmatrix} \begin{bmatrix} x \\ y \end{bmatrix} - (x^2 + y^2) \begin{bmatrix} x \\ y \end{bmatrix}. \tag{1.33}$$

The trivial solution $x \equiv y \equiv 0$ is possible for every λ. (We remark that $\lambda = 1$ indicates a closed region; $\lambda \leq 0$ indicates a region with a lot of interaction with adjacent regions.)

First consider a small perturbation $[\begin{smallmatrix} \delta x \\ \delta y \end{smallmatrix}]$ of the trivial solution. It approximately satisfies the linearized problem

$$\begin{bmatrix} \delta x \\ \delta y \end{bmatrix}_t = \begin{bmatrix} \lambda & -1 \\ 1 & \lambda \end{bmatrix} \begin{bmatrix} \delta x \\ \delta y \end{bmatrix}. \tag{1.34}$$

Let $A = [\begin{smallmatrix} \lambda & -1 \\ 1 & \lambda \end{smallmatrix}]$. The eigenvalue equation $Ax = \mu x$ is equivalent to $(\lambda - \mu)^2 + 1 = 0$, i.e., $\mu_\pm = \lambda \pm i$.

The eigenvalues are different; hence their eigenvectors are linearly independent. Therefore, A is diagonalizable; i.e., there exists a nonsingular (complex) matrix S such that

$$A = S^{-1}DS \text{ with } D = \begin{bmatrix} \mu_+ & 0 \\ 0 & \mu_- \end{bmatrix}. \tag{1.35}$$

From (1.34) and (1.35) it follows that

$$\left(S\begin{bmatrix} \delta x \\ \delta y \end{bmatrix}\right)_t = D\left(S\begin{bmatrix} \delta x \\ \delta y \end{bmatrix}\right). \tag{1.36}$$

The solution to (1.36) in $S\begin{bmatrix} \delta x \\ \delta y \end{bmatrix}$ is

$$S\begin{bmatrix} \delta x \\ \delta y \end{bmatrix} = \begin{bmatrix} c_1 e^{\mu_+ t} \\ c_2 e^{\mu_- t} \end{bmatrix} \quad (c_1, c_2 \text{ constants}).$$

The solution to (1.34), therefore, has the form

$$\begin{cases} \delta x &= a_1 e^{\mu_+ t} + a_2 e^{\mu_- t}, \\ \delta y &= b_1 e^{\mu_+ t} + b_2 e^{\mu_- t} \end{cases}$$

with complex constants a_1, a_2, b_1, b_2. For $\lambda < 0$ all solutions converge to 0. Hence the trivial solution $x \equiv y \equiv 0$ is stable if $\lambda < 0$ (at least for the linearized equation). For $\lambda = 0$ (1.34) reduces to

$$\begin{bmatrix} \delta x \\ \delta y \end{bmatrix}_t = \begin{bmatrix} 0 & -1 \\ 1 & 0 \end{bmatrix}\begin{bmatrix} \delta x \\ \delta y \end{bmatrix} \tag{1.37}$$

$\Longrightarrow (\delta x)_{tt} = (-\delta y)_t = -\delta x$ of $(\delta x)_{tt} + (\delta x) = 0$,

$$\begin{aligned} \delta x &= A \sin t + B \cos t, \\ \delta y &= -(\delta x)_t = -A \cos t + B \sin t, \end{aligned}$$

i.e., a periodic function.

Hence there is a loss of stability if λ crosses the origin and at the same time periodic orbits appear. We remark that A is the Jacobian matrix $G_{x,y}$ of the system

$$G(x,y) \equiv A\begin{bmatrix} x \\ y \end{bmatrix} - (x^2+y^2)\begin{bmatrix} x \\ y \end{bmatrix} = 0, \tag{1.38}$$

which describes the equilibria of (1.33). If λ crosses the origin, then a conjugate pair of complex eigenvalues of the Jacobian matrix crosses the imaginary axis.

We consider again the nonlinearized equations (1.33). The factor $x^2 + y^2$ suggests the introduction of polar coordinates

$$\begin{cases} x = \rho \cos \theta, \\ y = \rho \sin \theta. \end{cases}$$

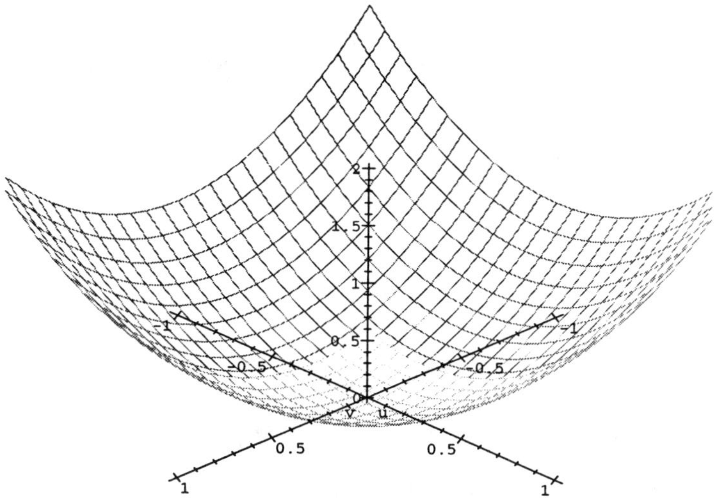

Figure 1.10: A paraboloid of periodic orbits.

Then (1.33) reduces to

$$\dot{\rho}\cos\theta - \dot{\theta}\rho\sin\theta = \lambda\rho\cos\theta - \rho\sin\theta - \rho^3\cos\theta, \tag{1.39}$$

$$\dot{\rho}\sin\theta + \dot{\theta}\rho\cos\theta = \rho\cos\theta + \lambda\rho\sin\theta - \rho^3\sin\theta. \tag{1.40}$$

By making suitable linear combinations we find

$$\dot{\rho} = \lambda\rho - \rho^3 = \rho(\lambda - \rho^2), \tag{1.41}$$

$$-\rho\dot{\theta} = -\rho. \tag{1.42}$$

Let (ρ_0, θ_0) describe an initial state. If $\rho_0 = 0$, then by (1.41) $\dot{\rho}_0 = 0$ and $\rho \equiv 0, \theta$ arbitrary is the general solution to (1.41), (1.42).

If $\rho_0 \neq 0$, then $\rho(t) \neq 0$ for all t since (1.41) would imply $\rho \equiv 0$.

Let us now assume $\rho_0 \neq 0$ so that, in fact, $\rho_0 > 0$ and $\rho(t) > 0$ for all t. From (1.42) it follows that

$$\theta(t) = \theta_0 + t \text{ for all } t.$$

If furthermore $\lambda > 0$ and $\rho(t) = \sqrt{\lambda}$ for a value of t, then by (1.41) $\rho \equiv \sqrt{\lambda}$. Hence the solution to (1.41)–(1.42) is a periodic circular movement.

Next assume that either $\lambda < 0$ or $0 < \lambda \neq \rho_0^2$. Then $\rho(\lambda - \rho^2) \neq 0$ for all t. From (1.41) it follows that

$$\frac{d\rho}{\rho(\lambda - \rho^2)} = dt \quad \text{or} \quad \frac{d\rho^2}{\rho^2(\lambda - \rho^2)} = 2\ dt.$$

After some manipulations with partial fractions we find

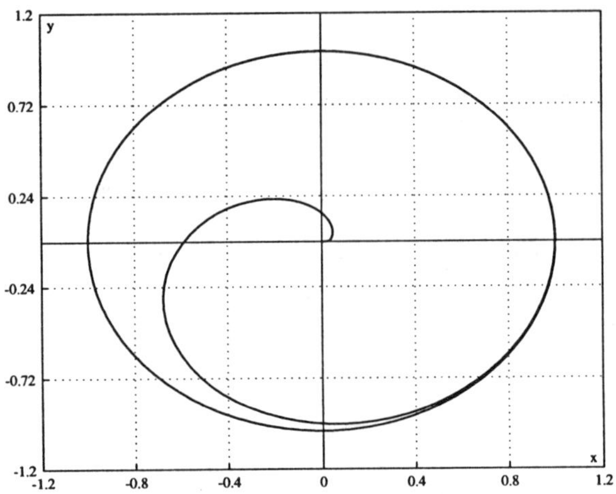

Figure 1.11: A representation of (1.44) in the x, y-plane for $C = 0.001$, $\lambda = 1$.

$$\left|\frac{\rho^2}{\lambda - \rho^2}\right| = Ce^{2\lambda t}, \tag{1.43}$$

where C is a positive constant, $C < 1$ if $\lambda < 0$. We have to distinguish two cases.

(i) $\lambda < 0$ or $0 < \lambda < \rho_0^2$. From (1.43) it follows that $\rho^2 = \lambda\frac{Ce^{2\lambda t}}{Ce^{2\lambda t}-1}$. Hence ρ converges to zero if $\lambda < 0$ and to $\sqrt{\lambda}$ if $\lambda > 0$.

(ii) $\rho_0^2 < \lambda$. From (1.43) it follows that

$$\rho^2 = \lambda\frac{Ce^{2\lambda t}}{Ce^{2\lambda t}+1}. \tag{1.44}$$

Hence ρ converges to $\sqrt{\lambda}$.

We remark that if $\lambda = 0$, then (1.41) reduces to $\dot{\rho} = -\rho^3$ with general solution $\rho(t) = \frac{1}{\sqrt{C+2t}}$. This function converges to zero for $t \to \infty$, hence, this case connects the cases $\lambda < 0$ and $\lambda > 0$.

We conclude that for $\lambda > 0$ the trivial state is unstable. For all starting values different from the trivial state the solution converges to a periodic movement with amplitude $\sqrt{\lambda}$; this orbit is therefore called a stable periodic orbit.

The stable periodic orbits form a paraboloid in (x, y, λ)-space. See Figure 1.10, where the λ-axis is vertical.

The above situation is a standard example of Hopf bifurcation where a branch of periodic solutions originates in a point on a branch of stationary solutions. The bifurcation point (here $(0, 0, 0)$) is called a Hopf point. In the present case the Hopf point separates stable and unstable equilibria; the stability of equilibria is taken over by the periodic solutions. For $\lambda < 0$, $\{0\}$ is called an attractor; for $\lambda > 0$ the periodic orbit is called an attractor. For this case an orbit that converges to the periodic orbit is presented in Figure 1.11.

We remark that for $\lambda = 0$ the behavior of the nonlinear system is qualitatively different from the linearized one. The linearized system has nontrivial periodic orbits; the nonlinear system has none.

1.3 An Example from Combustion Theory

Combustion problems lead to partial differential equations (PDEs) of the form

$$\frac{\partial u}{\partial t} = F(u, \alpha) \tag{1.45}$$

together with initial and boundary value conditions. Here u is a function of space and time and $F(u, \alpha)$ contains various space derivatives of u; it is based on the heat equation but can be a complicated function since it takes reaction effects into account. By space discretization (1.45) can in principle be reduced to a system of the form (1.4). The reduction is usually nontrivial, but we restrict it to the simplest setting.

Classical references to combustion theory are [41] and [242]. We consider a simple case where ignition and extinction can appear. The equilibrium equation ($F(u, \alpha) = 0$) consists of the one-dimensional boundary value problem

$$\frac{d^2u}{dx^2} + \lambda e^{\frac{u}{1+\mu u}} = 0 \tag{1.46}$$

with Dirichlet boundary conditions

$$u(0) = u(1) = 0. \tag{1.47}$$

This is a popular test example that describes an exothermic chemical reaction in an infinite slab; u denotes temperature scaled so that ambient temperature is 0. The function u depends on the one-dimensional state variable x. We refer to [93] for background and to [223], [246] for numerical studies. The two parameters λ, μ play a different role; λ is related to the thickness of the slab, while μ is related to properties of the reactants. Therefore, it makes sense to consider μ as fixed in a series of experiments and compare the influence of λ on the behavior of u.

For convenience we write (1.46) as

$$\frac{d^2u}{dx^2} + f(u, \lambda, \mu) = 0 \tag{1.48}$$

with Dirichlet conditions (1.47).

1.3.1 Finite Element Discretization

Consider the problem (1.47), (1.48) with

$$f(u, \lambda, \mu) \equiv \lambda e^{\frac{u}{1+\mu u}}. \tag{1.49}$$

We now describe a simple finite element method to reduce this problem to a finite-dimensional problem of the form in §1.1. The starting point is the observation that for every function $v(x)$ that is continuous in $[0,1]$, piecewise smooth, and zero at the endpoints $0, 1$, we have

$$\begin{aligned} 0 &= \int_0^1 (u'' + f(u,\lambda,\mu))v(x)dx \\ &= \int_0^1 v du' + \int_0^1 f(u,\lambda,\mu)v(x)dx \\ &= [vu']_0^1 - \int_0^1 u'v'dx + \int_0^1 f(u,\lambda,\mu)v(x)dx. \end{aligned}$$

Hence

$$0 = \int_0^1 (-u'v' + f(u,\lambda,\mu)v)dx. \tag{1.50}$$

Now we approximate u by functions in the finite-dimensional space generated by a set of functions called $\phi_0, \phi_1, \ldots, \phi_{n+1}$. In other words, we look for a function

$$u = \sum_{i=0}^{n+1} u_i\phi_i \tag{1.51}$$

and choose $u_i (0 \le i \le n+1)$ such that (1.50) holds for a class of functions v.

The base functions $\phi_0, \ldots, \phi_{n+1}$ are chosen by triangulating the integration domain of (1.50). In our one-dimensional setting we consider the points

$$x_0 = 0, \quad x_1 = h, \quad x_2 = 2h, \ldots, x_{n+1} = (n+1)h = 1$$

with $h = \frac{1}{n+1}$. We define ϕ_i by

$$\phi_i(x) = \begin{cases} \dfrac{x - x_{i-1}}{x_i - x_{i-1}} & \text{if} \quad x_{i-1} \le x \le x_i, \\ \dfrac{x_{i+1} - x}{x_{i+1} - x_i} & \text{if} \quad x_i \le x \le x_{i+1}, \\ 0 & \text{if} \quad x \notin [x_{i-1}, x_{i+1},], \end{cases} \tag{1.52}$$

where $x_{-1} = -h, x_{n+2} = 1 + h$ (see Figure 1.12).

If we require that $u = \sum_{i=0}^{n+1} u_i\phi_i$ satisfies the boundary conditions (1.47), then $u_0 = u_{n+1} = 1$. So we need n further conditions to determine $u_1, \ldots, u_n$.

In the Ritz–Galerkin method these are found by replacing v in (1.50) successively by $\phi_1, \ldots, \phi_n$. We obtain

$$-\sum_{i=0}^{n+1} u_i\langle\phi_i', \phi_j'\rangle + \int_0^1 f(u,\lambda,\mu)\phi_j\, dx = 0 \qquad (1 \le j \le n). \tag{1.53}$$

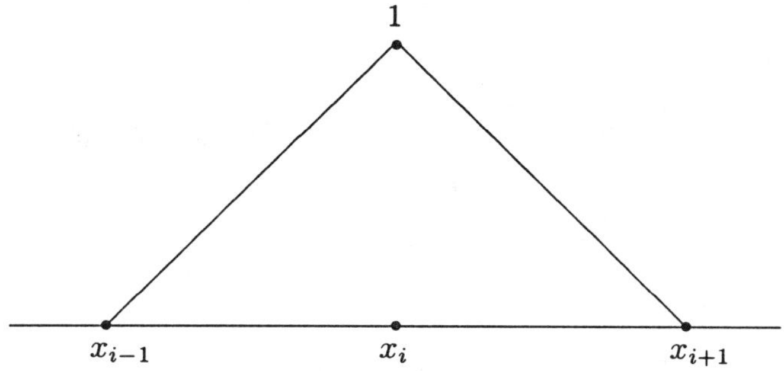

Figure 1.12: Picture of $\phi_i(x)$.

Clearly,

$$\phi_i' = \begin{cases} \frac{1}{h} \text{ if } x_{i-1} < x < x_i, \\ -\frac{1}{h} \text{ if } x_i < x < x_{i+1}, \\ 0 \text{ if } x \notin [x_{i-1}, x_{i+1}]. \end{cases} \tag{1.54}$$

Hence

$$\langle \phi_i', \phi_i' \rangle = \int_{x_{i-1}}^{x_{i+1}} \frac{1}{h}\frac{1}{h} dx = \frac{2}{h}, \tag{1.55}$$

$$\langle \phi_i', \phi_{i+1}' \rangle = \int_{x_i}^{x_{i+1}} \frac{-1}{h^2} dx = -\frac{1}{h}, \tag{1.56}$$

$$\langle \phi_i', \phi_j' \rangle = 0 \text{ if } \mid i - j \mid \geq 2. \tag{1.57}$$

Hence by multiplying (1.53) with h we obtain

$$-u_{j-1} + 2u_j - u_{j+1} - h \int_{x_{j-1}}^{x_{j+1}} f(u, \lambda, \mu)\phi_j dx = 0 \qquad (1 \leq j \leq n). \tag{1.58}$$

From this, one obtains equations in $u_j (1 \leq j \leq n)$ if the integral is replaced by an approximation in terms of $u_j (1 \leq j \leq n)$.

We remark that $u(x_j) = \sum_{i=0}^{n+1} u_i \phi_i(x_j) = u_j$. Hence the simplest approximation is $f(u, \lambda, \mu) \approx f(u_j, \lambda, \mu)$ in (1.58). Then

$$\int_{x_{j-1}}^{x_{j+1}} f(u, \lambda, \mu)\phi_j dx \approx f(u_j, \lambda, \mu) \int_{x_{j-1}}^{x_{j+1}} \phi_j dx = hf(u_j, \lambda, \mu). \tag{1.59}$$

From (1.58), (1.59), and $u_0 = u_{n+1} = 0$ it follows that

$$\begin{bmatrix} 2 & -1 & & & \\ -1 & 2 & -1 & & \\ & -1 & 2 & -1 & \\ & & & \ddots & -1 \\ & & & -1 & 2 \end{bmatrix} \begin{bmatrix} u_1 \\ u_2 \\ \vdots \\ \vdots \\ u_n \end{bmatrix} = \begin{bmatrix} c_1 \\ c_2 \\ \vdots \\ \vdots \\ c_n \end{bmatrix}, \tag{1.60}$$

where

$$c_j = h^2 f(u_j, \lambda, \mu). \tag{1.61}$$

The nonlinear system (1.60) determines $u_1, \ldots, u_n$ and hence u. A more sophisticated approach is to approximate $f(u, \lambda, \mu)$ in $[x_{j-1}, x_{j+1}]$ by an interpolation polynomial

$$f(u, \lambda, \mu) \sim \alpha x^2 + \beta x + \gamma, \tag{1.62}$$

whereby α, β, γ are determined so that

$$\alpha x_k^2 + \beta x_k + \gamma = f(u_k, \lambda, \mu) \qquad (k = j-1, j, j+1) \tag{1.63}$$

and we set

$$c_j = a^2 h \int_{x_{j-1}}^{x_{j+1}} (\alpha x^2 + \beta x + \gamma)\phi_j(x)dx. \tag{1.64}$$

By a translation of the x-axis we may assume that $x_{j-1} = -h, x_j = 0, x_{j+1} = h$. If, furthermore, we set $f_{-1} = f(u_{j-1}, \lambda, \mu), f_0 = f(u_j, \lambda, \mu), f_1 = f(u_{j+1}, \lambda, \mu)$, and if we obtain $\alpha_1, \beta_1, \gamma_1$ by solving

$$\left\{ \begin{array}{rcl} \alpha_1 h^2 - \beta_1 h + \gamma_1 &=& f_{-1}, \\ \gamma_1 &=& f_0, \\ \alpha_1 h^2 + \beta_1 h + \gamma_1 &=& f_1, \end{array} \right. \tag{1.65}$$

then $c_j \approx a^2 h \int_{-h}^{+h} (\alpha_1 x^2 + \beta_1 x + \gamma_1)\phi(x)dx$ where

$$\phi(x) \left\{ \begin{array}{l} = \dfrac{x+h}{h} \text{ if } -h < x < 0, \\ = \dfrac{h-x}{h} \text{ if } 0 < x < h, \\ = 0 \text{ if } x \notin [-h, +h]. \end{array} \right.$$

By a straightforward computation we find

$$\int_{-h}^{0} (\alpha_1 x^2 + \beta_1 x + \gamma_1) \left(\frac{x+h}{h} \right) dx = h \left(\frac{1}{12}\alpha_1 h^2 - \frac{1}{6}\beta_1 h + \frac{1}{2}\gamma_1 \right).$$

By a transformation $x \leftrightarrow (-x)$ it follows that

$$\int_{0}^{h} (\alpha_1 x^2 + \beta_1 x + \gamma_1) \left(\frac{h-x}{h} \right) dx = h \left(\frac{1}{12}\alpha_1 h^2 + \frac{1}{6}\beta_1 h + \frac{1}{2}\gamma_1 \right).$$

By a combination of both integrals one finds

$$c_j \approx a^2 h^2 \left(\frac{1}{6}\alpha_1 h^2 + \gamma_1 \right).$$

From (1.64) we obtain

$$\begin{array}{rcl} \gamma_1 &=& f_0, \\ \alpha_1 &=& \dfrac{1}{2h^2}(f_{-1} - 2f_0 + f_1), \end{array}$$

so that

$$\begin{aligned} c_j &= h^2\left(\frac{1}{12}(f_{-1}-2f_0+f_1)+f_0\right), && (1.66)\\ c_j &= \frac{h^2}{12}(f(u_{j-1},\lambda,\mu)+10f(u_j,\lambda,\mu)+f(u_{j+1},\lambda,\mu)). && (1.67)\end{aligned}$$

Here (1.67) is an improvement of (1.61).

1.3.2 Finite Difference Discretization

We now describe a simple finite difference method to discretize the problem (1.47), (1.48). The unknowns are the values of u in a number of points and we replace the derivatives by finite differences.

Consider the grid points

$$x_0 = 0, \quad x_1 = h, \quad x_2 = 2h, \ldots, x_{n+1} = (n+1)h = 1$$

with $h = \frac{1}{n+1}$.

We define $u_i = u(x_i)(i = 0, \ldots, n+1)$. By (1.47)

$$u_0 = u_{n+1} = 1. \tag{1.68}$$

For an arbitrary j we consider two Taylor expansions

$$\begin{aligned} u(x_{j+1}) &= u(x_j) + \frac{h}{1!}u'(x_j) + \frac{h^2}{2!}u''(x_j) + \frac{h^3}{3!}u'''(x_j) + \cdots\\ u(x_{j-1}) &= u(x_j) - \frac{h}{1!}u'(x_j) + \frac{h^2}{2!}u''(x_j) - \frac{h^3}{3!}u'''(x_j) + \cdots. \end{aligned}$$

By addition the odd powers of h vanish so that

$$u_{j+1} - 2u_j + u_{j-1} = 2\frac{h^2}{2!}u''(x_j) + 2\frac{h^4}{4!}u^{(4)}(x_j) + \cdots. \tag{1.69}$$

Substitution of (1.48) in (1.69) gives

$$-u_{j+1} + 2u_j - u_{j-1} = h^2 f(u_j, \lambda, \mu) - \frac{h^4}{12}u^{(4)}(x_j) + O(h^6). \tag{1.70}$$

Neglecting terms of order h^4, we obtain

$$-u_{j+1} + 2u_j - u_{j-1} = h^2 f(u_j, \lambda, \mu). \tag{1.71}$$

Together with (1.68) this leads to a nonlinear system that is formally identical to (1.60)–(1.61).

We have (replace u by u'')

$$-u''(x_{j+1}) + 2u''(x_j) - u''(x_{j-1}) = -h^2 u^{(4)}(x_j) + O(h^4), \tag{1.72}$$

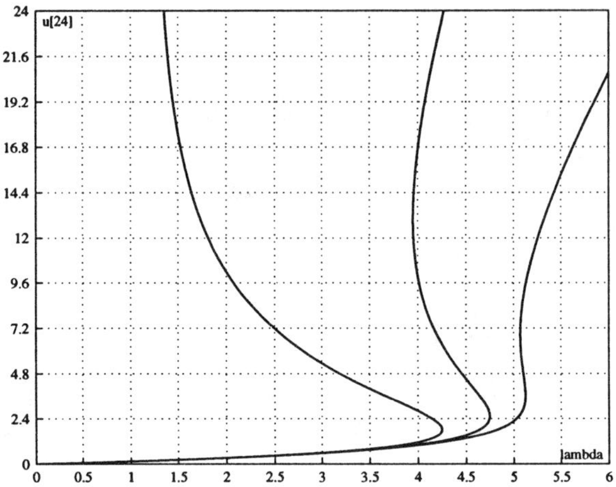

Figure 1.13: Solution branches of (1.46)–(1.47) for $\mu = 0.15, \mu = 0.21$, and 0.24.

which is similar to (1.69). Substituting (1.72) in (1.70) and using (1.48) we find

$$-u_{j+1} + 2u_j - u_{j-1} = \frac{h^2}{12}(f(u_{j-1},\lambda,\mu) + 10f(u_j,\lambda,\mu) + f(u_{j+1},\lambda,\mu)) + O(h^6), \quad (1.73)$$

which is analogous to (1.60)–(1.67). This discretization method is known as the Mehrstellenverfahren of Collatz.

1.3.3 Numerical Continuation: Motivation by an Example

The combustion example provides an interesting motivation for the use of numerical continuation methods in addition to the homotopy idea in §1.1.

The parameter μ is usually fixed during an experiment. For some values of μ ($\mu \sim 0.20$) two types of combustion are possible, namely, a cool solution and a hot solution. More interestingly, they are possible for the same value of λ. If one starts with low values of λ from the trivial initial state $\lambda = 0, u \equiv 0$, then with increasing λ a path of stable cool solutions can be computed. However, for a critical value λ_i of λ a turning point is reached, the cool solution loses stability, and λ decreases. At another critical value λ_e of λ another turning point is reached, the cool solution becomes stable again, and λ increases. So for values of λ between λ_e and λ_i three combustion equilibria are possible: two of them are stable (a cool one and a hot one) and a third is unstable. Physically λ_i corresponds to an ignition point (jump from a stable cool process to a stable hot process) and λ_e corresponds to an extinction point (jump from a stable hot process to a stable cool process).

The solution curves of (1.46)–(1.47) are presented in Figure 1.13 for $\mu = 0.15, 0.21$ and $\mu = 0.24$ from left to right. In the three cases $n = 48$, and the vertical axis is the component u_{24} of u. The curves were computed by a numerical continuation method (see Chapter 2) using the software package CONTENT [165].

We note that this approach to the study of equilibria of the combustion problem has some advantages. We mention two other approaches:

1. For given λ, μ one computes the time evolution of the combustion. In the limit one finds a stable state. Unstable states cannot be found. If $\lambda \approx \lambda_e$ or $\lambda \approx \lambda_i$, then convergence problems can arise.

2. For given λ, μ one computes the steady states only. Stable and unstable solutions may be found. If $\lambda \approx \lambda_e$ or $\lambda \approx \lambda_i$, then convergence problems are possible. Furthermore, it is usually hard to find good starting values.

The advantage of a numerical continuation method is that one can often start with a trivial solution (in our case $\lambda = 0, u \equiv 0$), convergence problems are unlikely, and both stable and unstable solutions can be computed.

1.4 An Example of Symmetry Breaking

In §1.2.3 we saw an example of a pitchfork bifurcation where a nontrivial branch originates from a trivial solution branch ($u \equiv 0$). This is a particular case of the phenomenon of symmetry-breaking bifurcation whereby a curve in a "nonsymmetric" subspace of $\mathbb{R}^n$ branches off from a curve in a "symmetric" subspace. "Symmetry" here means "invariance with respect to a group of orthogonal transformations of $\mathbb{R}^n$." If this group contains all orthogonal transformations, then the invariant subspace is the null space. For other groups more interesting situations may arise. We give a simple example where the group contains two elements only and is hence isomorphic to $\mathbb{Z}_2 = \{0, 1\}$ with addition modulo 2. In Chapter 8 we consider the situation in more detail.

Consider the boundary value problem

$$u'' + \lambda e^u = 0, \tag{1.74}$$

$$u(-1) = u(1) = 0 \tag{1.75}$$

in the interval $[-1, +1]$. (This is called Bratu's problem.) Let $\mathcal{L}$ be the space of all functions defined in $[-1, +1]$ that vanish in ± 1 and are at least twice continuously differentiable. $\mathcal{L}$ is invariant under the reflection S:

$$\begin{aligned} S: \ \mathcal{L} &\to \mathcal{L}, \\ u &\mapsto Su, \end{aligned}$$

whereby $(Su)(x) = u(-x)$. If I denotes the identity operator, then $\{I, S\}$ is a group under composition and isomorphic with $\mathbb{Z}_2$. Let C_0 be the space of all continuous functions on $[-1, +1]$. Then

$$\begin{aligned} D: \ \mathcal{L} &\to C_0, \\ u &\mapsto D(u) = u'' + \lambda e^u \end{aligned}$$

defines a nonlinear operator whose kernel is precisely the set of solutions to (1.74). We remark that

$$(DS)(u) = D(S(u)) = u''(-x) + \lambda e^{u(-x)} = SD(u);$$

i.e., D and S commute. Hence if $Du = 0$, then also $D(Su) = 0$; i.e., the reflection of a solution to (1.74), (1.75) is also a solution. The symmetric subspace of $\mathcal{L}$ is now defined as $\{u \in \mathcal{L} : Su = u\}$.

To obtain numerical results we discretize (1.74). In an extremely simple finite difference approximation we choose the grid points $x_0 = -1, x_1 = -\frac{1}{3}, x_2 = \frac{1}{3}, x_3 = 1$, so that $h = \frac{2}{3}$. The discrete equations in (u_1, u_2, λ) that replace (1.74) and (1.75) are now

$$F\left(\begin{pmatrix} u_1 \\ u_2 \end{pmatrix}, \lambda\right) \equiv \frac{1}{h^2}\begin{bmatrix} -2 & 1 \\ 1 & -2 \end{bmatrix}\begin{bmatrix} u_1 \\ u_2 \end{bmatrix} + \lambda \begin{bmatrix} e^{u_1} \\ e^{u_2} \end{bmatrix} = 0. \tag{1.76}$$

The discrete symmetry operator is

$$S = \begin{pmatrix} 0 & 1 \\ 1 & 0 \end{pmatrix} \qquad (S \neq I, S^2 = I), \tag{1.77}$$

$$F\left(S\begin{pmatrix} u_1 \\ u_2 \end{pmatrix}, \lambda\right) \equiv S\left(F\left(\begin{pmatrix} u_1 \\ u_2 \end{pmatrix}, \lambda\right)\right). \tag{1.78}$$

The symmetric space is

$$\mathcal{S} = \left\{\begin{pmatrix} u_1 \\ u_2 \end{pmatrix} : S\begin{pmatrix} u_1 \\ u_2 \end{pmatrix} = \begin{pmatrix} u_1 \\ u_2 \end{pmatrix}\right\} = \left\{\begin{pmatrix} u_1 \\ u_2 \end{pmatrix} : u_1 = u_2\right\}. \tag{1.79}$$

We parameterize $\mathcal{S}$ by introducing $u = u_1 = u_2$ as a variable. Equation (1.76) may be written in $\mathcal{S}$ as

$$f(u, \lambda) \equiv F\left(\begin{pmatrix} u \\ u \end{pmatrix}, \lambda\right) = \frac{-1}{h^2}u + \lambda e^u = 0 \tag{1.80}$$

(by (1.78) $F(\mathcal{S}) \subseteq \mathcal{S}$). The Jacobian matrix of $f(u, \lambda)$ now is

$$\left[\frac{\partial f(u,\lambda)}{\partial u}, \frac{\partial f(u,\lambda)}{\partial \lambda}\right] = \left[\frac{-1}{h^2} + \lambda e^u, e^u\right].$$

It has full rank 1 and the solution to (1.80) therefore is a smooth curve. (See §2; smoothness also follows from the fact that parameterization with u is possible.) Turning points of (1.80) are found for

$$\frac{-1}{h^2}u + \lambda e^u = 0, \tag{1.81}$$

$$\frac{-1}{h^2} + \lambda e^u = 0, \tag{1.82}$$

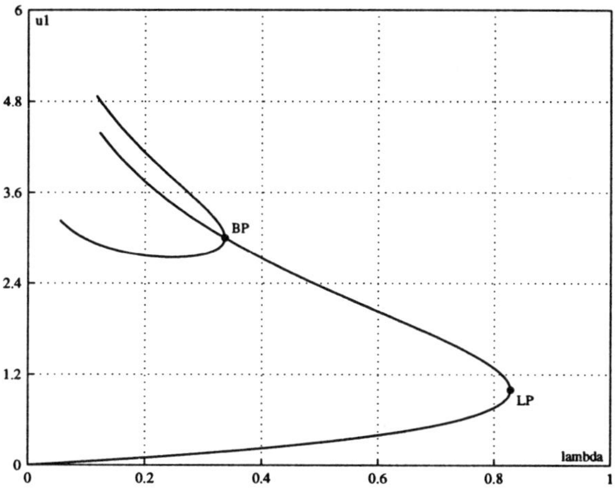

Figure 1.14: Symmetry-breaking bifurcation in (1.76).

i.e., for $u = 1, \lambda = \frac{1}{h^2 e}$. We now return to (1.76). Its Jacobian is

$$\frac{\partial F(u_1, u_2, \lambda)}{\partial(u_1, u_2, \lambda)} = \begin{bmatrix} \frac{-2}{h^2} + \lambda e^{u_1} & \frac{1}{h^2} & e^{u_1} \\ \frac{1}{h^2} & \frac{-2}{h^2} + \lambda e^{u_2} & e^{u_2} \end{bmatrix}.$$

The main part $\frac{\partial F(u_1,u_2,\lambda)}{\partial(u_1,u_2)}$ is singular if

$$\left(\frac{-2}{h^2} + \lambda e^{u_1}\right)\left(\frac{-2}{h^2} + \lambda e^{u_2}\right) - \frac{1}{h^4} = 0. \tag{1.83}$$

Solutions to (1.76), (1.83) are in $\mathcal{S}$ if

$$(-2 + \lambda h^2 e^u)(-2 + \lambda h^2 e^u) = 1, \tag{1.84}$$

$$\frac{-1}{h^2} u + \lambda e^u = 0, \tag{1.85}$$

$$u_1 = u_2 = u \tag{1.86}$$

hold. From (1.84) and (1.85) it follows that $u = 2 \pm 1$. Hence there are two solutions, namely, $(1, 1, \frac{1}{h^2 e})$ and $(3, 3, \frac{3}{h^2 e^3})$.

The first point is the turning point of the solution curve of (1.80) in $\mathcal{S}$. The Jacobian of (1.76) in that point is

$$\begin{bmatrix} \frac{-1}{h^2} & \frac{1}{h^2} & e \\ \frac{1}{h^2} & \frac{-1}{h^2} & e \end{bmatrix}$$

and has full rank. Hence the solution curve to (1.76) is smooth in $(1, 1, \frac{1}{h^2 e})$.

In $(3, 3, \frac{3}{h^2 e^3})$ the Jacobian is

$$\begin{bmatrix} \frac{1}{h^2} & \frac{1}{h^2} & e^3 \\ \frac{1}{h^2} & \frac{1}{h^2} & e^3 \end{bmatrix}$$

and has rank 1. It is possible to show that $(3, 3, \frac{3}{h^2 e^3})$ is indeed a bifurcation point of (1.76) where the symmetric branch (solution to (1.80)) intersects a nonsymmetric branch. The point $(3, 3, \frac{3}{h^2 e^3})$ is called a symmetry-breaking bifurcation point. We note that the singular vector at the turning point is in the symmetric space, while the singular vector at the symmetry-breaking bifurcation point is not.

Figure 1.14 graphically illustrates the phenomena (turning point LP and branching point BP) that we discussed in this section.

1.5 Linear and Nonlinear Stability

In §§1.2.1, 1.2.2, 1.2.4, and 1.3.3 we discussed stability in an informal way. It is now time to give precise definitions. Consider a nonlinear dynamical system

$$\dot{x} = G(x) \tag{1.87}$$

with $x, G(x) \in \mathbb{R}^N$.

Definition 1.5.1.

1. An equilibrium solution x_0 of (1.87) is stable if for each $\varepsilon > 0$ there exists a $\delta > 0$ such that if $x(t)$ $(t \geq 0)$ is a solution to (1.87) with $\|x(0) - x_0\| \leq \delta$, then $\|x(t) - x_0\| \leq \varepsilon$ for all $t \geq 0$.

2. An equilibrium solution x_0 to (1.87) is asymptotically stable if it is stable and furthermore there exists a $\delta \geq 0$ such that if $x(t)$ is a solution to (1.87) for $t \geq 0$ and $\|x(0) - x_0\| \leq \delta$, then $x(t) \to x_0$ for $t \to \infty$.

3. A solution to (1.87) is unstable if it is not stable.

The theory of stability of equilibria is discussed in all standard texts on dynamical systems. We restrict ourselves to a brief summary of the results.

The simplest case is linear stability where G is an affine mapping; i.e., (1.87) reduces to

$$\dot{x} = Ax - b \tag{1.88}$$

with A an $N \times N$ matrix. If x_0 is an equilibrium solution to (1.88), then $b = Ax_0$ and (1.88) is equivalent to

$$\frac{d}{dt}(x - x_0) = A(x - x_0).$$

Hence the study of (1.88) reduces to that of

$$\dot{x} = Ax \tag{1.89}$$

by a simple translation. It turns out that the stability of this solution is determined by the eigenvalues of A. We show this first in the simplest case where A has only real and distinct eigenvalues.

Assume that $Av_i = \lambda_i v_i$, $v_i \neq 0$ for $i = 1, \ldots, N$ with $\lambda_i \neq \lambda_j$ if $i \neq j$. Then the eigenvectors v_i are linearly independent and span $\mathbb{R}^N$. Hence a general solution of (1.89) can be written as $x(t) = x_1(t)v_1 + \cdots + x_N(t)v_N$; it satisfies

$$\dot{x}_1(t)v_1 + \cdots + \dot{x}_N(t)v_N = \lambda_1 x_1(t)v_1 + \cdots + \lambda_N x_N(t)v_N,$$

so that

$$\dot{x}_i(t) = \lambda_i x_i(t). \tag{1.90}$$

The general solution to (1.90) is

$$x_i(t) = C_i e^{\lambda_i t} \tag{1.91}$$

with C_i an integration constant.

Clearly $x(t)$ converges to zero if all eigenvalues of A are negative, and $x(t)$ converges to $+\infty$ or $-\infty$ if at least one eigenvalue λ_i is positive and the corresponding coefficient C_i is nonzero. In the first case zero is an asymptotically stable equilibrium of (1.89), in the second case it is an unstable equilibrium. If one eigenvalue is zero and all other eigenvalues are negative, then zero is a stable but not asymptotically stable equilibrium.

Now consider the case that all eigenvalues are distinct but some are complex. Assume, e.g., that λ_i, λ_j is a conjugate pair of complex eigenvalues $\lambda_i = \lambda^{(r)} + i\lambda^{(i)}$, $\lambda_j = \lambda^{(r)} - i\lambda^{(i)}$ with $\lambda^{(r)}, \lambda^{(i)}$ real and $\lambda^{(i)} \neq 0$. Furthermore, let $v_i = v^{(r)} + iv^{(i)}$ be the complex eigenvector corresponding to λ_i; i.e.,

$$A(v^{(r)} + iv^{(i)}) = (\lambda^{(r)} + i\lambda^{(i)})(v^{(r)} + iv^{(i)}).$$

Then $v_j = \bar{v_i} = v^{(r)} - iv^{(i)}$ is an eigenvector corresponding to λ_j.

In $\mathbb{C}^N$ every complex vector c can be written uniquely as

$$c = c_1 v_1 + \cdots + c_N v_N.$$

If c is real, then $c_1 v_1 + \cdots + c_N v_N = \bar{c}_1 \bar{v}_1 + \cdots + \bar{c}_N \bar{v}_N$; hence $c_i = \bar{c}_j$.

The general solution to (1.89) is $x(t) = x_1(t)v_1 + \cdots + x_N(t)v_N$ where $x_i(t) = x_i(0)e^{\lambda_i t}$. Since $x(0)$ is real, it follows that $x_i(0) = \overline{x_j(0)}$. Hence $x_i(0)v_i + x_j(0)\overline{v_i}$ is a real vector spanned by $v^{(r)}, v^{(i)}$. If we set

$$x_i^{(0)} v_i + x_j(0)\overline{v_i} = \alpha v^{(r)} + \beta v^{(i)}$$

(α, β real), then

$$2 \text{ Re } x_i^{(0)} v^{(r)} - 2 \text{ Im } x_i^{(0)} v^{(i)} = \alpha v^{(r)} + \beta v^{(i)};$$

i.e., $x_i^{(0)} = \frac{\alpha - i\beta}{2}$ and so $x_j^{(0)} = \overline{x_i^{(0)}} = \frac{\alpha + i\beta}{2}$.

The component of $x(t)$ in the space spanned by $v^{(r)}, v^{(i)}$ is therefore

$$\frac{\alpha - i\beta}{2} e^{(\lambda^{(r)} + i\lambda^{(i)})t}(v^{(r)} + iv^{(i)}) + \frac{\alpha + i\beta}{2} e^{(\lambda^{(r)} - i\lambda^{(i)})t}(v^{(r)} - iv^{(i)})$$

$$= e^{\lambda^{(r)}t} \left\{ \left(\frac{\alpha - i\beta}{2}(\cos \lambda^{(i)}t + i \sin \lambda^{(i)}t) + \frac{\alpha + i\beta}{2}(\cos \lambda^{(i)}t - i \sin \lambda^{(i)}t) \right) v^{(r)} \right.$$

$$+\left(\frac{\alpha - i\beta}{2}(\cos\lambda^{(i)}t + i\sin\lambda^{(i)}t) - \frac{\alpha + i\beta}{2}(\cos\lambda^{(i)}t - i\sin\lambda^{(i)}t)\right) iv^{(i)}\Big\}$$
$$= e^{\lambda^{(r)}t}\left\{(\alpha\cos\lambda^{(i)}t + \beta\sin\lambda^{(i)}t)v^{(r)} - (\alpha\sin\lambda^{(i)}t - \beta\cos\lambda^{(i)}t)v^{(i)}\right\}.$$

The factor between curly brackets is a periodic function with period $\frac{2\pi}{\lambda^{(i)}}$ and nonzero if α, β are not both zero. Hence the stability is determined completely by the factor $e^{\lambda^{(r)}t}$, i.e., by the real part of the complex eigenvalue.

If all eigenvalues are distinct (real or complex) and their real parts are all negative, then the zero solution to (1.89) is asymptotically stable. If at least one eigenvalue has a positive real part, then the zero solution is unstable. If some eigenvalues have a real part zero but no eigenvalue has a positive real part, then the zero solution is stable but not asymptotically stable.

For the case of multiple eigenvalues let us consider the simplest case, i.e., a real eigenvalue λ with geometric multiplicity 1 and algebraic multiplicity 2. So there exist generalized eigenvectors v_1, v_2 such that $Av_1 = \lambda v_1$, $Av_2 = \lambda v_2 + v_1$. The space spanned by v_1, v_2 (i.e., the generalized eigenspace of λ) is invariant under the dynamics of (1.89). If a solution to (1.89) has the form $x(t) = x_1(t)v_1 + x_2(t)v_2$, then we find that $x_1(t), x_2(t)$ must satisfy

$$\begin{cases} \dot{x}_1 = \lambda x_1 + x_2, \\ \dot{x}_2 = \lambda x_2. \end{cases}$$

From standard theory of linear differential equations it follows that

$$\begin{cases} x_1 = (C_1 t + C_2)e^{\lambda t}, \\ x_2 = C_1 e^{\lambda t} \end{cases} \tag{1.92}$$

for some constants C_1, C_2. Similar computations can be done in all other cases of multiple eigenvalues. It follows that the zero solution is still asymptotically stable if all eigenvalues have a negative real part. Also, it is unstable if at least one eigenvalue has a positive real part. If at least one nonsemisimple eigenvalue has real part zero, then the solution is also unstable (obvious in the case of (1.92)).

For a nonlinear system (1.87) with equilibrium solution x_0, we can consider the Taylor expansion

$$\frac{d}{dt}(x - x_0) = G(x) = G_x(x - x_0) + O(\|x - x_0\|^2). \tag{1.93}$$

It suggests that in a neighborhood of $x = x_0$ the solutions to (1.87) behave approximately like those of (1.89) with $A = G_x$ and in particular that the stability of the solutions to (1.87) is determined by the eigenvalues of G_x. It turns out that this is not always true. Nevertheless, the theorem of Hartman–Grobman (see, e.g., [191, §2.8]) guarantees that it is true if G_x has no eigenvalues with real part zero. In particular, x_0 is an asymptotically stable equilibrium if all eigenvalues of G_x have a negative real part; it is unstable if at least one eigenvalue has a positive real part.

Now suppose that G in (1.87) also depends on a parameter α. The set of eigenvalues of G_x is a continuous function of (x, α); in particular, it varies continuously along a solution branch of the equation $G(x, \alpha) = 0$. If the equilibrium solution is stable in certain parts of the solution branch, then stability may be lost if eigenvalues of G_x cross the imaginary axis. One possibility is that a real eigenvalue crosses the imaginary axis; another one is that a conjugate pair of complex eigenvalues crosses it. The first case generically leads to a turning point bifurcation, the second case to a Hopf bifurcation. Turning point bifurcations were found in §§1.2.2, 1.3.3, and 1.4. An example of Hopf bifurcation was discussed in §1.2.4.

We note that there are more complicated and degenerate cases of loss of stability. We continue their study (in particular numerical methods) throughout the book.

Remark. For an asymptotically stable equilibrium in a linear system the domain of attraction is the whole state space; i.e., for all starting points the system converges to the equilibrium. For a nonlinear system the domain of attraction is only a neighborhood of the asymptotically stable equilibrium. In some cases this neighborhood may be so small that for practical purposes (say, in a physical experiment) the equilibrium is unstable. A dynamical system is called linearly stable if the linearized system is stable; there is still a possibility that for practical purposes the full system is unstable.

1.6 Exercises

1. Consider the differential equation

$$\frac{dx}{dt} \equiv G(x, \alpha) = -x^3 + \alpha_1 x + \alpha_2$$

from §1.2.3, where x is a population density.

   (a) For what values of α_1, α_2 does this system have equilibria that are physically meaningful ($x \geq 0$) and realizable (i.e., stable)?

   (b) How many such equilibria exist for a given pair (α_1, α_2)? (Consider all cases.)

2. Consider the system in Exercise 1. Let (α_1, α_2) be a pair of parameters for which the equation $G(x, \alpha) = 0$ has three distinct real roots. Let x_0 be a nonequilibrium point. Prove that the solution $x(t)$ to

$$\frac{dx}{dt} \equiv G(x, \alpha) = -x^3 + \alpha_1 x + \alpha_2, x(0) = x_0$$

converges to a stable equilibrium if t tends to ∞.

3. Prove the same result as in Exercise 2 in the case that $G(x, \alpha) = 0$ has a real root and a conjugate pair of complex roots with nonzero real part.

4. The assumptions in (1.30) led to a simple system in which time-periodic behavior is possible but we had to allow negative values of x, y. Can you make other assumptions that allow time-periodic behavior with x, y strictly positive?

5. Consider the example in §1.2.4. Can you find equilibria that are stable but not asymptotically stable?

6. Consider the following Lotka–Volterra competitor–competitor model [42]:

$$\begin{cases} \dot{x} = x(r_1 - r_2 x - r_3 y), \\ \dot{y} = y(r_4 - r_5 x - r_6 y), \end{cases}$$

where $r_i > 0$ for $i = 1, \dots, 6$ and it is required that $x, y \geq 0$.

 (a) Prove that this system cannot have Hopf bifurcation points.

 (b) Find parameter values for which it has two distinct stable equilibria (the trivial equilibrium $x = y = 0$ is always unstable).

7. Consider the equation $\ddot{x} + kx = 0(x(t) \in \mathbb{R}, k \in \mathbb{R})$. Rewrite this as a dynamical system in $\binom{x}{y}$ where $y = \dot{x}$.

 (a) Assume $k > 0$. Is $\binom{0}{0}$ a stable, asymptotically stable, or unstable equilibrium?

 (b) The same question for $k < 0$.

 (c) Assume that we add a friction term so that the equation becomes $\ddot{x}+l\dot{x}+kx = 0$, where $l > 0$. Prove that $\binom{0}{0}$ is an asymptotically stable equilibrium for $k > 0$ and an unstable equilibrium for $k < 0$.

8. A fraction $x(t)$ $(0 \leq x(t) \leq 1)$ of a population is infected by a disease; the rest $y(t) = 1 - x(t)$ is not. Both infection and cure are possible, but the cure does not provide immunity. Let the time evolution of the process be described by the dynamical system

$$\dot{x} = \beta y x - \gamma x,$$

where $\beta > 0$ is the *transmission rate* and $\gamma > 0$ is the *recovery rate*. Discuss all possible stable and unstable equilibria for all possible combinations of β, γ.

9. Consider Bazykin's ecological model with two state variables x, y and four parameters α, β, γ, δ [26], [27]:

$$\begin{aligned} \dot{x} &= x - \tfrac{xy}{1+\alpha x} - \beta x^2, \\ \dot{y} &= -\gamma y + \tfrac{xy}{1+\alpha x} - \delta y^2. \end{aligned}$$

Fix the parameter values $\alpha = 0.3$, $\beta = 0.01$, $\gamma = 1.0$, $\delta = 0.4$. Use a numerical integrator code to compute an orbit starting with the values $x = 100$, $y = 10$. Find a stable equilibrium. Now repeat the experiment with $\delta = 0.2$. Find a stable periodic orbit.

Chapter 2

Manifolds and Numerical Continuation

2.1 Manifolds

2.1.1 Definitions

For the purposes of this book it is sufficient to consider only differentiable manifolds that are naturally embedded in finite-dimensional Euclidean spaces and can be defined locally as the regular null sets of continuously differentiable functions with values in other Euclidean spaces. For a more general (and more abstract) definition see [201, §4.1].

Definition 2.1.1. Let $k, N \in \mathbb{N}, N > k \geq 1$. A subset $\mathcal{S}$ of $\mathbb{R}^N$ is called a k-dimensional C^1-manifold if for every $\bar{x} \in \mathcal{S}$ there exist a neighborhood U of $\bar{x}$ in $\mathbb{R}^N$ and a C^1-function $G : U \to \mathbb{R}^{N-k}$ such that $G_x(\bar{x})$ has rank $N-k$ and $\{x \in U : G(x) = 0\} = \mathcal{S} \cap U$. We say that G is a local defining function for $\mathcal{S}$; if we want to stress that G takes values in a multidimensional space, then we call G a local defining system. The number $N-k$ is called the *codimension* of the manifold.

Note. It is a fundamental property that if G_1 is another $\mathbb{R}^{N-k}$-valued C^1-function defined in a neighborhood of $\bar{x}$ that locally vanishes on the manifold defined by G and has full linear rank $N-k$, then G_1 defines (locally) the same manifold. We will not prove this here; it is an easy corollary to Lemma 6.2.3. It will not be used in this chapter but will be used extensively in Chapter 5.

Definition 2.1.2. If $N > k \geq 1$ and $G : \mathbb{R}^N \to \mathbb{R}^{N-k}$ is a C^1-function, then G is called a global defining function for the manifold $\{x \in \mathbb{R}^N : G_x(x) \text{ has rank } N-k\}$. The points of this manifold are called ordinary points of the zero set of G.

Definitions 2.1.1 and 2.1.2 generalize in the obvious way to C^l-manifolds ($l > 1$) and to C^∞-manifolds. C^∞-manifolds will also be called smooth manifolds.

Let us assume without loss of generality that in Definition 2.1.1 the submatrix of $G_x(\bar{x})$ that is related to the first $N-k$ variables is nonsingular. Then we call the last k variables the basic variables and collect them in $x_B \in \mathbb{R}^k$; the others are called nonbasic variables and are collected in $x_N \in \mathbb{R}^{N-k}$. So formally $x = \binom{x_N}{x_B}$. The implicit function theorem tells us that there exists a continuously differentiable mapping G^i defined in a neighborhood of $\bar{x}_B \in \mathbb{R}^k$ and with values in $\mathbb{R}^{N-k}$ such that $G^i(\bar{x}_B) = \bar{x}_N$ and $G(G^i(x_B), x_B) \equiv 0$. The mapping G^i is locally unique and inherits the differentiability properties of G. If, in particular, G is a smooth function, then so is G^i. The map $\mathbb{R}^k \to \mathcal{S} : x_B \to \binom{G^i(x_B)}{x_B}$ is a homeomorphism from a neighborhood of $\bar{x}_B \in \mathbb{R}^k$ onto a neighborhood of $\bar{x} \in \mathcal{S}$. It is also a diffeomorphism; i.e., its Jacobian has full rank k.

2.1.2 The Tangent Space

Consider a C^1-map $G : \mathbb{R}^N \to \mathbb{R}^{N-k} (N > k \geq 1)$. Let $\xi(t)$ be a C^1-curve in the zero set $\mathcal{S}$ of G, i.e.,

$$G_i(\xi(t)) \equiv 0 \quad \text{for} \quad 1 \leq i \leq N-k. \tag{2.1}$$

By taking derivatives we obtain

$$\frac{\partial G_i}{\partial x_1}\dot{\xi}_1 + \frac{\partial G_i}{\partial x_2}\dot{\xi}_2 + \cdots + \frac{\partial G_i}{\partial x_N}\dot{\xi}_N \equiv 0. \tag{2.2}$$

Hence the tangent vector to each curve in $\mathcal{S}$ is in the kernel of the Jacobian in the point. We prove that the converse is true in a precise sense if the Jacobian has full rank.

Proposition 2.1.3. Let $G : \mathbb{R}^N \to \mathbb{R}^{N-k} (N > k \geq 1)$ be a C^1-map and suppose that $\bar{x} \in \mathbb{R}^N$, $G(\bar{x}) = 0$, and $G_x(\bar{x})$ has full rank $N-k$. If $p = (p_1, \ldots, p_N)^T$ is in the kernel of $J = G_x(\bar{x})$, then there exists a curve $x = \xi(t)$ in $\mathbb{R}^N$ which lies locally in $\mathcal{S} = \{x \in \mathbb{R}^N : G(x) = 0\}$ and for which $\xi(0) = \bar{x}$, $\dot{\xi}(0) = p$.

Proof. Without loss of generality we assume that the first $N-k$ columns of J are linearly independent. By the implicit function theorem there exist functions

$$F_i(x_{N-k+1}, \ldots, x_N) \;\; (1 \leq i \leq N-k)$$

such that

$$\bar{x}_i = F_i(\bar{x}_{N-k+1}, \ldots, \bar{x}_N) \;\; (1 \leq i \leq N-k)$$

and

$$G_i(F_1(x_{N-k+1}, \ldots, x_N), \ldots, F_{N-k}(x_{N-k+1}, \ldots, x_N), x_{N-k+1}, \ldots, x_N) \equiv 0 \tag{2.3}$$

in a neighborhood of $(\bar{x}_{N-k+1}, \ldots, \bar{x}_N)^T \in \mathbb{R}^k$.

Now define

$$\xi_i(t) = F_i(\bar{x}_{N-k+1} + tp_{N-k+1}, \ldots, \bar{x}_N + tp_N) \;\; (1 \leq i \leq N-k), \tag{2.4}$$

$$\xi_i(t) = \bar{x}_i + tp_i \;\; (N-k+1 \leq i \leq N). \tag{2.5}$$

These two equations determine a curve $\xi(t)$ in $\mathbb{R}^N$ with $\xi(0) = \bar{x}$ and locally $\xi(t) \in \mathcal{S}$. We now prove that $\dot{\xi}(0) = p$.

From (2.5) it follows that

$$\dot{\xi}_l(0) = p_l \;\; \text{if} \;\; N-k+1 \leq l \leq N. \tag{2.6}$$

By taking derivatives of (2.3) we find

$$\sum_{j\leq N-k}\frac{\partial G_i}{\partial x_j}\frac{\partial F_j}{\partial x_l}+\frac{\partial G_i}{\partial x_l}\equiv 0 \qquad (1\leq i\leq N-k, N-k+1\leq l\leq N). \tag{2.7}$$

After multiplication with $\dot{\xi}_l(t)$ and summation over l we obtain

$$\sum_{j\leq N-k}\frac{\partial G_i}{\partial x_j}\sum_{l>N-k}\frac{\partial F_j}{\partial x_l}\dot{\xi}_l+\sum_{l>N-k}\frac{\partial G_i}{\partial x_l}\dot{\xi}_l\equiv 0, \tag{2.8}$$

where the identity is with respect to t and $\dot{\xi}_l\equiv p_l \;\; (N=k+1\leq l\leq N)$.

From (2.4) we infer

$$\dot{\xi}_j(t)=\sum_{l>N-k}\frac{\partial F_j}{\partial x_l}\dot{\xi}_l \qquad (1\leq j\leq N-k). \tag{2.9}$$

Combining (2.8) and (2.9) we obtain

$$\sum_{j\leq N-k}\frac{\partial G_i}{\partial x_j}\dot{\xi}_j+\sum_{l>N-k}\frac{\partial G_i}{\partial x_l}\dot{\xi}_l\equiv 0. \tag{2.10}$$

On the other hand,

$$\sum_{j\leq N-k}\frac{\partial G_i}{\partial x_j}p_j+\sum_{l>N-k}\frac{\partial G_i}{\partial x_l}p_l\equiv 0. \tag{2.11}$$

Since $[\frac{\partial G_i}{\partial x_j}]_{1\leq i,j\leq N-k}$ has full rank, it follows from (2.10) and (2.11) that $\dot{\xi}_j(0)=p_j$ also if $1\leq j\leq N-k$. □

Definition 2.1.4. Let $k,N\in\mathbb{N}, N>k\geq 1$, $\mathcal{S}$ a k-dimensional C^1-manifold in $\mathbb{R}^N$, $\bar{x}\in\mathcal{S}$ and $G:U\to\mathbb{R}^{N-k}$ a local defining function of $\mathcal{S}$ at $\bar{x}$. Then the null space of $G_x(\bar{x})$ is called the tangent space of $\mathcal{S}$ at $\bar{x}$.

Proposition 2.1.3 proves that the above definition is independent of the choice of the local defining function G at $\bar{x}\in\mathcal{S}$.

2.1.3 Examples

Example 2.1.5. Let $N=3, k=2, R\in\mathbb{R}, R>0$. Define $G:\mathbb{R}^3\to\mathbb{R}, (x_1,x_2,x_3)^T\to x_1^2+x_2^2+x_3^2-R^2$. The zero set of G is the sphere with center in the origin and radius R; G_x has full rank 1 at every point of the sphere and so constitutes a global defining function.

Example 2.1.6. Let $N=3, k=2$. A torus in $\mathbb{R}^3$ has an obvious parametrization with two real parameters. Locally each part of the torus is diffeomorphic to an open set in $\mathbb{R}^2$. By local elimination of the parameters one obtains local defining functions. The existence of a global defining function is less obvious, but we can take

$$(x_1,x_2,x_3)^T\to(x_1^2+x_2^2-4)^2+x_3^2-1.$$

Example 2.1.7. Let $N = 2, k = 1$ and define $G : \mathbb{R}^2 \to \mathbb{R}, (x_1, x_2)^T \to x_1^2 - x_2^2$. The zero set of G consists of the two straight lines $x_1 = x_2$ and $x_1 = -x_2$. This set does not constitute a manifold since at $(0,0)^T$ the set of tangent vectors does not form a linear space. However, if we define

$$\mathcal{S} = \{(x_1, x_2)^T : x_1^2 - x_2^2 = 0, (x_1, x_2)^T \neq (0,0)^T\},$$

then $\mathcal{S}$ is a manifold with dimension 1 and G is a defining function at each point of $\mathcal{S}$. We remark that $\mathcal{S}$ is a disconnected set that consists of four connected parts (open half-lines).

Example 2.1.8. Let $N = 4, k = 3$ and identify $\mathbb{R}^4$ with the space of 2×2 matrices. Define $\det : \mathbb{R}^4 \to \mathbb{R}, (a_{11}, a_{12}, a_{21}, a_{22})^T \to a_{11}a_{22} - a_{12}a_{21}$. The zero set of det is the set of singular 2×2 matrices. The vectors $(1,0,0,0)^T$ and $(0,0,0,1)^T$ are both tangent vectors to curves in the zero set of det at the origin, but $(1,0,0,1)^T$ is not. Hence the zero set of det does not constitute a manifold. If we delete the origin, then

$$\mathcal{S} = \{x \in \mathbb{R}^4 : \det(x) = 0, x \neq 0\}$$

is the set of rank-1 matrices and constitues a three-dimensional differentiable manifold since det has a nonvanishing Jacobian at each point of $\mathcal{S}$.

Example 2.1.9. Let $N = 6, k = 4$ and identify $\mathbb{R}^6$ with the space of 3×2 matrices. Define

$$\det_1 : \mathbb{R}^6 \to \mathbb{R}, (a_{11}, a_{12}, a_{21}, a_{22}, a_{31}, a_{32})^T \to a_{11}a_{22} - a_{12}a_{21},$$

$$\det_2 : \mathbb{R}^6 \to \mathbb{R}, (a_{11}, a_{12}, a_{21}, a_{22}, a_{31}, a_{32})^T \to a_{11}a_{32} - a_{12}a_{31},$$

$$\det_3 : \mathbb{R}^6 \to \mathbb{R}, (a_{11}, a_{12}, a_{21}, a_{22}, a_{31}, a_{32})^T \to a_{21}a_{32} - a_{22}a_{31}.$$

Now

$$\mathcal{S} = \{x \in \mathbb{R}^6 : \det_1(x) = \det_2(x) = \det_3(x) = 0, x \neq 0\}$$

is the set of rank-1 3×2 matrices. If $x \in \mathcal{S}$, then at least one entry of x is nonzero. For example, suppose that $a_{11} \neq 0$. Then $(\det_1 = 0, \det_2 = 0)$ define $\mathcal{S}$ locally near x and the Jacobian of this system is nonsingular at x. So $\mathcal{S}$ is a four-dimensional manifold and local defining systems can be found by taking appropriate pairs from $(\det_1 = 0, \det_2 = 0, \det_3 = 0)$. We remark that each pair is actually a defining system for $\mathcal{S}$ in an open everywhere dense subset of $\mathbb{R}^6$.

2.2 Branches and Limit Points

The case of a one-dimensional manifold ($k = 1$) deserves special attention. It arises most naturally if one of the variables is singled out as a distinguished bifurcation parameter. Let us then subdivide the independent variables as (x, α), where $x \in \mathbb{R}^N, \alpha \in \mathbb{R}$. Let $G : \mathbb{R}^{N+1} \to \mathbb{R}^N$ be a C^1-function and let $\mathcal{Z}$ be the zero set of G, $\mathcal{S}$ the set of all ordinary points in $\mathcal{Z}$. Then $\mathcal{S}$ is a differentiable manifold with dimension 1. The maximal connected components of $\mathcal{S}$ are called branches; they are necessarily diffeomorphic with real intervals.

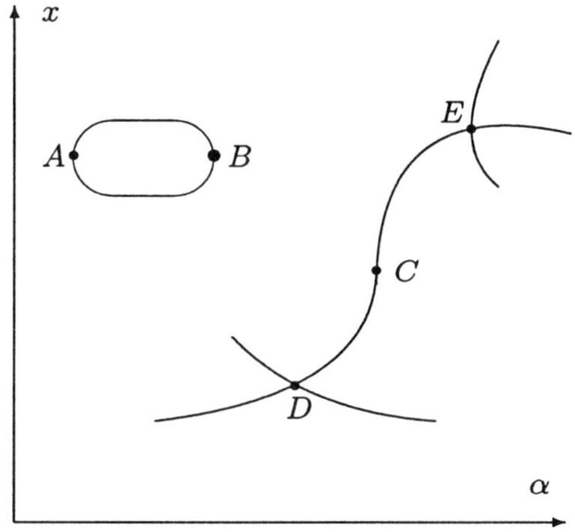

Figure 2.1: The zero set of a nonlinear function.

A possible picture of $\mathcal{Z}$ is given in Figure 2.1. Obviously D, E are not in $\mathcal{S}$. $\mathcal{S}$ has eight maximal connected components (branches). The point D is called a *transcritical bifurcation point*; E is a *pitchfork bifurcation*. The distinction between transcritical bifurcation points and pitchfork bifurcations exists only because of our choice of α as a distinguished bifurcation parameter.

A, B, C are in $\mathcal{S}$ but are special with respect to the parameter α because the tangents in these points are orthogonal to the α-axis. The parameter α cannot be used to parameterize $\mathcal{S}$ in a neighborhood of A or B. It can be used in a neighborhood of C, but the parameterization will then not be diffeomorphic as we would probably prefer it to be.

Definition 2.2.1. A point $(x, \alpha) \in \mathcal{S}$ is called a *limit point* or *fold* if G_x is singular. All other points of $\mathcal{S}$ are called *regular points.*

Proposition 2.2.2. In a limit point G_x has rank deficiency 1. There exists a vector $\psi \neq 0$, unique up to a scalar multiple, such that $\psi^T G_x = 0$. For this vector $\psi^T G_\alpha \neq 0$.

Proof. Since $[G_x \ G_\alpha]$ has rank N, G_x has rank at least $(N-1)$. □

Proposition 2.2.3. Let s be a diffeomorphic parameterization of a branch through the ordinary point (x, α). Then (x, α) is a limit point if and only if $\alpha_s \equiv \frac{d\alpha}{ds} = 0$.

Proof. From $G(x(s), \alpha(s)) \equiv 0$ it follows that $G_x x_s + G_\alpha \alpha_s \equiv 0$. If (x, α) is a limit point and ψ is the corresponding left singular vector of G_x, then

$$(\psi^T G_x)x_s + (\psi^T G_\alpha)\alpha_s = 0.$$

Hence $\alpha_s = 0$. Conversely, if $\alpha_s = 0$ then $G_x x_s = 0$. Since $(x_s^T, \alpha_s)^T \neq 0$ we infer $x_s \neq 0$ and hence G_x is singular. □

In the setting with a distinguished bifurcation parameter, limit points are often encountered as quadratic turning points. We saw this already in Chapter 1. To introduce it now formally, we start from the identity

$$G(x(s), \alpha(s)) \equiv 0$$

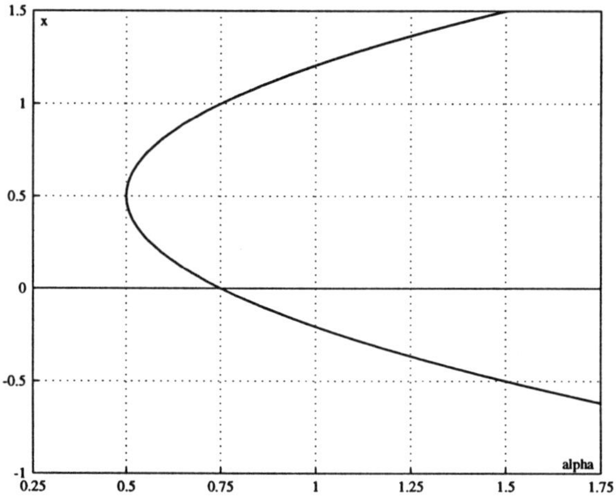

Figure 2.2: A quadratic turning point.

from which

$$G_x x_s + G_\alpha \alpha_s \equiv 0,$$

$$G_{xx} x_s x_s + G_{x\alpha} \alpha_s x_s + G_x x_{ss} + G_{x\alpha} x_s \alpha_s + G_{\alpha\alpha} \alpha_s \alpha_s + G_\alpha \alpha_{ss} \equiv 0.$$

In a limit point $\alpha_s = 0$; after multiplication with ψ^T we obtain

$$\psi^T G_{xx} x_s x_s + (\psi^T G_\alpha) \alpha_{ss} = 0.$$

Hence $\alpha_{ss} \neq 0$ if and only if $\psi^T G_{xx} x_s x_s \neq 0$. We remark that x_s is a nonzero right singular vector of G_x.

Definition 2.2.4. A limit point is a simple or quadratic turning point if $\alpha_{ss} \neq 0$.

It follows that $\alpha(s)$ locally behaves as cs^2 where $s = 0$ corresponds to the limit point and c is a nonzero constant. In particular, α_s changes sign. See Figure 2.2.

Remark 2.2.5. Of course one can define limit points of higher order. For example, a hysteresis point is defined by the conditions $\alpha_s = 0$, $\alpha_{ss} = 0, \alpha_{sss} \neq 0$, or, equivalently, by

$$\begin{cases} G_x \text{ singular}, \\ \psi^T G_{xx} x_s x_s = 0, \\ \psi^T G_{xxx} x_s x_s x_s + 3\psi^T G_{xx} x_s x_{ss} \neq 0 \end{cases}$$

(check this!). In transcritical and pitchfork bifurcation points one has $\psi^T G_\alpha = 0$. This case also deserves further study. See Chapter 7.

2.3 Numerical Continuation

2.3.1 Natural Parameterization

We consider the problem of computing a solution branch of the equation

$$G(x, \alpha) = 0 \quad (x \in \mathbb{R}^N, \alpha \in \mathbb{R}, G(x, \alpha) \in \mathbb{R}^N). \tag{2.12}$$

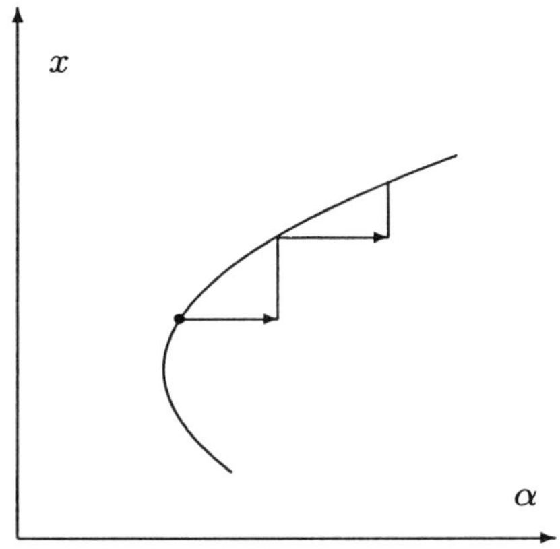

Figure 2.3: Continuation by natural parameterization.

Numerical continuation is a technique to find consecutive points of a solution branch to (2.12). We adapt our notation to this splitting of the variables, keeping in mind that the splitting is irrelevant in some cases.

Suppose that we already found a solution point $(x_0^T, \alpha_0)^T$. To find a new one, call it $(x_1^T, \alpha_1)^T$, we need a starting point $(\widehat{x}^T, \widehat{\alpha})^T$ and a strategy to determine $(x_1^T, \alpha_1)^T$.

One possibility is to choose $\widehat{\alpha}$ and to fix $\alpha_1 = \widehat{\alpha}$. This is the historical method [196]. (See Figure 2.3.) Then x_1 is determined by solving

$$G(x, \alpha_1) = 0. \tag{2.13}$$

Solving this system by Newton's method we construct a sequence $x^0, x^1, x^2, \ldots$ with $x^0 = \widehat{x} = x_0$ and

$$G_x(x^k, \alpha_1)\Delta x^k = -G(x^k, \alpha_1), \tag{2.14}$$

$$x^{k+1} = x^k + \Delta x^k \tag{2.15}$$

for $k = 0, 1, 2, \ldots$. Of course any other method to solve nonlinear systems can be used instead of (2.14), (2.15), in particular quasi-Newton methods.

Geometrically this amounts to approximating the curve first by a straight line (predictor step) and then correcting in a hyperplane $\alpha = \alpha_1$ (corrector step). See Figure 2.3.

This suggests another predictor, namely, along the tangent line. If we parameterize by α, then from the identity

$$G(x(\alpha), \alpha) = 0 \tag{2.16}$$

we infer that

$$G_x x'_\alpha = -G_\alpha. \tag{2.17}$$

Hence x'_α is found by solving a system similar to (2.14), and the predictor is $\widehat{x} = x_0 + (\widehat{\alpha} - \alpha_0)x'_\alpha(0)$. See Figure 2.4.

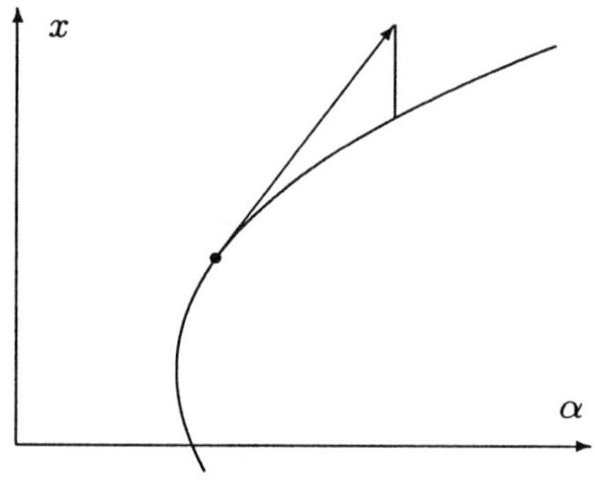

Figure 2.4: Prediction along the tangent.

2.3.2 Pseudoarclength Continuation

The methods proposed in §2.3.1 have difficulties if a branch of (2.12) contains limit points; α is not a good parameterization of the curve in the neighborhood of such points.

Suppose that a point $(x_1^T, \alpha_1)^T$ and a previous point $(x_0^T, \alpha_0)^T$ with tangent vector t_0 are known. To find a tangent vector t_1 at $(x_1^T, \alpha_1)^T$ we remark (as in §2.1.2) that

$$[G_x \quad G_\alpha]\, t_1 = 0. \tag{2.18}$$

This equation determines t_1 up to a scalar factor. To preserve the orientation of the branch we require

$$t_0^T\, t_1 = 1. \tag{2.19}$$

If we decompose t_1 in a natural way as $t_1 = \binom{t_1^{(x)}}{t_1^{(\alpha)}}$ and do similarly for t_0, then we can write (2.18) and (2.19) as

$$\begin{bmatrix} G_x & G_\alpha \\ t_0^{(x)T} & t_0^{(\alpha)} \end{bmatrix} \cdot \begin{bmatrix} t_1^{(x)} \\ t_1^{(\alpha)} \end{bmatrix} = \begin{bmatrix} 0 \\ 1 \end{bmatrix}, \tag{2.20}$$

where G_x is an $n \times n$ matrix, G_α and $t_0^{(x)}$ are n-vectors, and $t_0^{(\alpha)}$ is a scalar. A matrix with form

$$\begin{bmatrix} A & B \\ C^T & D \end{bmatrix} \tag{2.21}$$

with $A \in \mathbb{R}^{n\times n}$, $B,\ C \in \mathbb{R}^{n\times m}$, $D \in \mathbb{R}^{m\times m}$ will be called a bordered matrix with borderwidth m.

Suppose that a steplength Δs is chosen (this choice will be discussed further in §2.3.3). Then our predictor is

$$\widehat{x} = x_1 + \frac{\Delta s}{\|t_1\|} t_1^{(x)}, \quad \widehat{\alpha} = \alpha_1 + \frac{\Delta s}{\|t_1\|} t_1^{(\alpha)}. \tag{2.22}$$

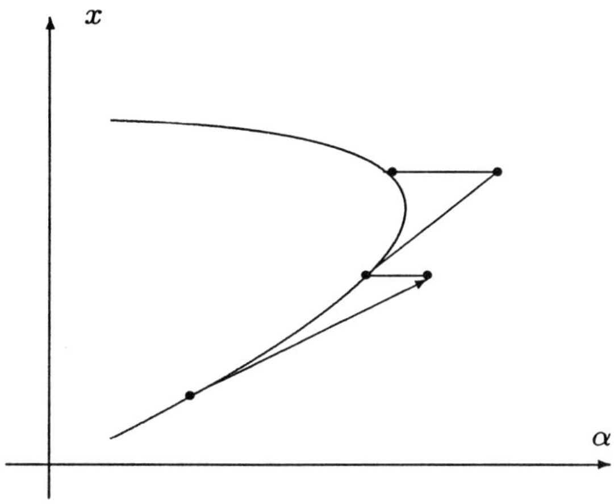

Figure 2.5: Continuation along a branch with limit points.

This is called a pseudoarclength predictor because Δs measures arclength along the tangent line and therefore approximates the arclength along the branch.

There are several possibilities for the corrector step; we discuss four of them.

1. A solution to (2.12) is sought in a coordinate hyperplane through $(\widehat{x}, \widehat{\alpha})$; i.e., an equation of the form $x_i = \widehat{x}_i$ ($i \in \{1, \ldots, n\}$) or $\alpha = \widehat{\alpha}$ is added. For a picture of the former case, see Figure 2.5. The latter case is the one discussed in §2.3.1, which fails if G_x is singular. In the former case the system is

$$G(x, \alpha) = 0, \tag{2.23}$$

$$e_i^T (x - \widehat{x}) = 0, \tag{2.24}$$

where $(e_i)_j = \delta_{ij}$ $(1 \leq j \leq n)$. The Jacobian of (2.23)–(2.24) is hence

$$M = \begin{bmatrix} G_x & G_\alpha \\ e_i^T & 0 \end{bmatrix}. \tag{2.25}$$

One has to choose i so that (2.25) is nonsingular. From a comparison of (2.25) and (2.18) it follows that M is singular if and only if the ith component of t_1 is 0.

The strategy proposed in [201] and implemented in the code PITCON (Pittsburgh Continuation Code; see §2.4) is, therefore, to choose that value of i for which $\frac{|t_{1i}|}{\|t_1\|}$ is maximal (with a small correction:

$$\text{if } \frac{|t_{1i}|}{\|t_1\|} < \frac{|t_{0i}|}{\|t_0\|} \text{ and } \frac{|t_{1j}|}{\|t_1\|} > \frac{|t_{0j}|}{\|t_0\|} \text{ whereby } |t_{1j}| \geq \mu |t_{1i}|$$

for a threshold value $\mu (0 < \mu < 1)$, then one chooses j instead of i).

2. A solution to (2.12) is sought in the hyperplane orthogonal to t_1. The system for the computation of the next point on the branch is, therefore,

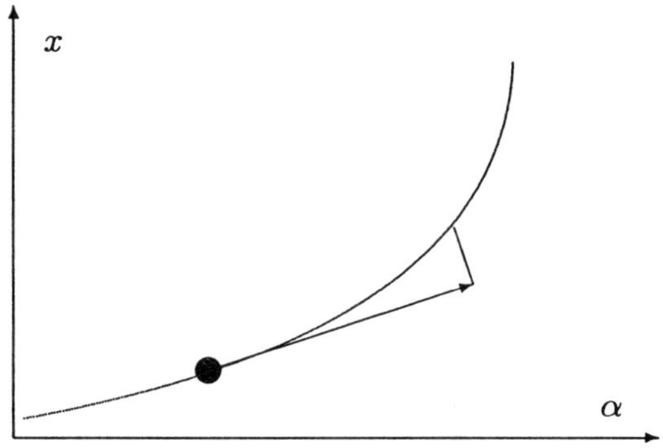

Figure 2.6: Correction in the hyperplane orthogonal to the tangent.

$$G(x, \alpha) = 0, \tag{2.26}$$

$$t_1^{(x)^T}(x - \widehat{x}) + t_1^{(\alpha)^T}(\alpha - \widehat{\alpha}) = 0 \tag{2.27}$$

with Jacobian

$$\begin{bmatrix} G_x & G_\alpha \\ t_1^{(x)^T} & t_1^{(\alpha)} \end{bmatrix}. \tag{2.28}$$

This is called *Keller's method* [151]. See Figure 2.6. We remark that orthogonality is not scale invariant. If x and α have different physical dimensions (as is often the case), then the meaning of orthogonality in $(x^T, \alpha)^T$-space is rather arbitrary. Therefore, in AUTO [80], [84] the possibility is provided to replace (2.27) with

$$\theta_x^2\, t_1^{(x)^T}(x - \widehat{x}) + \theta_\alpha^2\, t_1^{(\alpha)^T}(\alpha - \widehat{\alpha}) = 0 \tag{2.29}$$

with weights $\theta_x > 0$ and $\theta_\alpha > 0$.

3. In a variant of the previous method the tangent vector t_1 is replaced after every corrector step by the normalized singular vector of the Jacobian at the point from where the correction step was started. This is the strategy in CONTENT [165].

4. A solution to

$$G(\widehat{x} + \Delta x,\ \widehat{\alpha} + \Delta\alpha) = 0 \tag{2.30}$$

is sought so that $\|\binom{\Delta x}{\Delta \alpha}\|$ is minimal.

Since $G(x, \alpha)$ is nonlinear, this cannot be done exactly in one step. However, we can perform an iterative algorithm similar to Newton's method. More precisely, we choose $(x^0, \alpha^0) = (\widehat{x}, \widehat{\alpha})$ and $x^{k+1} = x^k + \Delta x$, $\alpha^{k+1} = \alpha^k + \Delta\alpha$, where Δx, $\Delta\alpha$ is the minimum-norm solution to

$$G_x(x^k, \alpha^k)\Delta x + G_\alpha(x^k, \alpha^k)\Delta\alpha = -G(x^k, \alpha^k). \tag{2.31}$$

The iterative scheme, which uses (2.31) as a step, is called the Gauss–Newton iteration [74], [75]. See Figure 2.7. It is analyzed, and quadratic convergence is proved in [74].

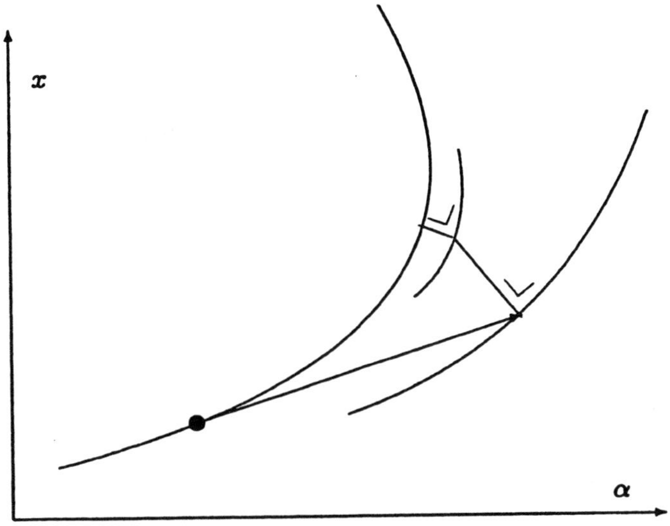

Figure 2.7: Gauss–Newton method.

We note that computing the minimum-norm solution to an underdetermined system of linear equations is a standard problem in numerical linear algebra; it is part of the more general problem of computing minimum norm least squares solutions. This in turn is the numerical procedure to apply the pseudoinverse or Moore–Penrose inverse of a given possibly nonsquare and possibly rank-deficient matrix. The recommended stable algorithm uses a QR decomposition with pivoting of the matrix [106], [230], [70]. However, in the context of continuation and bifurcation problems it is also natural to use a bordered matrix approach. We note that the solution to (2.31) is unique except for a multiple of the nullvector of $[G_x \;\; G_\alpha]$. Hence if one solves the systems

$$\begin{bmatrix} G_x & G_\alpha \\ t_1^{(x)^T} & t_1^{(\alpha)} \end{bmatrix} \begin{bmatrix} t^{(x)} \\ t^{(\alpha)} \end{bmatrix} = \begin{bmatrix} 0 \\ 1 \end{bmatrix}, \tag{2.32}$$

$$\begin{bmatrix} G_x & G_\alpha \\ t_1^{(x)^T} & t_1^{(\alpha)} \end{bmatrix} \begin{bmatrix} \Delta_1 x \\ \Delta_1 \alpha \end{bmatrix} = \begin{bmatrix} -G \\ 0 \end{bmatrix}, \tag{2.33}$$

then the general solution to (2.31) is given by

$$\left\{ \begin{array}{ll} \Delta x & = \Delta_1 x + \eta t^{(x)}, \\ \Delta\alpha & = \Delta_1\alpha + \eta t^{(\alpha)}, \end{array} \right. \tag{2.34}$$

where $\eta \in \mathbb{R}$ is arbitrary. The vector $[\begin{smallmatrix} \Delta x \\ \Delta\alpha \end{smallmatrix}]$ has minimal norm if

$$((\Delta x)^T \Delta\alpha) \begin{pmatrix} t^{(x)} \\ t^{(\alpha)} \end{pmatrix} = 0;$$

i.e.,

$$\eta = -\frac{(\Delta_1 x)^T t^{(x)} + (\Delta_1 \alpha) t^{(\alpha)}}{\|t\|^2}. \tag{2.35}$$

Extensions of this method to more general minimum norm and least squares problems can be found in [120].

Remark. In each of the four methods the solution of systems with G_x is replaced by the solution of systems with form

$$M = \begin{bmatrix} G_x & b \\ c^T & d \end{bmatrix},$$

i.e., bordered matrices with borderwidth 1. The underlying idea is that M can be nonsingular even if G_x is singular. We will discuss bordered matrices in detail in Chapter 3.

2.3.3 Steplength Control

Steplength control is an important part of a continuation method. If the steplength is too small, then a lot of unnecessary work is done. If it is too large, then the corrector algorithm may converge to a point on a different branch of (2.12) or not converge at all.

It is possible to make an a priori estimate of a good steplength, although such an estimate is never completely reliable. Suppose that we have computed a point (u_k, α_k) with normalized tangent vector t_k and the preceding point (x_{k-1}, α_{k-1}) with normalized tangent vector t_{k-1}. Let $t_k = (t_k^{(x)T}, t_k^{(\alpha)})^T$ and $t_{k-1} = (t_{k-1}^{(x)T}, t_{k-1}^{(\alpha)})^T$ be the natural splittings of t_k and t_{k-1}, respectively.

For a pseudoarclength Δs the predictor is

$$\widehat{x} = x_k + (\Delta s) t_k^{(x)}, \tag{2.36}$$

$$\widehat{\alpha} = \alpha_k + (\Delta s) t_k^{(\alpha)}. \tag{2.37}$$

Since the prediction is along the tangent line the error has order $(\Delta s)^2$ for small (Δs) (Taylor expansion!). An estimate of the error may be obtained by comparing the prediction with one of order $(\Delta s)^3$, namely,

$$q(\Delta s) = \begin{pmatrix} x_k \\ \alpha_k \end{pmatrix} + (\Delta s) t_k + \frac{1}{2} (\Delta s)^2 w_k, \tag{2.38}$$

whereby

$$w_k = \frac{1}{\Delta s_k}(t_k - t_{k-1}), \quad \Delta s_k = \left\| \begin{pmatrix} x_k \\ \alpha_k \end{pmatrix} - \begin{pmatrix} x_{k-1} \\ \alpha_{k-1} \end{pmatrix} \right\|.$$

Indeed, from

$$q(0) = \begin{pmatrix} x_k \\ \alpha_k \end{pmatrix}, \quad \frac{d\, q(\Delta s)}{d\, (\Delta s)}(0) = t_k, \quad \left(\frac{d\, q(\Delta s)}{d(\Delta s)}\right)(-\Delta s_k) = t_{k-1},$$

it follows that (2.38) is an estimate with order $\max\{|\Delta s_k||\Delta s|^2, |\Delta s|^3\}$ (check this!).

Suppose that a tolerance ε is chosen for the absolute error in $(\widehat{x}, \widehat{\alpha})$. The error in $q(\Delta s)$ is about $\frac{1}{2}(\Delta s)^2\|w_k\|$; therefore, the choice

$$\Delta s \simeq \sqrt{\frac{2\varepsilon}{\|w_k\|}} \tag{2.39}$$

is a reasonable one.

This choice is used (with some additional constraints) for steplength control in PITCON (see §2.4). The following corrector step can of course lead to a rejection of this stepsize and replacement by a smaller one.

It is also possible to base the steplength control completely on the convergence of Newton iterates in the corrector step. A simple way to do this is implemented in CONTENT. The convergence is declared successful if both $\|G\|$ and the length of the last correction vector are below certain user-chosen threshold values after no more than three correction steps; the steplength is then multiplied by 1.3. It is declared unsuccessful if the thresholds are not reached after a user-chosen number $I_{max} > 3$ of correction steps; the steplength is then halved and corrections are started again. In the intermediate case convergence is declared successful, but the steplength is left unchanged.

A more sophisticated method is used in AUTO [80], [84]. We discuss this further.

Suppose the corrector step for the computation of $(x_k^T, \alpha_k)^T$ starts from $(x_k^{(0)T}, \alpha_k^{(0)})^T = (\widehat{x_k}^T, \widehat{\alpha_k})^T$ and computes consecutively the values $(x_k^{(1)T}, \alpha_k^{(1)})^T, (x_k^{(2)T}, \alpha_k^{(2)})^T, \dots$ whereby $x_k^{(l+1)} = x_k^{(l)} + \Delta x^{(l+1)}, \alpha_k^{(l+1)} = \alpha_k^{(l)} + \Delta\alpha^{(l+1)}$. For every correction one computes the norm

$$\varepsilon^{(l+1)} = \|(\Delta x^{(l+1)T}, \Delta\alpha^{(l+1)})^T\|,$$

and the iteration is stopped if $\varepsilon^{(l+1)} < \mathcal{T}$ whereby $\mathcal{T}$ is a threshold value, e.g., $\mathcal{T} = 10^{-5}\|(x_{k-1}^T, \alpha_{k-1})^T\|$ if the machine precision is about 10^{-15}.

One decides that the iteration is unsuccessful if the condition

$$\varepsilon^{(l+1)} < \mathcal{T} \tag{2.40}$$

is not satisfied after 10 iterations or if $\varepsilon^{(l+1)} > \varepsilon^{(l)}$. In the case of unsuccessful iterations the steplength is halved; below a certain minimal threshold the continuation is interrupted.

Now let Δs_{k-1} be the steplength of the previous step; i.e.,

$$(\widehat{x_k}^T, \widehat{\alpha_k})^T = (x_{k-1}^T + \Delta s_{k-1} t_{k-1}^{(x)}, \alpha_{k-1} + \Delta s_{k-1} t_{k-1}^{(\alpha)})^T$$

is the point that is reached after successful correction step has yielded (x_k, α_k).

Let N_0 be the number of iterations that was needed to satisfy (2.40), and let us denote the norms of the corrections by $\varepsilon_{k-1}^{(n)}$ $(1 \le n \le N_0)$.

We now compute a new steplength Δs_k to find a suitable starting point $(\widehat{x}_{k+1}^T, \widehat{\alpha}_{k+1})^T = (x_k^T, \alpha_k)^T + \Delta s_k (t_k^{(x)T}, t_k^{(\alpha)})^T$.

Let us denote by $\varepsilon_k^{(n)}$ the (yet unknown) norms of the corrections that will be needed when starting from this point. From the hypothesis of quadratic convergence (theory of

Newton iterates) we find

$$\varepsilon_i^{(n)} = C(\varepsilon_i^{(n-1)})^2 = C^{1+2}(\varepsilon_i^{(n-2)})^4 \tag{2.41}$$
$$= C^{1+2+4}(\varepsilon_i^{(n-3)})^8 = \cdots = C^{(2^{n-1}-1)}(\varepsilon_i^{(1)})^{(2^{n-1})} \tag{2.42}$$

for $i = k - 1$ and for $i = k$.

We suppose further that

$$\varepsilon_i^{(1)} = \alpha(\Delta s_i) \quad (i = k-1, k). \tag{2.43}$$

A theoretical analysis, based on quadratic convergence, leads to an approximation of the form $\varepsilon_i^{(1)} = \alpha(\Delta s_i)^2$, but this is true only in the limit for small (Δs_i), and we have no reason to choose Δs smaller than necessary. The estimate (2.43) is in practice better if the steps are not very small.

We want to choose (Δs_k) so that (2.40) is satisfied after K steps whereby K is chosen heuristically. We always perform at least three Newton iterations, and ideally this number should be sufficient. So $N_0 \geq 3$ and we set

$$K = \begin{cases} 3 \text{ if } N_0 = 3, \\ N_0 - 1 \text{ if } N_0 > 3. \end{cases}$$

From (2.42) and (2.43) it follows that

$$\frac{\varepsilon_k^{(K)}}{\varepsilon_{k-1}^{(K)}} = \left(\frac{\Delta s_k}{\Delta s_{k-1}}\right)^{(2^{K-1})}.$$

If we require (as a safety measure) that $\varepsilon_k^{(K)} \simeq 0.1\mathcal{T}$ instead of $\varepsilon_k^{(K)} < \mathcal{T}$, then we get

$$\Delta s_k = \Delta s_{k-1} \cdot \left(\frac{0.1\mathcal{T}}{\varepsilon_{k-1}^{(K)}}\right)^{(2^{-K+1})}.$$

2.3.4 Convergence of Newton Iterates

Since convergence of Newton iterates is essential for most algorithms discussed in this book (not only in continuation), we take the opportunity to recall some basic facts. Suppose we are solving

$$F(x) = 0 \quad (x \in \mathbb{R}^n, F(x) \in \mathbb{R}^n), \tag{2.44}$$

using a starting point x^0 and the iterations

$$F_x(x^k)\Delta x^k = -F(x^k), \tag{2.45}$$

$$x^{k+1} = x^k + \Delta x^k. \tag{2.46}$$

An important property of this process is its affine invariance. Namely, if A is an arbitrary nonsingular matrix, then (2.44) is equivalent to the solution of

$$G(x) := AF(x) = 0. \tag{2.47}$$

The sequence $\{x^0, x^1, x^2, \ldots\}$ of Newton iterates is independent of A; from (2.45) it follows indeed that

$$G'(x^k)\triangle x^k = -G(x^k).$$

A convergence test of the form

$$\|\triangle x^{k+1}\| \leq \varepsilon$$

or monotonicity test

$$\|\triangle x^{k+1}\| \leq \theta \|\triangle x^k\| \tag{2.48}$$

for a fixed θ $(0 < \theta < 1)$ is also affine invariant. On the other hand, the apparently natural test

$$\|F(x^{k+1})\| \leq \theta \|F(x^k)\| \tag{2.49}$$

$(0 < \theta < 1)$ is not affine invariant. If $F(x^k)$ and $F(x^{k+1})$ are linearly independent vectors (as they usually are), then for any given $\theta > 0$ it is in fact possible to find a nonsingular matrix A such that

$$\|AF(x^{k+1})\| > \theta \|AF(x^k)\|.$$

On the other hand, an affine invariant convergence and monotonicity test can be based on $\|F_x(x^k)^{-1}F(x^{k+1})\|$. To be precise, the monotonicity test would be

$$\|F_x(x^k)^{-1}F(x^{k+1})\| \leq \theta \|F_x(x^k)^{-1}F(x^k)\| \tag{2.50}$$

$(0 < \theta < 1)$; affine invariance follows from

$$(AF_x(x^k))^{-1}(AF(x^{k+1})) = F_x(x^k)^{-1}F(x^{k+1}).$$

We remark that in (2.50) the computation of the right-hand side requires little additional work since $F_x(x^k)^{-1}F(x^k) = -\triangle x^k$ is already known. The left-hand side requires a new solution with the system $F_x(x^k)$. If these systems are solved by a factorization method (e.g., LU or QR), then the factorization of $F_x(x_k)$ is done already in the computation of $\triangle x_k$, and the additional work of solving a system is small. In [75] the reduced Newton correction $\bar{\triangle} x^{k+1}$ is defined by

$$F_x(x^k)\bar{\triangle} x^{k+1} = -F(x^{k+1}) \tag{2.51}$$

and the test (2.50) hence becomes

$$\|\bar{\triangle}(x^{k+1})\| \leq \theta \|\triangle x^k\|. \tag{2.52}$$

In [75] the value $\theta = 0.5$ is recommended in (2.52).

2.3.5 Some Practical Considerations

To start a continuation algorithm one needs a starting point $(u_0^T, \alpha_0)^T$, a tangent vector in that point, an initial steplength, and an orientation along the tangent.

A starting point is usually found by ad hoc methods. In many cases the problem has a trivial solution for certain parameter values. If only a guess is available, then we can try to improve it using a Newton iteration or any other solution method for nonlinear systems. A tangent vector is found by solving the system

$$\begin{bmatrix} G_u & G_\alpha \\ c^T & d \end{bmatrix} \begin{bmatrix} t^{(u)} \\ t^{(\alpha)} \end{bmatrix} = \begin{bmatrix} 0 \\ 1 \end{bmatrix}, \tag{2.53}$$

where $c \in \mathbb{R}^n, d \in \mathbb{R}$ are chosen fairly arbitrarily; it is only required that the matrix in (2.53) be nonsingular. An initial steplength is usually found by trial and error. For the orientation only two choices are possible and both may be worthwhile.

2.4 Notes and Further Reading

1. For a more complete treatment of numerical continuation methods we refer to [201], [74], [22] (code COLCON for boundary value problems), and [6].

Several good continuation codes are freely available. One of them is PITCON. For related literature see [201], [198], [73], [199], [200]. It can be obtained from the NETLIB library at http://www.netlib.org as contin/pcon61.f (single precision) or contin/dpcon61.f (double precision). For the computation of implicitly defined manifolds contin/manpak is also provided.

ALCON is another continuation code, developed at the Konrad Zuse Institut in Berlin. It can be obtained from http://elib.zib.de/pub/Packages/alcon_s. One of its main features is the use of Gauss–Newton iterations in the corrector stage.

An extensive list of continuation and related software is given in [9]. Unfortunately, such lists tend to become obsolete rather quickly.

2. Most codes work fine in simple low-dimensional problems as those we consider in the exercises for this chapter. On the other hand, some problems require a good continuation code and a careful handling of tolerance and threshold parameters. Examples in this book include Exercise 15, §5.6 and the computations in §7.7.

3. To detect bifurcations it is often necessary to monitor certain test functions while computing branches. We will see many examples in Chapters 4 and 5 and beyond. It has been suggested that these test functions can also be used in the choice of the steplength. Roughly the idea is to decrease the steplength if the behavior of the test functions suggests that a bifurcation point might be nearby. Alhough reasonable, this idea has not yet found its way into software.

2.5 Exercises

Note. CONTIN is a generic name for a continuation code, which the reader may write or, preferably, download from one of the sites mentioned in §2.4. The exercises that refer to CONTIN were tested with PITCON.

1. Prove that the set of rank-2 matrices in $\mathbb{R}^{3\times3}$ is a differentiable manifold with dimension 8 and give a global defining function.

2. Prove that the set of rank-1 matrices in $\mathbb{R}^{3\times3}$ is a differentiable manifold with dimension 5 and describe local defining systems for every matrix in the manifold.

3. Prove that the set of 2×2 matrices with exactly one zero eigenvalue is a differentiable manifold in $\mathbb{R}^4$. Give a local defining function for every matrix in the manifold. Is it also a global defining function?

4. The set of real normal 2×2 matrices ($AA^T = A^TA$) does not constitute a manifold. It is a finite disjoint union of manifolds of dimensions 1, 2, and 3. Prove this. (Such structures are called stratified sets; we omit the precise definition.)

5. Use CONTIN to compute 25 points of the curve $y = x^2$ starting with $x = 1, y = 1$ with initially increasing values of x.

6. Use CONTIN to find the smallest solution to the equation

   $x^2 - 3x + 2 = 0, x > 0$ (start $x = 0$) (*Hint:* $G(Y, X) \equiv Y - X^2 + 3X - 2 = 0$).

7. Compute $\sqrt{2}$ with CONTIN.

8. Use CONTIN to compute a number of solution points to $x^2 - y^3 = 0$, starting with $(x, y) = (-1, 1)$ with initially decreasing values of y. Can you continue the branch to points with $x > 0$? Then try again with $x^2 - y^3 + \epsilon = 0$, where ϵ is a positive or negative number. Explain what happens.

9. Find with CONTIN all real roots of the cubic equation $x^3 - 6x^2 + 11x - 6 = 0$.

10. Find with CONTIN all real roots of the equation $x^7 - 6x + 1 = 0$.

11. Continue with CONTIN the solution branch to the system

$$G(X(1), X(2), \lambda) = \frac{1}{h^2}\begin{bmatrix} -2 & 1 \\ 1 & -2 \end{bmatrix}\begin{bmatrix} X(1) \\ X(2) \end{bmatrix} + \lambda \begin{bmatrix} e^{X(1)} \\ e^{X(2)} \end{bmatrix}$$

   with $h = \frac{2}{3}$, starting point $(0, 0, 0)$ in the direction of initially decreasing λ-values (cf. §1.4). Questions:

   (a) Does one stay in the symmetric space $\mathcal{S} = \{(X(1), X(2), \lambda) : X(1) = X(2)\}$?

   (b) Is there a turning point with respect to λ? $(1, 1, -0.8277287)$

12. Find an asymmetric solution branch to the problem in the previous exercise by starting from a point in the neighborhood of the symmetry-breaking bifurcation point $(3, 3, -0.3360627)$.

13. Consider the map $G : \mathbb{R}^3 \to \mathbb{R}^2$, where

$$G(x) \equiv \begin{pmatrix} x_1 - x_2^3 + 5x_2^2 - 2x_2 + 34x_3 - 47 \\ x_1 + x_2^3 + x_2^2 - 14x_2 + 10x_3 - 39 \end{pmatrix}$$

[201, p. 146].

   (a) Which one of x_1, x_2, x_3 can be used easily to parameterize the solution set of the system? Why?

   (b) Give an explicit parameterization of the solution set by this variable.

   (c) Give a starting point with $x_2 = 0$.

   (d) Find with CONTIN a point with $x_2 > 0$ and $x_1 = 0$. Is there guaranteed to be one? Can you also find it with analytical methods?

   (e) Is there a point with $x_2 > 0$ and $x_3 = 0$?

   (f) Find with CONTIN a maximum of x_1 in the region $x_2 > 0$. Can you also find it with analytical methods?

14. Consider the following eutrophication model:

$$\begin{cases} \dot{x}_1 = & x_1(0.2(\lambda_1 - x_1 - x_2) - 0.445x_3 - 4), \\ \dot{x}_2 = & -0.0455x_2x_3 + 4x_1, \\ \dot{x}_3 = & \lambda_2(10 - x_3) - 2.67x_3(0.445x_1 + 0.0455x_2). \end{cases}$$

Here x_3 is the concentration of oxygen in the hypolimnion of a flat lake; x_1 and x_2 are the nutrient densities stored in biomass and detritus, respectively. The parameter λ_1 is the sum of all nutrients, including those in dissolved form and assumed to be constant; λ_2 determines the transportation of oxygen from epilimnion to hypolimnion (cf. Werner [237]).

   (a) Compute for $\lambda_2 = 0.7$ a curve of stationary solutions in $(x_1, x_2, x_3, \lambda_1)$-space starting from $(x_1 = 0, x_2 = 0, x_3 = 10, \lambda_1 = 42.25)$ in the direction of initially increasing x_1.

   (b) Find a turning point of this curve with respect to λ_1. Conclusion: There exist parameter pairs (λ_1, λ_2) for which two equilibrium states (x_1, x_2, x_3) are possible.

(*Remark*: You may have starting problems. See §7.6.2 for a study of the local behavior near the starting point.)

15. Consider the function

$$G(x, y) = -y^2 + x + x^2 + 0.1x^4 + 0.01x^6 - 0.001x^8$$

$(x, y \in \mathbb{R})$.

(a) Compute with CONTIN all turning points (with respect to either x or y) of the branch of $G(x, y) = 0$ through $(0, 0)$.

(b) Another solution branch goes through $(-3.570165, 4.442488)$. Compute also all turning points on this branch.

(c) Make a sketch of the computed solution branches to $G(x, y) = 0$ in $\mathbb{R}^2$.

(d) Prove that the solution set to $G(x, y) = 0$ is bounded.

16. Consider the function

$$G(x, y) = 7x^6 + x^3y + xy + y^2 - 8y^6.$$

(a) Compute all solutions of the equation $G(x, y) = 0$ that lie on the circle $\{(x, y) : x^2 + y^2 = 9\}$. (Hint: There is precisely one in each quadrant.)

(b) Do the computed points exhibit a form of symmetry? Was this to be expected?

(c) Compute for every i $(1 \leq i \leq 4)$ a solution branch of $G(x, y) = 0$ through P_i and follow it into the interior of the disc $\{(x, y) : x^2 + y^2 \leq 9\}$ until it leaves the disc again.

(d) Do all these branches pass through $(0, 0)$?

(e) Compute with CONTIN all limit points in the disc $\{(x, y) : x^2 + y^2 \leq 9\}$ with respect to either x or y.

(f) What can you guess from the computed points concerning the tangents at $(0, 0)$ to all branches?

(g) Compute the intersection points of all branches with the coordinate axes. Make a sketch of all solutions to the equation $G(x, y) = 0$ inside the disc $\{(x, y) : x^2 + y^2 \leq 9\}$. Indicate all the computed special points.

17. The 3-box Brusselator describes the Belusov–Zhabotinski reaction in a system of three cyclically coupled cells (see §4.2 for some background). In the ith cell a component with concentration X_i and one with concentration Y_i react in a process that involves the parameters A, B. Adjacent cells influence each other by diffusion, which is governed by diffusion coefficients D_1, D_2 and a coupling coefficient λ. The global equation of the so formed dynamical system is

$$\dot{X}_1 = D_1(X_2 - 2X_1 + X_3) + \lambda(A - (B+1)X_1 + X_1^2 Y_1),$$
$$\dot{Y}_1 = D_2(Y_2 - 2Y_1 + Y_3) + \lambda(BX_1 - X_1^2 Y_1),$$
$$\dot{X}_2 = D_1(X_3 - 2X_2 + X_1) + \lambda(A - (B+1)X_2 + X_2^2 Y_2),$$
$$\dot{Y}_2 = D_2(Y_3 - 2Y_2 + Y_1) + \lambda(BX_2 - X_2^2 Y_2),$$
$$\dot{X}_3 = D_1(X_1 - 2X_3 + X_2) + \lambda(A - (B+1)X_3 + X_3^2 Y_3),$$
$$\dot{Y}_3 = D_2(Y_1 - 2Y_3 + Y_2) + \lambda(BX_3 - X_3^2 Y_3).$$

We assume $D_1 = 1$, $D_2 = 10$, $A = 2.1$, $B = 3.5$.

(a) Write a driving program for CONTIN to compute a branch of equilibrium solutions with λ as the free parameter. Compute a number of points on a branch through the point with coordinates $X_1 = X_2 = 1.087734$, $Y_1 = Y_2 = 1.8488063$, $X_3 = 4.124552$, $Y_3 = 1.0389936$, $\lambda = 5$.

(b) Are the symmetry relations $X_1 = X_2$, $Y_1 = Y_2$ preserved along the branch?

(c) This question is similar to (a) for the branch through $X_1 = X_3 = 2.5975837$, $Y_1 = Y_3 = 1.4653965$, $X_2 = 1.1048325$, $Y_2 = 1.8634635$, $\lambda = 5$.

(d) This question is similar to (b) for the branch computed in (c).

(e) Does there exist a fully symmetric solution for $\lambda = 5$, i.e., a solution with $X_1 = X_2 = X_3$, $Y_1 = Y_2 = Y_3$? If so, then find for this solution the precise values X_1, Y_1. Describe the solution branch with free parameter λ through this point.

(f) Find (without computing) four not-yet-mentioned equilibrium solutions for $\lambda = 5$.

(g) Compute with CONTIN a branch of equilibrium solutions through the point with coordinates $X_1 = 1.1001799$, $Y_1 = 2.079928$, $X_2 = 1.9698811$, $Y_2 = 1.7320476$, $X_3 = 3.229939$, $Y_3 = 1.2280244$, $\lambda = 9$.

(h) Is the so-computed branch a closed one? Compute all limit points of this branch. How many do you find? How many points does the branch contain with $\lambda = 5$? How many with $\lambda = 9$?

(i) Can you find seven equilibrium points with $\lambda = 9$ that are not on the branch computed in (g)?

Chapter 3

Bordered Matrices

One of the most important objects in a nonlinear dynamical system is the Jacobian matrix of the system. Matrix constructions are at the heart of many methods. In Chapter 2 we saw that bordered matrices naturally arise in numerical continuation, especially in the case where the Jacobian is singular.

We recall that the rank r of a matrix $A \in \mathbb{R}^{m \times n}$ is the number of linearly independent columns (equivalently, rows). The number $\min(m, n) - r$ is called the *rank deficiency* of A. (*Rank defect* and *rank drop* are also common names.) In this chapter we study the construction and use of bordered matrices in a systematic way, in particular to obtain defining conditions for rank deficiencies in a matrix. This is essential in several methods that we will study later on. We assume that the reader has a basic knowledge of linear algebra, including singular values. Otherwise, this chapter is completely self-contained. By I_n we will denote the $n \times n$ identity matrix.

3.1 Introduction: Motivation by Cramer's Rule

Consider a matrix $A \in \mathbb{R}^{n \times n}$. We can think of A as a function of its entries. Suppose that we want conditions to express that A is singular. An obvious choice is $\det(A) = 0$. This choice has several disadvantages if n is not very small. First, the determinant is usually badly scaled as a function of the entries of A (since it is a polynomial of order n in these entries). To be computed exactly, it needs a large number of operations. In practice, it will usually be computed as a byproduct of an LU decomposition. This is inconvenient if we decide to solve the linear systems by an indirect (iterative) method. Also, the derivatives of the determinant function are even harder to compute; usually one will have to use finite difference approximations.

Bordered matrices allow us to find a substitute function of the determinant that avoids most of the above-mentioned problems. Consider a bordered extension M of A of the form

$$M = \begin{pmatrix} A & b \\ c^T & d \end{pmatrix},$$

where $b, c \in \mathbb{R}^n, d \in \mathbb{R}$. Suppose that M is nonsingular and we solve the system

$$M\begin{pmatrix} r \\ s \end{pmatrix} = \begin{pmatrix} 0_n \\ 1 \end{pmatrix},$$

where $r \in \mathbb{R}^n, s \in R$. Then by Cramer's rule we have

$$s = \frac{\det(A)}{\det(M)},$$

and so $\det(A) = 0$ if and only if $s(A) = 0$. Since $\det(M) \neq 0$ in some neighborhood of A, we can use the equation $s(A) = 0$ to characterize locally the matrices A that are singular. This approach avoids the scaling problems and is based on the solution of linear systems only. Also, we will show how to compute derivatives of $s(A)$. Since rectangular matrices and rank deficiencies higher than one are also important, we prefer to give precise results in a more general setting.

Consider a block matrix M with the simple form

$$M = \begin{pmatrix} A & B \\ C^T & D \end{pmatrix}, \tag{3.1}$$

where $A \in \mathbb{R}^{p\times q}$, $B \in \mathbb{R}^{p\times(n-q)}$, $C \in \mathbb{R}^{q\times(n-p)}$, $D \in \mathbb{R}^{(n-p)\times(n-q)}$.

Here $p < n$, $q < n$. If $(n-p), (n-q)$ are small then we call (3.1) a bordered matrix. This is sometimes numerically useful if A has a special structure or sparsity while the bordering matrices B, C^T, D have no particular structure. But in most cases the reason for introducing the block partition of M in (3.1) is that the different blocks each have a specific meaning in the theory in which M arises. So they deserve to be treated separately.

Our main use of bordered matrices will be to obtain conditions (smooth functions of the entries of M) that express that A has a specified rank deficiency r. The defining systems are local in the sense of being defined on the sets of a predefined cover of the manifold of all rank-r matrices with globally open and dense sets. The obtained systems have full linear rank; hence they define the underlying manifold in a regular way and can be used analytically to obtain local parameterizations or used numerically to set up Newton systems with local quadratic convergence. The final description of this is in §3.4. The main tool is the singular value inequality (Proposition 3.3.5).

3.2 The Construction of Nonsingular Bordered Matrices

In this section we prove some general results on matrices of the form (3.1). The first one is sometimes called Keller's lemma; actually it is an easy result even in the somewhat generalized form that we present. We will say that two linear subspaces of $\mathbb{R}^n$ are transversal if together they span $\mathbb{R}^n$.

Proposition 3.2.1. Let $A \in \mathbb{R}^{p\times q}$ and suppose that A has rank $r < \min(p,q)$. Let n be an integer with $n > p, n > q$ and let M be defined as in (3.1) with $B \in \mathbb{R}^{p\times(n-q)}$, $C \in \mathbb{R}^{q\times(n-p)}$, $D \in \mathbb{R}^{(n-p)\times(n-q)}$. Then the following conditions hold:

1. If $r + n < p + q$, then M is singular.

2. If $r+n = p+q$, then M is nonsingular if and only if B spans a subspace transversal to the range of A and C spans a subspace transversal to the range of A^T. In this case B and C have rank $n - q$ and $n - p$ (full column rank), respectively.

Proof. By padding a matrix with an additional row or column its rank increases by at most one. Hence the rank of M is at most $r + (n - q) + (n - p)$. If M is nonsingular, this implies $r + (n - q) + (n - p) \geq n$; i.e., $r + n \geq p + q$.

Now assume $r + n = p + q$ and that M is nonsingular. Then the columns of A and B together must span $\mathbb{R}^p$. Since B has $n - q = p - r$ columns and since the columns of A span an r-dimensional space only, this implies that B has full column rank and spans a subspace transversal to the range of A. The results on C are proved similarly.

Conversely, assume that $r + n = p + q$ and that B and C span subspaces transversal to the range of A and the range of A^T, respectively. Then the ranges of A and B have only the zero vector in common and so have the ranges of A^T and C. To prove that M is nonsingular, suppose that $Ax + By = 0, C^Tx + Dy = 0$, where $x \in \mathbb{R}^q, y \in \mathbb{R}^{n-q}$. From the first equality it follows that $Ax = By = 0$. Since B has rank $n-q$ this implies $y = 0$. Now we have $Ax = 0, C^Tx = 0$. Since A^T, C are transversal, this implies $x = 0$. □

In some applications $A \in \mathbb{R}^{p\times q}$ with rank r is given and one wants to choose a minimal n and $B \in \mathbb{R}^{p\times(n-q)}, C \in \mathbb{R}^{q\times(n-p)}, D \in \mathbb{R}^{(n-p)\times(n-q)}$ such that M is nonsingular. Proposition 3.2.1 shows that $n = p+q-r$ is minimal and that a generic choice of B, C, D will do. For practical purposes, "generic" here means that B, C, D could be generated by a random number generator. Of course one might want to choose B, C, D in such a way that M is optimal in the sense of having a minimal condition number. If the kernels of A and A^T are known, this can be accomplished. We recall that with each matrix, e.g., $A \in \mathbb{R}^{p\times q}$, we can associate its singular values $\sigma_1(A) \geq \sigma_2(A) \geq \cdots \geq \sigma_{\min(p,q)} \geq 0$. If A is square, then its spectral condition number $\kappa(A)$ is the ratio between the largest and smallest singular values. We have the following result.

Proposition 3.2.2. Suppose that $A \in \mathbb{R}^{p\times q}$ has rank $r < \min(p, q)$ and $n = p+q-r$. For any $B \in \mathbb{R}^{p\times(n-q)}, C \in \mathbb{R}^{q\times(n-p)}, D \in \mathbb{R}^{(n-p)\times(n-q)}$ with M defined by (3.1) we have

$$\sigma_1(M) \geq \sigma_1(A). \tag{3.2}$$

If $r > 0$, then also

$$\sigma_n(M) \leq \sigma_r(A) \tag{3.3}$$

and

$$\kappa(M) \geq \frac{\sigma_1(A)}{\sigma_r(A)}. \tag{3.4}$$

If $r > 0$ and we choose $D = 0$ and B, C such that their columns form orthogonal bases for the kernels of A^T, A, respectively, and the norms of the columns of B, C are between $\sigma_r(A)$ and $\sigma_1(A)$ (bounds included); then the lower bound $\kappa(M) = \frac{\sigma_1(A)}{\sigma_r(A)}$ is reached.

If $r = 0$ and we choose $D = 0$ and B, C as orthogonal matrices with dimensions $p \times p$ and $q \times q$, respectively, then $\kappa(M) = 1$.

Proof. The case $r = 0$ is trivial, so we assume $r > 0$. The spectral norm of a matrix does not decrease if we pad the matrix with additional rows or columns; hence (3.2) is obvious. Now by the definition of singular values there exist a $(q-r+1)$-dimensional subspace V of $\mathbb{R}^q$ such that $\|Av\| \le \sigma_r(A)\|v\|$ for all $v \in V$. Since C has only $n-p = q-r$ columns, there exists a nonzero $v \in V$ such that $C^T v = 0$. If we set

$$w = \begin{pmatrix} v \\ 0_{n-q} \end{pmatrix},$$

then $\|Mw\| = \|Av\| \le \sigma_r(A)\|v\| = \sigma_r(A)\|w\|$. Hence (3.3) follows. We now prove the statement of how to reach the lower bound of $\kappa(M)$ by an appropriate choice of B, C, D. First we remark that A has a singular value decomposition $A = U^T A_d V$, where $U \in \mathbb{R}^{p\times p}$ and $V \in \mathbb{R}^{q\times q}$ are orthogonal matrices and $A_d = \mathrm{Diag}_{p,q}(\sigma_1, \ldots, \sigma_r, 0, \ldots, 0)$. Then

$$\begin{pmatrix} U & 0_{p,n-p} \\ 0_{n-p,p} & I_{n-p,n-p} \end{pmatrix} \begin{pmatrix} A & B \\ C^T & D \end{pmatrix} \begin{pmatrix} V^T & 0_{q,n-q} \\ 0_{n-q,q} & I_{n-q,n-q} \end{pmatrix} = \begin{pmatrix} A_d & UB \\ C^T V^T & D \end{pmatrix}. \tag{3.5}$$

Now it is easy to see that the matrix on the right-hand side of (3.5) has condition number $\frac{\sigma_1(A)}{\sigma_r(A)}$ if we choose $D = 0$, and B, C so that UB, VC have columns with norms between $\sigma_r(A)$ and $\sigma_1(A)$, and the columns of UB, VC form orthogonal bases for the kernels of A_d^T and A_d, respectively. These conditions are equivalent to the conditions on B, C, D given in the proposition. □

3.3 The Singular Value Inequality

Suppose that a block matrix M of the form (3.1) is nonsingular. Can we manipulate M to obtain information on the rank of A? We shall prove a result in this direction, slightly generalizing the main result in [111] (where only the case of a square A is considered).

We first fix some notation. It is convenient to pad the singular values of every matrix with trailing zeros; i.e., $\sigma_i(A) = 0$ if $A \in \mathbb{R}^{m\times n}$ and $i > \min(m, n)$. This simplifies the statement of many results. It is useful to formulate the following elementary result first.

Lemma 3.3.1

1. Let $\lambda_i(X)$ be the eigenvalues of a nonnegative definite $k \times k$ matrix X, ordered $\lambda_1(X) \ge \cdots \lambda_k(X)$. Then
$$\sigma_i(A)^2 = \lambda_i(A^T A),$$
$$\sigma_i(A)^2 = \lambda_i(AA^T),$$
where the $\lambda_i(A^T A)$, $\lambda_i(AA^T)$ are padded with trailing zeros if necessary.

2. $\sigma_i(A) = \sigma_i(A^T)$ for all i.

3. For any i, $\sigma_i(A) = \min\{\|A - \tilde{A}\| : \tilde{A} \text{ has rank } i\}$.

4. The singular values of a matrix are unchanged by pre- or postmultiplication with a unitary matrix, by permuting or changing the sign of rows or columns, or by padding the matrix with zero rows or columns.

5. If B is a submatrix of A, then $\sigma_i(B) \le \sigma_i(A)$ for all i.

6. If $B = XAY$, then $\sigma_i(B) \leq \|X\|\|Y\|\sigma_i(A)$ for all i.

7. Let $A \in \mathbb{R}^{m\times n}$ and

$$B = \begin{pmatrix} I_n \\ A \end{pmatrix}.$$

Then $\sigma_i(B)^2 = 1 + \sigma_i(A)^2$ for $1 \leq i \leq n$.

Proof. Item 1 is standard and implies item 2. Item 3 is standard and implies items 4, 5, and 6. To prove item 7 we remark that $B^TB = I_n + A^TA$. The n eigenvalues of B^TB are thus obtained by adding 1 to those of A^T. Together with item 1 this implies item 7. □

It is convenient to prove the singular value inequality in three stages, in increasing order of generality. The first form is shown in Proposition 3.3.2.

Proposition 3.3.2. Let

$$M = \begin{pmatrix} A & B \\ C^T & D \end{pmatrix}$$

be a nonsingular block matrix with $A, B, C, D \in \mathbb{R}^{n\times n}$. Let the inverse

$$M^{-1} = \begin{pmatrix} P & Q \\ R^T & S \end{pmatrix}$$

be decomposed similarly. Then

$$\|M\|^{-2}\sigma_i(A) \leq \sigma_i(S) \leq \|M^{-1}\|^2\sigma_i(A) \tag{3.6}$$

for all i.

Proof. By the symmetry of the problem, it is sufficient to prove the second inequality in (3.6). Without loss of generality we may assume that A is nonsingular. Indeed, all quantities in (3.6) are continuous functions of the entries of A and an arbitrary small perturbation of a singular matrix can make it nonsingular.

Put $A^{-1}B = V$. Then

$$\begin{pmatrix} A^{-1} & 0 \end{pmatrix} M = \begin{pmatrix} I_n & V \end{pmatrix}. \tag{3.7}$$

From (3.7) and Lemma 3.3.1, items 5 and 6, we infer

$$\sigma_i(A^{-1}) \leq \|M^{-1}\|\sigma_i\begin{pmatrix} I_n & V \end{pmatrix} \tag{3.8}$$

for all i. Now

$$\left(\sigma_i\begin{pmatrix} I_n & V \end{pmatrix}\right)^2 = \left(\sigma_i\begin{pmatrix} I_n \\ V^T \end{pmatrix}\right)^2 \quad \text{[Lemma 3.3.1, item 2]}$$

$$= 1 + \sigma_i(V)^2 \quad \text{[Lemma 3.3.1, items 2 and 7]}$$

$$= \left(\sigma_i\begin{pmatrix} -V \\ I_n \end{pmatrix}\right)^2 \quad \text{[Lemma 3.3.1, items 4 and 7].}$$

Hence

$$\sigma_i \begin{pmatrix} I_n & V \end{pmatrix} = \sigma_i \begin{pmatrix} -V \\ I_n \end{pmatrix} \tag{3.9}$$

for all i. On the other hand,

$$M \begin{pmatrix} -V \\ I \end{pmatrix} = \begin{pmatrix} 0 \\ S^{-1} \end{pmatrix}. \tag{3.10}$$

By Lemma 3.3.1, items 4 and 6, we get

$$\sigma_i \begin{pmatrix} -V \\ I_n \end{pmatrix} \leq \|M^{-1}\| \sigma_i(S^{-1}). \tag{3.11}$$

Combining (3.8), (3.9), (3.11) we get

$$\sigma_i(A^{-1}) \leq \|M^{-1}\|^2 \sigma_i(S^{-1}). \tag{3.12}$$

The singular values of the inverse of a $k \times k$ matrix are, for $i \leq k$, the reciprocals of the singular values of the matrix in reverse order; using this and (3.12) the proposition now follows. □

Proposition 3.3.3. Let

$$M = \begin{pmatrix} A & B \\ C^T & D \end{pmatrix}$$

be a nonsingular $(n+m) \times (n+m)$ block matrix with $A \in \mathbb{R}^{n\times n}$, $B, C \in \mathbb{R}^{n\times m}$, $D \in \mathbb{R}^{m\times m}$. Let the inverse

$$M^{-1} = \begin{pmatrix} P & Q \\ R^T & S \end{pmatrix}$$

be decomposed similarly. Put $p = \min(n, m)$. Then

$$\|M\|^{-2}\sigma_{n-j}(A) \leq \sigma_{m-j}(S) \leq \|M^{-1}\|^2 \sigma_{n-j}(A) \qquad \text{for } 0 \leq j < p, \tag{3.13}$$

$$\|M\|^{-1} \leq \sigma_i(S) \leq \|M^{-1}\| \qquad \text{if } n < m \text{ and } 1 \leq i \leq m-n, \tag{3.14}$$

$$\|M^{-1}\|^{-1} \leq \sigma_i(A)\|M\| \qquad \text{if } n > m \text{ and } 1 \leq i \leq n-m. \tag{3.15}$$

Proof. For $m = n$ the result reduces to Proposition 3.3.2. Again, there is a striking symmetry in the statement of Proposition 3.3.3, and so it is sufficient to prove (3.13), (3.14) in the case $n < m$. We pad M to a matrix

$$\tilde{M} = \begin{pmatrix} \mu I_{m-n,m-n} & 0 & 0 \\ 0 & A & B \\ 0 & C^T & D \end{pmatrix},$$

where $\mu = \|M\|$. Then

$$\tilde{M}^{-1} = \begin{pmatrix} \mu^{-1} I_{m-n,m-n} & 0 & 0 \\ 0 & P & Q \\ 0 & R^T & S \end{pmatrix}.$$

Applying Proposition 3.3.2 we get

$$\|\tilde{M}\|^{-2}\sigma_i\begin{pmatrix} \mu I_{m-n,m-n} & 0 \\ 0 & A \end{pmatrix} \le \sigma_i(S) \le \|\tilde{M}^{-1}\|^2\sigma_i\begin{pmatrix} \mu I_{m-n,m-n} & 0 \\ 0 & A \end{pmatrix}. \tag{3.16}$$

Since $\|A\| \le \|M\| = \mu$, we have

$$\sigma_i\begin{pmatrix} \mu I_{m-n,m-n} & 0 \\ 0 & A \end{pmatrix} = \|M\| \qquad \text{for } 1 \le i \le m-n \tag{3.17}$$

$$= \sigma_{n-m+i}(A) \qquad \text{for } m-n < i \le m. \tag{3.18}$$

Next,

$$\|\tilde{M}\| = \max(\mu, \|M\|) = \|M\|, \tag{3.19}$$

$$\|\tilde{M}^{-1}\| = \max(\mu^{-1}, \|M^{-1}\|) = \|M^{-1}\|. \tag{3.20}$$

Combining (3.16), (3.18), (3.19), (3.20), we obtain

$$\|M\|^{-2}\sigma_{n-m+i}(A) \le \sigma_i(S) \le \|M^{-1}\|^2\sigma_{n-m+i}(A)$$

for $m-n < i \le m$. Put $j = m-i$ to get (3.13).

Combining (3.16), (3.17), (3.19), (3.20), we obtain

$$\|M\|^{-2}\|M\| \le \sigma_i(S) \le \|M^{-1}\|^2\|M\|;$$

replacing the second inequality by the obvious bound $\sigma_i(S) \le \|S\| \le \|M^{-1}\|$, we get (3.14). □

For our purposes the most important corollary follows.

Corollary 3.3.4. Let M be as in Proposition 3.3.3 and $p \le \min(m,n)$. Then A has rank deficiency p if and only if S has rank deficiency p.

In the case where $m << n$ and for small p we will often use this result to express that A has a desired rank deficiency. Of course Corollary 3.3.4. can be proved directly; however, Proposition 3.3.3 adds a quantitative feature to this rank property.

To finish this section we prove a generalization of Proposition 3.3.3 to the case where A is a rectangular block. We have the following proposition.

Proposition 3.3.5 (the singular value inequality). Let

$$M = \begin{pmatrix} A & B \\ C^T & D \end{pmatrix}$$

be a nonsingular $n \times n$ block matrix with $A \in \mathbb{R}^{p\times q}$, $B \in \mathbb{R}^{p\times(n-q)}$, $C \in \mathbb{R}^{q\times(n-p)}$, $D \in \mathbb{R}^{(n-p)\times(n-q)}$, where $p < n, q < n$. Let the inverse matrix

$$M^{-1} = \begin{pmatrix} P & Q \\ R^T & S \end{pmatrix}$$

be decomposed so that $P \in \mathbb{R}^{q\times p}$, $B \in \mathbb{R}^{q\times(n-p)}$, $C \in \mathbb{R}^{p\times(n-q)}$, $D \in \mathbb{R}^{(n-q)\times(n-p)}$. So M^{-1} is decomposed like M^T, not like M.

Put $s_A = \min(p,q), s_S = \min(n-p, n-q)$. Then

$$\|M\|^{-2}\sigma_{s_A-j}(A) \le \sigma_{s_S-j}(S) \le \|M^{-1}\|^2\sigma_{s_A-j}(A) \qquad \text{for } 0 \le j < \min(s_A, s_S). \tag{3.21}$$

Proof. We shall give the proof in the case where $p < q, p+q < n$. The other cases are similar. First we pad M to form the matrix

$$\tilde{M} = \begin{pmatrix} X & 0 & 0 & Y \\ 0 & A & B & 0 \\ 0 & C^T & D & 0 \end{pmatrix},$$

where $X \in \mathbb{R}^{(n-2p)\times(n-p-q)}$, $Y \in \mathbb{R}^{(n-2p)\times(q-p)}$ and

$$\begin{pmatrix} X & Y \end{pmatrix} = \mu I_{2n-p,2n-p},$$

$\mu = \|M\|$. It is easy to see that $\tilde{M}^{-1}$ has a decomposition

$$\tilde{M}^{-1} = \begin{pmatrix} Z & 0 & 0 \\ 0 & P & Q \\ 0 & R^T & S \\ U & 0 & 0 \end{pmatrix},$$

where $Z \in \mathbb{R}^{(n-p-q)\times(n-2p)}$, $U \in \mathbb{R}^{(q-p)\times(n-2p)}$, and

$$\begin{pmatrix} Z \\ U \end{pmatrix} = \mu^{-1} I_{2n-p,2n-p}.$$

Now $\|\tilde{M}\| = \|M\|$, $\|\tilde{M}^{-1}\| = \|M^{-1}\|$. By Proposition 3.3.3 we have

$$\|M\|^{-2}\sigma_{n-p-i}\begin{pmatrix} X & 0 \\ 0 & A \end{pmatrix} \le \sigma_{n-p-i}\begin{pmatrix} S \\ 0 \end{pmatrix} \le \|M^{-1}\|^2\sigma_{n-p-i}\begin{pmatrix} X & 0 \\ 0 & A \end{pmatrix} \tag{3.22}$$

for $0 \le i < n-p$.

It is not hard to see that

$$\begin{pmatrix} X & 0 \\ 0 & A \end{pmatrix}$$

has $n-p-q$ singular values equal to μ and $q-p$ singular values equal to zero in addition to the singular values $\sigma_1(A)\ldots\sigma_p(A)$. On the other hand,

$$\begin{pmatrix} S \\ 0 \end{pmatrix}$$

has $q-p$ singular values equal to zero in addition to the singular values $\sigma_1(S),\ldots,\sigma_{n-q}(S)$. By inserting these singular values into (3.22) and taking into account that $s_A = p, s_S = n-q$, and $\min(s_A, s_S) = p$ in our case, we obtain (3.21). $\square$

3.4 The Schur Inverse as Defining System for Rank Deficiency

We will obtain defining systems for the rank-r matrices $A \in \mathbb{R}^{p\times q}$ where $r < \min(p,q)$. A defining system is a collection of smooth (in our case, algebraic) functions of the entries of A whose null set is the set of rank-r matrices and which has full linear rank at every rank-r matrix.

Our defining systems will be local in the sense of being defined on certain open subsets of $\mathbb{R}^{p\times q}$ only. These subsets have the form $\mathcal{U}(B,C,D)$ with $B \in \mathbb{R}^{p\times(p-r)}$, $C \in \mathbb{R}^{q\times(q-r)}$, $D \in \mathbb{R}^{(q-r)\times(p-r)}$. Here

$$\mathcal{U}(B,C,D) = \{A \in \mathbb{R}^{p\times q} : M(A) \text{ is nonsingular}\}, \tag{3.23}$$

where

$$M(A) = \begin{pmatrix} A & B \\ C^T & D \end{pmatrix}. \tag{3.24}$$

The sets of the form $\mathcal{U}(B,C,D)$ form an open cover of the (open) set of matrices $A \in \mathbb{R}^{p\times q}$ with rank at least r; this follows from Proposition 3.2.1. This is still true if we restrict to matrices B, C with full rank. For this case we have the following proposition.

Proposition 3.4.1. Let $p > 0, q > 0, r < \min(p,q)$, and let $B \in \mathbb{R}^{p\times(p-r)}$ with rank $(p-r)$, $C \in \mathbb{R}^{q\times(q-r)}$ with rank $(q-r)$, $D \in \mathbb{R}^{(q-r)\times(p-r)}$. Then $\mathcal{U}(B,C,D)$ is an open dense subset of $\mathbb{R}^{p\times q}$.

Proof. First assume that $r = 0$. Then B, C are nonsingular square matrices. If $A \in \mathbb{R}^{p\times q}$ is not in the closure of $\mathcal{U}(B,C,D)$, then there exists a neighborhood U of 1 in $\mathbb{R}$ such that

$$\begin{pmatrix} \alpha A & B \\ C^T & D \end{pmatrix}$$

is singular for all $\alpha \in U$. This implies that $I_{q,q} - \alpha C^{-T} D B^{-1} A$ is singular for all $\alpha \in U$. Since $C^{-T} D B^{-1} A$ cannot have an infinite number of eigenvalues, this is impossible.

Next assume that $r > 0$. Since the rectangular matrix B has full column rank, it contains a nonsingular $(p-r)\times(p-r)$ block. Without loss of generality we can assume that this is the upper block. Now if the lower left $q \times q$ block in (3.24) is singular, we can make it nonsingular by an arbitrarily small perturbation of the last r rows of A. The result then follows by an argument similar to that of the case of $r = 0$. □

Now let B, C, D be as in Proposition 3.4.1, $A \in \mathcal{U}(B,C,D)$, and let $M(A)^{-1}$ be decomposed as in Proposition 3.3.5 with $n = p + q - r$. In this notation $s_S = \min(n-p, n-q) = \min(q-r, p-r) < \min(p,q) = s_A$. By the singular value inequality (Proposition 3.3.5) S vanishes if and only if $\sigma_{r+1}(A) = 0$; i.e., A has rank at most r. Since A has rank at least r, we conclude that

$$S(A) = 0 \tag{3.25}$$

is a set of $(p-r)\times(q-r)$ defining equations for the matrices with rank r in $\mathcal{U}(B,C,D)$.

If A is square and nonsingular, then $S = (D - C^T A^{-1} B)^{-1}$. In this case $D - C^T A^{-1} B$ is usually called the Schur complement of A in M. Therefore, we will call S the Schur inverse of A in M. It is important to keep in mind that the Schur inverse is defined also if A is nonsquare or square and singular.

Since the Schur inverse arises as a block in $M(A)^{-1}$ there are two obvious ways to compute it. Either we can solve

$$M(A)\begin{pmatrix} Q \\ S \end{pmatrix} = \begin{pmatrix} 0_{p\times(q-r)} \\ I_{q-r} \end{pmatrix}, \tag{3.26}$$

where $Q \in \mathbb{R}^{q\times(q-r)}$, or we can solve

$$\begin{pmatrix} W^T & S \end{pmatrix} M(A) = \begin{pmatrix} 0_{(p-r)\times q} & I_{p-r} \end{pmatrix}, \tag{3.27}$$

where $W \in \mathbb{R}^{p\times(p-r)}$.

In numerical applications it is often desirable to know the derivatives of $S(A)$. There is a now classical method to obtain these if both (3.26) and (3.27) are solved. Indeed, by taking derivatives of (3.26) we obtain

$$M(A)\begin{pmatrix} Q_z \\ S_z \end{pmatrix} + \begin{pmatrix} A_z & 0 \\ 0 & 0 \end{pmatrix}\begin{pmatrix} Q \\ S \end{pmatrix} = 0, \tag{3.28}$$

where z is any variable on which A depends (it might be just an entry of A). Multiplying (3.28) from the left with $\begin{pmatrix} W^T & S \end{pmatrix}$ and using (3.27) we find

$$S_z = -W^T A_z Q. \tag{3.29}$$

The relation (3.29) can also be used to prove the following result.

Proposition 3.4.2. The system (3.25) is regular; i.e., its Jacobian matrix with respect to the entries of A has full rank $(p-r)(q-r)$.

Proof. Suppose that $H = (h_{ij})_{ij} \in \mathbb{R}^{(p-r)\times(q-r)}$ is such that $\sum_{i,j} h_{ij}(S_z)_{ij} = 0$ whenever z denotes an entry of A. By (3.29) we have

$$\sum_{i,j} h_{ij}(W^T A_{a_{kl}} Q)_{ij} = 0$$

for every entry $a_{kl} \in A$. Hence

$$\sum_{i,j} h_{ij}(W^T)_{ik} Q_{lj} = 0$$

for all k, l. Equivalently, $QH^TW^T = 0$. Since Q, W are full-rank matrices by (3.26) and (3.27) whenever $S = 0$, this implies that $H = 0$. □

Remark 3.4.3. Proposition 3.4.2 implies the well-known fact that the set of rank-r matrices in $\mathbb{R}^{p\times q}$ ($r < \min(p, q)$) is a smooth manifold with dimension $pq-(p-r)(q-r) = r(p+q-r)$.

3.5 Invariant Subspaces of Parameter-Dependent Matrices

It is well known that the algebraically simple eigenvalues of a matrix and the associated left and right eigenvectors depend smoothly on the entries of the matrix. The generalization to invariant subspaces is also well known. Quantitative bounds can be found in [226] and [106]. For our purposes a qualitative understanding in terms of smooth perturbations of the matrix is more important. Since it is fundamental for a good understanding of the numerical methods in Chapter 5 and can be seen as a nice application of bordered matrices, we provide our own proof. We formulate the result for real matrices, but everything carries over to complex matrices.

Proposition 3.5.1. Let $A(\mu) \in \mathbb{R}^{n\times n}$ be a smooth family of matrices and denote $A(\mu_0) = A_0$. Let $\Sigma = \Sigma_1 \cup \cdots \cup \Sigma_k$ be a partition of the family Σ of all eigenvalues of A_0 such that eigenvalues that correspond to the same Jordan block are in the same group (so an eigenvalue with multiple geometric multiplicity may appear in more than one group). Let n_i be the sum of the algebraic multiplicities of the eigenvalues in Σ_i, so $n_1 + \cdots + n_k = n$. Then the following hold:

1. There exist $S_{i0} \in \mathbb{R}^{n\times n_i}$ and $A_{i0} \in \mathbb{R}^{n_i\times n_i}$ such that the columns of all S_{i0} together form a base of $\mathbb{R}^n$, $A_0S_{i0} = S_{i0}A_{i0}$ for all $i = 1, \dots, k$ and the eigenvalues of A_{i0} are precisely those of A_0 in Σ_i, with the same multiplicity.

2. For a given choice of S_{i0}, A_{i0} as above (these need not be unique) there exist smooth functions $S_i(\mu), A_i(\mu)$, defined in a neighborhood of μ_0 such that $S_i(\mu_0) = S_{i0}$, $A_i(\mu_0) = A_{i0}$, and $A(\mu)S_i(\mu) = S_i(\mu)A_i(\mu)$ for all i.

Proof. The first part (existence of S_{i0} and A_{i0} with the requested properties) follows from the existence of a Jordan decomposition of A_0; the conditions are indeed equivalent to the statement that $A_0 = S_0JS_0^{-1}$, where $S_0 = (S_{10}, \dots, S_{k0})$ and J is a block diagonal matrix with blocks $A_{10}, \dots, A_{k0}$ on the diagonal.

For the second part it is clearly sufficient to prove the result for $i = 1$. To simplify the notation we assume $n_1 = 2$; the generalization to an arbitrary n_1 is trivial.

Let c_1^T, c_2^T be the first two rows of S_0^{-1}, where $S_0 = (S_{10}, \dots, S_{k0})$. From the relation $A_0 = S_0JS_0S_0^{-1}$ we infer that

$$c_1^T A_0 = (A_{10})_{11}c_1^T + (A_{10})_{12}c_2^T, \tag{3.30}$$

$$c_2^T A_0 = (A_{10})_{21}c_1^T + (A_{10})_{22}c_2^T. \tag{3.31}$$

Now consider the system

$$\begin{cases} A(\mu)b_1 - x_1b_1 - x_2b_2 = 0, \\ A(\mu)b_2 - x_3b_1 - x_4b_2 = 0, \\ c_1^Tb_1 = 1, \\ c_1^Tb_2 = 0, \\ c_2^Tb_1 = 0, \\ c_2^Tb_2 = 1 \end{cases} \tag{3.32}$$

of $2n+4$ equations in the $2n+4$ unknowns that consist of n components of a vector b_1, n components of a vector b_2, and four scalar unknowns x_1, x_2, x_3, x_4. For $\mu = \mu_0$ this system has a solution: Take for b_1, b_2 the two columns of $S_{10} = (s_1, s_2)$ and for x_i the appropriate entries of A_{10}. So by the implicit function theorem system (3.32) defines functions $b_1(\mu), b_2(\mu), x_1(\mu), x_2(\mu), x_3(\mu), x_4(\mu)$ that reduce to the indicated solution for $\mu = \mu_0$ if the Jacobian of (3.32) with respect to the variables $(b_1, b_2, x_1, x_2, x_3, x_4)$ is nonsingular there.

So suppose that $p, q \in \mathbb{R}^n, r, s, t, u \in \mathbb{R}$ are such that

$$\begin{pmatrix} A_0 - (A_{10})_{11}I_n & -(A_{10})_{21}I_n & -s_1 & -s_2 & 0 & 0 \\ -(A_{10})_{12}I_n & A_0 - (A_{10})_{22}I_n & 0 & 0 & -s_1 & -s_2 \\ c_1^T & 0 & 0 & 0 & 0 & 0 \\ 0 & c_1^T & 0 & 0 & 0 & 0 \\ c_2^T & 0 & 0 & 0 & 0 & 0 \\ 0 & c_2^T & 0 & 0 & 0 & 0 \end{pmatrix} \begin{pmatrix} p \\ q \\ r \\ s \\ t \\ u \end{pmatrix} = \begin{pmatrix} 0 \\ 0 \\ 0 \\ 0 \\ 0 \\ 0 \end{pmatrix}. \tag{3.33}$$

Multiplying the first block row of (3.33) from the left with c_1^T and taking into account (3.30) and the third and fifth block row in (3.33) we get $c_1^T s_1 r + c_1^T s_2 s = 0$. Since $c_1^T s_1 = 1$, $c_1^T s_2 = 0$, this implies $r = 0$. By similar arguments it follows that s, t, u must also vanish.

Inserting this back in (3.33) we obtain

$$A_0(p, q) = (p, q)A_{10}. \tag{3.34}$$

The generalized eigenvectors of A_{10} span $\mathbb{R}^{n_1}$. So if p, q are not both zero, then there exist a $\lambda \in \Sigma_1$, $l \geq 1$, and $v \in \mathbb{R}^{n_1}$ such that $(A_{10} - \lambda I_{n_1})^l v = 0$ and $(p, q)v = 0$. By (3.34) this implies

$$(A_0 - \lambda)^l (p, q)v = 0. \tag{3.35}$$

Hence $(p, q)v$ is a generalized eigenvector of A_0 for an eigenvalue $\lambda \in \Sigma_1$. Hence it is in the span of s_1, s_2. But it is also orthogonal to c_1^T, c_2^T, which is a contradiction.

The result now follows by setting

$$B_1(\mu) = (b_1(\mu), b_2(\mu)),$$

$$A_{10}(\mu) = \begin{pmatrix} x_1(\mu) & x_3(\mu) \\ x_2(\mu) & x_4(\mu) \end{pmatrix}.$$

□

In the setting of Proposition 3.5.1 there must obviously be a neighborhood of μ_0 in which the columns of all $S_i(\mu)$ together also span $\mathbb{R}^n$. So Proposition 3.5.1 implies that partitions of $\mathbb{R}^n$ into invariant subspaces of a matrix with disjoint sets of eigenvalues locally vary smoothly under perturbations of the matrix.

3.6 Numerical Methods for Bordered Linear Systems

We consider linear systems of the form

$$M\begin{pmatrix} x \\ y \end{pmatrix} = \begin{pmatrix} f \\ g \end{pmatrix}, \tag{3.36}$$

where M has the bordered form

$$M = \begin{pmatrix} A & B \\ C^T & D \end{pmatrix} \tag{3.37}$$

with $A \in \mathbb{R}^{n\times n}$, $B, C \in \mathbb{R}^{n\times m}$, $D \in \mathbb{R}^{m\times m}$, $x, f \in \mathbb{R}^n$, $y, g \in \mathbb{R}^m$.

Such systems deserve special attention because they are very frequent in numerical continuation and bifurcation problems. Usually m is small. In fact, the most frequent situation is $m = 1$ and in §8.3.3 we describe a case with $m = 5$; cases with $m > 5$ really seem quite remote.

In typical applications, n is the number of state variables of a dynamical system. If n is reasonably small (say, $n \leq 20$), then for numerical purposes the block structure of (3.37) can best be ignored. On the other hand, large matrices A are usually sparse since they arise from a discretization process. In (3.37), B, C, D are usually dense, and so the sparsity structure of M is more complicated than that of A. This is clear in the case that A is a bandmatrix. To exploit the sparsity of M for a solver based on Gaussian elimination it is possible to do some clever surgery on the problem (3.36) to solve it without destroying much of the structure of A; see [152], [201].

In the search for a more general approach several authors have considered the question if (3.36) can be solved using solvers for the main block A; we refer to [150], [151], [152], [153], [51], [52]. This is easy if A and M are both well conditioned. Unfortunately, in many applications A is a nearly singular matrix so that a naive block elimination strategy to solve (3.36) leads to big errors. It was remarked in [205] that even this is often acceptable because (3.36) is usually solved only to compute Newton corrections. But of course this does not lead to very robust algorithms, and indeed it turns out to fail in complicated situations as we will consider in Chapters 6 and 7.

Several authors have developed algorithms to overcome this problem [50], [53], [188]. These methods require solvers not only for A but also for its transpose A^T. In typical applications this is usually not a big problem.

We will discuss a particular method, called block elimination mixed with wider borders (BEMW), developed in [112] and [113], for which software is freely available. For a good understanding we first recall some basic notions from numerical linear algebra.

3.6.1 Backward Stability

The importance of the notion of backward stability is well appreciated in the numerical linear algebra community since the work of J. H. Wilkinson; cf. [241]. For recent developments see [43], [135]. The idea is quite general, but we restrict ourselves to the solution of linear systems

$$Ax = b, \quad A \in \mathbb{R}^{n\times n}, x, b \in \mathbb{R}^n. \tag{3.38}$$

If A is invertible and a norm $\| \ \|$ is defined on $\mathbb{R}^n$ (in fact, one might have distinct norms on the domain and range spaces of A), then the condition number $\text{cond}(A) = \|A\| \|A^{-1}\|$ indicates how sensitive the solution of (3.38) is to perturbations of the right-hand side; if

$$A(x + \Delta x) = b + \Delta b, \tag{3.39}$$

then

$$\frac{\|\Delta x\|}{\|x\|} \le \text{cond}(A) \frac{\|\Delta b\|}{\|b\|}. \tag{3.40}$$

If A is also perturbed, i.e.,

$$(A + \Delta A)(x + \Delta x) = b + \Delta b \tag{3.41}$$

with $\|A^{-1}\Delta A\| \le 1$, then

$$\frac{\|\Delta x\|}{\|x\|} \le \frac{1}{1 - \|A^{-1}\Delta A\|} \text{cond}(A) \left(\frac{\|\Delta b\|}{\|b\|} + \frac{\|\Delta A\|}{\|A\|} \right). \tag{3.42}$$

Therefore, we also call $\text{cond}(A)$ the condition of the problem (3.38) itself. We note that it is not a property of a solution method. If the data of the problem, i.e., A, b are subject to relative errors of the order ϵ, then the computed solution by any method will typically have an error of order at least $\epsilon \ \text{cond}(A)$. In a practical application with floating point computations with relative precision u the computed solution is expected to have an error of order at least $u \ \text{cond}(A)$. We call this the inevitable error. If A is not invertible, then the condition number is said to be infinite.

A solution method for (3.38) is an operator $\mathcal{S}$ from $\mathbb{R}^n$ into itself for which we accept $\mathcal{S}(b)$ as an approximation to the solution x of (3.38). The method is called numerically stable if for all A, b the error in the solution is bounded by a small multiple of the inevitable error.

$\mathcal{S}$ is called *backward stable* if the computed solution is the exact solution to a problem with slightly perturbed data; i.e., $\mathcal{S}(b)$ satisfies

$$(A + \Delta A)\mathcal{S}(b) = b + \Delta b, \tag{3.43}$$

where

$$\|\Delta A\| \le uC_S \|A\|, \qquad \|\Delta b\| \le uC_S \|b\|, \tag{3.44}$$

where C_S is a modest real number that may depend on n. It is a useful exercise to show that in (3.43) we can assume that $\Delta b = 0$. Indeed we have

$$\|\Delta b\| \le uC_S \|b + \Delta b\| + uC_S \|\Delta b\|,$$

$$(1 - uC_S)\|\Delta b\| \le uC_S \|A + \Delta A\| \|\mathcal{S}(b)\|,$$

$$\|\Delta b\| \le u \frac{1 + uC_S}{1 - uC_S} C_S \|A\| \|\mathcal{S}(b)\|.$$

Setting $C_S^1 = \frac{1+uC_S}{1-uC_S} C_S$, we conclude from the last inequality that there exists a $\Delta_1 A \in \mathbb{R}^{n \times n}$, $\|\Delta_1 A\| \le uC_S^1 \|A\|$ such that $-\Delta b = \Delta_1 A \mathcal{S}(b)$. Hence

$$(A + \Delta A + \Delta_1 A)\mathcal{S}(b) = b,$$

where $\|\Delta A + \Delta_1 A\| \leq u(C_S + C_S^1)\|A\|$. Since C_S^1 is in all reasonable cases bounded by, say, $1.1C_S$, the result follows.

Backward stability is in a sense the ultimate desirable property for a linear solver. One reason is that the original data (A, b) usually contain errors; if instead of (3.38) we solve a problem sufficiently close to it, then the result may be the best that we can hope for anyway. The other argument is that backward stability together with a good conditioning of the problem guarantees an accurate solution by (3.42).

The solution of triangular systems by substitution is the ultimate example of a backward stable solver. Such systems can be arbitrarily ill conditioned, which shows clearly that conditioning and stability are completely different concepts. For practical purposes, Gaussian elimination with partial or complete pivoting are also backward stable and so are many iterative methods. We refer to [241], [43], and [135] for details.

3.6.2 Algorithm BEM for One-Bordered Systems

We consider (3.36) in the case $m = 1$. Since B, C are vectors and D is a scalar, we denote these by b, c, d, respectively. We will describe Algorithm BEM (block elimination mixed), which allows us to solve systems with M as in (3.36), (3.37) in a backward stable way if backward stable solvers are given for both A and A^T. As a bonus we obtain a solver for M^T as well; this turns out to be useful in the further extension to the case $m > 1$ in §3.6.3 and also in the applications in Chapters 6 and 7. We give the idea of why the method works but refer to [112] and [113] for the technical details of a full stability analysis.

Basically (3.36) can be solved by two different block LU factorizations of M. The first is Doolittle factorization

$$\begin{pmatrix} A & b \\ c^T & d \end{pmatrix} = \begin{pmatrix} I_n & 0 \\ w^T & 1 \end{pmatrix} \begin{pmatrix} A & b \\ 0 & \delta^* \end{pmatrix}. \tag{3.45}$$

The second is Crout block factorization

$$\begin{pmatrix} A & b \\ c^T & d \end{pmatrix} = \begin{pmatrix} A & 0 \\ c^T & \delta \end{pmatrix} \begin{pmatrix} I_n & v \\ 0 & 1 \end{pmatrix}. \tag{3.46}$$

These factorizations lead to the following respective algorithms:

Algorithm BED (block elimination Doolittle)

Step 1.	Solve	$A^T w = c$
Step 2.	Compute	$\delta^* = d - w^T b$
Step 3.	Compute	$y = (g - w^T f)/\delta^*$
Step 4.	Solve	$Ax = f - by$

Algorithm BEC (block elimination Crout)

Step 1.	Solve	$Av = b$
Step 2.	Compute	$\delta = d - c^T v$
Step 3.	Solve	$A\xi = f$
Step 4.	Compute	$y = (g - c^T \xi)/\delta$
Step 5.	Compute	$x = \xi - vy$

Algorithms BED and BEC work fine if A, M are both well conditioned and fail if M is well conditioned and A is nearly singular. Nevertheless, their numeric properties are quite different.

Consider the case of BED for M well conditioned and A nearly singular. Numerical tests in cases with backward stable solvers for A and A^T show that typically y is computed quite correctly, i.e., up to the inevitable order $u \, \|(x \; y)^T\| \mathrm{cond}(M)$. On the other hand, x usually has a large error; in fact, it often has no correct digit at all. It is easy to show why y is essentially correct if we assume for simplicity that the only errors that occur in BED are in the solutions of the systems with A and A^T; i.e., we ignore the roundoff error in the other computations. Then if $\bar{w}$, $\bar{\delta}^*$ and $\bar{y}$ denote the computed quantities in BED, we have

$$(A + \Delta A)^T \bar{w} = c, \quad \bar{\delta}^* = d - \bar{w}^T b, \quad \bar{\delta}^* \bar{y} = g - \bar{w}^T f, \quad \|\Delta A\| \leq u C_S \|A\| \tag{3.47}$$

with C_S a modest real number. Let us denote by x_1 the (unknown) exact solution of

$$(A + \Delta A)x = f - b\bar{y}. \tag{3.48}$$

Then (3.47) and (3.48) together imply

$$\begin{pmatrix} A + \Delta A & b \\ c^T & d \end{pmatrix} \begin{pmatrix} x_1 \\ \bar{y} \end{pmatrix} \begin{pmatrix} f \\ g \end{pmatrix}. \tag{3.49}$$

Since (3.49) is a small perturbation of the well-conditioned system (3.36) it follows that $\bar{y}$ must be a correspondingly good approximation of y.

Intuitively speaking, the computation of y by BED requires only one solve with A^T (and no solve with A); therefore, the backward errors can be pushed back into the well-conditioned matrix M.

Now consider the case of BEC for M well conditioned and A nearly singular. Numerical tests in cases with backward stable solvers for A show that typically both x and y have large errors; in fact, they often have no correct digit at all. Intuitively speaking, the computation of y depends on two solves with A, with different perturbations of A, and there is no way to push these back into M.

Numerical tests also show that if y is of the order of $u\|(x \; y)^T\|$, then BEC computes x accurately, i.e., with an error of order $u \, \mathrm{cond}(M)\|(x \; y)^T\|$. To explain this, we make again the assumption that the only errors that occur in BEC are in the solution of the systems with A. Let $\bar{v}$, $\bar{\delta}$, $\bar{\xi}$, $\bar{x}$, and $\bar{y}$ denote the quantities computed in BEC. We have

$$(A + \Delta_1 A)\bar{v} = b, \quad \bar{\delta} = d - c^T \bar{v}, \quad (A + \Delta_2 A)\bar{\xi} = f, \quad \bar{\delta}\bar{y} = g - c^T \bar{\xi}, \quad \bar{x} = \bar{\xi} - \bar{v}\bar{y} \tag{3.50}$$

with $\|\Delta_1 A\| \leq u C_S \|A\|$, $\|\Delta_2 A\| \leq u C_S \|A\|$. By some easy manipulations we find

$$\begin{pmatrix} A & b \\ c^T & d \end{pmatrix} \begin{pmatrix} \bar{x} \\ \bar{y} \end{pmatrix} = \begin{pmatrix} f - (\Delta_2 A)\bar{\xi} + (\Delta_1 A)\bar{v}\bar{y} \\ g \end{pmatrix}. \tag{3.51}$$

Under reasonable assumptions concerning bounds on $\|(A + \Delta_1 A)^{-1}\|$, $\|(A + \Delta_2 A)^{-1}\|$, one obtains

$$\left\| \begin{pmatrix} -(\Delta_2 A)\bar{\xi} + (\Delta_1 A)\bar{v}\bar{y} \\ 0 \end{pmatrix} \right\| \leq uC \left\| \begin{pmatrix} f \\ g \end{pmatrix} \right\| \tag{3.52}$$

for a modest constant C. The key idea is that $\|\xi\|$ is modest in spite of the near singularity of A, because f has the form $Ax+by$ where $|y|$ is small; on the other hand, $\|\bar{v}\|$ is indeed large (factor u^{-1}), but this is compensated for by the fact that $|\bar{y}|$ is small (see [113, Proposition 1] for details). Hence $\begin{pmatrix} \bar{x} & \bar{y} \end{pmatrix}^T$ approximates $\begin{pmatrix} x & y \end{pmatrix}^T$ with a relative precision of order u cond(M).

It is now natural to let BED compute y and do an iterative refinement step with BEC using the computed value of y and $x = 0$ as a first guess. In this manner we obtain the mixed block elimination method BEM.

Algorithm BEM

Step 1.	Solve	$A^T w = c$
Step 2.	Compute	$\delta^* = d - w^T b$
Step 3.	Solve	$Av = b$
Step 4.	Compute	$\delta = d - c^T v$
Step 5.	Compute	$y_1 = (g - w^T f)/\delta^*$
Step 6.	Compute	$f_1 = f - by_1$
Step 7.	Compute	$g_1 = g - dy_1$
Step 8.	Solve	$A\xi = f_1$
Step 9.	Compute	$y_2 = (g_1 - c^T\xi)/\delta$
Step 10.	Compute	$x = \xi - vy_2$
Step 11.	Compute	$y = y_1 + y_2$

The above discussion shows that in practically occurring cases, BEM produces accurate results if M is well conditioned even if A is nearly singular. It is possible to construct artificial cases where this fails to be true (see [112]), but these hardly arise in practice; the situation is comparable to Gaussian elimination with partial pivoting.

In fact, it is proved in [113, Proposition 1] that for practical purposes BEM is a backward stable solver for (3.36) if $m = 1$. In this result all error sources are taken into account.

In Algorithm BEM Steps 1–4 do not use the right-hand side. These steps form a preprocessing of M, comparable to LU decomposition in the case of Gaussian elimination. Steps 5–11 involve only one solve with A and some vector and scalar operations comparable to the solution stage in Gaussian elimination.

Interestingly, Steps 1–4 for M are the same as for M^T if one interchanges the roles of v, w and of δ, δ^*. Solving a linear system with M^T further requires only one solve with A^T and some vector and scalar operations.

3.6.3 Algorithm BEMW for Wider-Bordered Systems

Algorithm BEM has an obvious block analogue in which δ and δ^* are replaced by small $(m \times m)$ matrices. In exact arithmetic these are both equal to the Schur complement of A in M (see §3.4), i.e., the inverse of the Schur inverse. From Proposition 3.3.5 (the singular value inequality) it follows that for a well-conditioned M the condition of the Schur complement is determined by the m smallest singular values of A. In particular, the Schur complement can be arbitrarily ill conditioned. In the naive block form of

Algorithm BEM Steps 5 and 9 have to be replaced by solutions with δ and δ^*. So it is not surprising that this method fails if, e.g., $m = 2$ and A has near rank defect 1.

Algorithm BEMW proceeds in a different way, exploiting the backward stability of Algorithm BEM. The matrices B, C, D in (3.37) are split and recombined to form vectors $b_1, c_1 \in \mathbb{R}^n$, ..., $b_m, c_m \in \mathbb{R}^{n+m-1}$ and scalars $d_{11}, \ldots, d_{mm}$ as illustrated for the case $m = 3$:

$$\left(\begin{array}{cc|c|c} A & b_1 & b_2 & b_3 \\ \hline c_1^T & d_{11} & & \\ \hline \multicolumn{2}{c|}{c_2^T} & d_{22} & \\ \hline \multicolumn{3}{c|}{c_3^T} & d_{33} \end{array}\right). \tag{3.53}$$

We further denote by M_i $(i = 1, \ldots, m)$ the upper left $(n+i) \times (n+i)$ block in M, so $M_m = M$ and for $m = 3$ we have

$$M_1 = \left(\begin{array}{c|c} A & b_1 \\ \hline c_1^T & d_{11} \end{array}\right), \quad M_2 = \left(\begin{array}{c|c} M_1 & b_2 \\ \hline c_2^T & d_{22} \end{array}\right), \quad M_3 = M_m = M = \left(\begin{array}{c|c} M_2 & b_3 \\ \hline c_3^T & d_{33} \end{array}\right).$$

The input of Algorithm BEM consists of backward stable solvers for A and A^T plus the data for the borders. Then Steps 1–4 of BEM are performed, essentially to form backward stable solvers for M_1 and M_1^T. These solvers are used in a second application of Steps 1–4 of BEM to form backward stable solvers for M_2 and M_2^T and so on. Finally one obtains backward stable solvers for M and M^T, using m solves with A and m solves with A^T and some vector and scalar operations. To solve a linear system with M (respectively, M^T) one more solve with A (respectively, A^T) is needed.

The error analysis in [113] shows that BEMW is indeed backward stable with a stability constant that grows only linearly with m if the leading matrices $A, M_1, \ldots, M_m = M$ are not too ill conditioned. In practice, the natural perturbation by roundoff from a theoretically singular matrix is enough.

As formulated above, BEMW is a recursive algorithm. In [113] it is reformulated as an iterative algorithm that can easily be programmed in a recursion-unfriendly language such as Fortran 77.

The publicly distributed version of BEMW (dated September 13, 1993) in Fortran 77 can be obtained from http://www.netlib.org as the file linalg/bemw.tgr. There are single precision real, double precision real, complex, and real symmetric versions. An important feature is the reverse communication interface. This means that the subroutines called by BEMW return control to the driving program whenever they need to solve linear systems with A or A^T. These routines do not know anything about A, so the user is completely free to store or handle A in a problem-dependent way.

Tests with BEMW are described in [113] and [105]. This software was also intensively used in [159] to compute invariant manifolds of large dynamical systems. It is used systematically in the experiments described in Chapters 6, 7, and 8 for all computations that involve the continuous Brusselator model; the band structure of A is exploited by the use of LAPACK band solvers. This includes cases where A is singular to machine precision, even with rank defect 2 and values of m between 1 and 5.

3.7 Notes and Further Reading

Bordered matrices appear in numerical continuation and bifurcation theory from the beginning. Sometimes they seem to be merely self-suggesting computational tools; see, e.g., [150], [151], [152], [68], [51], [52]. However, their involvement is deeper, since they also constitute the numerical equivalent of the Liapunov–Schmidt reduction. This feature was stressed in particular in [146], [148]. A systematic use of bordered matrix methods was strongly promoted in [124] and [126]; these papers contain the fundamental ideas that we will develop further. However, quite similar systems were used in [1] and [194]. The papers [88] and [31] are particularly good references to the literature.

3.8 Exercises

1. Let n be a natural number and let $s_1, \dots, s_n$ be the real numbers with $1 \geq s_1 \geq s_2 \geq \cdots \geq s_n \geq 0$.

   (a) Prove that there exists at least one orthogonal matrix $U \in \mathbb{R}^{2n \times 2n}$ with block form
$$U = \begin{pmatrix} A & B \\ C^T & D \end{pmatrix}$$
   with $A, B, C, D \in \mathbb{R}^{n \times n}$ and $\sigma_i(A) = s_i$ for $i = 1, \dots, n$.

   (b) Prove that if a matrix U satisfies the requirements in (a), then $\sigma_i(D) = s_i$ for $i = 1, \dots, n$.

2. Consider a block matrix M
$$M = \begin{pmatrix} A & B \\ C^T & D \end{pmatrix}$$
   with $A \in \mathbb{R}^{n \times n}$, $B, C \in \mathbb{R}^{n \times m}$, $D \in \mathbb{R}^{m \times m}$. Let us denote by M^A the adjoint of M (whose (i, j)th entry is the minor of the (j, i)th element in M). Let M^A be partitioned as
$$M^A = \begin{pmatrix} P_1 & Q_1 \\ R_1^T & S_1 \end{pmatrix}$$
   in blocks with the same sizes as those of M.

   Prove that $\det(S_1) = \det(A) \det(M)^{m-1}$.

   (*Hint:* First prove a corresponding result for M^{-1} if both M and A are nonsingular; then apply a density argument.)

3. Let p, q, N be natural numbers, $1 \leq p, q \leq N$. Let $A \in \mathbb{R}^{p \times q}$, $B \in \mathbb{R}^{p \times (N-q)}$, $C \in \mathbb{R}^{q \times (N-p)}$ be such that
$$\begin{pmatrix} A & B \end{pmatrix}$$
   and
$$\begin{pmatrix} A \\ C^T \end{pmatrix}$$

both have full rank. Prove that there exists a $D \in \mathbb{R}^{(N-p)\times(N-q)}$ such that

$$\begin{pmatrix} A & B \\ C^T & D \end{pmatrix}$$

has full rank.

4. **Contact equivalence of Schur inverses.** Let $1 \le p, q \le N$, $A_0 \in \mathbb{R}^{p\times q}$, and assume that

$$M_1 = \begin{pmatrix} A & B_1 \\ C_1^T & D_1 \end{pmatrix}, \quad M_2 = \begin{pmatrix} A & B_2 \\ C_2^T & D_2 \end{pmatrix}$$

are nonsingular $N \times N$ matrices at $A = A_0$. Let $S_1(A)$, $S_2(A)$ be the Schur inverses of A in M_1 and M_2, respectively. Prove that locally near A_0 there exist smooth matrix functions $X(A) \in \mathbb{R}^{(N-q)\times(N-q)}$, $Y(A) \in \mathbb{R}^{(N-p)\times(N-p)}$, both nonsingular in a neighborhood of A_0, such that

$$S_2(A) = X(A)S_1(A)Y(A).$$

(*Hint:* Use the previous exercise to decompose the problem into two simpler ones.)

5. Let

$$f = \begin{pmatrix} f_1(x_1, \ldots, x_n) \\ \cdots \\ f_n(x_1, \ldots, x_n) \end{pmatrix}$$

be a smooth function defined in a neighborhood U of $a \in \mathbb{R}^n$ and with values in $\mathbb{R}^n$. A smooth function g defined on a neighborhood of $f(a)$ and with values in $\mathbb{R}$ is called a dependence function if its gradient at $f(a)$ is nonzero and

$$g(f_1(x_1, \ldots, x_n), \ldots, f_n(x_1, \ldots, x_n)) \equiv 0$$

in a neighborhood of a. A set of k dependence functions $g_1, \ldots, g_k$ is said to be independent if the k gradient vectors are linearly independent. Prove the following.

(a) If f has a dependence function, then the Jacobian $\frac{\partial f}{\partial x}$ is singular in a neighborhood of a.

(b) If f has k independent dependence functions $(1 \le k \le n)$, then in a neighborhood of a $\frac{\partial f}{\partial x}$ has rank at most $n - k$.

(c) If $\frac{\partial f}{\partial x}$ has rank $n-1$ in a neighborhood of a, then f has a dependence function.

(d) If $\frac{\partial f}{\partial x}$ has rank $n-k$ in a neighborhood of a $(1 \le k \le n)$, then f has k independent dependence functions.

(*Hint:* To prove (c) choose $b, c \in \mathbb{R}^n$, $d \in \mathbb{R}$, so that

$$\begin{pmatrix} f_x & b \\ c^T & d \end{pmatrix}$$

is nonsingular. Then define $F : U \times \mathbb{R} \to \mathbb{R}^{n+1}$ by

$$F_i(x_1, \ldots, x_n, x_{n+1}) = f_i(x_1, \ldots, x_n) + b_i x_{n+1}$$

for $i = 1, \ldots, n$ and

$$F_{n+1}(x_1, \ldots, x_n, x_{n+1}) = c_1 x_1 + \cdots + c_n x_n + d x_{n+1}.$$

Prove that F admits an inverse function G and that the last component G_{n+1} is independent of the $(n+1)$th variable. Then evaluate at $x_{n+1} = 0$.)

6. Prove the error bounds in (3.40) and (3.42).

7. Write a computer program that, for an input value σ, performs the following commands:

   (a) Construct a 50×50 diagonal matrix A_0 with diagonal elements $(1, 1, \ldots, 1, \sigma)$.

   (b) Compute $A = H_1 H_2 \ldots H_{10} A_0 H_{11} H_{12} \ldots H_{20}$, where each H_i is a Householder matrix with a random Householder vector.

   (c) Generate vectors $b, c, x_e \in \mathbb{R}^{50}$ and scalars $d, y_e \in \mathbb{R}$ with all components random in $[0, 1]$ (the distribution does not matter); compute $f = Ax_e + by_e$ and $g = c^T x_e + dy_e$.

   (d) Solve the system
   $$\begin{pmatrix} A & b \\ c^T & d \end{pmatrix} \begin{pmatrix} x \\ y \end{pmatrix} = \begin{pmatrix} f \\ g \end{pmatrix}$$
   using BED, BEC, and BEM with Gaussian elimination with partial pivoting as solver for A and A^T. (For example, use LAPACK routines.)

   Run this program for $\sigma = 0.1, 0.09, 0.08, \ldots, 0.0, \ldots, -0.1$ and compare the computed x, y with x_e, y_e. What do you observe?

8. Write a computer program that for input values σ_1, σ_2 performs the following commands:

   (a) Construct a 50×50 diagonal matrix A_0 with diagonal elements $(1, 1, \ldots, 1, \sigma_1, \sigma_2)$.

   (b) Compute $A = H_1 H_2 \ldots H_{10} A_0 H_{11} H_{12} \ldots H_{20}$, where each H_i is a Householder matrix with a random Householder vector.

   (c) Generate matrices $B, C \in \mathbb{R}^{50 \times 2}$, $D \in \mathbb{R}^{2 \times 2}$, vectors $x_e \in \mathbb{R}^{50}$ and $y_e \in \mathbb{R}^2$ with all components random in $[0, 1]$ (the distribution does not matter); compute $f = Ax_e + By_e$ and $g = C^T x_e + Dy_e$.

   (d) Solve the system
   $$\begin{pmatrix} A & B \\ C^T & D \end{pmatrix} \begin{pmatrix} x \\ y \end{pmatrix} = \begin{pmatrix} f \\ g \end{pmatrix}$$
   using the block analogue of BEM and Gaussian elimination with partial pivoting as solver for A and A^T. (For example, use LAPACK routines.)

   Run this program with $\sigma_1 = 1$, $\sigma_2 = 0.1, 0.09, 0.08, \ldots, 0.0, \ldots, -0.1$. Compare the computed x, y with x_e, y_e. Explain the results. Repeat the experiment with $\sigma_1 = 0$, $\sigma_2 = 0.1, 0.09, 0.08, \ldots, 0.0, \ldots, -0.1$. Explain again.

9. Repeat the previous exercise using BEMW instead of the block form of BEM. Do you see an improvement?

10. Let A be the 6×6 matrix

$$A = \begin{pmatrix} \epsilon & 2.0 & 0.986589 & 1.98767 & 6.23456 & 8.123876 \\ 0.0 & \epsilon & 0.89765 & 0.786987 & 7.7656765 & 8.765876 \\ 0.0 & 0.0 & \epsilon & 2.45 & 1.1117775 & 1.213243 \\ 0.0 & 0.0 & 0.0 & \epsilon & 3.56432 & 1.765456 \\ 9.1111111 & 1.78121 & 8.54321 & 0.9986543 & 1.654 & 3.0 \\ 2.223342 & 3.11432 & 1.7163 & 2.61543 & 3.333333 & 4.1234 \end{pmatrix}$$

in which ϵ is a parameter. Define $h = (1\ 2\ 3\ 4\ 5\ 6\ )^T$ and compute $b = Ah$. Then solve $Ax = b$ using BEMW with $n = 4, m = 2$ in Fortran single precision with substitution to solve the triangular systems (e.g., use LAPACK solvers). Do this for $\epsilon = 0.1, 0.01, 0.001, \ldots, 0.0000001$. Compare the computed value of x with h. For what values of ϵ is the result reasonably accurate? Explain this.

(*Hint:* Triangular systems can be very ill conditioned, but solution by substitution is always backward stable.)

Chapter 4

Generic Equilibrium Bifurcations in One-Parameter Problems

In Chapter 1 we discussed examples of quadratic turning points and Hopf bifurcations. In Chapter 2 quadratic turning points were again encountered. We now turn to a more detailed study of the numerical methods for detection, computation, and continuation of these bifurcations. This will lead us in a natural way back to the bordered matrices encountered in Chapter 3.

4.1 Limit Points

We consider the dynamical system

$$\dot{x} = G(x, \alpha), \tag{4.1}$$

where $x, G(x) \in R^N, \alpha \in \mathbb{R}$. A solution to the equilibrium equation

$$G(x, \alpha) = 0 \tag{4.2}$$

is called a limit point if G_x is singular with rank defect 1. Then G_x has a right singular vector $\phi \neq 0$ and a left singular vector $\psi \neq 0$, both unique up to a nonzero scalar factor. If (4.2) locally defines a smooth one-dimensional manifold, i.e., the Jacobian $[G_x, G_\alpha]$ has full rank N (cf. §2.1), then necessarily

$$\psi^T G_\alpha \neq 0 \tag{4.3}$$

(see §2.2). In this setting with the distinguished parameter α the limit point is called a simple (or quadratic) turning point if

$$\psi^T G_{xx}\phi\phi \neq 0 \tag{4.4}$$

(again, see §2.2). One of the most common problems in bifurcation theory is the computation of quadratic turning points. It is a fairly simple problem that can be solved in many ways. A survey is given in [201]. We will discuss the most widely used methods.

4.1.1 The Moore–Spence System for Quadratic Turning Points

The Moore–Spence system was developed in [215] and [187]. It is often used in software, e.g., [80], [84], [165]. In Chapter 7 we will see how it embeds in a more general singularity context.

The idea is to choose any vector $c \in \mathbb{R}^N$ with

$$c^T\phi \neq 0 \tag{4.5}$$

and to consider the extended system

$$\begin{aligned} &G(x,\alpha) = 0,\\ &G_x(x,\alpha)v = 0,\\ &c^Tv = 1, \end{aligned} \tag{4.6}$$

where $v \in \mathbb{R}^N$. This is a system of $2N+1$ equations in the $2N+1$ unknowns x, α, v. Clearly, if (x,α) is a quadratic turning point of (4.2), then (4.6) holds with $v = \phi/(c^T\phi)$. We note that in a continuation context an approximation to (x,α) is known and that the state component of the tangent vector of the equilibrium curve is an approximation of ϕ. So c can be chosen as this state component and $c/(c^Tc)$ as an approximation to ϕ. If (4.6) is a regular system in (x,α,v), then a Newton procedure with these starting values will converge to the quadratic turning point and right singular vector.

Proposition 4.1.1. If (x,α) is a quadratic turning point of (4.2) and if (4.5) holds, then (x,α,v) is a regular solution to (4.6) with $v = \phi/(c^T\phi)$.

Proof. It is sufficient to prove that

$$\begin{pmatrix} G_x & G_\alpha & 0\\ G_{xx}v & G_{x\alpha}v & G_x\\ 0 & 0 & c^T \end{pmatrix} \tag{4.7}$$

is nonsingular at the point that we want to compute. Multiplying the second block row from the left with ψ^T and adding this to the third block row we find that (4.7) is nonsingular if and only if

$$\begin{pmatrix} G_x & G_\alpha & 0\\ G_{xx}v & G_{x\alpha}v & G_x\\ \psi^TG_{xx}v & \psi^TG_{x\alpha}v & c^T \end{pmatrix} \tag{4.8}$$

is nonsingular. Now from (4.3) and (4.4) it follows that the $(N+1)\times(N+1)$ submatrix in (4.8), obtained by selecting the first $(N+1)$ columns, the first N rows, and the last row, is nonsingular; in addition we have that

$$\begin{pmatrix} G_x & G_\alpha\\ \psi^TG_{xx}v & \psi^TG_{x\alpha}v \end{pmatrix}\begin{pmatrix} vc^T\\ 0 \end{pmatrix} = \begin{pmatrix} 0\\ (\psi^TG_{xx}vv)c^T \end{pmatrix}. \tag{4.9}$$

Hence (4.8) is nonsingular if and only if

$$\begin{pmatrix} I & 0 & vc^T \\ G_{xx}v & G_{x\alpha}v & G_x \\ 0 & 1 & 0 \end{pmatrix} \tag{4.10}$$

is nonsingular. Multiplying the first block row of (4.10) with $G_{xx}v$, the third with $G_{x\alpha}v$, and subtracting this from the second block row, we find that (4.10) is nonsingular if and only if $G_x - G_{xx}vvc^T$ is nonsingular. Now suppose that $\zeta \in \mathbb{R}^N$ and $(G_x - G_{xx}vvc^T)\zeta = 0$. Multiplying this from the left with ψ^T we obtain $c^T\zeta = 0$. But then $G_x\zeta = 0$ as well; hence $\zeta = 0$. □

In a variant of (4.6) the third equation is replaced by

$$v^T v = 1. \tag{4.11}$$

A result analogous to Proposition 4.1.1 holds for this system with the same proof.

4.1.2 Quadratic Turning Points by Direct Bordering Methods

We consider again the problem of computing a quadratic turning point in the setting of (4.1)–(4.4). The most important property of a quadratic turning point is that G_x has rank deficiency 1. From Propositions 3.2.1 and 3.2.2 we know that there exist $b, c \in \mathbb{R}^N$ and $d \in \mathbb{R}$ such that

$$M(x,\alpha) = \begin{pmatrix} G_x(x,\alpha) & b \\ c^T & d \end{pmatrix} \tag{4.12}$$

is nonsingular at the turning point and hence in a neighborhood of it. Now we define the functions $v(x,\alpha), w(x,\alpha) \in \mathbb{R}^N$, $h(x,\alpha) \in \mathbb{R}$ by the equations

$$M \begin{pmatrix} v \\ h \end{pmatrix} = \begin{pmatrix} 0_n \\ 1 \end{pmatrix}, \tag{4.13}$$

$$\begin{pmatrix} w^T & h \end{pmatrix} M = \begin{pmatrix} 0_n^T & 1 \end{pmatrix}. \tag{4.14}$$

We note that (4.13) and (4.14) actually compute the last column and last row of M^{-1}, respectively; therefore, the two systems must produce the same value for h. By Corollary 3.3.4 G_x is singular if and only if $h = 0$. In this case v, w are nonzero scalar multiples of ϕ, ψ, respectively.

We propose the following defining system for a quadratic turning point:

$$\begin{aligned} G(x,\alpha) &= 0, \\ h(x,\alpha) &= 0. \end{aligned} \tag{4.15}$$

This is called a *minimally augmented system*, because the number of equations equals the number of free variables. There is a standard "trick" with which to compute the derivatives of h. If z is any state variable or parameter, then by taking derivatives of (4.13) with respect to z, multiplying from the left with $(w^T\ h)$, and using (4.14), we obtain

$$h_z + w^T G_{xz} v = 0. \tag{4.16}$$

In particular we have

$$h_x = -w^T G_{xx} v, \tag{4.17}$$

$$h_\alpha = -w^T G_{x\alpha} v. \tag{4.18}$$

Proposition 4.1.2. If (4.3) and (4.4) hold then the quadratic turning point (x, α) is a regular solution to (4.15).

Proof. It is sufficient to show that

$$\begin{pmatrix} G_x & G_\alpha \\ h_x & h_\alpha \end{pmatrix} \tag{4.19}$$

is nonsingular. By Proposition 3.2.1 it is sufficient (and necessary) that G_α is not in the range of G_x and h_x^T is not in the range of G_x^T. The first assertion is equivalent to (4.3); by (4.17) and the interpretation of v, w the second assertion is equivalent to (4.4). □

4.1.3 Detection of Quadratic Turning Points

Limit points are usually detected as quadratic turning points, all other cases being non-generic. Quadratic turning points are found easily by monitoring the parameter values along the computed curve. However, for an implementation in software it is desirable to have a test function, i.e., a function that changes sign at the quadratic turning point. By Proposition 4.1.2 the function $h(x, \alpha)$ defined by (4.13) or (4.14) is such a function.

A drawback of this approach is that one has to update the vectors b, c and scalar d used in (4.12) along the curve so that the matrix M in (4.12) remains nonsingular. If the linear systems are solved by a direct method that involves pivoting (partial or complete) of G_x, then the safest strategy is to use the pivot information by choosing $d = 0$ and taking for b, c scaled unit vectors where the nonzeros correspond to the row (respectively, the column), where the smallest pivot is found. If this is not feasible then one can choose $d = 0$, compute a reasonable choice of b, c in the initial point by some additional computations, and update b, c regularly by replacing

$$b := G_x^{-T} c, \qquad c := G_x^{-1} b$$

and scaling b, c appropriately. If G_x is well conditioned, then M will also be nonsingular; if G_x tends to singularity then b will tend to a left singular vector of G_x and c to a right singular vector so that M will again be nonsingular. Of course this is the asymptotic case; it might fail if, for example, the stepsize of the continuation is too large.

An alternative is to use the function $\det(G_x)$ if it is easily available. In §3.1 we saw indeed that $\det(G_x)$ necessarily changes sign together with $h(x, \alpha)$. We note that the determinant function is usually not very suitable for use in defining systems (because of scaling problems and because derivatives are hard to compute) but often is satisfactory as a test function.

Remark. It is clear that the determinant changes sign in a quadratic turning point if zero is an algebraically simple eigenvalue of G_x in that point; indeed the determinant is just the product of all eigenvalues and the critical eigenvalue has to change sign. However, the multiplicity of the zero eigenvalue does not really matter. For example, the system

$$\begin{cases} \dot{x} &= 0.5x^2 + y - \alpha, \\ \dot{y} &= 0.5x^2 + xy - \alpha \end{cases}$$

with two state variables x, y and one parameter α has a quadratic turning point at $(0, 0, 0)$ and the determinant of G_x changes sign. It does not matter that the zero eigenvalue has algebraic multiplicity 2 at $(0, 0, 0)$.

4.1.4 Continuation of Limit Points

The methods in §§4.1.1 and 4.1.2 can both be used for the numerical continuation of limit points if two parameters are free. It is advisable to adapt the auxiliary variables (c in the Moore–Spence system, b, c in the direct bordering method) along the computed branch. The obvious choice for c is a (scaled) right singular vector of G_x in a previously computed point, for b a (scaled) left singular vector.

A generic limit point satisfies (4.4), but a limit point curve generically contains isolated other points where (4.4) does not hold. See Chapter 6 for a full discussion. In the case where a distinguished bifurcation parameter is chosen (i.e., we consider the diagrams in state-distinguished parameter space) the (generic) quadratic turning point can generically degenerate in one of two ways: either a hysteresis point ((4.4) does not hold) or a transcritical bifurcation point ((4.3) does not hold). See Chapter 7 for a full discussion.

4.2 Example: A One-Dimensional Continuous Brusselator

4.2.1 The Model and Its Discretization

The Brusselator is a system of equations intended to model the Belusov–Zhabotinsky reaction. This is a system of reaction-diffusion equations that is known to exhibit oscillatory behavior. The history of the model goes back to [197]; the authors associated their model with the city where it was created.

There exist many variants of the Brusselator model. A discrete model with three cyclically coupled boxes was used in Exercise 2.5.17.

We will consider a continuous Brusselator with (for simplicity) a one-dimensional space variable. The system then exhibits only a $\mathbb{Z}_2$-symmetry, which nevertheless has profound consequences for the behavior of the system. In the present section symmetry plays no role, so we ignore it.

Our model is taken from [206] (that paper also gives references to previous work of the same and other authors). The unknowns are the concentrations $X(z,t)$, $Y(z,t)$, $A(z,t)$, $B(z,t)$ of four reactants. Here t denotes time and z is the one-dimensional space variable normalized so that $z \in [0, 1]$. The length L of the reactor is a parameter of the problem. For simplicity we assume that the diffusion coefficient of B is infinite so that B is equal to its imposed value at the endpoints; i.e., B is a parameter of the problem.

The time-evolution equations for X, Y, A are the reaction-diffusion equations

$$\frac{\partial X}{\partial t} = \frac{D_X}{L^2}\frac{\partial^2 X}{\partial z^2} - ((B+1)X - X^2Y - A), \tag{4.20}$$

$$\frac{\partial Y}{\partial t} = \frac{D_Y}{L^2}\frac{\partial^2 Y}{\partial z^2} - (X^2Y - BX), \tag{4.21}$$

$$\frac{\partial A}{\partial t} = \frac{D_A}{L^2}\frac{\partial^2 A}{\partial z^2} - A. \tag{4.22}$$

Here D_X, D_Y, D_A are the diffusion coefficients of X, Y, A, respectively.

The corresponding equilibrium equations are

$$\frac{D_X}{L^2}\frac{\partial^2 X}{\partial z^2} - ((B+1)X - X^2Y - A) = 0, \tag{4.23}$$

$$\frac{D_Y}{L^2}\frac{\partial^2 Y}{\partial z^2} - (X^2Y - BX) = 0, \tag{4.24}$$

$$\frac{D_A}{L^2}\frac{\partial^2 A}{\partial z^2} - A = 0. \tag{4.25}$$

We will consider equilibrium solutions with Dirichlet boundary conditions $A(0) = A(1) = A_0$, $X(0) = X(1) = A_0$, $Y(0) = Y(1) = B/A_0$. We note that this introduces the new parameter A_0 and that the boundary conditions respect the symmetry. Also (4.25) can be solved analytically and the resulting function

$$A(A_0, L, D_A, z) = A_0 \frac{1 - e^{-\frac{L}{\sqrt{D_A}}}}{e^{\frac{L}{\sqrt{D_A}}} - e^{-\frac{L}{\sqrt{D_A}}}} \left(e^{\frac{L}{\sqrt{D_A}}z} + e^{\frac{L}{\sqrt{D_A}}(1-z)}\right)$$

input into (4.23), (4.24). We note that $A(A_0, L, D_A, z)$ converges to A_0 uniformly in $[0, 1]$ if L tends to zero or D_A tends to ∞.

To simplify the notation we introduce the functions

$$f_1(D_X, D_Y, A_0, L, D_A, B, z, x, y) \equiv -\frac{L^2}{D_X}((B+1)x - x^2y - A(A_0, L, D_A, z)), \tag{4.26}$$

$$f_2(D_X, D_Y, A_0, L, D_A, B, z, x, y) \equiv -\frac{L^2}{D_Y}(x^2y - Bx). \tag{4.27}$$

If L, D_X, D_Y are all nonzero (this has to be the case if the model has a physical interpretation), then (4.23), (4.24) are equivalent to

$$\frac{d^2X}{dz^2} + f_1(D_X, D_Y, A_0, L, D_A, B, z, X, Y) = 0, \tag{4.28}$$

$$\frac{d^2Y}{dz^2} + f_2(D_X, D_Y, A_0, L, D_A, B, z, X, Y) = 0, \tag{4.29}$$

where the notation indicates that X, Y no longer depend on t.

These equations allow a trivial solution with constant $X \equiv A_0, Y \equiv B/A_0$ for $L = 0$ and arbitrary D_X, D_Y, A_0, D_A, B. Numerical continuation can start from the trivial solution, in the direction of positive values for L.

Now we have to discretize (4.28), (4.29). Adaptive mesh strategies are usually preferred for such purposes, at least if the solution varies much faster in some parts of the domain of definition than in other parts. Usually this does not happen in the case of the Brusselator, in spite of its complicated bifurcation behavior. Therefore and for simplicity reasons we will use equidistant meshes. To avoid spurious solutions (solutions that are induced by the discretization but do not actually correspond to solutions of the undiscretized problem) one can vary the number of mesh points. If the same solution curves are found for several discretizations, then we can assume (at least for practical purposes) that they correspond to solutions to the continuous problem.

The number of mesh points (n) is routinely set to 42 so that the discretization step is $h = 1/43$.

Let $z_i = hi(i = 1, \ldots, n)$ be the mesh points. For simplicity of notation we set $z_0 = 0, z_{n+1} = 1$. Let $u = (u_i)_{i=1,\ldots,2n}$ be the vector of unknowns where for each i, u_{2i-1} is the approximation to $X(z_i)$, and u_{2i} is the approximation to $Y(z_i)$. By repeating essentially the computations in §1.3.2 we find that

$$\frac{X_{i-1} - 2X_i + X_{i+1}}{h^2} + \left(\frac{1}{12} f_1|_{i-1} + \frac{10}{12} f_1|_i + \frac{1}{12} f_1|_{i+1}\right) = O(h^4), \tag{4.30}$$

$$\frac{Y_{i-1} - 2Y_i + Y_{i+1}}{h^2} + \left(\frac{1}{12} f_2|_{i-1} + \frac{10}{12} f_2|_i + \frac{1}{12} f_2|_{i+1}\right) = O(h^4). \tag{4.31}$$

Here X_j (respectively, Y_j) is a shorthand for $X(z_j)$ (respectively, $Y(z_j)$) $(j = i-1, i, i+1)$ and $f_k|_j$ is a shorthand for $f_k(D_X, D_Y, A_0, L, D_A, B, z_j, X_j, Y_j)$ $(j = i-1, i, i+1; k = 1, 2)$. We therefore consider the discretized equation

$$F(u, D_X, D_Y, A_0, L, D_A, B) = 0, \tag{4.32}$$

where F has $2n$ components, given by

$$F_{2i-1} = u_{2i-3} - 2u_{2i-1} + u_{2i+1} + h^2\left(\frac{1}{12} f_1|_{i-1} + \frac{10}{12} f_1|_i + \frac{1}{12} f_1|_{i+1}\right), \tag{4.33}$$

$$F_{2i} = u_{2i-2} - 2u_{2i} + u_{2i+2} + h^2\left(\frac{1}{12} f_2|_{i-1} + \frac{10}{12} f_2|_i + \frac{1}{12} f_2|_{i+1}\right) \tag{4.34}$$

for $i = 1, \ldots, n$ with the convention that $u_{-1} = u_{2n+1} = A_0$, $u_0 = u_{2n+2} = B/A_0$.

The relations (4.30), (4.31) express that our finite difference approximation is a method of order 4. We refer to [78] for background on this notion; we remark that the choice of an equidistant mesh allows to obtain this relatively high order of accuracy by fairly simple formulas.

The particular ordering of the variables was adopted to make the Jacobian matrix F_u as simple as possible. From (4.32) it is clear that F_u is a (nonsymmetric) seven-band matrix with three superdiagonals and three subdiagonals. In fact, the fine structure is

Table 4.1: Data for eight consecutive points on an equilibrium curve.

Counter	L
189	0.260590
190	0.260666
191	0.260725
192	0.260752
193	0.260749
194	0.260726
195	0.260675
196	0.260581

somewhat more specific since every row contains at most six nonzero elements (corresponding to the X- and Y-values in three adjacent discretization points).

We recall that a rescaling of the equations was performed to obtain (4.28), (4.29). This does not influence the structure of the set of equilibrium solutions (provided L, D_X, D_Y are nonzero).

4.2.2 Turning Points in the Brusselator Model

In the present computations we choose the parameters $D_X = 0.0016$, $D_Y = 0.008$, $A_0 = 2.0$, $D_A = 0.5$, and $B = 4.6$ fixed and L is the only variable of the problem. It is initially set to 0. We start with constant equilibrium values $u_i = A_0$ for i odd, $u_i = B/A_0$ for i even (we note that this is always a solution if $L = 0$). Since only values $L > 0$ have a physical meaning, we start by freeing L and follow a curve of equilibrium solutions in the direction of increasing L. From a linear algebra point of view this involves solutions with one-bordered extensions of the seven-band matrix F_u. These were solved using LAPACK band solvers [11] and the bordered matrix software described in §3.6. It turns out that L increases monotonously for some time until it reaches a value of about 0.260752. For eight consecutively computed points we give in Table 4.1 the value of L (the variable "Counter" keeps track of the number of computed points).

The data in Table 4.1 suggest that between the computed points 192 and 193 there is a turning point. We can locate it accurately by setting up the system (4.15). This requires the choice of the vectors b, c and scalar d in (4.12). We choose $c_i = b_i = i/\sqrt{\sum_{j=1}^{2n} j^2}$, $d = 0$. As a starting point for the Newton iteration we took the continuation point 192.

From a linear algebra point of view this again involves solutions with one-bordered extensions of the seven-band matrix F_u. These were again solved using LAPACK band solvers and the bordered matrix software described in §3.6.

The method converges well; the norms of the successive Newton corrections are 1.182797E-02, 4.133588E-04, 6.542175E-07, 2.137159E-12. The computed value of L turns out to be 0.26075331599577; for the values of u see Table 7.9.

4.3 Classical Methods for the Computation of Hopf Points

4.3.1 Hopf Points

We consider again the dynamical system

$$\dot{x} = G(x, \alpha), \tag{4.35}$$

where $x, G(x) \in \mathbb{R}^N, \alpha \in \mathbb{R}$. An ordinary solution point (x^0, α^0) of the equilibrium equation

$$G(x, \alpha) = 0 \tag{4.36}$$

is called a Hopf point if $G_x^0 = G_x(x^0, \alpha^0)$ has a conjugate pair of algebraically simple pure imaginary eigenvalues $\pm i\omega^0$, $\omega^0 > 0$. Then in a neighborhood of (x^0, α^0) G_x has the eigenvalues $\eta(x, \alpha) \pm i\omega(x, \alpha)$, with $\eta(x, \alpha), \omega(x, \alpha)$ smooth functions of (x, α) such that $\eta(x^0, \alpha^0) = 0$, $\omega(x^0, \alpha^0) = \omega^0$. We note that $\eta(x, \alpha)$ is usually nonzero in points different from (x^0, α^0).

Now (4.36) defines a smooth one-dimensional manifold through (x^0, α^0). On this manifold η, ω are smooth functions of the arclength s. We say that (x^0, α^0) is an isolated Hopf point if the transversality condition

$$\eta_s \neq 0 \tag{4.37}$$

holds in that point. The classical dynamic Hopf bifurcation theorem [139], [179], [132] states that a curve of periodic orbits originates in an isolated Hopf point if there are no other eigenvalues on the imaginary axis and a quantity ℓ_1, called the first Lyapunov coefficient, does not vanish. ℓ_1 depends on the second- and third-order derivatives of G with respect to the components of x; we refer to Chapter 9 for a definition. If the imaginary axis contains other eigenvalues or ℓ_1 vanishes, then the dynamic behavior of (4.35) is more complicated. We will discuss such situations in Chapters 5 and 9, respectively. However, this does not influence the computation or continuation of isolated Hopf points. On the other hand, condition (4.37) is essential.

There is an enormous literature on the detection, computation, and continuation of Hopf points. We will devote the rest of Chapter 4 to the computation by regular systems of equations. In the present section we discuss three approaches with systems of, respectively, $3N + 2$, $2N + 2$, and $N + 2$ equations. We note that the computational work for a smaller system is not necessarily less than for a larger system (this will become clear). However, smaller systems are at least from a theoretical point more satisfactory. In fact, ideally the system should contain only $N + 1$ equations since that is the total number of variables involved in the problem. This is indeed possible, but is postponed to §4.4 since it requires some preparation.

We fix some notation. Let $\pm i\omega$, $\omega > 0$ be a conjugate pair of algebraically simple eigenvalues of $A = G_x$ and let $p = p_1 + ip_2$, $q = q_1 + iq_2$ $(p_1, p_2, q_1, q_2 \in \mathbb{R}^n)$ be left and right eigenvectors of $i\omega$, respectively, i.e.,

$$A(q_1 + iq_2) = i\omega(q_1 + iq_2), \qquad (p_1 + ip_2)^H A = i\omega(p_1 + ip_2)^H. \tag{4.38}$$

Here p^H denotes the usual complex conjugate of a vector, i.e., $(p_1 + ip_2)^H = p_1^T - ip_2^T$. Hence (4.38) is equivalent to

$$Aq_1 = -\omega q_2, \quad Aq_2 = \omega q_1, \quad p_1^T A = \omega p_2^T, \quad p_2^T A = -\omega p_1^T. \tag{4.39}$$

Then

$$(A^2 + \omega)^2 q_1 = 0, \quad (A^2 + \omega^2) q_2 = 0, \quad p_1^T (A^2 + \omega^2) = 0, \quad p_2^T (A^2 + \omega^2) = 0. \tag{4.40}$$

$A^2 + \omega^2$ has rank deficiency 2 with a left singular space spanned by p_1, p_2 and a right singular space spanned by q_1 and q_2. These are called the left and right joint eigenspaces of the pair $\pm i\omega$. The following properties are standard and will be used freely.

Proposition 4.3.1.

1. There exists at least one pair p, q of left and right eigenvectors such that
$$p_i^T q_j = \delta_{ij} \; ; \; i, j \in \{1, 2\} \tag{4.41}$$
holds.

2. For every right eigenvector q there exists a unique left eigenvector p such that (4.41) holds (the converse holds similarly).

3. If p, q are left and right eigenvectors of A, respectively, then (4.41) holds if and only if $p^H q = 2$.

4. Every nonzero vector in the joint right eigenspace of $\pm i\omega$ is the real part of a unique right eigenvector of $i\omega$ (similarly in the left joint eigenspace).

5. There exists a (usually not unique) right eigenvector $q = q_1 + iq_2$ such that $q_1^T q_2 = 0$.

Proof. 1. The existence of at least one pair of left and right eigenvectors with the property (4.41) follows from the Jordan normal form of A (it is essential that the eigenvalue $i\omega$ is algebraically simple).

2. Now if $q^1 = q_1^1 + iq_2^1$ is another right eigenvector, then there exist $r_1, r_2 \in \mathbb{R}$, not both zero, such that $q_1^1 + iq_2^1 = (q_1 + iq_2)(r_1 + ir_2)$. Every left eigenvector has the form $p^1 = p_1^1 + ip_2^1 = (p_1 + ip_2)(r_1^1 + ir_2^1)$. Then (4.41) holds for p^1, q^1 if and only if $r_1^1 + ir_1^1 = (r_1 + ir_2)/(r_1^2 + r_2^2)$.

3. Next, (4.41) clearly implies $p^H q = 2$. Conversely, if $q^1 = q_1^1 + iq_2^1 = (q_1 + iq_2)(r_1 + ir_2)$ and $p^1 = p_1^1 + ip_2^1 = (p_1 + ip_2)(r_1^1 + ir_2^1)$ are such that $p^{1H} q^1 = 2$, then $(r_1 - ir_2)(r_1^1 + ir_2^1) = 2$. Hence $r_1^1 + ir_2^1 = (r_1 + ir_2)/(r_1^2 + r_2^2)$; i.e., (4.41) holds for p^1, q^1.

Statements 4 and 5 are now straightforward. □

Proposition 4.3.2. Let A depend smoothly on a parameter s and have the algebraically simple eigenvalue $\lambda(s)$ in a neighborhood of $s = s^0$. Let p, q be corresponding left and right eigenvectors. Then

$$p^H A_s q = \lambda_s p^H q. \tag{4.42}$$

Proof. Take derivatives of the identity $Aq = \lambda q$, multiply from the left with p^H, and apply $p^H A = \lambda p^H$. □

In the case of a Hopf point in the setting of (4.36) we can think of x, α as implicitly depending on a single parameter s; with p, q chosen to satisfy (4.38) we have in particular

$$p_1^T(G_x)_s q_1 + p_2^T(G_x)_s q_2 = 2\eta_s, \tag{4.43}$$

$$p_1^T(G_x)_s q_2 - p_2^T(G_x)_s q_1 = 2\omega_s. \tag{4.44}$$

4.3.2 Regular Systems with $3N + 2$ Equations

The natural analogue of (4.6) for Hopf points is the complex system

$$\begin{aligned} &G(x, \alpha) = 0, \\ &(G_x(x, \alpha) - i\omega I_N)q = 0, \\ &c^H q = 1, \end{aligned} \tag{4.45}$$

where c is a complex vector not orthogonal to q. To obtain a real system we set $q = q_1 + iq_2$, $c = c_1 + ic_2$ and take real and imaginary parts of the equations in (4.45). We find

$$\begin{aligned} &G(x, \alpha) = 0, \\ &G_x(x, \alpha)q_1 + \omega q_2 = 0, \\ &G_x(x, \alpha)q_2 - \omega q_1 = 0, \\ &c_1^T q_1 + c_2^T q_2 = 1, \\ &-c_2^T q_1 + c_1^T q_2 = 0. \end{aligned} \tag{4.46}$$

Now (4.46) is a system of $3N + 2$ equations in $3N + 2$ unknowns $x, \alpha, q_1, q_2, \omega$. We will prove that this system is regular at the Hopf point if and only if (4.37) holds; however, it is convenient to generalize the system somewhat to include other cases that have also been proposed in the literature. First, we remark that the condition $c^H q \neq 0$ is equivalent to the requirement that

$$\begin{pmatrix} c_1^T & c_2^T \\ -c_2^T & c_1^T \end{pmatrix} \begin{pmatrix} q_1 & -q_2 \\ q_2 & q_1 \end{pmatrix}$$

has full rank. Second, if $Q = Q_1 + iQ_2$ is another right eigenvector of G_x, then there exist $\beta, \gamma \in \mathbb{R}$ such that $(Q_1 + iQ_2) = (q_1 + iq_2)(\beta + i\gamma)$ so that

$$\begin{pmatrix} Q_1 & -Q_2 \\ Q_2 & Q_1 \end{pmatrix} = \begin{pmatrix} q_1 & -q_2 \\ q_2 & q_1 \end{pmatrix} \begin{pmatrix} \beta & -\gamma \\ \gamma & \beta \end{pmatrix}. \tag{4.47}$$

Hence for $c_1, c_2, c_3, c_4 \in \mathbb{R}^n$ the condition that

$$\begin{pmatrix} c_1^T & c_2^T \\ c_3^T & c_4^T \end{pmatrix} \begin{pmatrix} q_1 & -q_2 \\ q_2 & q_1 \end{pmatrix} \tag{4.48}$$

has full rank is independent of the choice of the right eigenvector $q_1 + iq_2$ of G_x.

Proposition 4.3.3. Assume that $[G_x, G_\alpha]$ has full rank and that $\pm i\omega$ is a pair of algebraically simple eigenvalues of G_x. If for $c_1, c_2, c_3, c_4 \in \mathbb{R}^N$ the matrix in (4.48) is nonsingular and r_1, r_2 are real numbers that are not both zero, then

$$\begin{aligned} &G(x,\alpha) = 0, \\ &G_x(x,\alpha)q_1 + \omega q_2 = 0, \\ &G_x(x,\alpha)q_2 - \omega q_1 = 0, \\ &c_1^T q_1 + c_2^T q_2 = r_1, \\ &c_3^T q_1 + c_4^T q_2 = r_2 \end{aligned} \tag{4.49}$$

holds for a unique right eigenvector $q_1 + iq_2$ of G_x for the eigenvalue $i\omega$; the system is regular if and only if (4.37) holds.

Proof. The existence of a unique solution to (4.49) follows from (4.47) and the nonsingularity of (4.48). Now assume that (4.37) holds and suppose that $U \in \mathbb{R}^N$, $a \in R$, $Q_1, Q_2 \in \mathbb{R}^N$, $\Omega \in \mathbb{R}$ are such that

$$\begin{pmatrix} G_x & G_\alpha & 0 & 0 & 0 \\ G_{xx}q_1 & G_{x\alpha}q_1 & G_x & \omega & q_2 \\ G_{xx}q_2 & G_{x\alpha}q_2 & -\omega & G_x & -q_1 \\ 0 & 0 & c_1^T & 0 & 0 \\ 0 & 0 & c_2^T & 0 & 0 \end{pmatrix} \begin{pmatrix} U \\ a \\ Q_1 \\ Q_2 \\ \Omega \end{pmatrix} = 0. \tag{4.50}$$

From the first block of equations in (4.50) it follows that there exists a $\beta \in \mathbb{R}$ such that $U = \beta x_s$, $a = \beta\alpha_s$. Substituting this in the second and third blocks of equations we obtain

$$\begin{aligned} &\beta G_{xs}q_1 + G_x Q_1 + \omega Q_2 + q_2\Omega = 0, \\ &\beta G_{xs}q_2 - \omega Q_1 + G_x Q_2 - q_1\Omega = 0. \end{aligned} \tag{4.51}$$

Now choose a left eigenvector $p = p_1 + ip_2$ so that (4.41) holds. We multiply the first equation in (4.51) from the left with p_1^T and add to the second multiplied with p_2^T. By (4.39), (4.41), and (4.43) we get $2\beta\eta_s = 0$. By (4.37) this implies $\beta = 0$; i.e., $U = 0, a = 0$. Substituting this back into (4.51) we obtain after an easy manipulation that

$$(G_x^2 + \omega^2)Q_1 + 2\omega q_1\Omega = 0. \tag{4.52}$$

Multiplying (4.52) from the left with p_1^T and applying (4.40) and (4.41) we find that $\Omega = 0$. Substituting this in (4.49) we see that $Q_1 + iQ_2$ must be an eigenvector of G_x for the eigenvalue $i\omega$, so it must have the form $(q_1 + iq_2)(\beta + i\gamma)$; the last two equations in (4.49) and the nonsingularity of (4.48) imply that $\beta = \gamma = 0$, i.e., $Q_1 = Q_2 = 0$.

Conversely, if (4.37) does not hold, then it is easily seen that (4.49) admits the nonzero solution $U = x_s$, $a = \alpha_s$, $Q_1 = q_{1s}$, $Q_2 = q_{2s}$, $U = \omega_s$ if Q_1, Q_2 is the (unique) right eigenvector of G_x for the eigenvalue $i\omega$ that satisfies the two scalar equations in (4.49). □

The system (4.49) includes (4.46) as a special case. Also, there are some obvious alternatives for the two scalar conditions in (4.46), e.g., $q_1^T q_1 = 1$, $L^T q_1 = 0$, where L is any real vector not orthogonal to the right joint eigenspace of $\pm i\omega$; or $q^H q = 1$, $L^H q = 0$ with L a complex vector that is not orthogonal to q.

Proposition 4.3.3 generalizes immediately to such situations. Such variants were proposed for the computation of Hopf points in [144] and [123] and were routinely used in AUTO97 [85] and its precursors.

4.3.3 Regular Systems with $2N + 2$ Equations

Two two-dimensional subspaces V, W are said to be orthogonal if V contains a nonzero vector orthogonal to W or, equivalently, if W contains a nonzero vector orthogonal to V. For example, by Proposition 4.3.1 the left and right joint eigenspaces of the algebraically simple eigenvalues $\pm i\omega$ are not orthogonal.

Let $c_1, c_2 \in \mathbb{R}^N$ span a two-dimensional space not orthogonal to the joint right eigenspace of $\pm i\omega$, and let r_1, r_2 be any two numbers, not both zero. There exists a unique vector $v = q_1$ in the right eigenspace of $\pm i\omega$ such that $c_1^T v = r_1, c_2^T v = r_2$. We can avoid the introduction of a corresponding q_2 by considering

$$\begin{aligned} &G(x,\alpha) = 0, \\ &(G_x^2(x,\alpha) + k)v = 0, \\ &c_1^T v = r_1, \\ &c_2^T v = r_2, \end{aligned} \tag{4.53}$$

where implicitly $k = \omega^2$. Now (4.53) can be seen as a system of $2N + 2$ equations in $2N+2$ unknowns x, α, v, k. Variants with slightly different normalizations were proposed for the computation of Hopf points in [138], [160], and [204] and implemented in [165].

Proposition 4.3.4. Assume that $[G_x, G_\alpha]$ has full rank and that $\pm i\omega$ is a pair of algebraically simple eigenvalues of G_x. If c_1, c_2 span a space that is not orthogonal to the joint right eigenspace of $\pm i\omega$, then (4.53) is a regular system if and only if (4.37) holds.

Proof. Assume that (4.37) holds and suppose that $U \in \mathbb{R}^N$, $a \in R$, $Q_1 \in \mathbb{R}^N$, $\Omega \in \mathbb{R}$ are such that

$$\begin{pmatrix} G_x & G_\alpha & 0 & 0 \\ (G_x)_x^2 v & (G_x)_\alpha^2 v & G_x^2 + k & v \\ 0 & 0 & c_1^T & 0 \\ 0 & 0 & c_2^T & 0 \end{pmatrix} \begin{pmatrix} U \\ a \\ Q_1 \\ \Omega \end{pmatrix} = 0 \tag{4.54}$$

holds. From the first block of equations in (4.54) it follows that there exists a $\beta \in \mathbb{R}$ such that $U = \beta x_s$, $a = \beta \alpha_s$. Substituting this in the second block of equations we obtain

$$\beta(G_x G_{xs} + G_{xs} G_x)v + (G_x^2 + k)Q_1 + v\Omega = 0. \tag{4.55}$$

Now choose a left eigenvector $p = p_1 + ip_2$ so that (4.41) holds with $v = q_1$. Multiplying (4.55) from the left with p_2^T and using (4.39), (4.41), and (4.43) we get $-2\beta\omega\eta_s = 0$. By (4.37) this implies $\beta = 0$, i.e., $U = 0, a = 0$. Substituting this back into (4.55), multiplying from the left with p_1^T, and applying (4.41) we find that $\Omega = 0$. Substituting this also in (4.55) and taking the two bottom rows in (4.54) we find also $Q_1 = 0$.

Conversely, suppose that (4.37) does not hold. We define $v(s) \in \mathbb{R}^n$, $k(s) \in \mathbb{R}$, $l(s) \in \mathbb{R}$, where s denotes arclength along the solution branch to (4.36) by requiring that they reduce to v, k, and 0, respectively, at the Hopf point and satisfy

$$\begin{aligned} &(G_x^2(s) + k(s))v(s) + l(s)q_2 = 0, \\ &c_1^T v(s) = r_1, \\ &c_2^T v(s) = r_2. \end{aligned} \tag{4.56}$$

Here q_2 is fixed and corresponds to $q_1 = v$ at the Hopf point. The system (4.56) is regular at the Hopf point. Taking derivatives with respect to s we find

$$(G_{xs}G_x + G_xG_{xs})v + k_sv + (G_x^2 + k)v_s + l_sq_2 = 0. \tag{4.57}$$

Multiplying this from the left with p_2^T we obtain

$$-\omega(p_2^TG_{xs}q_2 + p_1^TG_{xs}q_1) + l_s = 0. \tag{4.58}$$

By (4.43) this implies that $l_s = 0$. Hence (4.54) admits the nonzero solution $U = x_s$, $a = \alpha_s$, $Q_1 = q_{1s}$, $\Omega = k_s$ and therefore is a singular system. □

Again, there are some obvious alternatives for the two scalar conditions in (4.53), e.g., $q_1^Tq_1 = 1$, $L^Tq_1 = 0$, where L is any vector not orthogonal to the right joint eigenspace of $\pm i\omega$. Proposition 4.3.4 remains valid.

4.3.4 Regular Systems with $N + 2$ Equations

The Newton systems that arise by a straightforward application of the methods described in §§4.3.2 and 4.3.3 require the solution of linear systems of order $3N + 2$ or $2N + 2$, respectively. However, the Jacobian matrices (appearing, e.g., in (4.50) and (4.54)) have a special structure and it was soon remarked (see [123] and [204]) that a solution of these systems can be reduced to a solution with a few smaller systems, typically involving $G_x^2 + kI_n$. This suggests looking straightaway for smaller systems in which $G_x^2 + kI_n$ is involved, and these were indeed found; see [31], [57], [237].

The underlying idea is to express that in the Hopf point $G_x^2 + k$ has rank defect 2. Let $B, C \in \mathbb{R}^{N\times 2}$, $D \in \mathbb{R}^{2\times 2}$ be such that

$$M(x,\alpha,k) = \begin{pmatrix} G_x^2 + kI_n & B \\ C^T & D \end{pmatrix} \tag{4.59}$$

is nonsingular. (By Proposition 3.2.1 it is sufficient that B spans a space that is not orthogonal to the left singular space of $Gx^2 + kI_n$, C spans a space not orthogonal to the right singular space of $G_x^2 + kI_n$, and $D = 0$.) Now choose any real numbers r_1, r_2, not both zero, and define $v(x,\alpha) \in \mathbb{R}^N$, $h_1(x,\alpha), h_2(x,\alpha) \in \mathbb{R}$ by

$$M\begin{pmatrix} v \\ h_1 \\ h_2 \end{pmatrix} = \begin{pmatrix} 0_n \\ r_1 \\ r_2 \end{pmatrix}. \tag{4.60}$$

By Corollary 3.3.4 we have

$$\begin{aligned} &G(x,\alpha) = 0, \\ &h_1(x,\alpha,k) = 0, \\ &h_2(x,\alpha,k) = 0 \end{aligned} \tag{4.61}$$

at the Hopf point. To compute the derivatives h_{1z}, h_{2z} where z is in x, α, k we may first solve

$$\begin{pmatrix} w_1^T & h_1 & h_3 \\ w_2^T & h_2 & h_4 \end{pmatrix} M = \begin{pmatrix} 0_n^T & 1 & 0 \\ 0_n^T & 0 & 1 \end{pmatrix}, \tag{4.62}$$

where $w_1, w_2 \in \mathbb{R}^N$. By a standard argument we find

$$\begin{aligned} h_{1z} &= -w_1^T(G_x^2)_z v, \\ h_{2z} &= -w_2^T(G_x^2)_z v, \\ h_{1k} &= -w_1^T v, \\ h_{2k} &= -w_2^T v. \end{aligned} \tag{4.63}$$

Proposition 4.3.5. Assume that $[G_x, G_\alpha]$ has full rank and that $\pm i\omega$ is a pair of algebraically simple eigenvalues of G_x. If b_1, b_2, c_1, c_2 are such that M is nonsingular at the Hopf point, then (4.63) is a regular system if and only if (4.37) holds.

Proof. Assume that (4.37) holds and suppose that $U \in \mathbb{R}^N$, $a \in R$, $K \in \mathbb{R}$ are such that

$$\begin{pmatrix} G_x & G_\alpha & 0 \\ h_{1x} & h_{1\alpha} & h_{1k} \\ h_{2x} & h_{2\alpha} & h_{2k} \end{pmatrix} \begin{pmatrix} U \\ a \\ K \end{pmatrix} = 0 \tag{4.64}$$

holds. From the first block of equations in (4.64) it follows that there exists a $\beta \in \mathbb{R}$ such that $U = \beta x_s$, $a = \beta\alpha_s$. Substituting this in the last two rows and using (4.63) we find

$$\begin{aligned} -\beta w_1^T(G_x G_{xs} + G_{xs} G_x)v - (w_1^T v)K = 0, \\ -\beta w_2^T(G_x G_{xs} + G_{xs} G_x)v - (w_2^T v)K = 0. \end{aligned} \tag{4.65}$$

By (4.60) v is in the joint right eigenspace of the eigenvalues $\pm i\omega$ and by (4.62) w_1, w_2 span the joint left eigenspace. So by making linear combinations of the two equations in (4.65) we may assume that

$$\begin{aligned} \beta p_1^T(G_x G_{xs} + G_{xs} G_x)q_1 + (p_1^T q_1)K = 0, \\ \beta p_2^T(G_x G_{xs} + G_{xs} G_x)q_1 + (p_2^T q_1)K = 0, \end{aligned} \tag{4.66}$$

where p_1, p_2, q_1, q_2 satisfy (4.39), (4.41), and (4.43). Hence $\beta = 0$, $K = 0$, and so $X = 0$, $a = 0$.

Conversely, assume that (4.37) does not hold. To prove that (4.64) admits a nonzero solution it is sufficient to prove that there exists a $K \in \mathbb{R}$ such that

$$\begin{aligned} w_1^T(G_x^2)_s v + (w_1^T v)K = 0, \\ w_2^T(G_x^2)_s v + (w_2^T v)K = 0, \end{aligned} \tag{4.67}$$

or, equivalently, such that

$$\begin{aligned} p_1^T(G_x^2)_s q_1 + (p_1^T q_1)K = 0, \\ p_2^T(G_x^2)_s q_1 + (p_2^T q_1)K = 0, \end{aligned} \tag{4.68}$$

where p_1, p_2, q_1, q_2 satisfy (4.39), (4.41), (4.43), and (4.44). Clearly (4.68) holds if we choose $K = 2\omega\omega_s$. □

4.3.5 Zero-Sum Pairs of Real Eigenvalues

It is not hard to see that the system (4.53) can also have solutions if $k < 0$. Suppose that λ_1 and $\lambda_2 = -\lambda_1$, $\lambda_1 > 0$ are both algebraically simple eigenvalues of G_x. Such points are called *neutral saddle points* Let p_1, p_2 be corresponding left eigenvectors and q_1, q_2 corresponding right eigenvectors, normalized so that (4.41) holds. In the context of following a branch one easily finds that (4.43) is replaced by

$$p_1^T G_{xs} q_1 + p_2^T G_{xs} q_2 = (\lambda_1 + \lambda_2)_s. \tag{4.69}$$

Now consider the system

$$\begin{aligned} &G(x, \alpha) = 0, \\ &(G_x^2(x, \alpha) + k)v = 0, \\ &c_1^T v = r_1, \\ &c_2^T v = r_2, \end{aligned} \tag{4.70}$$

which is formally identical to (4.53) and admits as a solution the point with zero-sum eigenvalues for $k = -\lambda_1^2$ and v an appropriate linear combination of q_1, q_2. To prove an analogue of Proposition 4.3.4 we have to replace (4.37) by

$$(\lambda_1 + \lambda_2)_s \neq 0. \tag{4.71}$$

Another, less obvious modification is also needed.

Proposition 4.3.6. Assume that $[G_x, G_\alpha]$ has full rank and that λ_1, λ_2 are algebraically simple eigenvalues of G_x, $\lambda_1 > 0$, $\lambda_2 = -\lambda_1$. Let p_1, p_2, q_1, q_2 be as above. If c_1, c_2 span a space that is not orthogonal to the joint right eigenspace of $\pm i\omega$, then the following are equivalent:

1. (4.70) is a regular system;
2. (4.71) holds and $r_1 c_2^T q_1 - r_2 c_1^T q_1 \neq 0$, $r_1 c_2^T q_2 - r_2 c_1^T q_2 \neq 0$.

Proof. Assume that (4.71) holds, together with the two inequalities in the proposition and suppose that $U \in \mathbb{R}^N$, $a \in R$, $Q_1 \in \mathbb{R}^N$, $\Omega \in \mathbb{R}$ are such that

$$\begin{pmatrix} G_x & G_\alpha & 0 & 0 \\ (G_x)_x^2 v & (G_x)_\alpha^2 v & G_x^2 + k & v \\ 0 & 0 & c_1^T & 0 \\ 0 & 0 & c_2^T & 0 \end{pmatrix} \begin{pmatrix} U \\ a \\ Q_1 \\ \Omega \end{pmatrix} = 0 \tag{4.72}$$

holds. From the first block of equations in (4.72) it follows that there exists a $\beta \in \mathbb{R}$ such that $U = \beta x_s$, $a = \beta \alpha_s$. Substituting this in the second block of equations we obtain

$$\beta(G_x G_{xs} + G_{xs} G_x)v + (G_x^2 + k)Q_1 + v\Omega = 0. \tag{4.73}$$

Now there exist real numbers d_1, d_2 such that $v = d_1 q_1 + d_2 q_2$. The two required inequalities imply $d_1 \neq 0$, $d_2 \neq 0$. Multiplying (4.73) from the left with $d_2 p_1^T - d_1 p_2^T$ we get

$$2\beta d_1 d_2 \lambda_1 (\lambda_1 + \lambda_2)_s = 0. \tag{4.74}$$

By (4.71) this implies $\beta = 0$, i.e., $U = 0, a = 0$. Substituting this back into (4.73) and multiplying from the left with $d_1 p_1^T + d_2 p_2^T$ we find that $\Omega = 0$. Substituting this also in (4.73) and taking the two bottom rows in (4.72) we find also $Q_1 = 0$.

The converse is proved as in Proposition 4.3.4. We recall it briefly in the case $r_1 c_2^T q_1 - r_2 c_1^T q_1 = 0$. In the solution to (4.70) we then have $v = d_1 q_1$ with $d_1 \neq 0$.

We define $v(s) \in \mathbb{R}^n$, $k(s) \in \mathbb{R}$, $l(s) \in \mathbb{R}$, where s denotes arclength along the solution branch to (4.36) by requiring that they reduce to v, k, and 0, respectively, at the zero-sum point and satisfy

$$\begin{aligned} (G_x^2(s) + k(s))v(s) + l(s)q_2 &= 0, \\ c_1^T v(s) &= r_1, \\ c_2^T v(s) &= r_2. \end{aligned} \tag{4.75}$$

The system (4.75) is regular. Taking derivatives with respect to s we find

$$(G_{xs}G_x + G_x G_{xs})v + k_s v + (G_x^2 + k)v_s + l_s q_2 = 0. \tag{4.76}$$

Multiplying this from the left with p_2^T we obtain $l_s = 0$. Hence (4.72) admits the nonzero solution $U = x_s$, $a = \alpha_s$, $Q_1 = v_s$, $\Omega = k_s$ and therefore is a singular system. □

The border case $k = 0$ corresponds to Bogdanov–Takens (BT) points where G_x has an eigenvalue with algebraic multiplicity 2 and geometric multiplicity 1. We return to this (more complicated) situation in §5.1.

System (4.61) similarly admits neutral saddle points as solutions for $k < 0$ and BT points for $k = 0$. Proposition 4.3.6 carries over with the same restrictions on c_1, c_2, now interpreted as the columns of the matrix C in (4.59).

Finally, we note that in a practical application the choice of c_1, c_2, r_1, r_2 is usually easy. Indeed, the choice $c_1 = p_1, c_2 = p_2$, $r_1 = 1, r_2 = 1$ is good and approximations to p_1 and p_2 can usually be obtained easily.

4.3.6 Hopf Points by Complex Arithmetic

It is a natural idea to express directly that $G_x - i\omega I_N$ is a singular matrix. Let $b, c \in \mathbb{C}^N$, $d \in \mathbb{C}$ be chosen so that

$$M(x, \alpha, \omega) = \begin{pmatrix} G_x - i\omega I_N & b \\ c^T & d \end{pmatrix} \tag{4.77}$$

is nonsingular. (By Proposition 3.2.1 it is sufficient that b is not orthogonal to the left singular space of $Gx - i\omega I_N$, c is not orthogonal to the right singular space of $G_x - i\omega I_N$, and $d = 0$.) Now define $v(x, \alpha) \in \mathbb{C}^N$, $h(x, \alpha) \in \mathbb{C}$ by

$$M \begin{pmatrix} v \\ h \end{pmatrix} = \begin{pmatrix} 0_N \\ 1 \end{pmatrix}. \tag{4.78}$$

The Hopf point is then characterized by $h = 0$. The further details (e.g., computation of derivatives) are similar to the case treated in §4.1.2. This method has not attracted much attention, presumably because of the obvious disadvantage that it requires complex arithmetic. Nevertheless, our experience (limited to some test computations) indicates that the convergence behavior (domain of attraction for the Newton method) compares favorably with other methods. A clear numerical advantage is that in the case of large systems it better preserves the sparsity of the Jacobian matrix. See §4.6.3 for details.

4.4 Tensor Products and Bialternate Products

We will now introduce some matrix constructions that are not usually found in textbooks on linear or numerical linear algebra. However, [169] gives an introduction to tensor products. Our aim is to obtain defining functions for Hopf bifurcations that involve only the entries of the matrix (i.e., no external objects such as the Hopf eigenvalues or Hopf eigenvectors). In this way the numerical methods to compute Hopf points are linked directly to the structure of the manifold of matrices with Hopf pairs. In Chapter 5 we will use these methods also to compute more complicated bifurcation points; therefore, our treatment is more extensive than strictly necessary for the present purposes.

The origins of the notion of a bialternate product go back to the paper [225] of Stéphanos (1900); Stéphanos's term is "composition bialternée." The English translation "bialternate product" is sometimes abbreviated to "biproduct." An extensive treatment is given in [96]; bialternate products were considered in [129] and [164] to compute Hopf bifurcations; their importance for multiple Hopf was established in [119].

In this section we will prove the fundamental properties of bialternate products and their relation to eigenvalues and the Jordan structure.

We will consider only the case where all matrices are square and have the same dimension; the theory of tensor products extends naturally to rectangular matrices; cf. [169].

Since we are interested in complex eigenvalues and since the complex Jordan normal form of a matrix is somewhat simpler than the real form, we prefer to formulate the results in this section for general complex matrices. In this way we avoid some awkward notational problems. However, when applied to real matrices all the matrix constructions that we consider will again yield real matrices. Also, the rank of a real matrix (over the real numbers) is the same as its rank as a complex matrix (over the complex numbers). Hence the results that we obtain on ranks, in particular in Proposition 4.4.27, hold also for real matrices.

4.4.1 Tensor Products

Tensor products arise naturally when matrices are multiplied in a specific order. Consider the matrices $A, B \in \mathbb{C}^{n\times n}$. They define a mapping

$$\mathbb{C}^{n\times n} \to \mathbb{C}^{n\times n}, \quad X \to Y = AXB^T. \tag{4.79}$$

Denoting the elements of a matrix by lowercase letters we find

$$y_{ij} = \sum_{k=1}^{n}\sum_{l=1}^{n} a_{ik}b_{jl}x_{kl}. \tag{4.80}$$

Now X, Y may also be interpreted as vectors if we order their entries sequentially. In this way (4.80) is the explicit form of a matrix–vector multiplication in which the matrix $\mathcal{A}$ is in $\mathbb{C}^{n^2\times n^2}$ and has entries $\mathcal{A}_{((i,j),(k,l))} = a_{ik}b_{jl}$.

The formal definition is as follows. Let $\{e_i : 1 \le i \le n\}$ be the canonical base of unit vectors in $\mathbb{C}^n$. Then the tensor product $\mathbb{C}^n \otimes \mathbb{C}^n$ is the space $\mathbb{C}^{n^2}$ with formal base

$\{e_i \otimes e_j : 1 \le i, j \le n\}$. If $x = \sum_{i=1}^n x_i e_i \in \mathbb{C}^n$, $y = \sum_{i=1}^n y_i e_i \in \mathbb{C}^n$, then we define $x \otimes y = \sum_{i=1}^n \sum_{j=1}^n x_i y_j e_i \otimes e_j \in \mathbb{C}^{n^2}$. The mapping $(x, y) \to x \otimes y$ is clearly bilinear; i.e., $x \otimes (\alpha y_1 + \beta y_2) = \alpha(x \otimes y_1) + \beta(x \otimes y_2)$ and $(\alpha x_1 + \beta x_2) \otimes y = \alpha(x_1 \otimes y) + \beta(x_2 \otimes y)$.

Two matrices $A, B \in \mathbb{C}^{n \times n}$ can be identified with linear mappings $\mathbb{C}^n \to \mathbb{C}^n$. The tensor product $A \otimes B \in \mathbb{C}^{n^2 \times n^2}$ is then defined as the linear mapping from $\mathbb{C}^n \otimes \mathbb{C}^n$ into itself for which $(A \otimes B)(e_k \otimes e_l) = Ae_k \otimes Be_l = \sum_{i,j=1}^n a_{ik} b_{jl} (e_i \otimes e_j)$. It is easily seen that

$$(A \otimes B)(x \otimes y) = Ax \otimes By \tag{4.81}$$

for all $x, y \in \mathbb{C}^n$. Denoting by $(A \otimes B)_{(i,j),(k,l)}$ the $(e_i \otimes e_j)$-component of $(A \otimes B)(e_k \otimes e_l)$ we have

$$(A \otimes B)_{(i,j),(k,l)} = a_{ik} b_{jl}. \tag{4.82}$$

For numerical work we can use (4.82); the coordinate-free notation (4.81) is more useful for theoretical proofs.

Proposition 4.4.1. Let $A, A_1, A_2, B, B_1, B_2 \in \mathbb{C}^{n \times n}$. Then

1. $A \otimes (B_1 + B_2) = A \otimes B_1 + A \otimes B_2$;
2. $(A_1 + A_2) \otimes B = A_1 \otimes B + A_2 \otimes B$;
3. $(A_1 \otimes B_1)(A_2 \otimes B_2) = (A_1 A_2) \otimes (B_1 B_2)$;
4. $I_n \otimes I_n = I_{n^2}$;
5. if A, B are nonsingular, then so is $A \otimes B$ and $(A \otimes B)^{-1} = A^{-1} \otimes B^{-1}$.

Proof. The proof is obvious from (4.81).

4.4.2 Condensed Tensor Products

Tensor products of matrices appear in a natural way in the transformation of higher-order derivatives when a function of several variables is subject to a change of variables. Since mixed derivatives do not change if the order of derivation is changed (at least under the appropriate smoothness conditions), this leads in a natural way to the idea of a condensed tensor product.

For our applications in Chapter 6 we need only the case of one function of two variables, so we restrict ourselves to this simple situation.

Proposition 4.4.2. Let $A \in \mathbb{C}^{2 \times 2}$. Let V be the set of all $(x_{(k,l)})_{1 \le k,l \le 2}$ such that $x_{(k_1,l_1)} = x_{(k_2,l_2)}$ if (k_2, l_2) is a permutation of (k_1, l_1). Then V is a three-dimensional subspace of $\mathbb{C}^{2 \times 2}$ that is invariant under the action of $A \otimes A$.

Proof. Obviously V is three-dimensional. If $(x_{(k,l)}) \in V$, then

$$((A \otimes A)(x_{(k,l)}))_{(i,j)} = \Sigma_{k,l} a_{ik} a_{jl} x_{(k,l)},$$

$$\begin{aligned} ((A \otimes A)(x_{(k,l)}))_{(j,i)} &= \Sigma_{k,l} a_{jk} a_{il} x_{(k,l)} \\ &= \Sigma_{k,l} a_{jl} a_{ik} x_{(l,k)} = \Sigma_{k,l} a_{jl} a_{ik} x_{(k,l)}. \end{aligned}$$

So $(A \otimes A)(x_{(k,l)})$ is invariant under a permutation of indices, i.e., is again in V. □

Definition 4.4.3. Let $A \in \mathbb{C}^{2\times 2}$. The condensed tensor product of order 2, denoted as $\otimes_2(A)$, is the matrix that represents the restriction of $A \otimes A$ to its invariant subspace V as defined in Proposition 4.4.2 with respect to the basis (x^{11}, x^{12}, x^{22}), where $x^{11}_{(1,1)} = x^{12}_{(1,2)} = x^{12}_{(2,1)} = x^{22}_{2,2} = 1$ and all other entries are zero.

The general form of an entry of $\otimes_2(A)$ is

$$(\otimes_2(A))_{(i,j),(k,l)} = \Sigma\{a_{ik_1}a_{jl_1} : (k_1, l_1) \text{ is a permutation of } (k, l)\}.$$

Explicitly we have that

$$\otimes_2(A) = \begin{pmatrix} a_{11}^2 & 2a_{11}a_{12} & a_{12}^2 \\ a_{11}a_{21} & a_{11}a_{22} + a_{12}a_{21} & a_{12}a_{22} \\ a_{21}^2 & 2a_{21}a_{22} & a_{22}^2 \end{pmatrix}.$$

Proposition 4.4.4. If A is nonsingular, then so is $\otimes_2(A)$.

Proof. This follows from Proposition 4.4.1(5) and the fact that a restriction of a one-to-one mapping is one to one. Of course it can also be proved directly; by a simple computation one finds that $\det(\otimes_2(A)) = (\det(A))^3$. □

The operation of forming tensor products is associative if we allow a small abuse of notation. More precisely, for any three matrices $A, B, C \in \mathbb{C}^{n\times n}$ we can define the tensor product matrix $A\otimes B\otimes C$ as an operator from $\mathbb{C}^{n^3}$ into itself by $(A\otimes B\otimes C)_{(i,j,k),(l,m,p)} = a_{il}b_{jm}c_{kp}$.

As in the previous case, the tensor product operator $A \otimes A \otimes A$ has an invariant subspace consisting of all vectors $(x_{(l,m,p)})$ whose component values are invariant under a permutation of the indices. This invariant subspace has dimension 4 and is spanned by the vectors $x^{111}, x^{112}, x^{122}, x^{222}$, where $x^{ijk}_{(l,m,p)}$ is equal to one if (l, m, p) is a permutation of (i, j, k) and zero in all other entries.

The condensed tensor product $\otimes_3(A)$ is the restriction of $A \otimes A \otimes A$ to V. With respect to the given basis its entries are

$$\otimes_3(A)_{(i,j,k),(l,m,p)} = \Sigma\{(a_{il_1}a_{jm_1}a_{kp_1}) : (l_1, m_1, p_1) \text{ is a permutation of } (l, m, p)\}.$$

Explicitly we have

$$\otimes_3(A) = \begin{pmatrix} a_{11}^3 & 3a_{11}^2a_{12} & 3a_{11}a_{12}^2 & a_{12}^2 \\ a_{11}^2a_{21} & a_{11}^2a_{22} + 2a_{11}a_{12}a_{21} & a_{12}^2a_{21} + 2a_{11}a_{12}a_{22} & a_{12}^2a_{22} \\ a_{11}a_{21}^2 & a_{21}^2a_{12} + 2a_{11}a_{21}a_{22} & a_{22}^2a_{11} + 2a_{12}a_{21}a_{22} & a_{12}a_{22}^2 \\ a_{21}^3 & 3a_{21}^2a_{22} & 3a_{21}a_{22}^2 & a_{22}^3 \end{pmatrix}.$$

Proposition 4.4.5. If A is nonsingular, then so is $\otimes_3(A)$.

Proof. This follows from Proposition 4.4.1(5) and the fact that a restriction of a one-to-one mapping is one to one. □

We now come to the connection with transformation of independent variables. Let $f(x, y)$ be a sufficiently smooth function of two variables (x, y) and let $g(x, y)$ be defined by $g(x, y) = f(X(x, y), Y(x, y))$, where $X(x, y), Y(x, y)$ are also sufficiently smooth. Let $J = \frac{\partial(X,Y)}{\partial(x,y)}$ denote the Jacobian of the transformation. We then have Proposition 4.4.6.

Proposition 4.4.6. There exists smooth 4×2 and 4×1 matrix functions $M_{4,2}(x,y)$ and $M_{4,1}(x,y)$, respectively, such that

$$\begin{pmatrix} g_{xx} \\ g_{xy} \\ g_{yx} \\ g_{yy} \end{pmatrix} = (J^T \otimes J^T) \begin{pmatrix} f_{xx} \\ f_{xy} \\ f_{yx} \\ f_{yy} \end{pmatrix} + M_{4,2} \begin{pmatrix} f_x \\ f_y \end{pmatrix} + M_{4,1} f,$$

where the functions $f, f_x, f_y, f_{xx}, f_{xy}, f_{yx}, f_{yy}$ are evaluated in $(X(x,y), Y(x,y))$.

Proof. The proposition is proved by straightforward verification. □

Now if $f(x,y)$ is sufficiently smooth, then the vector

$$\begin{pmatrix} f_{xx} \\ f_{xy} \\ f_{yx} \\ f_{yy} \end{pmatrix} \tag{4.83}$$

is in the invariant subspace of $J^T \otimes J^T$ considered in the definition of $\otimes_2(J^T)$. Hence we have the following result.

Proposition 4.4.7. There exists smooth 3×2 and 3×1 matrix functions $M^1_{3,2}(x,y)$ and $M^1_{3,1}(x,y)$, respectively, such that

$$\begin{pmatrix} g_{xx} \\ g_{xy} \\ g_{yy} \end{pmatrix} = (\otimes_2(J^T)) \begin{pmatrix} f_{xx} \\ f_{xy} \\ f_{yy} \end{pmatrix} + M^1_{3,2} \begin{pmatrix} f_x \\ f_y \end{pmatrix} + M^1_{3,1} f, \tag{4.84}$$

where the functions $f, f_x, f_y, f_{xx}, f_{xy}, f_{yy}$ are evaluated in $(X(x,y), Y(x,y))$.

Proof. The proof is straightforward.

The above results can easily be extended to higher-order derivatives. For example, in the case of third order we have Proposition 4.4.8.

Proposition 4.4.8. There exists smooth 8×4, 8×2, and 8×1 matrix functions $M_{8,4}(x,y)$, $M_{8,2}$, and $M_{4,1}(x,y)$, respectively, such that

$$\begin{pmatrix} g_{xxx} \\ g_{xxy} \\ \cdots \\ g_{yyx} \\ g_{yyy} \end{pmatrix} = (J^T \otimes J^T \otimes J^T) \begin{pmatrix} f_{xxx} \\ f_{xxy} \\ \cdots \\ f_{yyx} \\ f_{yyy} \end{pmatrix} + M_{8,4} \begin{pmatrix} f_{xx} \\ f_{xy} \\ f_{yx} \\ f_{yy} \end{pmatrix} + M_{8,2} \begin{pmatrix} f_x \\ f_y \end{pmatrix} + M_{8,1} f, \tag{4.85}$$

where the functions $f, f_x, \ldots, f_{yyy}$ are evaluated in $(X(x,y), Y(x,y))$.

Proof. The proposition is proved by straightforward verification or inductively from Proposition 4.4.6. □

Proposition 4.4.9. There exists smooth 4×3, 4×2, and 4×1 matrix functions $M^1_{4,3}(x,y)$, $M^1_{4,2}(x,y)$, and $M^1_{4,1}(x,y)$, respectively, such that

$$\begin{pmatrix} g_{xxx} \\ g_{xxy} \\ g_{xyy} \\ g_{yyy} \end{pmatrix} = (\otimes_3(J^T)) \begin{pmatrix} f_{xxx} \\ f_{xxy} \\ f_{xyy} \\ f_{yyy} \end{pmatrix} + M^1_{4,3} \begin{pmatrix} f_{xx} \\ f_{xy} \\ f_{yy} \end{pmatrix} + M^1_{4,2} \begin{pmatrix} f_x \\ f_y \end{pmatrix} + M^1_{4,1} f, \tag{4.86}$$

where the functions $f, f_x, \ldots, f_{yyy}$ are evaluated in $(X(x,y), Y(x,y))$.

Proof. The proof is straightforward. □

4.4.3 The Natural Involution in $\mathbb{C}^n \times \mathbb{C}^n$

There exists a natural linear mapping $\sigma : \mathbb{C}^n \otimes \mathbb{C}^n \to \mathbb{C}^n \otimes \mathbb{C}^n$ defined by

$$\sigma(e_i \otimes e_j) = e_j \otimes e_i (1 \leq i, j \leq n). \tag{4.87}$$

If $x = \sum_{i=1}^n x_i e_i, y = \sum_{j=1}^n y_j e_j$, then $\sigma(x \otimes y) = \sigma \sum_{i,j=1}^n x_i y_j e_i \otimes e_j = \sum_{i,j=1}^n x_i y_j e_j \otimes e_i$ $= y \otimes x$, i.e.,

$$\sigma(x \otimes y) = y \otimes x \tag{4.88}$$

for all $x, y \in \mathbb{C}^n$.

The map σ is an involution, i.e., $\sigma^2 = I_{\mathbb{C}^n \otimes \mathbb{C}^n}$. In particular it is a bijection. There is also a natural connection to tensor products of matrices, given in the next proposition.

Proposition 4.4.10. For all $A, B \in \mathbb{C}^{n \times n}$ the relation

$$\sigma(A \otimes B) = (B \otimes A)\sigma \tag{4.89}$$

holds.

Proof. For every pair $i, j \in \{1, \ldots, n\}$ we have $\sigma(A \otimes B)(e_i \otimes e_j) = \sigma(Ae_i \otimes Be_j) = Be_j \otimes Ae_i = (B \otimes A)\sigma(e_i \otimes e_j)$. □

The spectrum (set of eigenvalues) of σ is remarkable. For every pair of indices (i, j) with $1 \leq i, j \leq n$ we define the vectors

$$\zeta_{ij} = e_i \otimes e_j - e_j \otimes e_i,$$

$$\eta_{ij} = e_i \otimes e_j + e_j \otimes e_i.$$

We denote by E_a and E_s, respectively, the subspaces of $\mathbb{C}^n \otimes \mathbb{C}^n$ spanned by all vectors of the form ζ_{ij}, respectively, η_{ij}. Then the following holds.

Proposition 4.4.11. The operator σ has two eigenvalues, namely, the eigenvalue -1 with algebraic and geometric multiplicity $\frac{n(n-1)}{2}$ and the eigenvalue $+1$ with algebraic and geometric multiplicity $\frac{n(n+1)}{2}$. The eigenspace corresponding to -1 is E_a and has a base consisting of all vectors $\zeta_{ij}(n \geq i > j \geq 1)$; the eigenspace corresponding to $+1$ is E_s and has a base consisting of all vectors $\eta_{ij}(n \geq i \geq j \geq 1)$.

Proof. We first remark that $\zeta_{ji} = -\zeta_{ij}$. Hence $\zeta_{ii} = 0$ and $(\zeta_{ij})_{ij}$ span a space with dimension $\frac{n(n-1)}{2}$. On the other hand, $(\eta_{ij})_{ij}$ spans a space with dimension $\frac{n(n+1)}{2}$. Since these two spaces have a trivial intersection and together span $\mathbb{C}^n \otimes \mathbb{C}^n$, the Proposition follows by noting that $\sigma(\zeta_{ij}) = -\zeta_{ij}$, $\sigma(\eta_{ij}) = \eta_{ij}$. □

4.4.4 The Bialternate Product of Matrices

Let $A, B \in \mathbb{C}^{n \times n}$. We define the bialternate product by

$$A \odot B = \frac{1}{2}(A \otimes B + B \otimes A). \tag{4.90}$$

Some properties of bialternate products are immediate.

Proposition 4.4.12. Let $A, B, B_1, B_2 \in \mathbb{C}^{n\times n}$. Then

1. $A \odot B = B \odot A$;

2. $A \odot (B_1 + B_2) = A \odot B_1 + A \odot B_2$;

3. $A \odot A = A \otimes A$;

4. if A is nonsingular, then so is $A \odot A$ and $(A \odot A)^{-1} = A^{-1} \odot A^{-1}$;

5. $\sigma(A \odot B) = (A \odot B)\sigma$.

Proof. The first four claims follow immediately from (4.90). The last one follows from $\sigma(A \odot B) = \frac{1}{2}\sigma(A \otimes B + B \otimes A) = \frac{1}{2}(B \otimes A + A \otimes B)\sigma = (A \odot B)\sigma$. □

Proposition 4.4.13. For every $A, B \in \mathbb{C}^{n\times n}$ the spaces E_a and E_s are invariant subspaces of $A \odot B$.

Proof. If $f \in E_a$ then $\sigma(A \odot B)f = (A \odot B)\sigma f$ (by Proposition 4.4.4). Since $\sigma f = -f$ it follows that $\sigma(A \odot B)f = -(A \odot B)f$; that is, $(A \odot B)f \in E_a$. The proof for E_s is similar. □

For our purposes only the restriction of $(A \odot B)$ to E_a is important. For the numerical applications we need a representation of this operator with respect to a suitable base of E_a. From Proposition 4.4.11 it follows that $(\zeta_{ij})_{i>j}$ is such a base. The representation of $A \odot B$ is then given by the following proposition.

Proposition 4.4.14. With respect to the base $(\zeta_{ij})_{i>j}$ the restriction of $A \odot B$ to E_a is represented by

$$(A \odot B)_{(i,j),(k,l)} = \frac{1}{2}\left\{\begin{vmatrix} a_{ik} & a_{il} \\ b_{jk} & b_{jl} \end{vmatrix} + \begin{vmatrix} b_{ik} & b_{il} \\ a_{jk} & a_{jl} \end{vmatrix}\right\}. \tag{4.91}$$

Proof. Let $n \geq k > l \geq 1$. Then

$$(A \odot B)\zeta_{kl} = \frac{1}{2}(A \otimes B + B \otimes A)(e_k \otimes e_l - e_l \otimes e_k)$$

$$= \frac{1}{2}(Ae_k \otimes Be_l + Be_k \otimes Ae_l - Ae_l \otimes Be_k - Be_l \otimes Ae_k)$$

$$= \frac{1}{2}\left\{\sum_{i,j=1}^{n} a_{ik}b_{jl}e_i \otimes e_j + \sum_{i,j=1}^{n} b_{ik}a_{jl}e_i \otimes e_j - \sum_{i,j=1}^{n} a_{il}b_{jk}e_i \otimes e_j - \sum_{i,j=1}^{n} b_{il}a_{jk}e_i \otimes e_j\right\}$$

$$= \frac{1}{2}\left\{\sum_{i,j=1}^{n} a_{ik}b_{jl}e_i \otimes e_j + \sum_{i,j=1}^{n} b_{ik}a_{jl}e_i \otimes e_j - \sum_{i,j=1}^{n} a_{jl}b_{ik}e_j \otimes e_i - \sum_{i,j=1}^{n} b_{jl}a_{ik}e_j \otimes e_i\right\}$$

$$= \frac{1}{2}\left\{\sum_{i,j=1}^{n} a_{ik}b_{jl}(e_i \otimes e_j - e_j \otimes e_i) + \sum_{i,j=1}^{n} b_{ik}a_{jl}(e_i \otimes e_j - e_j \otimes e_i)\right\}$$

$$= \frac{1}{2}\left\{\sum_{i,j=1}^{n}(a_{ik}b_{jl} + b_{ik}a_{jl})\zeta_{ij}\right\}.$$

In this summation the terms with $i = j$ vanish since $\zeta_{ii} = 0$. From $\zeta_{ij} = -\zeta_{ji}$ it follows that

$$(A \odot B)\zeta_{kl} = \frac{1}{2}\left\{\sum_{i>j}(a_{ik}b_{jl} + b_{ik}a_{jl})\zeta_{ij} - \sum_{i>j}(a_{jk}b_{il} + b_{jk}a_{il})\zeta_{ij}\right\}$$
$$= \frac{1}{2}\sum_{i>j}(a_{ik}b_{jl} + b_{ik}a_{jl} - a_{jk}b_{il} - b_{jk}a_{il})\zeta_{ij}.$$

This implies (4.91). □

Convention. With an abuse of notation we will usually identify $A \odot B$ with its restriction to the invariant subspace E_a and take the matrix representation on this invariant subspace in Proposition 4.4.14 as the standard one. The dimension $\frac{n(n-1)}{2}$ will be abbreviated as $b(n)$.

The special case of a bialternate product of the form $2A \odot I_n$ is so important that we simply call it the bialternate product of A. From Proposition 4.4.14 we obtain the explicit form of this bialternate product. We find

$$(2A \odot I_n)_{(i,j),(k,l)} = \begin{cases} -a_{il} & \text{if } k = j, \\ a_{ik} & \text{if } k \neq i \text{ and } l = j, \\ a_{ii} + a_{jj} & \text{if } k = i \text{ and } l = j, \\ a_{jl} & \text{if } k = i \text{ and } l \neq j, \\ -a_{jk} & \text{if } l = i, \\ 0 & \text{else.} \end{cases} \tag{4.92}$$

To check results in simple cases it is helpful to have the bialternate products explicit for a few small values of n. With the lexicographic ordering of indices we have for a general 2×2 matrix A

$$2A \odot I_2 = \begin{pmatrix} a_{11} + a_{22} \end{pmatrix} \tag{4.93}$$

and for a general 3×3 matrix A

$$2A \odot I_3 = \begin{pmatrix} a_{11} + a_{22} & a_{23} & -a_{13} \\ a_{32} & a_{11} + a_{33} & a_{12} \\ -a_{31} & a_{21} & a_{22} + a_{33} \end{pmatrix}. \tag{4.94}$$

Structural zeros appear first in the 4×4 case where the bialternate product is

$$\begin{pmatrix} a_{11} + a_{22} & a_{23} & -a_{13} & a_{24} & -a_{14} & 0 \\ a_{32} & a_{11} + a_{33} & a_{12} & a_{34} & 0 & -a_{14} \\ -a_{31} & a_{21} & a_{22} + a_{33} & 0 & a_{34} & -a_{24} \\ a_{42} & a_{43} & 0 & a_{11} + a_{44} & a_{12} & a_{13} \\ -a_{41} & 0 & a_{43} & a_{21} & a_{22} + a_{44} & a_{23} \\ 0 & -a_{41} & -a_{42} & a_{31} & a_{32} & a_{33} + a_{44} \end{pmatrix}. \tag{4.95}$$

If A is a large matrix, then $2A \odot I_n$ has $O(n^4)$ entries but only $O(n^3)$ nonzero elements, so it is fairly sparse.

It follows from (4.91) that $(A \odot B)^T = A^T \odot B^T$. In particular $(2A \odot I_n)^T = (2A^T \odot I_n)$.

4.4.5 The Jordan Structure of the Bialternate Product Matrix

A remarkable property of the bialternate product matrix is that its eigenvalues and Jordan structure are completely determined by those of the original matrix. To be precise, we have the following proposition.

Proposition 4.4.15. Let $A, B \in \mathbb{C}^{n\times n}$ be two similar matrices and let P be a nonsingular matrix such that $B = PAP^{-1}$. Then $P \odot P$ is nonsingular and

$$(P \odot P)(2A \odot I_n)(P \odot P)^{-1} = 2B \odot I_n.$$

Proof. From Proposition 4.4.12 it follows that $P \odot P$ is nonsingular. So it is sufficient to prove that

$$(P \odot P)(2A \odot I_n) = (2B \odot I_n)(P \odot P).$$

The left-hand side in this expression is the restriction to E_a of

$$(P \otimes P)(A \otimes I_n + I_n \otimes A) = PA \otimes P + P \otimes PA.$$

The right-hand side is the restriction to E_a of

$$(PAP^{-1} \otimes I_n + I_n \otimes PAP^{-1})(P \otimes P) = PA \otimes P + P \otimes PA.$$

Since both sides are equal the result follows. □

We first consider eigenvalues. Here we have the following proposition.

Proposition 4.4.16. Suppose that A has eigenvalues $\lambda_1, \dots, \lambda_n$ ordered in an arbitrary order. Then $2A \odot I_n$ has eigenvalues $(\lambda_i + \lambda_j)_{n\geq i>j\geq 1}$.

Proof. The eigenvalues of a matrix are preserved by similarity transforms. Hence by Proposition 4.4.15 we may assume that A is in upper triangular form (by reducing to the Jordan or Schur form if necessary). Then the eigenvalues of A are its diagonal elements. Now we apply (4.92) in the case of an upper triangular matrix. It is easily seen that $2A \odot I_n$ is an upper triangular matrix if its indices are ordered lexicographically. In addition, its diagonal elements are given by $(2A \odot I_n)_{(i,j),(i,j)} = a_{ii} + a_{jj}$. This proves the result. □

We recall the convention that we only consider the restriction of the bialternate product matrix to its invariant subspace E_a for which we obtained an explicit representation in Proposition 4.4.14. If $v, w \in \mathbb{C}^n$ then clearly $w \otimes v - v \otimes w \in E_a$. Furthermore, with respect to the basis $(\zeta_{ij})_{n\geq i>j\geq 1}$ in §4.4.3 we have $w \otimes v - v \otimes w = \sum_{n\geq i>j\geq 1}(v_j w_i - v_i w_j)\zeta_{ij}$. This naturally leads to the following definition.

Definition 4.4.17. Let $v, w \in \mathbb{C}^n$. The wedge product of v and w is the vector $v \wedge w \in \mathbb{C}^{\frac{n(n-1)}{2}}$ with components $(v \wedge w)_{i,j} = v_j w_i - v_i w_j$ $(n \geq i > j \geq 1)$.

We can visualize the components of $v \wedge w$ as determinants of 2×2 blocks in the $n \times 2$ matrix with columns v, w. Formally, $v \wedge w$ is the representation of $w \otimes v - v \otimes w$ with respect to the canonical base of E_a.

Obviously, the wedge product $v \wedge w$ is linear with respect to both v and w and vanishes if and only if v, w are linearly dependent. It is antisymmetric ($v \wedge w + w \wedge v = 0$) and is determined up to a scalar multiple by the two-dimensional space that contains v, w. Conversely, if it is nonzero, then it defines this space completely.

Proposition 4.4.18. If $(v_i)_{i=1}^{k}$, $k \leq n$, are linearly independent vectors in $\mathbb{C}^n$, then $(v_i \wedge v_j)_{1 \leq j < i \leq k}$ are linearly independent in $\mathbb{C}^{b(n)}$. In particular, if $k = n$, then they form a base of $\mathbb{C}^{b(n)}$.

Proof. First assume that $k = n$ and let e_i denote the ith unit vector in $\mathbb{C}^n$ ($1 \leq i \leq n$). Since every e_i is in the span of the $(v_i)_{1 \leq i \leq n}$ it follows that each $e_i \wedge e_j$ ($1 \leq j < i \leq n$) is in the span of the $v_i \wedge v_j$. Since the $e_i \wedge e_j$ obviously form a base for $\mathbb{C}^{b(n)}$ the result follows.

The case $k < n$ then follows from the fact that a linearly independent set of vectors can be extended to a base. □

We now prove the basic relation between bialternate product matrices and wedge products of vectors.

Proposition 4.4.19. Let $A, B \in \mathbb{C}^{n \times n}$ and $v, w \in \mathbb{C}^n$. Then

$$(A \odot B)(v \wedge w) = \frac{1}{2}(Av \wedge Bw - Aw \wedge Bv). \tag{4.96}$$

In particular,

$$(2A \odot I_n)(v \wedge w) = Av \wedge w + v \wedge Aw. \tag{4.97}$$

Proof. We provide a proof in the canonical base of E_a; a coordinate-free proof can also be given easily. Let i, j be integers with $n \geq i > j \geq 1$. Then

$$\begin{aligned}
[(2A \odot B)(v \wedge w)]_{i,j} &= \Sigma_{n \geq k > l \geq 1}(a_{ik}b_{jl} - a_{il}b_{jk} + a_{jl}b_{ik} - a_{jk}b_{il})(v_l w_k - v_k w_l) \\
&= \Sigma_{n \geq k, l \geq 1}(a_{ik}b_{jl} + a_{jl}b_{ik})(v_l w_k - v_k w_l) \\
&= (Bv)_j(Aw)_i - (Bw)_j(Av)_i + (Bw)_i(Av)_j - (Bv)_i(Aw)_j \\
&= (Av \wedge Bw)_{i,j} - (Aw \wedge Bv)_{i,j}.
\end{aligned}$$

This implies the proposition. □

We note that (4.96) completely defines $A \odot B$. The next result is a natural complement to Proposition 4.4.16.

Proposition 4.4.20. Let $A \in \mathbb{C}^{n \times n}$ and let λ_1, λ_2 be eigenvalues of A with corresponding eigenvectors v_1, v_2. If v_1, v_2 are linearly independent, then $v_1 \wedge v_2$ is an eigenvector of $2A \odot I_n$ for the eigenvalue $\lambda_1 + \lambda_2$.

Proof. By Proposition 4.4.19 we have

$$\begin{aligned}
(2A \odot I_n)(v_1 \wedge v_2) &= Av_1 \wedge v_2 - Av_2 \wedge v_1 \\
&= \lambda_1 v_1 \wedge v_2 - \lambda_2 v_2 \wedge v_1 \\
&= (\lambda_1 + \lambda_2)(v_1 \wedge v_2).
\end{aligned}$$

This completes the proof. □

It is useful to consider the Jordan structures in more detail. First we have Proposition 4.4.21.

Proposition 4.4.21. Let $A \in \mathbb{C}^{n\times n}$ and let λ_1, λ_2 be two eigenvalues of A that correspond to different Jordan blocks with dimensions n_1 and n_2, respectively. If no other pair of eigenvalues of A has sum $\lambda_1 + \lambda_2$, then $\lambda_1 + \lambda_2$ is an eigenvalue of $2A \odot I_n$ with algebraic multiplicity $n_1 n_2$. Furthermore,

1. If either $n_1 = 1$ or $n_2 = 1$, then $\lambda_1 + \lambda_2$ has geometric multiplicity 1.
2. If $n_1 = n_2 = 2$, then $\lambda_1 + \lambda_2$ has geometric multiplicity 2; the two Jordan blocks have sizes 1 and 3, respectively.
3. If $n_1 = 2$, $n_2 = 3$, then $\lambda_1 + \lambda_2$ has geometric multiplicity 2; the two Jordan blocks have sizes 2 and 4, respectively.

Proof. Let us start with the second situation, the first being similar but easier. Suppose that $Av_1 = \lambda_1 v_1$, $Av_2 = \lambda_1 v_2 + v_1$, $Aw_1 = \lambda_2 w_1$, $Aw_2 = \lambda_2 w_2 + w_1$. Applying (4.97) several times we find that

$$(2A \odot I_n - (\lambda_1 + \lambda_2) I_{b(n)})(v_2 \wedge w_2) = v_1 \wedge w_2 + v_2 \wedge w_1,$$

$$(2A \odot I_n - (\lambda_1 + \lambda_2) I_{b(n)})(v_1 \wedge w_2 + v_2 \wedge w_1) = 2v_1 \wedge w_1,$$

$$(2A \odot I_n - (\lambda_1 + \lambda_2) I_{b(n)})(v_1 \wedge w_1) = 0,$$

$$(2A \odot I_n - (\lambda_1 + \lambda_2) I_{b(n)})(v_1 \wedge w_2 - v_2 \wedge w_1) = 0.$$

Together with Proposition 4.4.18 (independence of the eigenvectors $v_1 \wedge w_1$ and $v_1 \wedge w_2 - v_2 \wedge w_1$) this completes the proof of the second statement.

For the third statement we start from the vector $v_2 \wedge w_3$, which leads to $v_1 \wedge w_3 + v_2 \wedge w_2$, $2v_1 \wedge w_2 + v_2 \wedge w_1$, and $3v_1 \wedge w_1$, and also from $2v_1 \wedge v_3 - v_2 \wedge w_2$, which leads to $v_1 \wedge w_2 - v_2 \wedge w_1$. Since $v_1 \wedge w_1$ and $v_1 \wedge w_2 - v_2 \wedge w_1$ are linearly independent eigenvectors of $2A \odot I_{b(n)}$ for the eigenvalue $\lambda_1 + \lambda_2$ the result follows. □

Proposition 4.4.21 covers the cases that are of immediate interest. However, it is natural to wonder what happens for general values of n_1, n_2. We need the following definition.

Definition 4.4.22. For $1 \le l \le n_1 \le n_2$, we define $C(n_1, n_2, l)$ as the $l \times l$ matrix whose (i,j)th entry is the binomial coefficient $\binom{n_1+n_2-2l}{n_1-l+i-j}$ with the understanding that this entry vanishes if $n_1 - l + i - j$ is not in the range $[0, n_1 + n_2 - 2l]$.

Proposition 4.4.23. Every matrix of the form $C(n_1, n_2, l)$ $(1 \le l \le n_1)$ as defined in Definition 4.4.22 is nonsingular.

For a proof we refer to [178, §I.3, Example 4] where a more general result is obtained. In [119] the special case $n_1 = n_2$ is proved by a direct argument. See also the exercises.

Proposition 4.4.24. Let $A \in \mathbb{C}^{n\times n}$ and let λ_1, λ_2 be two eigenvalues of A that belong to different Jordan blocks with dimensions n_1 and n_2, respectively, with $1 \le n_1 \le n_2$. Then the matrix $2A \odot I_n$ has n_1 Jordan blocks for the eigenvalue $\lambda_1 + \lambda_2$, one each of size $1 + n_2 - n_1$, $3 + n_2 - n_1$, ..., $2n_1 - 1 + n_2 - n_1$, corresponding to this eigenvalue pair (total algebraic multiplicity of $\lambda_1 + \lambda_2$ for this eigenvalue pair: $n_1 n_2$).

Proof. For simplicity of notation we prove this in the case $n_1 = 4$, $n_2 = 6$, the generalization being obvious. Let $v_1, \ldots, v_4$ be the generalized eigenvectors associated with λ_1; i.e., $(A - \lambda_1 I_n)v_i = v_{i-1}$ for $i = 2, 3, 4$ and $(A - \lambda_1 I_n)v_1 = 0$. Let $w_1, \dots, w_6$ be the generalized eigenvectors associated with λ_2 in the same way. The generalized eigenspace of the eigenvalue $\lambda_1 + \lambda_2$ has dimension $n_1 n_2 = 24$ and is spanned by the vectors $V_{ij} = v_i \wedge w_j$ where $1 \le i \le 4$, $1 \le j \le 6$ (the linear independence of these vectors follows from Proposition 4.4.18). It is convenient to order them as follows:

$$\begin{array}{ccccccccc}
 & & & V_{46} & & & & & \\
 & & V_{36} & & V_{45} & & & & \\
 & V_{26} & & V_{35} & & V_{44} & & & \\
V_{16} & & V_{25} & & V_{34} & & V_{43} & & \\
 & V_{15} & & V_{24} & & V_{33} & & V_{42} & \\
 & & V_{14} & & V_{23} & & V_{32} & & V_{41} \\
 & & & V_{13} & & V_{22} & & V_{31} & \\
 & & & & V_{12} & & V_{21} & & \\
 & & & & & V_{11} & & &
\end{array} \tag{4.98}$$

The picture illustrates the action of $B = (2A \odot I_n) - (\lambda_1 + \lambda_2)I_{N_b}$ on the generalized eigenspace of $\lambda_1 + \lambda_2$ since V_{ij} is transformed into $V_{i-1,j} + V_{i,j-1}$ with the convention that such a vector vanishes if it is not in (4.98). If B acts p times on a vector in (4.98) then the resulting vector is a linear combination of vectors p rows down in (4.98); by induction we have

$$B^p V_{ij} = \sum_{r=0}^{p} \binom{p}{r} V_{i-p+r,j-r}. \tag{4.99}$$

Now denote

$$E_k = \text{Span}\,\{V_{ij} | i + j = k\}$$

for $k = 2, 3, \dots, 10$.

Clearly B has four linearly independent singular vectors, namely, V_{11}, $V_{12} - V_{21}$, $V_{13} - V_{22} + V_{31}$, and $V_{14} - V_{23} + V_{32} - V_{41}$ in E_2, E_3, E_4, E_5, respectively.

Now B^{12-2k} is a linear map from E_{12-k} to E_k for $k = 2, 3, 4, 5$. The result follows if these maps are all onto. From (4.99) we can compute an explicit representation of B^{12-2k} in terms of the basis $V_{6-k,6}, \dots, V_{4,8-k}$ of E_{12-k} and the basis $V_{1,k-1}, \dots, V_{k-1,1}$ of E_k. One finds

$$B^{12-2k} V_{5-k+j,7-j} = \sum_{r=0}^{12-2k} \binom{12-2k}{r} V_{-7+k+j+r,7-j-r}$$

for $j = 1, \dots, k-1$. By requiring that $-7+k+j+r = i$, i.e., $r = 7-k+i-j$, we find that that B^{12-2k} is represented by a matrix whose (i, j)th entry is $\binom{12-2k}{7-k+i-j} = \binom{10-2l}{4-l+j-i}$, where we have set $l = k - 1$. So B^{12-2k} is represented by the square matrix $C(4, 6, l)^T$. Since this matrix is nonsingular by Proposition 4.4.23, the proof is complete. □

In Propositions 4.4.21 and 4.4.24 it does not matter if λ_1, λ_2 are equal or not. If another pair of Jordan blocks leads to the same eigenvalue sum $\lambda_1 + \lambda_2$, then this eigenvalue of $2A \odot I_n$ has the two collections of Jordan blocks; there is no interaction since the respective generalized eigenspaces have only the zero vector in common.

The case of eigenvalues within the same Jordan block must be considered separately. The simplest cases are discussed in Proposition 4.4.25.

Proposition 4.4.25. Let $A \in \mathbb{C}^{n\times n}$ and let λ be an eigenvalue of A in a Jordan block with dimension k. Assume that no other pair of Jordan blocks has eigenvalues with sum 2λ. Let $(A-\lambda)v_k = v_{k-1}, \ldots, (A-\lambda)v_1 = 0$.

1. If $k = 2$, then 2λ is an eigenvalue of $2A \odot I_n$ with geometric multiplicity 1 with eigenvector $v_1 \wedge v_2$.

2. If $k = 3$, then 2λ is an eigenvalue of $2A \odot I_n$ with algebraic multiplicity 3 and geometric multiplicity 1. More precisely,

$$(2A \odot I_n - 2\lambda)(v_2 \wedge v_3) = v_1 \wedge v_3,$$

$$(2A \odot I_n - 2\lambda)(v_1 \wedge v_3) = v_1 \wedge v_2,$$

$$(2A \odot I_n - 2\lambda)(v_1 \wedge v_2) = 0.$$

Proof. The proof is now a straightforward computation. □

In general, if λ is an eigenvalue of A with algebraic multiplicity k and geometric multiplicity 1, then clearly 2λ is an eigenvalue of $2A \odot I_n$ with algebraic multiplicity $\frac{1}{2}k(k-1)$. By Proposition 4.4.18 the $\frac{1}{2}k(k-1)$ vectors of the form $v_i \wedge v_j$ where $1 \le j < i \le k$ always span the generalized eigenspace of the eigenvalue 2λ.

Proposition 4.4.26. Let $A \in \mathbb{C}^{n\times n}$ and let λ be an eigenvalue of A in a Jordan block with dimension k. Let $v_1, \ldots, v_k$ be a base of the generalized eigenspace that corresponds to λ, i.e., $(A-\lambda)v_k = v_{k-1}, \ldots, (A-\lambda)v_1 = 0$. Then

1. If $k = 2l, l \ge 1$, then the matrix $2A \odot I_n$ has l Jordan blocks for the eigenvalue 2λ. The vectors $v_1 \wedge v_2$, $v_1 \wedge v_4 - v_2 \wedge v_3$, ..., $v_1 \wedge v_{2l} - v_2 \wedge v_{2l-1} + \cdots + (-1)^{l+1} v_l \wedge v_{l+1}$ are linearly independent eigenvectors of $2A \odot I_n$ and correspond to Jordan blocks with sizes $4l-3, 4l-7, \ldots, 1$, respectively.

2. If $k = 2l+1, l \ge 1$, then the matrix $2A \odot I_n$ has l Jordan blocks for the eigenvalue 2λ of $2A \odot I_n$. The vectors $v_1 \wedge v_2$, $v_1 \wedge v_4 - v_2 \wedge v_3$, ..., $v_1 \wedge v_{2l} - v_2 \wedge v_{2l-1} + \cdots + (-1)^{l+1} v_l \wedge v_{l+2}$ are linearly independent eigenvectors of $2A \odot I_n$ and correspond to Jordan blocks with sizes $4l-1, 4l-5, \ldots, 3$, respectively.

Proof. We will prove the first statement in the case $l = 4$; the same argument applies for other values of l as well.

The generalized eigenspace corresponding to 2λ has dimension $\frac{7\times 8}{2} = 28$ and is spanned by the vectors $V_{ij} = v_i \wedge v_j$ with $1 \le i < j \le 8$ (these vectors are linearly independent by Proposition 4.4.18). It is natural to order them in the following scheme:

$$\begin{array}{ccccccc}
 & & & & & & V_{78} \\
 & & & & & V_{68} & \\
 & & & & V_{58} & & V_{67} \\
 & & & V_{48} & & V_{57} & \\
 & & V_{38} & & V_{47} & & V_{56} \\
 & V_{28} & & V_{37} & & V_{46} & \\
V_{18} & & V_{27} & & V_{36} & & V_{45} \\
 & V_{17} & & V_{26} & & V_{35} & \\
 & & V_{16} & & V_{25} & & V_{34} \\
 & & & V_{15} & & V_{24} & \\
 & & & & V_{14} & & V_{23} \\
 & & & & & V_{13} & \\
 & & & & & & V_{12}
\end{array} \tag{4.100}$$

The action of $B = 2A \odot I_n - 2\lambda I_{N_b}$ on the vectors V_{ij} can be read from (4.100) since this matrix transforms V_{ij} into $V_{i-1,j} + V_{i,j-1}$ with the convention that vectors not represented in (4.100) are zero.

Let us define

$$E_s = \text{Span}\,\{V_{ij} | i + j = s\}$$

for $s = 3, \ldots, 15$. It is clear that $V_{12} \in E_3$, $V_{14} - V_{23} \in E_5$, $V_{16} - V_{25} + V_{34} \in E_7$, and $V_{18} - V_{27} + V_{36} - V_{45} \in E_9$ are null vectors of B.

The result follows if we prove that B^{12} maps E_{15} onto E_3, B^8 maps E_{13} onto E_5, B^4 maps E_{11} onto E_7, and I_{N_b} maps E_9 onto E_9. We will prove the result for the case of B^8 acting on E_{13}, the other cases being similar.

For $i = 1, \ldots, 4$ let c_i be the number of downward paths in (4.100) that connect V_{58} with $V_{i,9-i}$, or, obviously equivalently, $V_{i,9-i}$ with V_{14}. Similarly, let d_i be the number of downward paths in (4.100) that connect V_{67} with $V_{i,9-i}$, or, obviously equivalently, $V_{i,9-i}$ with V_{23}. With respect to the basis vectors in (4.100) B^4 as a map from E_{13} to E_9 is represented by the matrix

$$M_4 = \begin{pmatrix} c_1 & d_1 \\ c_2 & d_2 \\ c_3 & d_3 \\ c_4 & d_4 \end{pmatrix}.$$

We note that M_4 has full rank 2 because B is one to one on E_{13}, E_{12}, E_{11}, and E_{10}. Furthermore, with respect to the basis vectors in (4.100) B^4 as a map from E_9 to E_5 is represented by the matrix M_4^T. Hence, with respect to the basis vectors in (4.100) B^8 as a map from E_{13} to E_5 is represented by the matrix $M_4^T M_4$; since M_4 has full rank, so has $M_4^T M_4$ by elementary linear algebra.

The second statement can be proved similarly. □

If there are two Jordan blocks with the same eigenvalue, then Proposition 4.4.26 applies separately to each block. The eigenvalue 2λ of $2A \odot I_n$ has the two collections of Jordan blocks. There is also an interaction between the blocks that is described by

Proposition 4.4.24, leading to still more Jordan blocks for the eigenvalue 2λ. This obviously generalizes easily to any number of pairs of Jordan blocks with the same eigenvalue sum.

Since $(2A \odot I_n)^T = (2A^T \odot I_n)$ all results obtained in Propositions 4.4.24 and 4.4.26 concerning the right eigenspaces of $2A \odot I_n$ carry over to the left eigenspaces, using left eigenvectors of A instead of right eigenvectors.

The next result allows us to compute in most practically occurring cases the algebraic and geometric multiplicities of the zero eigenvalue of $2A \odot I_n$ in the case that A has Hopf pairs.

Proposition 4.4.27. Suppose that A has p different conjugate pairs of pure imaginary eigenvalues $\pm i\omega_1, \pm i\omega_2, \ldots, \pm i\omega_p$, $\omega_j > 0$ for all j. Suppose that $i\omega_j$ has algebraic multiplicity n_j and geometric multiplicity 1. Suppose also that A has no other pairs of eigenvalues with sum zero.

Then zero is an eigenvalue of $2A \odot I_n$ with algebraic multiplicity $n_1^2 + n_2^2 + \cdots + n_p^2$ and geometric multiplicity $n_1 + n_2 + \cdots + n_p$. Corresponding to each $j (1 \le j \le p)$ the Jordan form of $2A \odot I_n$ contains n_j Jordan blocks with eigenvalue zero and with dimensions $1, 3, \ldots, 2n_j - 1$, respectively.

Proof. This follows from Proposition 4.4.24, case $n_1 = n_2$ (there is no interaction between different Hopf pairs). □

The simplest and most frequently occurring cases are the following ones.

1. A has one conjugate pair of pure imaginary eigenvalues with algebraic multiplicity 1 and no other pair of eigenvalues with sum zero. Then we can apply Proposition 4.4.27 with $p = 1$, $n_1 = 1$; $2A \odot I_n$ has eigenvalue zero with algebraic and geometric multiplicity 1.

2. A has two conjugate pairs of pure imaginary eigenvalues with algebraic multiplicity 1 and no other pair of eigenvalues with sum zero. We can apply Proposition 4.4.27 with $p = 2, n_1 = 1, n_2 = 1$. Then $2A \odot I_n$ has eigenvalue zero with algebraic and geometric multiplicity 2. We remark that in this case $(2A \odot I_n)^2$ has rank deficiency 2.

3. A has one conjugate pair of pure imaginary eigenvalues with algebraic multiplicity 2 and geometric multiplicity 1; in addition it has no other pair of eigenvalues with sum zero. Then $2A \odot I_n$ has eigenvalue zero with algebraic multiplicity 4 and geometric multiplicity 2. Its Jordan form contains two Jordan blocks with zeros on the diagonal, namely, one with dimension 1 and one with dimension 3. In this case $(2A \odot I_n)^2$ has rank deficiency 3.

We remark that the three cases can be distinguished by the rank deficiencies of $(2A \odot I_n)$ and $(2A \odot I_n)^2$ without computing eigenvalues.

4.5 Hopf Points with Bialternate Product Methods

We return to the dynamic problem (4.1) and its equilibrium solution (4.2). In Hopf bifurcation problems we are interested in the question whether a real matrix A has a conjugate pair of pure imaginary eigenvalues $\pm i\omega, \omega > 0$. A common method to detect Hopf points is to compute all eigenvalues; if a conjugate pair of complex eigenvalues crosses the imaginary axis, then one of the methods in §§4.3.2, 4.3.3, or 4.3.4 can be used to compute and continue (if desired) the Hopf point. For large problems, typically discretized PDEs the size of the Jacobian may make this impossible; see §§10.2 and 10.3 for a further discussion.

The bialternate matrix product offers another possibility for small systems. From Proposition 4.4.16 we infer that in a Hopf point $2A \odot I_N$ has an eigenvalue $i\omega - i\omega = 0$; i.e., $2A \odot I_N$ is a singular matrix.

For this criterion it is not necessary to compute the eigenvalues; by expressing, e.g., that the determinant of $2A \odot I_N$ is zero, one finds (using (4.92)) a condition that is purely an algebraic function of the coefficients of A.

As in the turning point case, a bordered matrix approach is more convenient for computation and continuation than the determinant function.

The basic idea is similar to that in §4.1.2 except that we border the bialternate product $2A \odot I_N$ instead of A itself. If (as we generically assume) A has only one zero-sum pair of eigenvalues, then by Proposition 4.4.16, $2A \odot I_N$ has zero as an algebraically simple eigenvalue. In particular, this matrix has rank deficiency 1. For a generic choice of vectors $b, c \in \mathbb{R}^{b(N)}$ and scalar $d \in \mathbb{R}$ the matrix

$$M(x, \alpha) = \begin{pmatrix} 2A(x, \alpha) \odot I_N & b \\ c^T & d \end{pmatrix} \tag{4.101}$$

is nonsingular; see Proposition 3.2.1. We define $q(x, \alpha) \in \mathbb{R}^{b(N)}, s(x, \alpha) \in \mathbb{R}$ by

$$M(x, \alpha) \begin{pmatrix} q(x, \alpha) \\ s(x, \alpha) \end{pmatrix} = \begin{pmatrix} 0_{b(N)} \\ 1 \end{pmatrix}. \tag{4.102}$$

Now $s(x, \alpha) = 0$ defines (locally) the matrices A for which $2A \odot I_N$ is singular (cf. Corollary 3.3.4 and Proposition 3.4.2). In particular, this function can be used to detect, compute, and continue matrices with a zero-sum pair of eigenvalues. For detection alone, the use of $\det(2A \odot I_N)$ is an alternative. For computation and continuation it is useful to have also the derivatives of $s(x, \alpha)$. Let z be any variable in x, α. By taking derivatives of (4.102) we find that

$$M \begin{pmatrix} q_z \\ s_z \end{pmatrix} + \begin{pmatrix} (2A_z \odot I_N)q \\ 0 \end{pmatrix} = 0. \tag{4.103}$$

This allows us to compute s_z. If several derivatives are desired, then as in §4.1.2 it is useful to solve the system

$$\begin{pmatrix} w^T(x, \alpha) & s(x, \alpha) \end{pmatrix} M = \begin{pmatrix} 0^T_{b(N)} & 1 \end{pmatrix}, \tag{4.104}$$

which is a natural adjoint to (4.102). Multiplying (4.103) from the left with

$$\begin{pmatrix} w^T(x, \alpha) & s(x, \alpha) \end{pmatrix}$$

we obtain

$$s_z + w^T(2A_z \odot I_N)q = 0. \tag{4.105}$$

A generic choice of b, c, d may cause trouble when following curves with fixed b, c, d if M becomes singular at certain points. Switching to a different choice is a natural remedy to this problem. On the other hand, it is natural to adapt b, c, d during continuation to avoid the problem altogether and to make M as well conditioned as possible. The ideal choice for b, c, d (for singular $2A \odot I_N$) is to set $d = 0$ and take for b, c scaled versions of w, q, respectively. This is another good reason to solve both (4.102) and (4.104). The scaling should be so that b, c have approximately the same size as $\|2A(x, \alpha) \odot I_N\|$ (cf. Proposition 3.2.2). Of course, q, w are not yet available at a point that is to be computed. Since we are in a continuation context, the natural solution is to use the computed values of q, w in the previous computed point.

We note that q is the right singular vector of $2A \odot I_N$ and that w is the right singular vector of $(2A \odot I_N)^T = 2A^T \odot I_N$. By Proposition 4.4.20 q (respectively, w) is the wedge product of any two base vectors in the joint right (respectively, left) eigenspace of A with respect to the zero-sum eigenvalues (generalized eigenspace in the case of two zero eigenvalues and up to a scalar factor).

4.5.1 Reconstruction of the Eigenstructure

If a curve of Hopf points is computed using the bialternate product method, then it may be necessary to reconstruct the eigenstructure of the zero-sum eigenvalues, e.g., to compute the first Lyapunov coefficient. We now show how the right eigenstructure can be recovered from q. The left eigenstructure can be obtained in the same way, starting from w.

First, to construct a base for the right eigenspace we consider an index pair $(i, j), i > j$ such that $q_{i,j}$ is nonzero (in practice, we would take a pair for which $|q_{i,j}|$ is maximal). We may assume that $q_{i,j} = 1$. The projection of the right eigenspace onto the (i, j) plane is onto. Hence there exist v_1, v_2 in the eigenspace such that $v_1 \wedge v_2 = q$, $v_{1j} = v_{2i} = 1, v_{1i} = v_{2j} = 0$. The other components then follow from the definition of the wedge product in Definition 4.4.17. In fact we find that $v_{1k} = q_{i,k}$ for $k < i$, $v_{1k} = -q_{k,i}$ for $k > i$, $v_{2k} = -q_{j,k}$ for $k < j$, and $v_{2k} = q_{k,j}$ for $k > j$.

Second, we consider the action of A on the space spanned by v_1, v_2. Let $Av_1 = c_{11}v_1 + c_{21}v_2, Av_2 = c_{12}v_1 + c_{22}v_2$. It is easily seen that the coefficients $c_{11}, c_{12}, c_{21}, c_{22}$ can be obtained from the equations

$$\begin{pmatrix} v_1^T v_1 & v_1^T v_2 \\ v_2^T v_1 & v_2^T v_2 \end{pmatrix} \begin{pmatrix} c_{11} \\ c_{21} \end{pmatrix} = \begin{pmatrix} v_1^T A v_1 \\ v_2^T A v_1 \end{pmatrix}, \tag{4.106}$$

$$\begin{pmatrix} v_1^T v_1 & v_1^T v_2 \\ v_2^T v_1 & v_2^T v_2 \end{pmatrix} \begin{pmatrix} c_{12} \\ c_{22} \end{pmatrix} = \begin{pmatrix} v_1^T A v_2 \\ v_2^T A v_2 \end{pmatrix}. \tag{4.107}$$

The square matrix in (4.106) and (4.107) is nonsingular because v_1, v_2 are linearly independent.

The local representation of A in the v_1, v_2 plane with coordinates β, γ is now given by

$$\begin{pmatrix} \beta \\ \gamma \end{pmatrix} \to \begin{pmatrix} c_{11} & c_{12} \\ c_{21} & c_{22} \end{pmatrix} \begin{pmatrix} \beta \\ \gamma \end{pmatrix}. \tag{4.108}$$

Hence the zero-sum eigenvalues are obtained by solving the equation

$$\lambda^2 - (c_{11} + c_{22})\lambda + (c_{11}c_{22} - c_{12}c_{21}) = 0. \tag{4.109}$$

Since the sum of the solutions to (4.109) must be zero, we have in fact $c_{11} + c_{22} = 0$. From (4.106), (4.107), (4.109) it follows easily that

$$\lambda^2 + \frac{(v_1^T A v_1)(v_2^T A v_2) - (v_2^T A v_1)(v_1^T A v_2)}{(v_1^T v_1)(v_2^T v_2) - (v_1^T v_2)^2} = 0. \tag{4.110}$$

Hence it is not necessary to solve (4.106), (4.107) if only the zero-sum eigenvalues are needed.

If λ is a solution to (4.110), then the corresponding eigenvector of A has the form $\beta v_1 + \gamma v_2$, where β, γ are not both zero and satisfy

$$\begin{pmatrix} c_{11} - \lambda & c_{12} \\ c_{21} & c_{22} - \lambda \end{pmatrix} \begin{pmatrix} \beta \\ \gamma \end{pmatrix} = \begin{pmatrix} 0 \\ 0 \end{pmatrix}. \tag{4.111}$$

If in particular the eigenvalues are $\pm i\omega, \omega > 0$, then we may assume without loss of generality that v_1 is the real part of the (complex) eigenvector. The eigenvector itself is $v_1 - \frac{i}{\omega} A v_1$ (unique up to multiplication with a nonzero complex number).

4.5.2 Double Borders and Detection of Double Hopf Points

On a path of Hopf points (more generally, of points with a zero-sum pair of eigenvalues) a point may be encountered where the Jacobian has a second zero-sum pair of eigenvalues. In such a point $2A \odot I_N$ has rank deficiency 2 and the matrix in (4.101) is singular so that the defining system for single Hopf based on the use of (4.101) is undefined. This problem can be avoided by using a doubly bordered extension of $2A \odot I_N$, i.e., a nonsingular matrix

$$M^e(x, \alpha) = \begin{pmatrix} 2A(x, \alpha) \odot I_N & b_1 & b_2 \\ c_1^T & d_{11} & d_{12} \\ c_2^T & d_{21} & d_{22} \end{pmatrix}.$$

We define $Q(x, \alpha) \in \mathbb{R}^{b(N) \times 2}, S(x, \alpha) \in \mathbb{R}^{2 \times 2}$ by

$$M^e(x, \alpha) \begin{pmatrix} Q(x, \alpha) \\ S(x, \alpha) \end{pmatrix} = \begin{pmatrix} 0_{b(N) \times 2} \\ I_{2 \times 2} \end{pmatrix}.$$

Now $\det(S(x, \alpha)) = 0$ defines (locally) the matrices A for which $2A \odot I_N$ is singular (cf. Corollary 3.3.4 and Proposition 3.4.2). To describe a good adaptation strategy for the borders we use the notation

$$b_2^e = \begin{pmatrix} b_2 \\ d_{12} \end{pmatrix}, \quad c_2^e = \begin{pmatrix} c_2 \\ d_{21} \end{pmatrix}, \quad M = \begin{pmatrix} 2A \odot I_N & b_1 \\ c_1^T & d_{11} \end{pmatrix}.$$

Table 4.2: Coordinates of an equilibrium point in the catalytic oscillator model.

	State	Value	Param	Value
1	x	0.0014673	q_1	2.5
2	y	0.826167	q_2	1.92373
3	s	0.123119	q_3	10
4			q_4	0.0675
5			q_5	1
6			q_6	0.1
7			k	0.4

We always choose $d_{11} = d_{22} = 0$. Furthermore, b_1, c_1 are adapted along the Hopf curve as approximations to the left and right singular vectors of $2A \odot I_N$, respectively. At an initial single Hopf point b_2^e is chosen as a nonzero vector with norm approximately equal to $\|A\|$ and c_2^e is a similarly scaled multiple of $M^{-1}b_2^e$. In this way M^e is initially nonsingular. Next at regular intervals b_2^e is replaced by a multiple of $M^{-T}c_2^e$ and c_2^e is replaced by a multiple of $M^{-1}b_2^e$. In this way M^e remains nonsingular along the Hopf curve; in a point with another zero-sum pair of eigenvalues, b_2^e tends to a left singular vector of M and c_2^e tends to a right singular vector. Let σ be a regular parameterization of the Hopf curve. In a point where another pair of eigenvalues λ_1, λ_2 has sum zero we have $g_{22} = 0$ and $\det(M) = 0$. If

$$\frac{d(\lambda_1 + \lambda_2)}{d\sigma} \neq 0 \tag{4.112}$$

in this point, then $g_{22\sigma} \neq 0$; i.e., g_{22} changes sign and is a test function for the occurrence of two zero-sum pairs of eigenvalues (we leave this as an exercise to the reader). By the argument in §3.1 $\det(M)$ also changes sign and thus also provides a test function.

4.6 Computation of Hopf Points: Examples

4.6.1 Zero-Sum Pairs of Eigenvalues in the Catalytic Oscillator Model

The following chemical model of a catalytic reaction of CO-oxidation was studied in [45] and [154]:

$$\begin{cases} \dot{x} &= 2\,q_1\,z^2 - 2\,q_5\,x^2 - q_3\,x\,y, \\ \dot{y} &= q_2\,z - q_6\,y - q_3\,x\,y, \\ \dot{s} &= q_4\,z - k\,q_4\,s, \end{cases} \tag{4.113}$$

where $z = 1 - x - y - s$.

In this low-dimensional problem a curve of equilibrium solutions was traced by a continuation method as described in Chapter 2, starting from the initial values in Table 4.2 and with q_2 as a free parameter. We used the software package CONTENT [165], [166], [173].

This branch is represented in Figure 4.1 as the curve that starts in the bottom right corner. During the computation the functions $\det(A)$ and $\det(2A \odot I_3)$ were monitored;

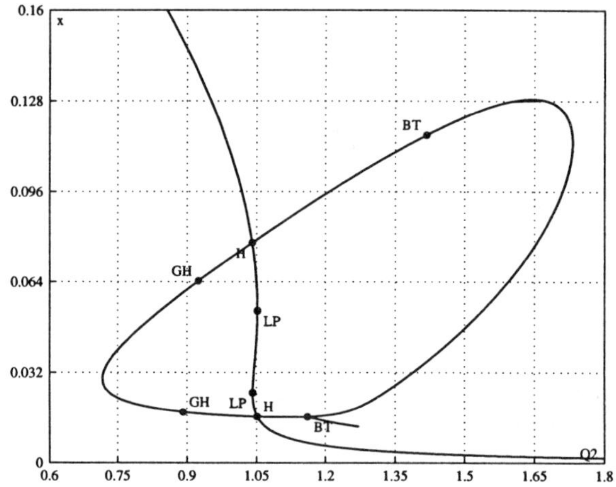

Figure 4.1: Bifurcation diagram of the catalytic oscillator model.

Table 4.3: Coordinates of a Hopf point in the catalytic oscillator model.

	State	Value	Param	Value
1	x	0.07792759	q_1	2.5
2	y	0.2330654	q_2	1.040992
3	s	0.4921479	q_3	10
4			q_4	0.0675
5			q_5	1
6			q_6	0.1
7			k	0.4

sign changes of these functions allow us to detect quadratic turning points and Hopf points, respectively. In this way two quadratic turning points (indicated LP) and two Hopf points (indicated H) were found. The coordinates of the second Hopf point are given in Table 4.3.

Using this as the starting point a Hopf curve was continued with free parameters q_2, k by the method described in §4.3.3. We found a closed curve of equilibria with a zero-sum pair of eigenvalues that connects the two Hopf points found on the equilibrium curve. A projection on the (q_2, x)-space, produced by CONTENT, is shown in Figure 4.1.

The symbol BT on Figure 4.1 denotes Bogdanov–Takens points. The part of the curve to the right of the BT points corresponds to neutral saddles.

The two GH points in Figure 4.1 are generalized Hopf (or Bautin) points, i.e., points where the first Lyapunov coefficient changes sign (see Chapter 9 for more details).

4.6.2 The Clamped Hodgkin–Huxley Equations

The Hodgkin–Huxley equations model the electrochemical activity in the giant axon of a squid under experimental conditions. An axon is part of a nerve cell; in the case of the squid it is unusually large and therefore easier to handle in an experimental setting. We refer to [137], [203], [168], [109], and [190, §6.5] for background information. We

Table 4.4: Coordinates of an equilibrium point in the Hodgkin–Huxley equations.

	State	Value	Param	Value
1	V	3.79763	C	1
2	M	0.0819466	I	6.09423
3	N	0.377125	T	6.3
4	H	0.460421	$\bar{g}_{Na}$	120
5			V_{Na}	115
6			$\bar{g}_K$	36
7			V_K	−12
8			$\bar{g}_l$	0.3
9			V_l	10.559

reproduce the results in part in [244].

The electrochemical activity of the axon is modeled by a dynamical system with four state variables V, M, N, H and nine parameters $C, I, T, \bar{g}_{Na}, V_{Na}, \bar{g}_K, V_K, \bar{g}_L, V_L$. The system is given by

$$\begin{aligned} \frac{dV}{dt} &= \frac{I - G(V,M,N,H)}{C}, \\ \frac{dM}{dt} &= \Phi(T)((1-M)\alpha_M(V) - M\beta_M(V)), \\ \frac{dN}{dt} &= \Phi(T)((1-N)\alpha_N(V) - N\beta_N(V)), \\ \frac{dH}{dt} &= \Phi(T)((1-H)\alpha_H(V) - H\beta_H(V)), \end{aligned} \tag{4.114}$$

where the following functions are used:

$$G(V,M,N,H) = \bar{g}_{Na}M^3H(V - V_{Na}) + \bar{g}_K N^4(V - V_K) + \bar{g}_L(V - V_L),$$

$$\begin{aligned} &\alpha_M(V) = \Psi(\tfrac{25-V}{10}), && \alpha_N(V) = 0.1\Psi(\tfrac{10-V}{10}), \\ &\alpha_H(V) = 0.07e^{-V/20}, && \beta_M(V) = 4e^{-V/18}, \\ &\beta_N(V) = 0.125e^{-V/80}, && \beta_H(V) = (1 + e^{(30-V)/10})^{-1}, \end{aligned}$$

$$\Phi(T) = 3^{\frac{T-6.3}{10}}, \ \Psi(x) = \frac{x}{e^x - 1}.$$

All variables are dimensionless and M, N, H must lie in $[0, 1]$. In Table 4.4 we give the data for an equilibrium solution to the equations in (4.114).

A path of equilibrium solutions was computed with I as a free parameter. On this path two Hopf points were detected; see Figure 4.2. Starting from the first Hopf point we then computed a Hopf curve using the method in §4.5 (bordered bialternate product) with free parameters I, T. This curve connects the two Hopf points and again contains a GH point. The data of the latter point are given in Table 4.5.

4.6.3 Discretization and Generalized Eigenvalue Problems

In §4.2.2 we saw that in the case of discretized boundary value problems the sparsity of the Jacobian can be exploited in the continuation of equilibria and in the computation (and continuation) of quadratic turning points. This is also true in the case of Hopf

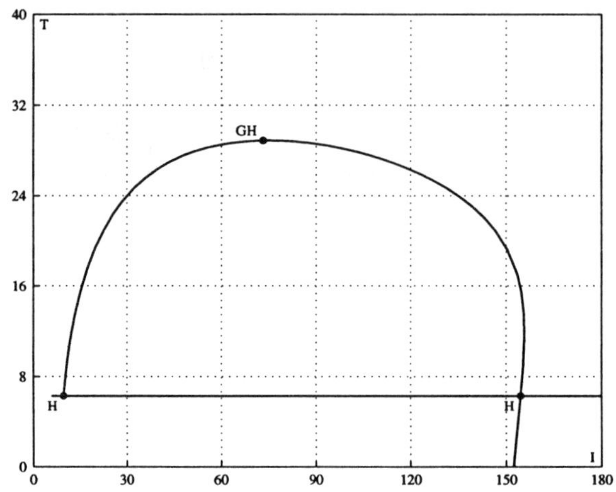

Figure 4.2: Bifurcation diagram of the Hodgkin–Huxley equations.

Table 4.5: Coordinates of a GH point in the Hodgkin–Huxley equations.

	State	Value	Param	Value
1	V	16.16858	C	1
2	M	0.2764981	I	73.10221
3	N	0.5674828	T	28.8525
4	H	0.134613	$\bar{g}_{Na}$	120
5			V_{Na}	115
6			$\bar{g}_K$	36
7			V_K	−12
8			$\bar{g}_l$	0.3
9			V_l	10.559

bifurcation points, but the situation is somewhat more complicated. We discuss this in an example, but the idea applies to a wide range of problems.

Consider again the continuous Brusselator model in (4.20)–(4.22). To simplify the discussion we assume that A is an equilibrium solution to (4.22). Since the dynamic behavior of A is independent of X, Y, we may consider (4.20) and (4.21) as a dynamical system with fixed space-dependent function A.

With the discretization and notation of §4.2.1 we get

$$\frac{L^2}{D_X}\dot{X}_i = X_i^{(2)} + f_1|_i, \tag{4.115}$$

$$\frac{L^2}{D_Y}\dot{Y}_i = Y_i^{(2)} + f_2|_i. \tag{4.116}$$

As in §1.3.2 this implies

$$\frac{L^2}{D_X}\dot{X}_i = \frac{X_{i+1} - 2X_i + X_{i-1}}{h^2} - \frac{1}{12}h^2 X_i^{(4)} + f_1|_i + O(h^4), \tag{4.117}$$

$$\frac{L^2}{D_Y}\dot{Y}_i = \frac{Y_{i+1} - 2Y_i + Y_{i-1}}{h^2} - \frac{1}{12}h^2 Y_i^{(4)} + f_2|_i + O(h^4). \tag{4.118}$$

Omitting terms of order h^2 we obtain a dynamical system in X_i, Y_i whose Jacobian matrix is a band matrix with bandwidth 5. To obtain a higher accuracy we can, as in §1.3.2 substitute the relations

$$X_i^{(4)} = \frac{X_{i+1}^{(2)} - 2X_i^{(2)} + X_{i-1}^{(2)}}{h^2} + O(h^2), \tag{4.119}$$

$$Y_i^{(4)} = \frac{Y_{i+1}^{(2)} - 2Y_i^{(2)} + Y_{i-1}^{(2)}}{h^2} + O(h^2) \tag{4.120}$$

in (4.117) and (4.118), use (4.115) and (4.116) again, and find

$$\frac{L^2}{D_X}\frac{1}{12}(\dot{X}_{i+1}+10\dot{X}_i+\dot{X}_{i-1}) = \frac{X_{i+1} - 2X_i + X_{i-1}}{h^2} + \frac{1}{12}(f_1|_{i+1}+10f_1|_i+f_1|_{i-1})+O(h^4), \tag{4.121}$$

$$\frac{L^2}{D_Y}\frac{1}{12}(\dot{Y}_{i+1}+10\dot{Y}_i+\dot{Y}_{i-1}) = \frac{Y_{i+1} - 2Y_i + Y_{i-1}}{h^2} + \frac{1}{12}(f_2|_{i+1}+10f_2|_i+f_2|_{i-1})+O(h^4). \tag{4.122}$$

Let u be defined as in §4.2.1. Let M be the $2n \times 2n$ matrix whose only nonzero elements are $(M)_{2i-1,2i-3} = \frac{1}{12}$, $(M)_{2i-1,2i-1} = \frac{10}{12}$, $(M)_{2i-1,2i+1} = \frac{1}{12}$, $(M)_{2i,2i-2} = \frac{1}{12}$, $(M)_{2i,2i} = \frac{10}{12}$, $(M)_{2i,2i+2} = \frac{1}{12}$ for $i = 1, \ldots, n$ whenever the given indices are between 1 and $2n$. Then (4.121) and (4.122) may be written in the compact form

$$M\dot{u} = F^*(u, D_X, D_Y, A_0, L, D_A, B) + O(h^4), \tag{4.123}$$

where $(F^*)_{2i-1} = \frac{D_X}{L^2h^2}F_{2i-1}$, $(F^*)_{2i} = \frac{D_Y}{L^2h^2}F_{2i}$ for $i = 1, \ldots, n$, and F is given by (4.33), (4.34).

In (4.123) M is a band matrix with bandwidth 5. It is also a strictly diagonally dominant and therefore nonsingular matrix. If we omit in (4.123) the $O(h^4)$ terms, then we obtain a dynamical system

$$\dot{u} = M^{-1}F^*(u, D_X, D_Y, A_0, L, D_A, B) \tag{4.124}$$

in u with Jacobian matrix $M^{-1}F_u^*$.

So (λ, v) is an eigenpair of the problem if and only if it is a solution to the generalized eigenvalue problem

$$F_u^* v = \lambda M v. \tag{4.125}$$

The methods in §§4.3.2, 4.3.3, and 4.3.4 can be applied if one is willing to compute $M^{-1}F_u^*$ explicitly, i.e., to lose the advantage that can be gained from the sparsity structure of F_u^*.

The method in §4.3.6 is superior in this respect. We have formally to solve systems of the form

$$\begin{pmatrix} M^{-1}F_u^* - i\omega I_{2n} & b \\ c^T & d \end{pmatrix}\begin{pmatrix} v \\ h \end{pmatrix} = \begin{pmatrix} 0_{2n} \\ 1 \end{pmatrix}.$$

Multiplying from the left with the nonsingular matrix

$$\begin{pmatrix} M & 0_{2n} \\ 0_{2n}^T & 1 \end{pmatrix}$$

we obtain the equivalent system

$$\begin{pmatrix} F_u^* - i\omega M & Mb \\ c^T & d \end{pmatrix} \begin{pmatrix} v \\ h \end{pmatrix} = \begin{pmatrix} 0_{2n} \\ 1 \end{pmatrix},$$

which preserves all the sparsity in F_u^*.

4.7 Notes and Further Reading

1. The question under what conditions a given polynomial has only roots in the negative half-plane of $\mathbb{C}$ is an old one and was extensively studied and solved in the second half of the nineteenth century. For a survey see [97, II, Ch. XV] or [98, Ch. V]. It turns out that this is equivalent to the condition that a number of determinants of matrices with increasing sizes (whose entries are coefficients of the polynomial or zeros) are positive. In particular, Hopf bifurcation can be detected by the vanishing of the determinant of the Routh–Hurwitz matrix whose entries are coefficients of the characteristic polynomial of the Jacobian matrix. This idea was used in LOCBIF [156]. Mathematically this is equivalent to the determinant of the bialternate product matrix. However, the use of the determinant of the Routh–Hurwitz matrix requires a careful handling because of scaling problems. Also, this approach does not extend naturally to the case of multiple Hopf points as the bialternate product matrix does (see the next chapter).

2. The problem of detecting (as distinguished from computing) Hopf bifurcation points has attracted considerable attention. We refer in particular to [191] and [117] for methods that try to detect Hopf points in combination with obtaining an approximation to the Hopf eigenvalue. Although these methods lack the elegance and simplicity of the methods that use the bialternate product matrix, they work well in test cases that involve discretized one-dimensional PDEs. For larger problems some form of subspace reduction is probably necessary anyway. See Chapter 10 for further remarks.

3. In §4.3.1 we noted that in a one-parameter problem a family of periodic orbits originates in every isolated Hopf point where the Jacobian has no other eigenvalues on the imaginary axis and $\ell_1 \neq 0$. The computation and numerical continuation of such periodic orbits is a well-understood problem. We refer to [83] for a description of the method now implemented in AUTO and CONTENT. Basically, one deals with a boundry value problem with periodic boundary conditions and a phase condition. The discretization method is the method of orthogonal collocation with piecewise polynomials; see [17] for details and the implementation in COLSYS. The method was analyzed in [66] and [209]. An important feature in the software implementation for periodic orbits is the efficient handling of the structured sparse linear systems that naturally arise.

4. The bifurcation behavior of periodic orbits is largely determined by the eigenvalues of a so-called monodromy matrix. Therefore, most of the methods for the computation of bifurcations of equilibria carry over to bifurcations of periodic orbits. The analogue

of a turning point bifurcation is a *turning point of periodic orbits*; we give examples in §§9.4.1 and 9.4.3. The analogue of a Hopf bifurcation is *torus bifurcation* where a diffeomorphic image of a torus forms an invariant set of the flow of (4.1). A new codimension-1 phenomenon is the *period doubling bifurcation.*

4.8 Exercises

1. Let $A, B \in \mathbb{C}^{n\times n}$. Prove that $\text{rank}(A \otimes B) = \text{rank}(A).\text{rank}(B)$.

2. Let $A \in \mathbb{C}^{n\times n}$. How many entries of $2A \odot I_n$ (representation following Proposition 4.4.14) can be nonzero?

3. Give a coordinate-free proof of Proposition 4.4.19.

4. Prove directly from the representations in (4.93) and (4.94) that in the cases $n = 2$ and $n = 3$, $2A \odot I_n$ is singular if and only if A has two eigenvalues with sum zero.

5. Suppose that $A \in \mathbb{C}^{n\times n}$ has a conjugate pair of pure imaginary eigenvalues with algebraic and geometric multiplicity 2. Suppose also that there are no other pairs of eigenvalues with sum zero. What is the algebraic and geometric multiplicity of the eigenvalue zero of $2A \odot I_n$? Is it possible to distinguish this case from the three cases discussed at the end of §4.3 merely by computing the rank deficiencies of $2A \odot I_n$ and its powers?

6. Let $A, B \in \mathbb{C}^{n\times n}$. Prove that the n^2 eigenvalues of $A \otimes B$ are precisely all products of the form $\lambda_i\mu_j$, where λ_i is an eigenvalue of A and μ_j is an eigenvalue of B. Prove (as a consequence) that $\det(A \otimes B) = (\det(A).\det(B))^n$.

   (*Hint:* Prove this first in the case that no two products of the form $\lambda_i\mu_j$ are the same, then apply a density argument in an appropriate space of matrices.)

7. Suppose that $A \in \mathbb{C}^{n\times n}$ has eigenvalues $\lambda_1, \ldots, \lambda_n$. Prove that the eigenvalues of $A \odot A$ are the products $\lambda_i\lambda_j$ with $i > j$. Prove (as a consequence) that $\det(A \odot A) = (\det(A))^{n-1}$.

   (*Hint:* Use a reduction to triangular form.)

8. Let $A \in \mathbb{R}^{n\times n}$ have a unique pair λ_1, λ_2 of eigenvalues with $\lambda_1\lambda_2 = 1$ (either of the form $e^{\pm i\omega}$ or two real eigenvalues). Prove

   (a) $A \odot A - I_{b(n)}$ is singular.

   (b) If q is a nonzero right singular vector of $A \odot A - I_{N_b}$, then q may be written as $q = v_1 \wedge v_2$ with $v_1, v_2 \in \mathbb{R}^n$.

   (c) In this case
   $\frac{1}{2}(\lambda_1 + \lambda_2)$
   $$= \frac{1}{2}\frac{(v_1^Tv_1)(v_2^TAv_2) + (v_2^Tv_2)(v_1^TAv_1) - (v_1^Tv_2)(v_2^TAv_1) - (v_2^Tv_1)(v_1^TAv_2)}{(v_1^Tv_1)(v_2^Tv_2) - (v_1^Tv_2)^2}.$$
   (In the case of complex eigenvalues this is the real part of the eigenvalues.)

9. Let $A, B \in \mathbb{C}^{n\times n}$. Prove that $(AB) \odot (AB) = (A \odot A)(B \odot B)$.

10. Let $n \geq 3$, $A \in \mathbb{C}^{n\times n}$ and set $m = n(n-1)(n-2)/6$. A matrix $\odot_3(A) \in \mathbb{C}^{m\times m}$ is defined by indexing its rows and columns by triples (i, j, k), where $n \geq i > j > k \geq 1$ and setting

$$(\odot_3(A))_{(i,j,k),(l,m,n)} = \det \begin{pmatrix} a_{il} & a_{im} & a_{in} \\ a_{jl} & a_{jm} & a_{jn} \\ a_{kl} & a_{km} & a_{kn} \end{pmatrix}.$$

Prove the following:

(a) $\odot_3(I_n) = I_m$.

(b) $\odot_3(AB) = \odot_3(A) \odot_3 (B)$ for all $A, B \in \mathbb{C}^{n\times n}$.

(c) If $A \in \mathbb{C}^{n\times n}$ is nonsingular, then so is $\odot_3(A)$ and $(\odot_3(A))^{-1} = \odot_3(A^{-1})$.

(d) If $A, B \in \mathbb{C}^{n\times n}$ are similar, then so are $\odot_3(A)$ and $\odot_3(B)$.

(e) If $\lambda_1, \ldots, \lambda_n$ are the eigenvalues of $A \in \mathbb{C}^{n\times n}$ in any given order, then the eigenvalues of $\odot_3(A)$ are precisely all products of the form $\lambda_i\lambda_j\lambda_k$ with $n \geq i > j > k \geq 1$.

(f) $\det(\odot_3(A)) = (\det(A))^{(n-1)(n-2)/2}$.

11. Generalize the preceding exercise to the case of $\odot_4(A)$ (and beyond).

12. Prove that $A \odot A = 0$ if and only if A has rank less than 2. Generalize this to $\odot_3(A)$ and $\odot_4(A)$ as defined in the preceding exercises.

13. Let $A \in \mathbb{C}^{10\times 10}$. Assume that A has eigenvalue -1 with algebraic multiplicity 3 and geometric multiplicity 1, eigenvalue 1 with algebraic multiplicity 2 and geometric multiplicity 1, eigenvalue -2 with algebraic multiplicity 3 and geometric multiplicity 1, eigenvalue 2 with algebraic multiplicity 2 and geometric multiplicity 1. Compute all eigenvalues of $2A \odot I_{10}$ and give a complete description of their Jordan structures.

14. Let $A \in \mathbb{C}^{6\times 6}$. Assume that A has eigenvalue -3 with algebraic multiplicity 3 and geometric multiplicity 1, eigenvalue 3 with algebraic multiplicity 3 and geometric multiplicity 1. Compute all eigenvalues of $2A \odot I_6$ and give a complete description of their Jordan structures.

15. Let $A \in \mathbb{C}^{9\times 9}$. Assume that A has eigenvalue $-i$ with algebraic multiplicity 4 and corresponding Jordan blocks of sizes 2,2. Assume that A also has eigenvalue i with algebraic multiplicity 5 and corresponding Jordan blocks of sizes 2, 3. Compute all eigenvalues of $2A \odot I_9$ and give a complete description of their Jordan structures.

16. Prove Proposition 4.4.5 directly, using the explicit form of $\otimes_3(A)$ given in §4.5.2.

17. Prove directly that all matrices of the form $C(n_1, n_2, l)$, $1 \leq n_1$, as defined in Definition 4.4.22 are nonsingular if $n_1 = 2$, 3, or 4.

18. Prove directly that all matrices of the form $C(n_1, n_1, l)$, $1 \le n_1$, as defined in Definition 4.4.22, are nonsingular. (*Hint:* Use an argument similar to that in Proposition 4.4.26.)

19. Prove directly that all matrices of the form $C(n_1, n_1 + 1, l)$, $1 \le n_1$, as defined in Definition 4.4.22 are nonsingular.

20. Give necessary and sufficient conditions on the Jordan structure of a real matrix $A \in \mathbb{R}^{n\times n}$ for $2A \odot I_n$ to have rank defect 1.

21. The continuous Brusselator model in §4.2.1 is studied in [109, §7.5] in the limit case $D_A \to \infty$. See also [54] for a recent numerical study. This case is incorporated in our model in §4.2.1 if the function $A(A_0, L, D_A, z)$ is the constant A_0. Then there is a trivial branch $X \equiv A_0, Y \equiv B/A_0$. It is shown in [109] that in such trivial solution points the Jacobian matrix has a zero eigenvalue if

$$B = 1 + \frac{D_X}{D_Y}A_0^2 + \frac{\pi^2 k^2}{L^2}D_X + \left(\frac{L}{\pi k}\right)^2 \frac{A_0^2}{D_Y},$$

where $k = 1, 2, \ldots$ and a Hopf pair of eigenvalues if

$$B = 1 + A_0^2 + \frac{\pi^2 k^2}{L^2}(D_X + D_Y),$$

where again $k = 1, 2, \ldots$ (this is proved by analytical means, using a spectral decomposition method, which is irrelevant to our present purposes). In particular, one can consider a setting with fixed D_X, D_Y, A_0, B and treat $L > 0$ as a bifurcation parameter. The above formulas can then be used to compute values of L for which bifurcations occur. Prove the following:

(a) Zero eigenvalues occur only if $r \equiv B - 1 - \frac{D_X}{D_Y}A^2 > 0$ and $r^2 \ge 4D_X A^2/D_Y$. If this is the case, then they occur for $L = k\pi\sqrt{\zeta_\pm}$, where $k = 1, 2, \ldots$ and

$$\zeta_\pm = 2(r \pm \sqrt{r^2 - 4D_X A^2/D_Y})D_Y/A^2.$$

(b) Hopf pairs occur only if $B - 1 - A^2 > 0$. If this is the case, then they occur for $L = k\eta$ where $k = 1, 2, \ldots$ and

$$\eta = \pi\sqrt{\frac{D_X + D_Y}{B - 1 - A^2}}.$$

(c) Consider the case $D_X = 0.1, D_Y = 10.9, A_0 = 2, B = 6$. Then zero eigenvalues occur for $L_1 = 0.4462607518$, $L_2 = 11.54500979$, and positive integer multiples of these values. In particular, the first 20 values to be encountered are $L = L_1, 2L_1, 3L_1, \ldots, 20L_1 = 8.925215036$. The first Hopf pair is found for $L = 10.41948408$; all others are positive integer multiples of this one.

(d) Consider the case $D_X = 10, D_Y = 1, A_0 = 2, B = 6$. Then no zero eigenvalues occur. The first Hopf pair is found for $L = 10.41948408$; all others are positive integer multiples of this one.

22. Consider the problem described in Exercise 21(c). Now discretize it numerically as in §4.2.1, using 42 equidistant internal mesh points. The problem is somewhat simpler than the one in §4.2.1 since $A \equiv A_0$. Since the discrete problem has only 84 eigenvalues, we cannot hope to match all the eigenvalues of the continuous problem. But you can do some numerical comparisons.

    (a) Compute all eigenvalues of the Jacobian of (4.32) at the trivial solution $X \equiv A_0$, $Y \equiv B/A_0$ for the parameter values $D_X = 0.1$, $D_Y = 10.9$, $A_0 = 2$, $B = 6$, and $L = 0.1, 0.2, 0.3, \ldots, 50.0$.

    (b) Check that initially all eigenvalues have negative real part and that the number of eigenvalues with positive real part increases monotonously. Show that the first 15 jumps occur precisely in the intervals predicted by the theory for the continuous problem and with the right amount (one real eigenvalue at a time), namely, in the intervals that contain the first 15 multiples of $L_1 = 0.4462607518$. The only exception is the thirteenth point, where indeed $13 \times 0.4462607518 = 5.801389773$ is close to an interval boundary, providing a scientific proof of superstition.

    (c) Try to locate the values of L for which the discretized problem has zero eigenvalues by using a direct bordering method as described in §4.1.2 to set up systems similar to the one used in §4.2.2. You should be able to find the L values 0.44626072605149 (very close to L_1 in Exercise 21(c)) and 11.545009096514 (very close to L_2 in Exercise 21(c)) but convergence to the points with L values near $2L_1$, $3L_1$, etc., may fail. Is this consistent with Proposition 4.1.2?

    (d) Repeat the previous series of computations using a mesh with 20 internal mesh points. Compare the accuracy of the obtained approximations to L_1 and L_2 with the previous ones. (11.544997724366 is now the approximation to L_2.)

23. Consider the problem described in Exercise 21(d).

    (a) Discretize it numerically as in §4.6.3, using 42 equidistant internal mesh points. Use the formulae (4.121)–(4.122) with left-hand sides replaced by $\frac{L^2}{D_X}\dot{X}_i$ and $\frac{L^2}{D_Y}\dot{Y}_i$, respectively (this somewhat simplifies the computations). Compute the eigenvalues at the trivial equilibrium $X \equiv A_0$, $Y \equiv B/A_0$ in the case $D_X = 10$, $D_Y = 1$, $A_0 = 2$, $B = 6$ for $L = 0.1, 0.2, 0.3, \ldots, 50.0$. Check that initially all eigenvalues have negative real part and that the number of eigenvalues with positive real part grows monotonously. Show that there are precisely four jumps, each with the amount predicted by the theory for the continuous problem (each time a Hopf pair). They occur between 10.4–10.5, 20.7–20.8, 31.3–31.4, and 41.3–41.4. For the continuous problem the theory predicts 10.41948408, 20.83896816, 31.25845224, and 41.67793632, respectively.

    (b) Repeat the computations using the generalized eigenvalue method described in §4.6.3. Do you find the jumps in the same intervals? (We did.)

    (c) Repeat the last series of computations using a discretization with 80 internal mesh points. You should find that the jumps are now closer to those of the continuous problem.

24. Consider the Brusselator problem as discussed in §4.6.3. Set $D_X = 10$, $D_Y = 1$, $A_0 = 2$, $D_A = 0.5$, $B = 6$. Continue a curve of equilibrium solutions starting from $L = 0$. Compute at all continuation points the generalized eigenvalues of the problem (4.125) (there is a LAPACK routine for that). Check that stability is lost through a Hopf bifurcation between the L values 6.389787 and 6.421930. Compute the Hopf bifurcation point using the method suggested at the end of §4.6.3 that allows us to exploit the sparsity of F_u^*.

25. Repeat the computations for the catalytic oscillator model described in §4.6.1. Find some stable periodic orbits in this model by time integration.

26. Prove that the functions g_{22} and $\det(M)$ introduced in §4.5.2 have nonzero derivatives in the point with two zero-sum pairs of eigenvalues if (4.112) holds.

Chapter 5

Bifurcations Determined by the Jordan Form of the Jacobian

We consider the dynamical system

$$\dot{x} = G(x, \alpha) \quad (x \in \mathbb{R}^N, \alpha \in \mathbb{R}^m, G(x, \alpha) \in \mathbb{R}^N) \tag{5.1}$$

and its equilibrium solutions for which

$$G(x, \alpha) = 0. \tag{5.2}$$

In Chapter 4 we considered the generic equilibrium bifurcations in one-parameter problems. We now consider the case of two parameters with some extensions to cases with more parameters. We restrict ourselves, however, to bifurcations that are determined by the structure of the Jacobian G_x (the nonlinear terms in the Taylor expansion of G may be used in the nondegeneracy conditions).

For systematic studies of manifolds of matrices with given Jordan structures we refer to [236] and [72] (but see also [12]).

5.1 Bogdanov–Takens Points and Their Generalizations

5.1.1 Introduction

A solution point to (5.2) is called a Bogdanov–Takens (BT) point if G_x has an eigenvalue zero with algebraic multiplicity 2 and geometric multiplicity 1; i.e., the characteristic polynomial

$$p(\lambda) = \det(G_x - \lambda I) \tag{5.3}$$

has a double root (but no triple root) zero and G_x has rank $N-1$. Equivalently, the Jordan normal form of G_x must contain a block of the form

$$\begin{pmatrix} 0 & 1 \\ 0 & 0 \end{pmatrix}$$

and no other block with zeros on the diagonal. The dynamic behavior of (5.1) near a BT point was first studied in [36] and [229]. Standard references are [127, §7.3] and [164, §8.4]. We note only that generically paths of Hopf points and paths of homoclinic orbits originate at BT points.

BT points can be expected generically on paths of fold points (when a second real eigenvalue crosses the imaginary axis) and on paths of Hopf points (when the two conjugate pure imaginary eigenvalues coalesce). They are therefore fairly common in two-parameter problems. See §5.1.3 for details.

Important papers on the numerical computation of BT points are [126], [114], and [245]. Our approach is simpler but leads to essentially the same equations as [114]. It is convenient to generalize the notion of a BT point as follows.

A solution point to (5.2) is called a BT point of order n if G_x has an eigenvalue 0 with algebraic multiplicity n and geometric multiplicity 1; i.e., the characteristic polynomial (5.3) has an n-tuple zero root and G_x has rank $N-1$. Equivalently, the Jordan normal form of G_x must contain a Jordan block of order n and no other zeros on the diagonal. The case $n = 2$ is the original BT situation; the case $n = 3$ is sometimes called a triple-point bifurcation point; cf. [155].

5.1.2 Numerical Computation of BT Points

It is natural to consider the general case of a matrix $A \in \mathbb{R}^{N\times N}$ with rank defect at most 1 for which

$$p_A(\lambda) = \det(A - \lambda I) \tag{5.4}$$

has precisely n zero eigenvalues ($1 \le n \le N$). We will then say that A is a BT matrix of order n.

In the sequel the dependence on A will usually be suppressed. We choose $b \in \mathbb{R}^N, c \in \mathbb{R}^N, d \in \mathbb{R}$ so that

$$M = \begin{pmatrix} A & b \\ c^T & d \end{pmatrix} \tag{5.5}$$

is nonsingular and define $q_A(\lambda) \in \mathbb{R}^N$, $w_A(\lambda) \in \mathbb{R}^N$, $s_A(\lambda) \in \mathbb{R}$ for λ in a neighborhood of $0 \in \mathbb{R}$ by

$$M(\lambda)\begin{pmatrix} q_A(\lambda) \\ s_A(\lambda) \end{pmatrix} = \begin{pmatrix} 0_N \\ 1 \end{pmatrix}, \tag{5.6}$$

$$\begin{pmatrix} w_A^T(\lambda) & s_A(\lambda) \end{pmatrix} M(\lambda) = \begin{pmatrix} 0_N^T & 1 \end{pmatrix}, \tag{5.7}$$

where

$$M(\lambda) = \begin{pmatrix} A-\lambda I & b \\ c^T & d \end{pmatrix}. \tag{5.8}$$

We note that there is a close relation between $p(\lambda)$ and $s(\lambda)$. First, we obviously have

$$p(\lambda) = \det(M(\lambda))s(\lambda). \tag{5.9}$$

Taking first, second, third, ... derivatives of this identity we find

$$\begin{pmatrix} p \\ p_\lambda \\ \cdots \\ p_{\lambda^k} \end{pmatrix} = T \begin{pmatrix} s \\ s_\lambda \\ \cdots \\ s_{\lambda^k} \end{pmatrix} \tag{5.10}$$

for every $k = 1, 2, \ldots$; here T is a lower triangular matrix with $\det(M(\lambda))$ in every diagonal entry. Hence this matrix is nonsingular for $\lambda = 0$. It follows that A is a BT matrix of order n if and only if $s(0) = s_\lambda(0) = s_{\lambda^{n-1}}(0) = 0$, $s_{\lambda^n}(0) \neq 0$.

To compute derivatives of the form $s_{\lambda^k}(0)$ we take derivatives of (5.6) and (5.7) and evaluate them at $\lambda = 0$. We obtain

$$M \begin{pmatrix} q_\lambda \\ s_\lambda \end{pmatrix} = \begin{pmatrix} q \\ 0 \end{pmatrix}, \tag{5.11}$$

$$\begin{pmatrix} w_\lambda^T & s_\lambda \end{pmatrix} M = \begin{pmatrix} w^T & 0 \end{pmatrix}, \tag{5.12}$$

and by induction

$$M \begin{pmatrix} q_{\lambda^k} \\ s_{\lambda^k} \end{pmatrix} = k \begin{pmatrix} q_{\lambda^{k-1}} \\ 0 \end{pmatrix}, \tag{5.13}$$

$$\begin{pmatrix} w_{\lambda^k}^T & s_{\lambda^k} \end{pmatrix} M = k \begin{pmatrix} w_{\lambda^{k-1}}^T & 0 \end{pmatrix}. \tag{5.14}$$

In view of these relations it is natural to rescale the defining equations for a BT point of order n by introducing

$$q_k(\lambda) = \frac{q_{\lambda^k}}{k!}, \quad w_k(\lambda) = \frac{w_{\lambda^k}}{k!}, \quad s_k(\lambda) = \frac{s_{\lambda^k}}{k!}$$

for $k = 0, 1, \ldots$. We then have

$$M \begin{pmatrix} q_0 \\ s_0 \end{pmatrix} = \begin{pmatrix} 0_N \\ 1 \end{pmatrix}, \tag{5.15}$$

$$\begin{pmatrix} w_0^T & s_0 \end{pmatrix} M = \begin{pmatrix} 0_N^T & 1 \end{pmatrix}, \tag{5.16}$$

and by induction

$$M \begin{pmatrix} q_k \\ s_k \end{pmatrix} = \begin{pmatrix} q_{k-1} \\ 0 \end{pmatrix}, \tag{5.17}$$

$$\begin{pmatrix} w_k^T & s_k \end{pmatrix} M = \begin{pmatrix} w_{k-1}^T & 0 \end{pmatrix} \tag{5.18}$$

for $k = 1, 2, \ldots$. We infer that

$$M^{k+1} \begin{pmatrix} q_k \\ s_k \end{pmatrix} = \begin{pmatrix} 0_N \\ 1 \end{pmatrix}, \tag{5.19}$$

$$\begin{pmatrix} w_k^T & s_k \end{pmatrix} M^{k+1} = \begin{pmatrix} 0_N^T & 1 \end{pmatrix}. \tag{5.20}$$

In a typical application (e.g., $A = G_x$ with G as in (5.1)) A depends on certain parameters, and we want to compute parameter values for which A is a BT matrix of a given order n, i.e., solutions to a system of the form

$$s_0(0) = s_1(0) = \cdots = s_{n-1}(0) = 0. \tag{5.21}$$

It is important to note that (5.21) implies that

$$\begin{aligned} Aq_0 = 0, Aq_1 = q_0, \ldots, Aq_{n-1} = q_{n-2}, \\ w_0^T A = 0, w_1^T A = w_0^T, \ldots, w_{n-1}^T A = w_{n-2}^T. \end{aligned} \tag{5.22}$$

Hence the vectors $q_0, q_1, \ldots, q_{n-1}$ are linearly independent and span the n-dimensional right generalized eigenspace of the eigenvalue 0. Similarly, the vectors $w_0, w_1, \ldots, w_{n-1}$ are linearly independent and span the n-dimensional left generalized eigenspace of the eigenvalue 0.

Let z be any parameter on which A depends (in the setting of (5.1) this would be a component of x or α). To compute s_{kz} we take the derivative of (5.20) with respect to z and obtain

$$\begin{aligned} &\begin{pmatrix} w_{kz}^T & s_{kz} \end{pmatrix} M^{k+1} + \begin{pmatrix} w_k^T & s_k \end{pmatrix} \\ &\times \left\{ \begin{pmatrix} A_z & 0 \\ 0 & 0 \end{pmatrix} M^k + M \begin{pmatrix} A_z & 0 \\ 0 & 0 \end{pmatrix} M^{k-1} + \cdots + M^k \begin{pmatrix} A_z & 0 \\ 0 & 0 \end{pmatrix} \right\} = 0. \end{aligned} \tag{5.23}$$

Multiplying from the right with

$$\begin{pmatrix} q_k \\ s_k \end{pmatrix} \tag{5.24}$$

and applying (5.17) and (5.18) several times we obtain

$$s_{kz} = -w_k^T A_z q_0 - w_{k-1}^T A_z q_1 - \cdots - w_0^T A_z q_k. \tag{5.25}$$

This formula allows us to solve system (5.21) for a BT matrix of order n by Newton's method. Finally, we prove that (5.21) has full linear rank in the space of all $N \times N$ matrices. Suppose that $r_1, \ldots, r_n$ are scalars such that

$$r_1 s_{0\zeta} + \cdots + r_n s_{(n-1)\zeta} = 0 \tag{5.26}$$

for all variables ζ on which A may depend. By using (5.25) and replacing ζ successively by all entries of A we obtain

$$r_1 q_0 w_0^T + r_2(q_0 w_1^T + q_1 w_0^T) + \cdots + r_n(q_0 w_{n-1}^T + \cdots + q_{n-1} w_0^T) = 0. \tag{5.27}$$

Multiplying (5.27) from the left with A^{n-1} and using (5.22) we obtain $r_n q_0 w_0^T = 0$. Since q_0, w_0 are nonzero, this implies $r_n = 0$. Inserting this into (5.27) and multiplying (5.27) from the left with A^{n-2} we then find $r_{n-1} = 0$. This procedure can be repeated to prove that all coefficients r_k must vanish.

The above results imply that the BT matrices of order n constitute a smooth manifold with dimension $N^2 - n$.

5.1.3 Local Analysis of BT Matrices

We now study the equilibrium surface of (5.1) near a BT point in some more detail; it is hard to get a good understanding of the numerical methods without such background knowledge. We need a few basic facts from unfolding theory for matrices (see [12] for more details).

First, the group $\mathbf{GL}(N)$ of all nonsingular real $N \times N$ matrices acts on the set $\mathcal{M}_N$ of all $N \times N$ matrices in the sense that every $G \in \mathbf{GL}(N)$ defines a mapping $A \to GAG^{-1}$. This is called a *group action* because the action of the product G_1G_2 is the same as the composition of the actions of G_1 and G_2; i.e., the multiplicative group $\mathbf{GL}(N)$ is mapped to the permutation group of $\mathcal{M}_N$. The *group orbit* $\{GAG^{-1} : G \in \mathbf{GL}(N)\}$ is precisely the set of matrices similar to A. If $A_0 \in \mathcal{M}_N$ and $A(\lambda)$ is a smooth family in $\mathcal{M}_N$, defined for all λ in a neighborhood of $0 \in \mathbb{R}^k$ such that $A(0) = A_0$, then $A(\lambda)$ is called an unfolding of A_0. An unfolding $A(\lambda)$ of A_0 is called versal with respect to the group action (we will say $\mathbf{GL}(N)$ versal if confusion is possible) if for every other unfolding $B(\mu)$ of A_0 ($\mu \in \mathbb{R}^l$ where l may be different from k) there exist smooth functions $\phi(\mu) \in \mathbb{R}^k$ and $C(\mu) \in \mathbf{GL}(N)$ such that $\phi(0) = 0$, $C(0) = I_N$, and $B(\mu) = C(\mu)A(\phi(\mu))C^{-1}(\mu)$. The complete unfolding where perturbations of all entries of the matrix are introduced as parameters is obviously always versal. A versal unfolding is called universal if the mapping $\phi(\mu)$ is unique as a germ (i.e., unique in a sufficiently small neighborhood of 0). It is called miniversal if the number of parameters in the unfolding is minimal. In all cases of interest to us we will see that universality and miniversality are equivalent (see also Exercise 4, §5.6). For BT matrices the following result is fundamental.

Proposition 5.1.1. Let $A_0 \in \mathbb{R}^{k\times k}$ be a Jordan block matrix with real eigenvalue a:

$$A_0 = \begin{pmatrix} a & 1 & 0 & \dots & 0 \\ 0 & a & 1 & 0 & \vdots \\ \vdots & \vdots & \ddots & \ddots & 0 \\ 0 & \dots & 0 & a & 1 \\ 0 & \dots & \dots & 0 & a \end{pmatrix}. \tag{5.28}$$

Then the following hold:

1. $$A(\delta_1, \dots, \delta_k) = \begin{pmatrix} a & 1 & 0 & \dots & 0 \\ 0 & a & 1 & 0 & \vdots \\ \vdots & \vdots & \ddots & \ddots & 0 \\ 0 & \dots & \dots & a & 1 \\ \delta_1 & \delta_2 & \dots & \dots & a+\delta_k \end{pmatrix} \tag{5.29}$$

   is a versal unfolding of A_0.

2. This unfolding is universal and miniversal.

3. An unfolding of A_0 is universal if and only if it is versal and miniversal.

Proof. To prove that (5.29) is a versal unfolding of (5.28) it is sufficient to prove that the complete unfolding

$$A(\alpha_{11},\ldots,\alpha_{kk}) = A_0 + \begin{pmatrix} \alpha_{11} & \cdots & \alpha_{kk} \\ \vdots & \ddots & \vdots \\ \alpha_{k1} & \cdots & \alpha_{kk} \end{pmatrix} \tag{5.30}$$

factors through $A(\delta_1,\ldots,\delta_k)$ as in the definition of versality. To simplify notation, we assume $k = 3$, the general case being similar. By a first similarity transformation we obtain

$$\begin{pmatrix} 1 & 0 & 0 \\ \frac{\alpha_{11}}{1+\alpha_{12}} & 1 & 0 \\ 0 & 0 & 1 \end{pmatrix} A(\alpha_{11},\ldots,\alpha_{33}) \begin{pmatrix} 1 & 0 & 0 \\ \frac{-\alpha_{11}}{1+\alpha_{12}} & 1 & 0 \\ 0 & 0 & 1 \end{pmatrix} = \begin{pmatrix} a & 1+\beta_{12}^1 & \beta_{13}^1 \\ \beta_{21}^1 & a+\beta_{22}^1 & 1+\beta_{23}^1 \\ \beta_{31}^1 & \beta_{32}^1 & a+\beta_{33}^1 \end{pmatrix},$$

where $\beta_{12}^1,\ldots,\beta_{33}^1$ are smooth functions of $\alpha_{11},\ldots,\alpha_{33}$, which all vanish if $\alpha_{11} = \cdots = \alpha_{33} = 0$.

By a second similarity transformation we obtain

$$\begin{pmatrix} 1 & 0 & 0 \\ 0 & 1 & \frac{\beta_{13}^1}{1+\beta_{12}^1} \\ 0 & 0 & 1 \end{pmatrix} \begin{pmatrix} a & 1+\beta_{12}^1 & \beta_{13}^1 \\ \beta_{21}^1 & a+\beta_{22}^1 & 1+\beta_{23}^1 \\ \beta_{31}^1 & \beta_{32}^1 & a+\beta_{33}^1 \end{pmatrix} \begin{pmatrix} 1 & 0 & 0 \\ 0 & 1 & \frac{-\beta_{13}^1}{1+\beta_{12}^1} \\ 0 & 0 & 1 \end{pmatrix}$$
$$= \begin{pmatrix} a & 1+\beta_{12}^2 & 0 \\ \beta_{21}^2 & a+\beta_{22}^2 & 1+\beta_{23}^2 \\ \beta_{31}^2 & \beta_{32}^2 & a+\beta_{33}^2 \end{pmatrix},$$

where $\beta_{12}^2,\ldots,\beta_{33}^2$ are smooth functions of $\alpha_{11},\ldots,\alpha_{33}$, which all vanish if $\alpha_{11} = \cdots = \alpha_{33} = 0$.

By a third similarity transformation we obtain

$$\begin{pmatrix} 1 & 0 & 0 \\ 0 & 1 & 0 \\ \frac{\beta_{21}^2}{1+\beta_{23}^2} & 0 & 1 \end{pmatrix} \begin{pmatrix} a & 1+\beta_{12}^2 & 0 \\ \beta_{21}^2 & a+\beta_{22}^2 & 1+\beta_{23}^2 \\ \beta_{31}^2 & \beta_{32}^2 & a+\beta_{33}^2 \end{pmatrix} \begin{pmatrix} 1 & 0 & 0 \\ 0 & 1 & 0 \\ \frac{-\beta_{21}^2}{1+\beta_{23}^2} & 0 & 1 \end{pmatrix}$$
$$= \begin{pmatrix} a & 1+\beta_{12}^3 & 0 \\ 0 & a+\beta_{22}^3 & 1+\beta_{23}^3 \\ \beta_{31}^3 & \beta_{32}^3 & a+\beta_{33}^3 \end{pmatrix},$$

where $\beta_{12}^3,\ldots,\beta_{33}^3$ are smooth functions of $\alpha_{11},\ldots,\alpha_{33}$, which all vanish if $\alpha_{11} = \cdots = \alpha_{33} = 0$.

By a fourth similarity transformation we obtain

$$\begin{pmatrix} 1 & 0 & 0 \\ 0 & 1 & 0 \\ 0 & \frac{\beta_{22}^3}{1+\beta_{23}^3} & 1 \end{pmatrix} \begin{pmatrix} a & 1+\beta_{12}^3 & 0 \\ 0 & a+\beta_{22}^3 & 1+\beta_{23}^3 \\ \beta_{31}^3 & \beta_{32}^3 & a+\beta_{33}^3 \end{pmatrix} \begin{pmatrix} 1 & 0 & 0 \\ 0 & 1 & 0 \\ 0 & \frac{-\beta_{22}^3}{1+\beta_{23}^3} & 1 \end{pmatrix}$$
$$= \begin{pmatrix} a & 1+\beta_{12}^4 & 0 \\ 0 & a & 1+\beta_{23}^4 \\ \beta_{31}^4 & \beta_{32}^4 & a+\beta_{33}^4 \end{pmatrix},$$

where $\beta_{12}^4, \ldots, \beta_{33}^4$ are smooth functions of $\alpha_{11}, \ldots, \alpha_{33}$, which all vanish if $\alpha_{11} = \cdots = \alpha_{33} = 0$.

Finally, if p, q, r are any nonzero real numbers, then

$$\begin{pmatrix} 1/p & 0 & 0 \\ 0 & 1/q & 0 \\ 0 & 0 & 1/r \end{pmatrix} \begin{pmatrix} a & 1+\beta_{12}^4 & 0 \\ 0 & a & 1+\beta_{23}^4 \\ \beta_{31}^4 & \beta_{32}^4 & a+\beta_{33}^4 \end{pmatrix} \begin{pmatrix} p & 0 & 0 \\ 0 & q & 0 \\ 0 & 0 & r \end{pmatrix}$$
$$= \begin{pmatrix} a & \frac{q}{p}(1+\beta_{12}^4) & 0 \\ 0 & a & \frac{r}{q}(1+\beta_{23}^4) \\ \frac{p}{r}\beta_{31}^4 & \frac{q}{r}\beta_{32}^4 & a+\beta_{33}^4 \end{pmatrix}.$$

So by setting $r = 1, q = 1 + \beta_{23}^4, p = q(1 + \beta_{12}^4)$, we obtain the desired form (5.29) using only smooth similarity transformations.

To prove that the unfolding in (5.29) is universal we must prove that $\delta_1, \ldots, \delta_k$ are unique. Now it is easy to see that

$$\det(A(\delta_1, \ldots, \delta_k) - \lambda I_k) = (-1)^{k+1}(\delta_1 + \delta_2(\lambda - a) + \cdots + \delta_k(\lambda - a)^{k-1} - (\lambda - a)^k).$$

Since the characteristic polynomial of a matrix is invariant under similarity transformations, the universality follows.

To prove the miniversality consider any versal unfolding $A(\mu), \mu \in \mathbb{R}^l$ of A_0. So there exists smooth mappings ϕ (respectively, ψ) defined on neighborhoods of $0 \in \mathbb{R}^l$ (respectively, $0 \in \mathbb{R}^k$) and with values in $\mathbb{R}^k$ (respectively, $\mathbb{R}^l$) such that $A(\phi(\mu))$ is similar to $A(\mu)$ (respectively, $A(\psi(\lambda))$ is similar to $A(\lambda)$). By the universality, the composition $\phi\psi$ must be the identity mapping. Hence $\phi_\mu \psi_\lambda$ is the identity matrix; so the rank of ϕ_μ must be at least k, which necessitates $l \geq k$.

The third statement is proved similarly. □

In a BT point the Jordan normal form of G_x contains a Jordan block

$$\begin{pmatrix} 0 & 1 \\ 0 & 0 \end{pmatrix}. \tag{5.31}$$

By Proposition 5.1.1 the family of matrices

$$\begin{pmatrix} 0 & 1 \\ \delta_1 & \delta_2 \end{pmatrix} \tag{5.32}$$

with parameters δ_1, δ_2 is a universal unfolding of (5.31) with respect to the group action of **GL**(2). If a BT point is encountered in a two-parameter problem, then by the universality property we expect generically that in a neighborhood of this point the eigenvalue structure is the same as that of (5.32). We assume this for the moment; a precise formulation is postponed to §5.1.4.

Now the eigenvalues of (5.32) satisfy

$$\lambda^2 - \delta_2 \lambda - \delta_1 = 0. \tag{5.33}$$

Solutions to (5.33) are presented in Figure 5.1. We note in particular that the axis $\delta_1 = 0$ corresponds to a fold curve and the axis $\delta_2 = 0$ to a Hopf–BT-neutral saddle curve.

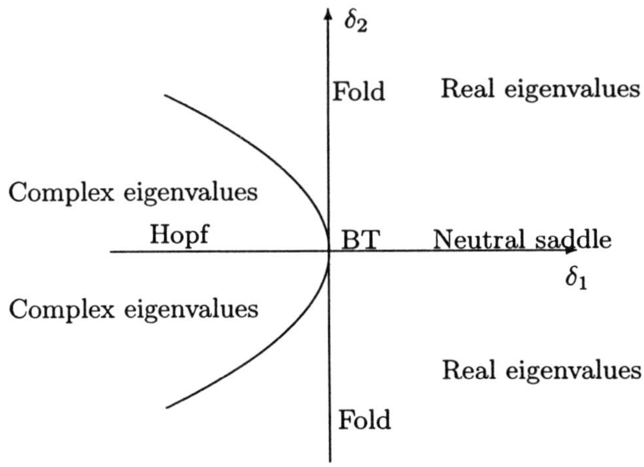

Figure 5.1: Unfolding of a BT matrix.

Now consider a triple-point bifurcation point. In a generic three-parameter unfolding with parameters $\delta_1, \delta_2, \delta_3$ the unfolded Jordan block is

$$\begin{pmatrix} 0 & 1 & 0 \\ 0 & 0 & 1 \\ \delta_1 & \delta_2 & \delta_3 \end{pmatrix} \tag{5.34}$$

and the eigenvalues are the solutions to

$$\lambda^3 - \delta_3\lambda^2 - \delta_2\lambda - \delta_1 = 0. \tag{5.35}$$

The plane $\delta_1 = 0$ now collects the fold points and the line $\delta_1 = \delta_2 = 0$ is a line of BT points. Zero-sum eigenvalues occur if

$$2A \odot I_3 = \begin{pmatrix} 0 & 1 & 0 \\ \delta_2 & \delta_3 & 1 \\ -\delta_1 & 0 & \delta_3 \end{pmatrix} \tag{5.36}$$

is singular; here A is (5.34) and we have used (4.94). Hence the zero-sum eigenvalues are found on the quadratic surface

$$\delta_1 + \delta_2\delta_3 = 0, \tag{5.37}$$

which has a saddle point at the origin; see Figure 5.2.

The plane $\delta_2 = 0$ intersects the surface of zero-sum points precisely in the BT curve and separates Hopf points from neutral saddles. For $\delta_2 < 0$ we find Hopf points; for $\delta_2 > 0$ we find neutral saddles.

We note that the intersection of the fold and zero-sum surfaces has two branches, namely, the BT points ($\delta_1 = \delta_2 = 0$) and another branch determined by ($\delta_1 = \delta_3 = 0$). The latter one is a branch of zero-Hopf (ZH) points. These will be considered in §5.2.

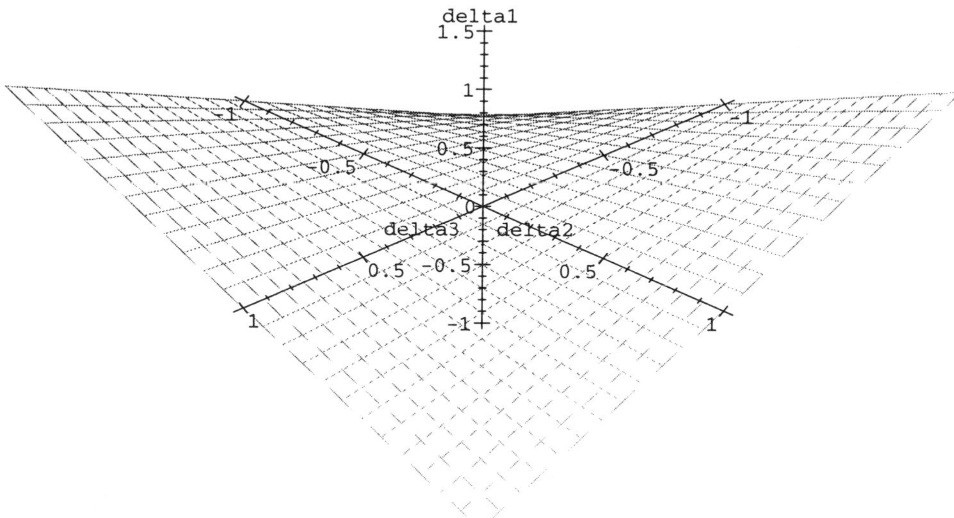

Figure 5.2: Matrices with zero-sum eigenvalue pairs.

The branches of BT points and of ZH points intersect in the triple-point bifurcation point.

In the simplest case ($n = 1$) the unfolding by a natural parameter is obviously universal if and only if the zero eigenvalue has a nonzero derivative as a function of the arclength of the solution curve to (5.2). In §5.1.4 we will see that this is precisely equivalent to the condition that the fold point is a quadratic turning point.

5.1.4 Transversality and Genericity

We now explain the genericity assumption in §5.1.3. We consider the case of a BT point in a two-dimensional family of matrices $A(\beta) = A(\beta_1, \beta_2)$. By Proposition 3.5.1 the eigenvalue structure of $A(\beta)$ for eigenvalues in a neighborhood of 0 is that of an unfolding $B(\beta)$ of (5.31). By the universality of (5.32) there exist smooth functions $\delta_1(\beta), \delta_2(\beta)$ such that $B(\beta)$ is similar to (5.32) with δ_1, δ_2 replaced by $\delta_1(\beta), \delta_2(\beta)$ and the similarity matrix a smooth function of β. The set of functions $\delta_1(\beta), \delta_2(\beta)$ has a nonsingular Jacobian if and only if $B(\beta)$ is a universal unfolding of (5.31).

Two submanifolds of an Euclidean space are called transversal at an intersection point if the union of their tangent spaces spans the whole space. If the sum of their dimensions equals the dimension of the whole space, then this amounts to saying that they have only the null vector in common. We will show that the universality of the unfolding $B(\beta)$ is equivalent to the transversality of the similarity orbit of (5.31) with the manifold $B(\beta)$. We first note that δ_1, δ_2 are functions of four parameters, namely, the the entries of $B(\beta)$. If we denote these collectively by ζ, then the Jacobian matrix

$$J_\zeta = \frac{\partial(\delta_1, \delta_2)}{\partial \zeta} \tag{5.38}$$

has full rank 2 because it contains a 2×2 identity block as a submatrix. Now the similarity orbit of (5.31) is defined by

$$\delta_1(\zeta) = \delta_2(\zeta) = 0$$

and so it is a manifold with codimension 2. Its tangent space is the two-dimensional null space of J_ζ. By the chain rule we have

$$J_\zeta \frac{\partial \zeta}{\partial \beta} = \frac{\partial(\delta_1, \delta_2)}{\partial \beta}. \tag{5.39}$$

From this it follows that $\frac{\partial(\delta_1,\delta_2)}{\partial \beta}$ is nonsingular if and only if the tangent space to the manifold $B(\beta)$ contains no tangent vector of the group orbit; i.e., the manifolds are transversal. The same property is proved in the same way in the space of $N \times N$ matrices; i.e., the unfolding of $B(\beta)$ is universal if and only if the $\mathbf{GL}(N)$ orbit of A in the BT point is transversal to the manifold $A(\beta)$.

In §5.1.3 it was proved that the tangent space to the $\mathbf{GL}(N)$ orbit of a BT matrix of order 2 is the space orthogonal to the two vectors $s_{0\zeta}, s_{1\zeta}$, where the index ζ runs through all entries of an $\mathbb{R}^{N\times N}$ matrix. Now assume in particular that the matrix arises as the Jacobian matrix of a solution to (5.2) with two free parameters α_1, α_2 in a regular point of this manifold, i.e., where $(G_x\ G_\alpha)$ has full rank N. Let $\phi_1, \phi_2 \in \mathbb{R}^{N+2}$ span the two-dimensional right singular space of $(G_x\ G_\alpha)$. Then the $N \times N$ matrices $A_1 = (G_{xx}, G_{x\alpha})\phi_1$, $A_2 = (G_{xx}, G_{x\alpha})\phi_2$ span the tangent space to the manifold of matrices G_x in solution points to (5.2). Also, the vectors $h_{0\zeta}$, $h_{1\zeta}$ with ζ running through all entries of an $N \times N$ matrix span the orthogonal complement of the tangent space to the $\mathbf{GL}(N)$ orbit. We note that if ζ corresponds to the entry (i,j) then $s_{0\zeta} = -w_0(i)q_0(j)$, $s_{1\zeta} = -w_1(i)q_0(j) - w_0(i)q_1(j)$ by (5.25). So the transversality condition in the space of $N \times N$ matrices is equivalent to the nonsingularity of the 2×2 matrix

$$\begin{pmatrix} A_1^T s_{0\zeta} & A_1^T s_{1\zeta} \\ A_2^T s_{0\zeta} & A_2^T s_{1\zeta} \end{pmatrix}, \tag{5.40}$$

where the inner products are summations over the $N \times N$ entries of a matrix. Now (5.40) precisely expresses that the defining system

$$\begin{aligned} G(x,\alpha) &= 0, \\ s_0(x,\alpha) &= 0, \\ s_1(x,\alpha) &= 0 \end{aligned} \tag{5.41}$$

of the BT point has full linear rank. In other words, the unfolding of the BT point by the natural parameters of the problem is universal if and only if the defining system (5.41) has full linear rank.

This result obviously carries over to BT points of any order, in particular to triple-point bifurcation points. In the simplest case where $n = 1$ we find that the unfolding by one natural parameter is universal if and only if the fold point is a quadratic turning point (cf. §4.1).

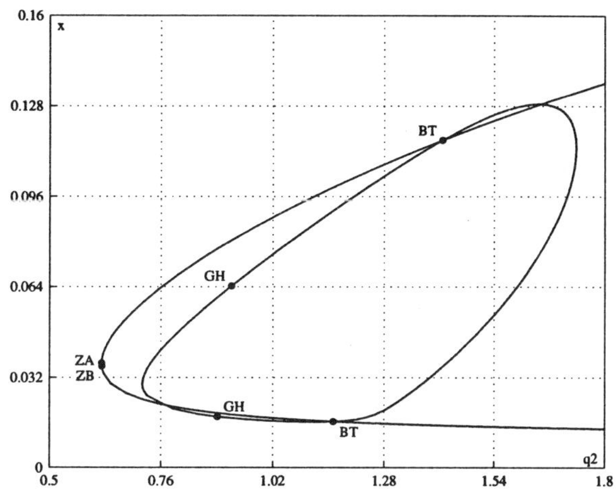

Figure 5.3: Zero sum and BT points in the catalytic oscillator model.

5.1.5 Test Functions for BT Points

BT points are typically found either on curves of limit points or on curves of points with zero-sum eigenvalues, in particular, Hopf points. On a curve of limit points, the function s_1 as obtained in (5.17) is a test function for BT. On a curve of points with zero-sum eigenvalues the product of the eigenvalues can be used as a test function. For the methods in §§4.3.2, 4.3.3, and 4.3.4 this product is easily available. If the bialternate matrix product is used, then the product of the eigenvalues can be obtained from (4.110). In the method that uses complex arithmetic (§4.3.6) ω itself is a test function.

5.1.6 Example: A Curve of BT Points in the Catalytic Oscillator Model

In §4.6.1 two BT points were found on a closed curve of Hopf points in the catalytic oscillator model. Taking the lower BT point in Figure 4.1 as a starting point we computed a curve of BT points with free parameters q_1, q_2, k. Figure 5.3 presents the Hopf curve and the curve of BT points. We note that the BT curve connects the two BT points on the Hopf curve. Also, it contains two close but distinct degeneracies denoted on Figure 5.3 by ZA and ZB. These are degeneracies in the nonlinear terms; the dynamical behavior of (5.1) is more complicated in the neighborhood of such points than in the neighborhood of a BT point with no such degeneracies. Degeneracies in the nonlinear terms will be discussed in Chapter 9.

5.2 ZH Points and Their Generalizations

A solution point to (5.2) is called a ZH point if G_x has an eigenvalue zero and a conjugate pair of pure imaginary eigenvalues $\pm i\omega$, $\omega > 0$. Standard references to the (usually quite complicated) dynamic behavior of (5.1) near a ZH point (also called *Gavrilov–Guckenheimer point*) are [127, §7.5] and [164, §8.5].

The numerical computation of ZH points is less exciting. It suffices to express that G_x is singular and has a Hopf pair of eigenvalues. The methods discussed in Chapter 4 can be applied.

The case of BT + Hopf or (BT of higher order) + Hopf can be treated similarly by combining methods for Hopf (Chapter 4) and methods for BT or higher-order BT (§5.1).

More attention is required for the interaction of ZH or BT–Hopf points with other bifurcation points. We saw in §5.1.3 that in a triple point bifurcation point one can generically jump to a branch of ZH points. Conversely, on a branch of ZH points a triple-point bifurcation might be found and so one could jump to a curve of BT points. A branch of ZH points might also contain a BT + Hopf point from which one might start a double Hopf (DH) curve (see §5.3).

5.2.1 Transversality and Genericity for Simple Hopf

In §4.5 a family $A(\alpha)$ of $N \times N$ matrices was given and we obtained a function $s(\alpha) \in \mathbb{R}$ so that

$$s(\alpha) = 0 \tag{5.42}$$

characterizes the matrices with a zero-sum pair of eigenvalues locally near a matrix with precisely one such pair of eigenvalues.

Proposition 5.2.1. The $N \times N$ matrices with precisely one zero-sum pair of eigenvalues form a manifold with dimension $N^2 - 1$. In the setting of §4.7 they all satisfy (5.42) for a choice of b, c, d that makes (4.101) nonsingular. Furthermore, (5.42) is then locally a regular defining function for the manifold.

Proof. If a matrix has precisely one zero-sum pair of eigenvalues, then it has a neighborhood (in the space of all matrices) in which every matrix has at most one zero-sum pair of eigenvalues. Then (5.42) further implies (at least in another, possibly smaller neighborhood) that there is such a pair. So only the regularity of the defining function remains to be shown. We will show that it is actually sufficient to consider only the diagonal entries. Since 0 is an algebraically simple eigenvalue of $2A \odot I_N$ we must have $w^T q \neq 0$ where w, q are the left and right singular vectors of $2A \odot I_N$, respectively (notation of §4.5). If s is not a regular defining function then by (4.105) we have $w^T(2A_{ii} \odot I_N)q = 0$ for $i = 1, \dots, N$. Now it is easily observed that $\Sigma_{i=1,\dots,N}(2A_{ii} \odot I_N) = (N-1)I_N$ and a contradiction follows. □

We note that (5.42) holds for an appropriate nonsingular matrix in (4.101) if and only if $2A \odot I_N$ has rank defect 1. This includes not only the points of the manifold in Proposition 5.2.1 but some other points as well. In fact, from Propositions 4.4.2.1, 4.4.24–26 it follows that $2A \odot I_N$ also has rank defect 1 in the following cases (and no other ones):

1. A has eigenvalue zero with algebraic multiplicity 3 and geometric multiplicity 1.

2. A has eigenvalues $\pm\lambda$ for a $\lambda > 0$ for which both have geometric multiplicity 1 and one of them has algebraic multiplicity 1.

For unfolding results, we have to adapt Proposition 5.1.1 to the case of complex eigenvalues. If the complex Jordan form of a real matrix $A \in \mathbb{R}^{2k\times 2k}$ contains a Jordan block J_k with complex eigenvalue $a+ib$,

$$J_k = \begin{pmatrix} a+ib & 1 & 0 & \dots & 0 \\ 0 & a+ib & 1 & 0 & \vdots \\ \vdots & \vdots & \ddots & \ddots & 0 \\ 0 & \dots & 0 & a+ib & 1 \\ 0 & \dots & \dots & 0 & a+ib \end{pmatrix}, \tag{5.43}$$

then it must also contain the block $\overline{J_k}$ and the corresponding real form is

$$\begin{pmatrix} \frac{1}{2}(J_k+\overline{J_k}) & \frac{1}{2i}(J_k-\overline{J_k}) \\ -\frac{1}{2i}(J_k-\overline{J_k}) & \frac{1}{2}(J_k+\overline{J_k}) \end{pmatrix} = \begin{pmatrix} \frac{1}{\sqrt{2}}I_k & \frac{1}{\sqrt{2}}I_k \\ \frac{i}{\sqrt{2}}I_k & \frac{-i}{\sqrt{2}}I_k \end{pmatrix} \begin{pmatrix} J_k & 0 \\ 0 & \overline{J_k} \end{pmatrix} \begin{pmatrix} \frac{1}{\sqrt{2}}I_k & \frac{-i}{\sqrt{2}}I_k \\ \frac{1}{\sqrt{2}}I_k & \frac{i}{\sqrt{2}}I_k \end{pmatrix}. \tag{5.44}$$

Proposition 5.2.2. Let $A_0 \in \mathbb{R}^{2k\times 2k}$ have the complex eigenvalues $a \pm ib$ with a complex Jordan block of the form J_k as in (5.43) and suppose that A_0 is in real normal Jordan form (5.44) (we represent the case $k=3$):

$$A_0 = \begin{pmatrix} a & 1 & & b & & \\ & a & 1 & & b & \\ & & a & & & b \\ -b & & & a & 1 & \\ & -b & & & a & 1 \\ & & -b & & & a \end{pmatrix}. \tag{5.45}$$

Then the following hold:

1.
$$A(\delta_1,\eta_1,\dots,\delta_k,\eta_k) = \begin{pmatrix} a & 1 & & b & & \\ & a & 1 & & b & \\ \delta_1 & \dots & a+\delta_k & \eta_1 & \dots & b+\eta_k \\ -b & & & a & 1 & \\ & -b & & & a & 1 \\ -\eta_1 & \dots & -b-\eta_k & \delta_1 & \dots & a+\delta_k \end{pmatrix} \tag{5.46}$$

is a $\mathbf{GL}(2k)$ versal unfolding of A_0.

2. This unfolding is universal and miniversal.

3. An unfolding of A_0 is universal if and only if it is versal and miniversal.

Proof. We prove the $\mathbf{GL}(2k)$ versality of the unfolding $A(\delta_1,\eta_1,\dots,\delta_k,\eta_k)$; for the other statements the proofs in Proposition 5.1.1 carry over easily.

Let $A(\alpha), \alpha \in \mathbb{R}^{2k\times 2k}$ be the complete real unfolding of A_0. By the complex variant of Proposition 3.5.1 there exists matrix functions $B(\alpha) \in \mathbb{R}^{2k\times k}$ and $J(\alpha) \in \mathbb{R}^{k\times k}$ such that $A(\alpha)B(\alpha) = B(\alpha)J(\alpha)$ with

$$B(0) = \begin{pmatrix} I_k \\ iI_k \end{pmatrix}$$

and $J(0) = J_k$. By the complex variant of Proposition 5.1.1 there exist a smooth complex matrix function $C(\alpha) \in \mathbb{R}^{k\times k}$ and complex numbers $\delta_i(\alpha) + i\eta_i(\alpha)$ $(1 \le i \le k)$ such that $C(0) = I_k$, $\delta_i(0) = \eta_i(0) = 0$ for all i and $J(\alpha) = C(\alpha)J_p(\alpha)C(\alpha)^{-1}$, where

$$J_p(\alpha) = \begin{pmatrix} a+ib & 1 & 0 & \dots & 0 \\ 0 & a+ib & 1 & 0 & \vdots \\ \vdots & \vdots & \ddots & \ddots & 0 \\ 0 & \dots & \dots & a+ib & 1 \\ \delta_1 + i\eta_1 & \delta_2 + i\eta_2 & \dots & \dots & a+ib+\delta_k+i\eta_k \end{pmatrix}.$$

Setting $B^1(\alpha) = B(\alpha)C(\alpha)$ we find $A(\alpha)B^1(\alpha) = B^1(\alpha)J_p(\alpha)$. Taking complex conjugates we find that

$$A(\alpha)\begin{pmatrix} B^1(\alpha) & \overline{B^1(\alpha)} \end{pmatrix} = \begin{pmatrix} B^1(\alpha) & \overline{B^1(\alpha)} \end{pmatrix}\begin{pmatrix} J_p(\alpha) & 0 \\ 0 & \overline{J_p(\alpha)} \end{pmatrix}.$$

Now set

$$M(\alpha) = \begin{pmatrix} B^1(\alpha) & \overline{B^1(\alpha)} \end{pmatrix}\begin{pmatrix} \frac{1}{\sqrt{2}}I_k & \frac{-i}{\sqrt{2}}I_k \\ \frac{1}{\sqrt{2}}I_k & \frac{i}{\sqrt{2}}I_k \end{pmatrix}.$$

Then $M(\alpha)$ is a real matrix and

$$A(\alpha)M(\alpha) = M(\alpha)A(\delta_1, \eta_1, \dots, \delta_k, \eta_k).$$

Since $M(0) = I_{2k}$ the $\mathbf{GL}(2k)$ versality follows. □

Consider again the Hopf case where the complex Jordan normal form of G_x contains a block

$$\begin{pmatrix} ri \end{pmatrix} \tag{5.47}$$

with $r > 0$. The real Jordan normal form then contains the block

$$\begin{pmatrix} 0 & r \\ -r & 0 \end{pmatrix} \tag{5.48}$$

and by Proposition 5.2.2 has the $\mathbf{GL}(2)$ universal unfolding

$$\begin{pmatrix} \delta & r+\eta \\ -r-\eta & \delta \end{pmatrix} \tag{5.49}$$

with real parameters δ, η.

Similarity transforms are not the only transformations that leave the Hopf structure (number of eigenvalues with zero real part and their multiplicities) invariant. Multiplication with positive real scalars has the same effect. Let $\mathbb{R}^+_\cdot$ be the multiplicative group of positive real numbers. The product group $\mathbb{R}^+_{,\cdot}\times$ $\mathbf{GL}(N)$ acts on $\mathcal{M}_N$ by $A \to rGAG^{-1}$ for $(r, G) \in \mathbb{R}^+_{,\cdot}\times$ $\mathbf{GL}(N)$. The orbit of (5.49) with respect to this group contains

$$\begin{pmatrix} \beta & 1 \\ -1 & \beta \end{pmatrix}. \tag{5.50}$$

So (5.50) is a one-parameter versal unfolding of (5.47) with respect to the group action of $\mathbb{R}^+_{,\cdot}\times$ $\mathbf{GL}(2)$. The unfolding parameter β in (5.50) is uniquely determined if the eigenvalues of (5.50) are known up to multiplication with the same positive real scalar. Hence (5.50) is a universal unfolding of (5.50) with respect to the group action of $\mathbb{R}^+_{,\cdot}\times$ $\mathbf{GL}(2)$.

Now assume that the Hopf point is encountered in a one-parameter problem, say, with parameter α. As in §5.1.4 it follows that the natural unfolding is transversal to the Hopf surface if and only if it corresponds to a universal unfolding (5.50), i.e.,

$$\frac{d\beta}{d\alpha} \neq 0$$

or, equivalently, if and only if (4.37) holds. Transversality is equivalent to the regularity of the defining system

$$\begin{aligned} G(x, \alpha) &= 0, \\ s(x, \alpha) &= 0. \end{aligned} \tag{5.51}$$

5.2.2 Transversality and Genericity for ZH

The manifold of matrices with precisely one zero eigenvalue and one zero-sum pair of eigenvalues can be determined by the two conditions $h_0 = 0, s = 0$, where h_0 is determined as in (4.13) and s as in (4.102). It is easy to see (by reduction to a base that is provided by the Jordan normal form) that these two equations form a regular defining system for a manifold with dimension $N^2 - 2$. In a two-parameter problem results on transversality and genericity of the natural unfolding are similar to those obtained in §§5.1.4 and 5.2.1 and can be proved by similar arguments. Also, they are related to the regularity of the defining system in a similar way.

We note that the natural unfolding is universal if there is a block

$$\begin{pmatrix} \beta & 1 & 0 \\ -1 & \beta & 0 \\ 0 & 0 & \gamma \end{pmatrix} \tag{5.52}$$

in the real Jordan normal form of the Jacobian (scaled so that the pure imaginary eigenvalues have imaginary part ± 1) with (β, γ) a full-rank system in the natural parameters.

The above results extend naturally and by similar arguments to the case of BT + Hopf and (BT of higher order) + Hopf.

5.2.3 Detection of ZH Points

ZH points appear generically on curves of limit points and Hopf points. As a test function on a curve of limit points any method for detecting Hopf points (§4.5) can be used.

However, a BT point typically also satisfies the Hopf conditions; therefore, a test function for BT should be monitored at the same time. Conversely, on a curve of Hopf points any test function for limit points can be used, but the case of a BT point should be excluded by an appropiate test condition.

5.3 DH Points and Resonant DH Points

5.3.1 Introduction

A solution point to (5.2) is called a DH point if G_x has two pairs $\pm i\omega_1, \pm i\omega_2$ ($\omega_1 > 0, \omega_2 > 0$) of pure imaginary eigenvalues. The special case $\omega_1 = \omega_2$ is called a one-to-one resonant DH point. It presents special numerical features and will be treated in more detail in §5.3.3.

The dynamic behavior of (5.1) near DH points can be quite complicated even if there is no resonance. Standard references are [127, §7.5] and [164, §8.6]. We note only that homoclinic, quasi-periodic, and chaotic motions exist near DH points. The case of one-to-one resonance is even more complicated.

We note that two-to-one, three-to-one, etc., resonances also influence the dynamic behavior profoundly (cf. [127], [164]). However, their computation in the set of equilibria does not present features different from expressing a "resonance" of, say, 1.3 to 1, which is dynamically uninteresting. We will not discuss this situation.

Double Hopf points are common in two-parameter problems. They can be expected generically on paths of Hopf points. In §4.5.2 we saw how they can be detected there.

5.3.2 Defining Functions for Multiple Hopf Points

To characterize, detect, compute, and continue DH points we shall combine bialternate product methods (§§4.6 and 4.7) with bordered matrix techniques (Chapter 3). We recall that $b(N) = N(N-1)/2$.

Suppose that in a point (x, α) the matrix $A = G_x(x, \alpha)$ has eigenvalues $\pm i\omega_1, \pm i\omega_2$, where $\omega_1 > 0$, $\omega_2 > 0$, $\omega_1 \neq \omega_2$. If there are no other zero-sum eigenvalue pairs, then by Proposition 4.4.27 the matrix $2A \odot I_N$ has rank defect 2. We choose $B, C \in \mathbb{R}^{b(N)\times 2}$, $D \in \mathbb{R}^{2\times 2}$ such that

$$M(x, \alpha) = \begin{pmatrix} 2A \odot I_N & B \\ C^T & D \end{pmatrix} \tag{5.53}$$

is nonsingular at the DH point. Then we define the $b(N) \times 2$ matrix $Q(x, \alpha)$ and the 2×2 matrix $S(x, \alpha)$ by

$$M(x, \alpha) \begin{pmatrix} Q(x, \alpha) \\ S(x, \alpha) \end{pmatrix} = \begin{pmatrix} 0_{b(N),2} \\ I_{2,2} \end{pmatrix}. \tag{5.54}$$

By Corollary 3.3.4 the four entries of $S(x, \alpha)$ vanish together if and only if $2A \odot I_N$ has rank defect 2.

The derivatives of S can be obtained in the now familar way. Define the $b(N) \times 2$ matrix $W(x, \alpha)$ by

$$\begin{pmatrix} W^T(x,\alpha) & S(x,\alpha) \end{pmatrix} M(x,\alpha) = \begin{pmatrix} 0_{2\times b(N)} & I_{2,2} \end{pmatrix}. \tag{5.55}$$

Then S_z can be obtained from

$$S_z(x,\alpha) + W^T(2A_z(x,\alpha) \odot I_n)Q = 0_{2,2}. \tag{5.56}$$

We note that $S(x, \alpha)$ has four components, while intuitively a DH point is a codimension-2 phenomenon only (we will see that it is). So we can suspect that the four resulting equations are not independent. We will discuss this in a somewhat more general setting, but we first prove Lemma 5.3.1 to elucidate the meaning of (5.56).

Lemma 5.3.1. Let $v_1, v_2, w_1, w_2 \in \mathbb{C}^N$ and let $1 \le i, j \le N$. Then for any matrix $A \in \mathbb{C}^{N\times N}$ we have

$$\begin{aligned} &(w_1 \wedge w_2)^T(2A \odot I_N)(v_1 \wedge v_2) \\ &= (w_1^T A v_1)(w_2^T v_2) + (w_2^T A v_2)(w_1^T v_1) - (w_1^T A v_2)(w_2^T v_1) - (w_2^T A v_1)(w_1^T v_2). \end{aligned} \tag{5.57}$$

In particular,

$$(w_1 \wedge w_2)^T(v_1 \wedge v_2) = ((w_1^T v_1)(w_2^T v_2) - (w_2^T v_1)(w_1^T v_2)). \tag{5.58}$$

If $1 \le i, j \le N$ and z denotes the (i, j)th entry of A, then

$$\begin{aligned} &(w_1 \wedge w_2)^T(2A_z \odot I_N)(v_1 \wedge v_2) \\ &= w_{1i}v_{1j}w_2^T v_2 - w_{1i}v_{2j}w_2^T v_1 - w_{2i}v_{1j}w_1^T v_2 + w_{2i}v_{2j}w_1^T v_1. \end{aligned} \tag{5.59}$$

Proof. We prove (5.57); the other statements follow easily. We have

$$(w_1 \wedge w_2)^T(2A \odot I_N)(v_1 \wedge v_2)$$

$$= \Sigma_{1\le j<i\le N}(w_1 \wedge w_2)_{i,j}(Av_1 \wedge v_2 + v_1 \wedge Av_2)_{i,j}$$

(by Proposition 4.4.19)

$$= \Sigma_{1\le j<i\le N}(w_{1j}w_{2i} - w_{2j}w_{2i})((Av_1)_j v_{2i} - (Av_1)_i v_{2j} + v_{1j}(Av_2)_i - v_{1i}(Av_2)_j)$$

$$= \frac{1}{2}\Sigma_{1\le i,j\le N}(w_{1j}w_{2i} - w_{2j}w_{2i})((Av_1)_j v_{2i} - (Av_1)_i v_{2j} + v_{1j}(Av_2)_i - v_{1i}(Av_2)_j)$$

$$= (w_1^T A v_1)(w_2^T v_2) + (w_2^T A v_2)(w_1^T v_1) - (w_1^T A v_2)(w_2^T v_1) - (w_2^T A v_1)(w_1^T v_2). \qquad \square$$

Proposition 5.3.2. In the nonresonant DH situation the four gradient vectors contained in (5.56) span a two-dimensional space.

Proof. Using essentially the notation of §4.4 we let $q_1^j + iq_2^j$ denote the right eigenvector that corresponds to $i\omega_j$ $(j = 1, 2)$ and $p_1^j + ip_2^j$ $(j = 1, 2)$ the left eigenvector. Since ω_1, ω_2 are positive and not equal we necessarily have

$$(p_i^j)^T q_k^l = 0 \qquad (j \ne l), \tag{5.60}$$

while by an appropriate choice of the vectors we may assume (cf. Proposition 4.4.1)

$$(p_i^j)^T q_k^j = \delta_{ik} \qquad (j = 1, 2). \tag{5.61}$$

From the proof of Proposition 4.4.21 it is clear that $2A \odot I_N$ has a two-dimensional right singular space spanned by the (complex) vectors $V_1 = (q_1^1 + iq_2^1) \wedge (q_1^1 - iq_2^1)$ and $V_2 = (q_1^2 + iq_2^2) \wedge (q_1^2 - iq_2^2)$; its left singular space is spanned by the (complex) vectors $W_1 = (p_1^1 + ip_2^1) \wedge (p_1^1 - ip_2^1)$ and $W_2 = (p_1^2 + ip_2^2) \wedge (p_1^2 - ip_2^2)$.

Now consider the four gradient vectors defined in the space of all $N \times N$ matrices by (5.56). Obviously their span is the same as that of the four vectors $W_i^T(2A_z \odot I_N)V_j$ for $i, j = 1, 2$. But by Lemma 5.3.1 and (5.60) we have $W_i^T(2A_z \odot I_N)V_j = 0$ if $i \neq j$. So we are left with the two gradient vectors $W_i^T(2A_z \odot I_N)V_i$ for $i = 1, 2$.

From (5.61) it follows that $(p_1^1 + ip_2^1)^T(q_1^1 + iq_2^1) = 0$, $(p_1^1 + ip_2^1)^T(q_1^1 - iq_2^1) = 2$, $(p_1^1 - ip_2^1)^T(q_1^1 + iq_2^1) = 2$, $(p_1^1 - ip_2^1)^T(q_1^1 - iq_2^1) = 0$. By Lemma 5.3.1 we infer that

$$\begin{aligned}(W_1^T(2A_{ij} \odot I_N)V_1) &= -2(p_1^1 + ip_2^1)_i(q_1^1 - iq_2^1)_j - 2(p_1^1 - ip_2^1)_i(q_1^1 + iq_2^1)_j \\ &= -4(p_1^1)_i(q_1^1)_j - 2(p_2^1)_i(q_2^1)_j.\end{aligned}$$

A similar formula holds of course for the other gradient vector. To prove that the two are linearly independent, suppose that there exist α_1, α_2 such that

$$\alpha_1((p_1^1)_i(q_1^1)_j + (p_2^1)_i(q_2^1)_j) + \alpha_2((p_1^2)_i(q_1^2)_j + (p_2^2)_i(q_2^2)_j) = 0 \tag{5.62}$$

for all i, j. Multiplying (5.62) with $(q_1^1)_i$ we find after summation over i that $\alpha_1(q_1^1)_j = 0$ for all j. This implies $\alpha_1 = 0$. Similarly, $\alpha_2 = 0$. □

Proposition 5.3.3. For each nonresonant DH matrix $A_0 \in \mathbb{R}^{N\times N}$ there exists a neighborhood (in the space of all matrices $A \in \mathbb{R}^{N\times N}$) in which the DH matrices form a manifold with codimension 2 and are characterized by the fact that $2A \odot I_N$ has rank defect 2.

Proof. Let $A_0 = CJ_0C^{-1}$ be the Jordan decomposition of A_0. Obviously, J_0 has one-element Jordan blocks of the form $i\omega_1$, $i\omega_2$, $-i\omega_1$, $-i\omega_2$. By Proposition 5.2.2 there is a neighborhood of A_0 in which every matrix has corresponding Jordan blocks of the form $i\omega_1 + \delta_1 + i\delta_2$, $i\omega_2 + \delta_3 + i\delta_4$, $-i\omega_1 + \delta_1 - i\delta_2$, $-i\omega_2 + \delta_3 - i\delta_4$, where δ_1, δ_2, δ_3, δ_4 are smooth functions of the matrix entries and vanish at A_0. Clearly, $2A \odot I_N$ has rank defect 2 if and only if A is a DH matrix if and only if $\delta_1 = \delta_3 = 0$. So it is sufficient to show that δ_1, δ_3 form a regular set of functions of the entries of A, i.e., that their $2 \times N^2$ Jacobian

$$\begin{pmatrix} (\delta_1)_A \\ (\delta_3)_A \end{pmatrix} \tag{5.63}$$

has full rank 2 at $A = A_0$. Now consider the real two-parameter unfolding $A(\alpha_1, \alpha_2) = CJ(\alpha_1, \alpha_2)C^{-1}$ where $J(\alpha_1, \alpha_2)$ is obtained by replacing in J_0 $\pm i\omega_1$ by $\pm i\omega_1 + \alpha_1$ and $\pm i\omega_2$ by $\pm i\omega_2 + \alpha_2$. Obviously the mapping $(\alpha_1, \alpha_2)^T \to (\delta_1(A(\alpha_1, \alpha_2)), \delta_3(A(\alpha_1, \alpha_2)))^T$ is the identity mapping; by the chain rule for derivatives this implies that (5.63) must have full rank. □

The above result naturally generalizes to three other cases.

Proposition 5.3.4. Let $A_0 \in \mathbb{R}^{N\times N}$ be a matrix of one of the following four types.

1. A is nonresonant DH (type DH).
2. A has a pair of algebraically simple eigenvalues of the form $a+ib$ and $-a-ib$ with a, b both real and nonzero (rectangular eigenvalues point RT).
3. A has four different nonzero real algebraically simple eigenvalues $-\lambda_1, -\lambda_2, \lambda_1, \lambda_2$ (double neutral saddle point DN).
4. A has algebraically simple eigenvalues $\pm i\omega$, $\pm\lambda$ with $\omega, \lambda > 0$ (Hopf neutral saddle point HN).

Then there exists a neighborhood of A_0 (in the space of all matrices $A \in \mathbb{R}^{N\times N}$) in which the matrices of the same type form a manifold with codimension 2 and are characterized by the fact that $2A \odot I_N$ has rank defect 2.

Proof. The first case is Proposition 5.3.3; the other cases can be proved similarly. □

By Propositions 5.3.2 and 5.3.3 in the DH case two of the four functions contained in the system

$$S(x, \alpha) = 0, \tag{5.64}$$

where $S(x, \alpha)$ is defined by (5.54), form a defining system for the two-dimensional manifold of DH matrices near a given one. One might also consider any two linear combinations of these four functions; it is sufficient that the gradient vectors are linearly independent. In fact, Proposition 5.3.2 provides sufficient information to make an a priori choice based on local information in the matrix under consideration.

In practice it is convenient to compute the four gradient vectors with respect to the variables in x, α and choose the two components of $S(x, \alpha)$ that span a space as close as possible to the orthogonal complement of the space spanned by the rows of the Jacobian of $G(x, \alpha)$ since this optimizes the condition number of the system obtained by linearizing (5.4) and (5.64).

One way to achieve this starts with the computation of a column base

$$V_N = \begin{pmatrix} V_1 & V_2 & V_3 \end{pmatrix}$$

for the three-dimensional singular space of (G_z, G_α). Then we compute $(S_{ij})_z$ with z either a state variable or a parameter. In this way, we obtain four vectors S_{11z}, S_{12z}, S_{21z}, and S_{22z} in $\mathbb{R}^{N+3}$. The tangent vector to the DH curve is necessarily in the space spanned by V_N and is orthogonal to all vectors S_{ijz}. We project the four of them orthogonally onto the span of V_N and obtain

$$\left\{ \begin{array}{lcl} S_{11}^\star & = & V_N\,(V_N^T V_N)^{-1} V_N^T\, S_{11z}, \\ S_{12}^\star & = & V_N\,(V_N^T V_N)^{-1} V_N^T\, S_{12z}, \\ S_{21}^\star & = & V_N\,(V_N^T V_N)^{-1} V_N^T\, S_{21z}, \\ S_{22}^\star & = & V_N\,(V_N^T V_N)^{-1} V_N^T\, S_{22z}. \end{array} \right. \tag{5.65}$$

So in the absence of roundoff and truncation errors the four vectors in (5.65) span only a two-dimensional space. We choose two index pairs so that the corresponding vectors span this space in a numerically optimal way. First we compute $\|S_{11}^\star\|$, $\|S_{12}^\star\|$, $\|S_{21}^\star\|$, and $\|S_{22}^\star\|$; the largest norm gives the first index pair (i_1, j_1). After that, $S_{i_1,j_1}^\star$

is normalized. To find the second index pair (i_2, j_2), we project $S^\star_{ij}$ onto the orthogonal complement of $S^\star_{i_1j_1}$ for all $(i,j) \neq (i_1, j_1)$; i.e.,

$$S^{\star\star}_{ij} = S^\star_{ij} - \langle S^\star_{ij}, S^\star_{i_1j_1}\rangle S^\star_{i_1j_1} \quad \text{with } (i,j) \neq (i_1, j_1).$$

Again, norms are calculated and the index with the largest norm gives (i_2, j_2).

We note that an attractive alternative is to develop a generalization of continuation methods to overdetermined systems; this would avoid the elimination of two equations.

The preceding remarks apply equally in the three other cases described in Proposition 5.3.4

5.3.3 Branch Switching at a DH Point

We next consider following a curve of single Hopf points through a DH point. The defining function based on simple bordering is not defined at a DH point. However, if we border the bialternate product matrix with two rows and columns instead of one, the defining function for the single Hopf point is the determinant of $S(x, \alpha)$. This defining system still can be computed at a DH point. We detect the DH point by monitoring the entries of $S(x, \alpha)$ (all must be zero at the DH point and generically they will all change sign). To locate the DH point one can use two entries of S instead of its determinant; the choice can be done as above.

If a DH point is detected and computed, then the tangent directions of the curves of single Hopf points through this point can be computed from the doubly bordered bialternate product matrix. Let $S_z(x, \alpha)$ be the matrix given by (5.56). Below, we identify the row indices of the entries of Q_0 with the corresponding element symbols in (x, α).

Proposition 5.3.5. Suppose that the DH point (x^0, α^0) is an ordinary point of the solution manifold of (5.2); i.e., two parameters are active and $[G_x, G_\alpha]$ has full rank. Let C_1, D_1 be $N \times 2$ and 2×2 matrices, respectively, such that

$$M_0 = \begin{pmatrix} G_x(x^0, \alpha^0) & G_\alpha(x^0, \alpha^0) \\ C_1^T & D_1 \end{pmatrix}$$

is nonsingular and compute the $(N+2) \times 2$ matrix V_0 by solving

$$M_0 V_0 = \begin{pmatrix} 0_{N,2} \\ I_{2,2} \end{pmatrix}.$$

Let $S_{V_{01}}$ and $S_{V_{02}}$ be the 2×2 matrices defined by

$$S_{V_{0i}} = \sum \{ S_z(x, \alpha) V_0(z, i) : z \in (x, \alpha) \}$$

for $i = 1, 2$.

Then the tangent directions of the Hopf curves through (x^0, α^0) are given by

$$V_0 \begin{pmatrix} a_1 \\ a_2 \end{pmatrix},$$

where $(a_1, a_2) \in R^2$ satisfy the homogeneous and quadratic equation

$$\det(a_1 S_{V_{01}} + a_2 S_{V_{02}}) = 0.$$

Proof. The columns of V_0 span the tangent space of the equilibrium manifold at (x^0, α^0). Also $S_{V_{01}}$ and $S_{V_{02}}$ are the directional derivatives of $S(x, \alpha)$ along the column vectors in V_0. If σ parameterizes a curve of single Hopf points through (x^0, α^0), then $\det(S(x(\sigma), \alpha(\sigma)) \equiv 0$. By taking the second derivative of this identity and evaluating it at the double Hopf point we find that $\det(S_\sigma(x^0, \alpha^0)) = 0$. The directional derivative of $S(x, \alpha)$ in the direction of $V_0\binom{a_1}{a_2}$ is equal to $a_1 S_{V_{01}} + a_2\ S_{V_{02}}$. If this direction is the tangent direction to a Hopf curve, then $\det(a_1 S_{V_{01}} + a_2 S_{V_{02}}) = 0$. □

5.3.4 Resonant DH Points

We now consider the one-to-one resonant case where A has double eigenvalues $\pm i\omega$, $\omega > 0$ with geometric multiplicity 1. By Proposition 4.4.27 $2A \odot I_N$ has eigenvalue zero with two Jordan blocks with sizes 1 and 3, respectively. Hence $(2A \odot I_N)^2$ has rank defect 3. Since this case presents a certain analogy to the BT situation (§5.1) it is natural to define the $b(N) \times 2$ matrix $Q_1(x, \alpha)$ and the 2×2 matrix $S_1(x, \alpha)$ by solving

$$M(x,\alpha)\begin{pmatrix} Q_1(x,\alpha) \\ S_1(x,\alpha) \end{pmatrix} = \begin{pmatrix} Q(x,\alpha) \\ S(x,\alpha) \end{pmatrix}, \tag{5.66}$$

where $M(x, \alpha)$ is defined in (5.53) and $Q(x, \alpha), S(x, \alpha)$ obtained from (5.54). Obviously we also have

$$\begin{pmatrix} (2A\odot I_N)^2 + BC^T & B_1 \\ C_1^T & D_1 \end{pmatrix}\begin{pmatrix} Q_1(x,\alpha) \\ S_1(x,\alpha) \end{pmatrix} = \begin{pmatrix} 0_{b(N)\times 2} \\ I_{2\times 2} \end{pmatrix}, \tag{5.67}$$

where $B_1, C_1 \in \mathbb{R}^{b(N)\times 2}$ and $D_1 \in \mathbb{R}^{2\times 2}$. Since BC^T has rank at most 2, it follows that $(2A \odot I_N)^2 + BC^T$ has rank defect at least 1, so by Corollary 3.3.4 S_1 is singular. On the other hand, S_1 cannot be zero, since then (5.66) would imply that $2A \odot I_N$ has Jordan blocks with sizes 2,2 instead of 1,3. So it is natural to add the condition

$$\det(S_1(x,\alpha)) = 0 \tag{5.68}$$

to the conditions for DH to obtain conditions for one-to-one resonant DH. Before dealing with the regularity of this system we note that the derivatives of S_1 can be obtained in the now familar way. Define the $b(N) \times 2$ matrix $W_1(x, \alpha)$ by

$$\begin{pmatrix} W_1^T(x,\alpha) & S_1(x,\alpha) \end{pmatrix} M(x,\alpha) = \begin{pmatrix} W^T(x,\alpha) & S(x,\alpha) \end{pmatrix}. \tag{5.69}$$

Then S_{1z} can be obtained from

$$S_{1z}(x,\alpha) + W_1^T(2A_z(x,\alpha)\odot I_N)Q + W^T(2A_z(x,\alpha)\odot I_N)Q_1 = 0_{2,2}. \tag{5.70}$$

It is convenient to deal with the regularity issue in a more general situation.

Proposition 5.3.6. Consider the following four types of matrices $A \in \mathbb{R}^{N\times N}$.

1. A is resonant double Hopf and the Hopf eigenvalues have geometric multiplicity 1 (resonant DH point RDH).

2. A has real eigenvalues $\pm\lambda$, $\lambda > 0$, each with algebraic multiplicity 2 and geometric multiplicity 1 (resonant double neutral saddle RDN).

3. A has algebraically simple eigenvalues $\pm\lambda$, $\lambda > 0$ and eigenvalue zero with algebraic multiplicity 2 and geometric multiplicity 1 (BT neutral saddle point BTN).

4. A has algebraically simple eigenvalues $\pm i\omega$, $\omega > 0$ and eigenvalue zero with algebraic multiplicity 2 and geometric multiplicity 1 (BT–Hopf point BTH).

The matrices of each type form a manifold of codimension 3 in the space of all $\mathbb{R}^{N\times N}$ matrices. If A_0 is in one of these classes, then $2A_0 \odot I_N$ has rank defect 2. Furthermore, there exists a neighborhood of A_0 (in the space of all matrices $A \in \mathbb{R}^{N\times N}$) in which the matrices for which $2A \odot I_N$ has rank defect 2 form a manifold with codimension 2. In the case of RDH this neighborhood contains only matrices of the types RDH, DH, and RT. For RDN it is RDN, RT, and DN. For BTN it is BTN, DN, and HN. For BTH it is BTH, HN, and DH. In each case a regular set of defining functions for the manifold of codimension 2 is obtained by taking two of the functions contained in (5.64) for which the gradient system has full rank 2. In the cases RDN and RDH a regular set of defining functions for the manifold of codimension 3 is obtained by adding the condition (5.68). In the cases BTN and BTH one obtains a regular set of defining equations by adding the condition $s_f(A) = 0$, where $s_f(A)$ is obtained by solving

$$\begin{pmatrix} A & b_f \\ c_f^T & d_f \end{pmatrix} \begin{pmatrix} v_f \\ s_f \end{pmatrix} = \begin{pmatrix} 0_N \\ 1 \end{pmatrix} \tag{5.71}$$

with $v_f \in \mathbb{R}^N$; $b_f, c_f \in \mathbb{R}^N$ and $d_f \in \mathbb{R}$ are fixed and chosen in such a way that the square matrix in (5.71) is nonsingular in the codimension-3 point.

Proof. We first consider the case RDN in some detail. There exist vectors v_1, v_2, v_3, v_4 and w_1, w_2, w_3, w_4, all in $\mathbb{R}^N$, such that $Av_1 = \lambda v_1$, $Av_2 = \lambda v_2 + v_1$, $Av_3 = -\lambda v_3$, $Av_4 = -\lambda v_4 + v_3$, $w_1^T A = \lambda w_1^T$, $w_2^T A = \lambda w_2^T + w_1^T$, $w_3^T A = -\lambda w_3^T$, $w_4^T A = -\lambda w_4 + w_3^T$. We have $w_i^T v_j = 0$ if $i \in \{1,2\}, j \in \{3,4\}$ or $i \in \{3,4\}, j \in \{1,2\}$ (different eigenvalues). Furthermore, we may assume that $w_i^T v_j = 0$ if $i = j$ and $w_i^T v_j = 1$ if $i \neq j$ and i,j correspond with the same eigenvalue.

By Proposition 4.4.24 $2A \odot I_N$ has the linearly independent right singular vectors $v_1 \wedge v_3$ and $v_1 \wedge v_4 - v_2 \wedge v_3$. Furthermore, $v_1 \wedge v_3$ is in the range of $2A \odot I_N$; in fact, $(2A \odot I_N)(\frac{1}{2}(v_1 \wedge v_4 + v_2 \wedge v_3)) = v_1 \wedge v_3$. Similar relations hold for $(2A \odot I_N)^T$ if we replace every v_i by a w_i.

Let us first determine the dimensions of the relevant manifolds. In the Jordan form of A a diagonal block of the form

$$\begin{pmatrix} \lambda & 1 & 0 & 0 \\ 0 & \lambda & 0 & 0 \\ 0 & 0 & -\lambda & 1 \\ 0 & 0 & 0 & -\lambda \end{pmatrix} \tag{5.72}$$

appears with real universal unfolding

$$\begin{pmatrix} \lambda & 1 & 0 & 0 \\ \delta_1 & \lambda+\delta_2 & 0 & 0 \\ 0 & 0 & -\lambda & 1 \\ 0 & 0 & \delta_3 & -\lambda+\delta_4 \end{pmatrix}. \tag{5.73}$$

Universality is understood with respect to the group of similarity transformations. By the same argument as in Proposition 5.3.3 one proves that the system $(\delta_i(A))_{1\le i\le 4}$ has full rank 4. Furthermore, it is not hard to check that for matrices with a diagonal block (5.73) the matrix $2A \odot I_N$ has rank defect 2 if and only if $\delta_3 = -\delta_1, \delta_4 = -\delta_2$. This proves the assertion concerning the manifold with codimension 2.

Next, a matrix of the form (5.73) for which $2A \odot I_N$ has rank defect 2 is either of type RT (if $\delta_2^2 + 4\delta_1 < 0$) or type DN (if $\delta_2^2 + 4\delta_1 > 0$) or type RDN (if $\delta_2^2 + 4\delta_1 = 0$).

Since the system of three conditions for RDN ($\delta_1 + \delta_3 = 0$, $\delta_2 + \delta_4 = 0$, $\delta_2^2 + 4\delta_1 = 0$) has a Jacobian with full rank 3 at the origin, the assertion concerning the manifold with codimension 3 follows.

We now turn to the regularity of the systems obtained by the bordered matrix methods. First note that the columns of Q span the right singular space of $2A \odot I_N$, i.e., the same space as $v_1 \wedge v_3$ and $v_1 \wedge v_4 - v_2 \wedge v_3$. Similarly, the columns of W span the same space as $w_1 \wedge w_3$ and $w_1 \wedge w_4 - w_2 \wedge w_3$.

By Lemma 5.3.1 and the relations between the vectors v, w we have for $1 \le i, j \le n$ that

$$(w_1 \wedge w_3)^T (2A_{i,j} \odot I_N)(v_1 \wedge v_3) = 0, \tag{5.74}$$

$$(w_1 \wedge w_3)^T (2A_{i,j} \odot I_N)(v_1 \wedge v_4 - v_2 \wedge v_3) = w_{1i}v_{1j} - w_{3i}v_{3j}, \tag{5.75}$$

$$(w_1 \wedge w_4 - w_2 \wedge w_3)^T (2A_{i,j} \odot I_N)(v_1 \wedge v_3) = w_{1i}v_{1j} - w_{3i}v_{3j}, \tag{5.76}$$

$$(w_1 \wedge w_4 - w_2 \wedge w_3)^T (2A_{i,j} \odot I_N)(v_1 \wedge v_4 - v_2 \wedge v_3) = -w_{1i}v_{2j} - w_{4i}v_{3j} - w_{2i}v_{1j} - w_{3i}v_{4j}. \tag{5.77}$$

Obviously these four gradient vectors span at most a two-dimensional space. Multiplying (5.75) with v_{2i} and summing over i we obtain $-(w_1^T v_2)v_{1j}$ for all j; this implies that the gradient vector in (5.75) is nonzero. On the other hand, multiplying it with v_{1i} and summing over i we find zero; multiplying (5.77) also with v_{1i} and summing over i we obtain $-w_2^T v_1 v_{1j}$ for all j; hence the two gradient vectors in (5.75) and (5.77) are linearly independent and span a two-dimensional space. So the equations corresponding to the second and fourth gradient vectors determine the two-dimensional manifold of matrices that we are considering.

Now let us look at the resonance condition $\det(S_1) = 0$. We know already that S_1 has rank 1 at the resonant point. Set

$$S_1 = \begin{pmatrix} s_{11} & s_{12} \\ s_{21} & s_{22} \end{pmatrix}.$$

We first consider the special case that $s_{12} = s_{21} = s_{22} = 0$. Then necessarily $s_{11} \neq 0$ and $\det_z(S_1) = s_{11}(s_{22})_z$. Also,

$$(s_{22})_z = -W_1^{2T}(2A_z \odot I_N)Q^2 - W^{2T}(2A_z \odot I_N)Q_1^2, \tag{5.78}$$

where the upper index 2 in each case indicates that we take the second column out of a matrix with 2 columns. By the assumptions on S_1 we must necessarily have that

$$Q_1^2 = \alpha_r v_1 \wedge v_3 + \beta_r(v_1 \wedge v_4) + \gamma_r(v_2 \wedge v_3)$$

with $\beta_r + \gamma_r \neq 0$ and

$$Q^2 = (2A \odot I_N)Q_1^2 = (\beta_r + \gamma_r)(v_1 \wedge v_3),$$

$$W_1^2 = \alpha_l w_1 \wedge w_3 + \beta_l(w_1 \wedge w_4) + \gamma_l(w_2 \wedge w_3)$$

with $\beta_l + \gamma_l \neq 0$ and

$$W^2 = (2A \odot I_N)^T W_1^2 = (\beta_l + \gamma_l)(w_1 \wedge w_3).$$

Comparing (5.78) with (5.74) through (5.77) we note that it is natural to rewrite Q_1^2 and W_1^2 as

$$Q_1^2 = \alpha_r v_1 \wedge v_3 + \beta_r(v_1 \wedge v_4 - v_2 \wedge v_3) + (\beta_r + \gamma_r)(v_2 \wedge v_3),$$

$$W_1^2 = \alpha_l w_1 \wedge w_3 + \beta_l(w_1 \wedge w_4 - w_2 \wedge w_3) + (\beta_l + \gamma_l)(w_2 \wedge w_3).$$

Inserting these expressions into (5.78) we obviously find a linear combination of the expressions in (5.75) and (5.76) plus an additional nonzero multiple of

$$(w_2 \wedge w_3)^T(2A_{i,j} \odot I_N)(v_1 \wedge v_3) + (w_1 \wedge w_3)^T(2A_{i,j} \odot I_N)(v_2 \wedge v_3) = 2w_{3i}v_{3j}. \quad (5.79)$$

By multiplying (5.75), (5.77), and (5.79) with v_{1i} and v_{2i} and summing over i it follows easily that these gradient vectors are linearly independent, so the result follows under the assumptions we made concerning S_1.

In the general case there exists by the Jordan decomposition theorem a 2×2 nonsingular matrix X such that $S_{1J} = X^{-1}S_1X$ has at resonance a nonzero element in the upper left entry and zeros everywhere else.

Now define $B_J = BX, C_J = CX^{-T}, D_J = X^{-1}DX$. Then the matrix

$$M_J = \begin{pmatrix} 2A \odot I_N & B_J \\ C_J^T & D_J \end{pmatrix}$$

is obviously nonsingular and by trivial computations we find that

$$(M_J)^2 \begin{pmatrix} Q_1X \\ X^{-1}S_1X \end{pmatrix} = \begin{pmatrix} 0 \\ I_{2,2} \end{pmatrix}$$

in a neighborhood of the resonant matrix.

In this neigbborhood we have that $S_{1J}(A) = XS_J(A)X^{-1}$ and hence $\det(S_1(A)) = \det(S_{1J}(A))$. So the result follows from the special case that we considered first.

This proves the case RDN. The case RDH is similar. Now consider the case BTN. There exist a $\lambda > 0$ and vectors v_1, v_2, v_3, v_4 and w_1, w_2, w_3, w_4 in $\mathbb{R}^n$ such that $Av_1 = 0$, $Av_2 = v_1$, $Av_3 = \lambda v_3$, $Av_4 = -\lambda v_4$, $w_1^T A = 0$, $w_2^T A = w_1^T$, $w_3^T A = \lambda w_3^T$, $w_4^T A = -\lambda w_4^T$.

We may assume that $w_i^T v_j = 0$ for $i, j = 1, 2, 3, 4$ except for the cases $w_1^T v_2 = w_2^T v_1 = w_3^T v_3 = w_4^T v_4 = 1$. The left singular vectors of $2A \odot I_N$ are the vectors $w_1 \wedge w_2$ and $w_3 \wedge w_4$. The right singular vectors are $v_1 \wedge v_2$ and $v_3 \wedge v_4$.

The universal unfolding of the Jordan form now has the form

$$\begin{pmatrix} 0 & 1 & 0 & 0 \\ \delta_1 & \delta_2 & 0 & 0 \\ 0 & 0 & \lambda + \delta_3 & 0 \\ 0 & 0 & 0 & -\lambda + \delta_4 \end{pmatrix}. \tag{5.80}$$

For a matrix with a diagonal block (5.80) the bialternate product matrix has rank defect 2 if and only if $\delta_2 = \delta_3 + \delta_4 = 0$; this proves the claim concerning the codimension-2 manifold. Furthermore, it is a BTN matrix if and only if $\delta_1 = \delta_2 = \delta_3 + \delta_4 = 0$. This proves the claim concerning the codimension-3 manifold.

Now we consider the regularity of the defining systems. By Lemma 5.3.1 and the relations between the vectors v, w we have for $1 \le i, j \le n$ that

$$(w_1 \wedge w_2)^T (2A_{i,j} \odot I_N)(v_1 \wedge v_2) = -w_{2i} v_{1j} - w_{1i} v_{2j}, \tag{5.81}$$

$$(w_3 \wedge w_4)^T (2A_{i,j} \odot I_N)(v_3 \wedge v_4) = w_{3i} v_{3j} + w_{4i} v_{4j}. \tag{5.82}$$

The two other candidates for a gradient vector are zero. Furthermore, the gradient that corresponds to the condition $s_f(A) = 0$ is given by

$$-w_1^T A_{ij} v_1 = -w_{1i} v_{1j}. \tag{5.83}$$

It is easy to prove that the three gradient vectors with components given by (5.81)–(5.83) are linearly independent (multiply successively with w_{1j}, w_{2j}, and w_{3j} and sum over j).

Finally, the case BTH is similar to BTN. □

Figure 5.4 presents the eight types of matrices described in Propositions 5.3.4 and 5.3.6 and their possible interactions. As a typical application, one might compute a curve of points in a three-parameter problem, expressing the requirement that $2A \odot I_N$ has rank defect 2. Then in the scheme of Figure 5.4 we expect to move from each type to one of the two adjacent types, the types at the corners generically being isolated points on the computed curve. The numbers between brackets indicate the sizes of the Jordan blocks of $2A \odot I_N$ for the zero eigenvalue. We note that all these cases can be found in real examples, e.g., in the LP neuron that we will describe in §5.4 and where we will compute some of the most interesting cases.

We note that there are other matrices A for which the bialternate product has rank defect 2. However, they have to lie on certain manifolds with codimension higher than 3 and the systems that we obtained may not be regular in such points. For example, consider the case where A has the eigenvalue 1 with algebraic multiplicity 2 and geometric multiplicity 1 and the eigenvalue -1 with algebraic multiplicity 3 and geometric multiplicity 1. Then by Proposition 4.4.21 $2A \odot I_N$ has rank defect 2 and its square has rank defect 4. In the setting of the preceding methods we have $S = S_1 = 0$. So the gradient vector of $\det(S_1)$ vanishes.

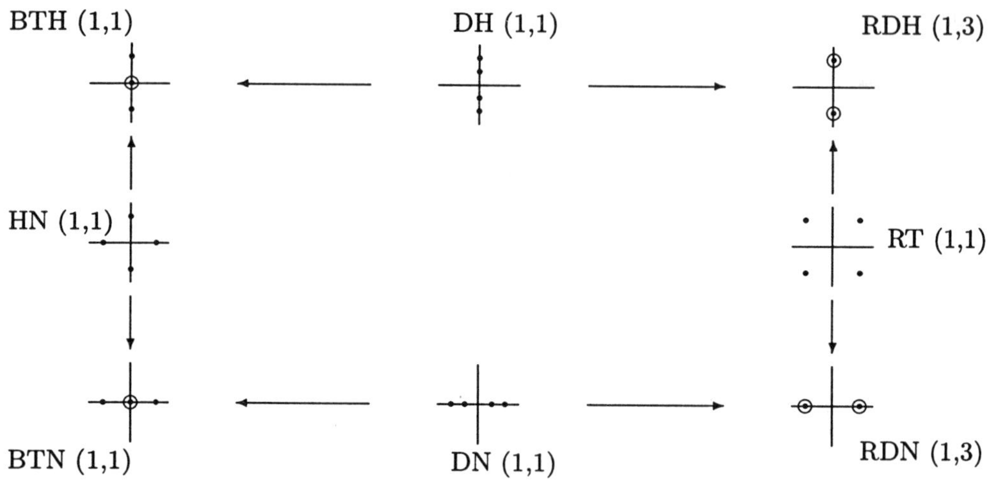

Figure 5.4: Eight types of matrices A where $2A \odot I_N$ has rank defect 2.

5.3.5 The Stratified Set of Hopf Points Near a Point with One-to-One Resonance

We saw in §4.4.5 and explained by an unfolding argument in §5.1.3 that curves of single Hopf points typically end when a BT point is encountered but have natural continuations as curves of equilibrium points with a zero-sum pair of real eigenvalues. Numerically one observes (see the example in §5.4) that a curve of double Hopf points similarly ends when a one-to-one resonant DH point is encountered. The natural continuation then is a curve of equilibrium points with two opposite conjugate pairs of complex eigenvalues. From a theoretical point of view this is to be expected from the analysis in §5.3.4; cf. Figure 5.4.

It is also observed numerically that curves of Hopf points are hard to compute near one-to-one resonant DH points. To be precise, the stepsize of the continuation code based on techniques described in §4.6 is forced down by sharp turning points. Furthermore, self-crossings are observed and pairs of eigenvalues swap their places while the loop is traversed.

To understand these phenomena we show that the Hopf points near a one-to-one resonant DH point in a general three-dimensional parameter space form a Whitney umbrella. Up to smooth equivalence, the Whitney umbrella in (x, y, z) space is the surface parameterized by $x = u, z = v^2, y = uv$ with (u, v) in a neighborhood of $0 \in R^2$. Equivalently, $y^2 = zx^2$. Smooth equivalence means that we allow a regular (but possibly nonlinear) transformation of the range space onto itself. For a picture of the Whitney umbrella, see Figure 5.5.

We note that this is not a manifold but rather a union of several manifolds of different dimensions (0, 1, and 2). Such objects are called stratified sets (we omit the precise definition, which is irrelevant for our purposes).

The appearance of Whitney umbrellas is more common than one might suspect. In a

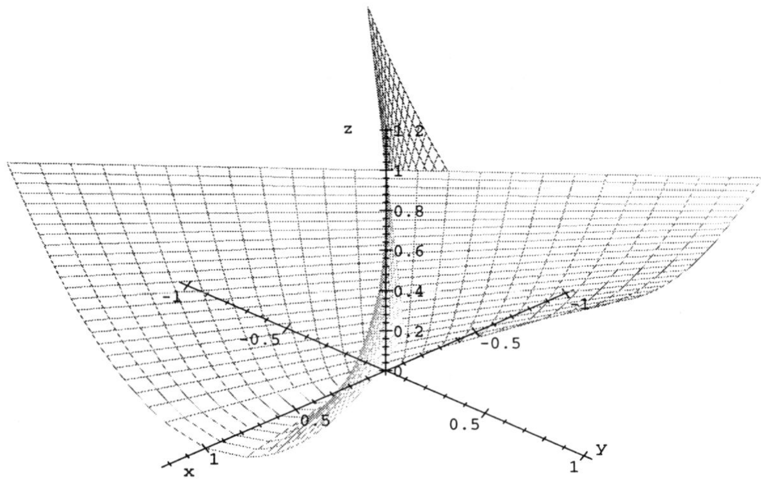

Figure 5.5: The Whitney umbrella.

sense, it is the most common organizing center for the singular behavior of mappings from $\mathbb{R}^2$ into $\mathbb{R}^3$. The study of such objects is far beyond the scope of this book, but we refer to Golubitsky and Guillemin [108], where this singularity is introduced as the simplest example of a cross cap. On the other hand, the reader can convince himself or herself by defining in a random way some complicated mappings from $\mathbb{R}^2$ into $\mathbb{R}^3$ and looking at the two-dimensional surfaces created in this way (presumably using mathematical software such as MATHEMATICA, MAPLE, or others). The reader then will probably remark that the self-intersections of such surfaces are either infinite or end in points where the surface looks like a Whitney umbrella.

To explain that in general the Hopf points form a Whitney umbrella near a one-to-one resonant DH point we will again use the notion of transversality.

Without loss of generality we assume that the imaginary part of the resonant Hopf value is $\pm i$. Therefore, in the case of geometric multiplicity 1 the complex Jordan normal form of G_x contains a Jordan block

$$J_2^c = \begin{pmatrix} i & 1 \\ 0 & i \end{pmatrix}. \tag{5.84}$$

If we denote by J_2^r and J_2^i the real and imaginary parts of J_2^c, respectively, then by (5.44) the real Jordan normal form of G_x contains the 4×4 block

$$\begin{pmatrix} J_2^r & J_2^i \\ -J_2^i & J_2^r \end{pmatrix}. \tag{5.85}$$

We note that there is in general a one-to-one linear relation between complex $k \times k$ matrices of the form $A + iB$, where A, B are real and, on the other hand, real $2k \times 2k$ matrices of the form

$$\begin{pmatrix} A & B \\ -B & A \end{pmatrix};$$

moreover, this one-to-one relation preserves matrix addition and multiplication, identity matrices, and similarity transformations.

Let $\mathbf{GL}(n, \mathbb{C})$ be the group of nonsingular complex $n \times n$ matrices. By Proposition 5.2.2 the universal unfolding of (5.85) with respect to real similarity transformations corresponds to the universal unfolding of (5.84) with respect to $\mathbf{GL}(2, \mathbb{C})$, i.e.,

$$\begin{pmatrix} i & 1 \\ \delta_1 + i\delta_2 & i + \delta_3 + i\delta_4 \end{pmatrix} \tag{5.86}$$

with four real parameters (i.e., the complex variant of Proposition 5.1.1).

By the identity

$$\begin{aligned} &\begin{pmatrix} 1 & 0 \\ \beta_1 + i\beta_2 & 1 \end{pmatrix} \begin{pmatrix} i + \beta_1 + i\beta_2 & 1 \\ \gamma_1 + i\gamma_2 & i + \beta_1 + i\beta_2 \end{pmatrix} \begin{pmatrix} 1 & 0 \\ -\beta_1 - i\beta_2 & 1 \end{pmatrix} \\ &\quad = \begin{pmatrix} i & 1 \\ \gamma_1 + i\gamma_2 - (\beta_1 + i\beta_2)^2 & i + 2(\beta_1 + i\beta_2) \end{pmatrix} \end{aligned} \tag{5.87}$$

and since the mapping

$$(\gamma_1, \gamma_2, \beta_1, \beta_2) \to (\gamma_1 - \beta_1^2 + \beta_2^2, \gamma_2 - 2\beta_1\beta_2, 2\beta_1, 2\beta_2)$$

has full linear rank at $(0, 0, 0, 0)$ we infer that the family

$$\begin{pmatrix} i + \beta_1 + i\beta_2 & 1 \\ \gamma_1 + i\gamma_2 & i + \beta_1 + i\beta_2 \end{pmatrix} \tag{5.88}$$

is a versal unfolding of (5.84) with respect to the action of $\mathbf{GL}(2, \mathbb{C})$.

As in §5.2 we consider the action of the product group $\mathbb{R}^+_{,\cdot} \times \mathbf{GL}(2, \mathbb{C})$ on $\mathcal{M}_2$ given by $A \to rGAG^{-1}$ $(r, G) \in \mathbb{R}^+_{,\cdot} \times \mathbf{GL}(2, \mathbb{C})$. The orbit of (5.88) with respect to this group contains

$$\begin{pmatrix} i + \beta_1/(1 + \beta_2) & 1/(1 + \beta_2) \\ (\gamma_1 + i\gamma_2)/(1 + \beta_2) & i + \beta_1/(1 + \beta_2) \end{pmatrix}. \tag{5.89}$$

By a similarity transformation with matrix

$$\begin{pmatrix} 1 & 0 \\ 0 & (1 + \beta_2)^{-1} \end{pmatrix}$$

one shows that it also contains a matrix with the form

$$\begin{pmatrix} i + \beta & 1 \\ \mu_1 + i\mu_2 & i + \beta \end{pmatrix}, \tag{5.90}$$

where $\beta = \beta_1/(1 + \beta_2)$, $\mu_1 = \gamma_1/(1 + \beta_2)^2$, $\mu_2 = \gamma_2/(1 + \beta_2)^2$.

The versality condition for the group $\mathbb{R}^+_{,\cdot} \times \mathbf{GL}(2, \mathbb{C})$

$$B(\mu) = r(\mu)C(\mu)A(\phi(\mu))C^{-1}(\mu)$$

with $r(0) = 1$, $C(0) = I_N$, $\phi(0) = 0$. So (5.90) is a versal unfolding of (5.84) with respect to the group action of $\mathbb{R}^+_{,\cdot} \times \mathbf{GL}(2, \mathbb{C})$. Clearly the unfolding parameters β, μ_1, μ_2

in (5.90) are uniquely determined if the two eigenvalues of (5.90) are known up to multiplication with the same positive real scalar. Hence (5.90) is really a universal unfolding of (5.84) with respect to the group action of $\mathbb{R}^+_{,\cdot}\times$ $\mathbf{GL}(2,\mathbb{C})$. The matrix (5.90) has a pure imaginary eigenvalue if and only if there exists a real number δ such that $(i(\delta-1)-\beta)^2-(\mu_1+i\mu_2)=0$. Eliminating δ we find

$$\beta^2(-4\mu_1+4\beta^2)=\mu_2^2. \tag{5.91}$$

If we set $x=\beta, y=\mu_2, z=-4\mu_1+4\beta^2$, then (5.91) becomes $y^2=zx^2$; i.e., the stratified set of Hopf points is a Whitney umbrella.

Now consider a three-dimensional family of matrices $A(\alpha)=A(\alpha_1,\alpha_2,\alpha_3)$ that contains a one-to-one resonant DH point. By taking restrictions to a four-dimensional generalized eigenspace we find that the Hopf structure of $A(\alpha)$ is locally the same as that of an unfolding $B(\alpha)$ of (5.84). By the universality of (5.90) there exist smooth functions $\beta(\alpha),\mu_i(\alpha)$, $i=1,2$ such that $B(\alpha)$ is (5.90) with β,μ_i replaced by $\beta(\alpha),\mu_i(\alpha)$. The set of functions $\beta(\alpha),\mu_i(\alpha)$, $i=1,2$ has a nonsingular Jacobian if and only if $B(\alpha)$ is a universal unfolding of (5.84).

We will show that the universality of the unfolding $B(\alpha)$ is equivalent to the transversality of the $\mathbb{R}^+_{,\cdot}\times$ $\mathbf{GL}(2,\mathbb{C})$ orbit of (5.84) with the manifold $B(\alpha)$. We first note that the β,μ_1,μ_2 are functions of eight real parameters, namely, the real and imaginary parts of the entries of a 2×2 complex matrix. If we denote these collectively by ζ, then the Jacobian matrix

$$J_\zeta=\frac{\partial(\beta,\mu_1,\mu_2)}{\partial\zeta} \tag{5.92}$$

has full rank 3 because it contains a 3×3 identity block as a submatrix. Now the $\mathbb{R}^+_{,\cdot}\times$ $\mathbf{GL}(2,\mathbb{C})$ orbit of (5.84) is defined by

$$\beta(\zeta)=\mu_1(\zeta)=\mu_2(\zeta)=0,$$

and so it is a manifold with codimension 3. Its tangent space is the five-dimensional null space of J_ζ. By the chain rule we have

$$J_\zeta\frac{\partial\zeta}{\partial\alpha}=\frac{\partial(\beta,\mu_1,\mu_2)}{\partial\alpha}. \tag{5.93}$$

From this it follows that $\frac{\partial(\beta,\mu_1,\mu_2)}{\partial\alpha}$ is nonsingular if and only if the tangent space to the manifold $B(\alpha)$ contains no tangent vector of the group orbit; i.e., the manifolds are transversal. In this case the set of Hopf points in α-space is a Whitney umbrella.

The DH points of (5.90) form the half-line $\beta=\mu_2=0,\mu_1\le 0$. Along the continuation $\mu_1>0$ of this half-line the eigenvalues of (5.90) are given by $\lambda=i\pm\sqrt{\mu_1}$. Remark that μ_1 represents the arclength along the curve of DH points with origin in the one-to-one resonant point.

Figure 5.5 presents a picture of the Whitney umbrella $y^2=zx^2$. A section through the origin is typically a cusp curve. To see this, substitute $z=rx+sy$ where r,s are parameters. The equation in the (x,y)-plane is then

$$F(x,y)\equiv y^2-rx^3-syx^2=0.$$

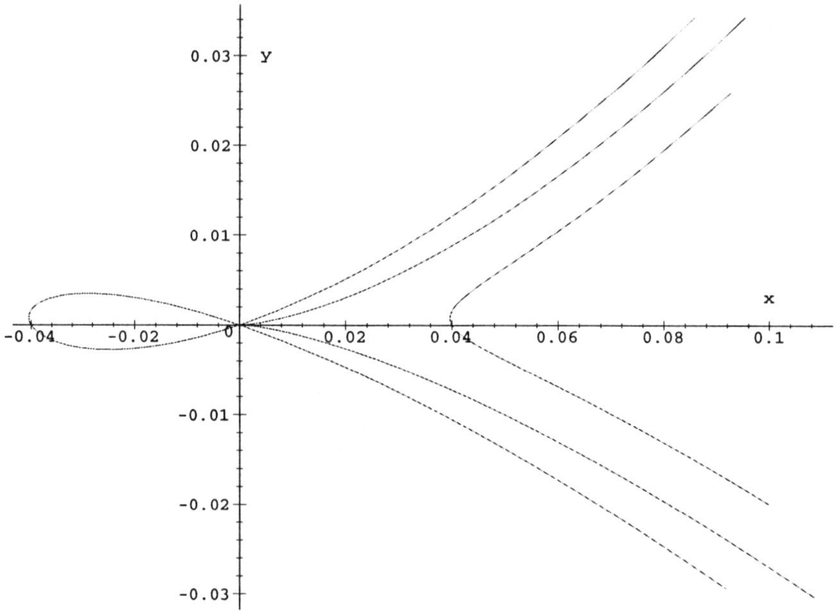

Figure 5.6: Self-crossing ($t = 0.04$), cusp ($t = 0$), and sharp turning ($t = -0.04$).

From the classification results in §6.3.3 it follows that this represents a cusp curve if $r \neq 0$. Nearby sections defined by $z = x + y + t$ with $t < 1$ can be parameterized by $\gamma = \delta - 1$, where δ is the parameter that was eliminated to obtain (5.91). One easily finds that $x = \frac{4\gamma^2 - t}{1-2\gamma}$, $y = -2\gamma x, z = 4\gamma^2$, For $t > 0$ the section has a self-crossing point at $(0,0,0)$, which reduces to a cusp point for $t = 0$. For $t < 0$ only a sharp turning point remains. See Figure 5.6.

Now consider the behavior of the eigenvalues on a loop with self-crossing; i.e., $t > 0$. The Hopf eigenvalue is then parameterized by γ as $\lambda_1(\gamma) = i(1+\gamma)$, the other eigenvalue is $\lambda_2(\gamma) = 2\frac{4\gamma^2-t}{1-2\gamma} + i(1-\gamma)$. So for $\gamma = \frac{\sqrt{t}}{2}$ we have $\lambda_1 = i(1 + \frac{\sqrt{t}}{2})$, $\lambda_2 = i(1 - \frac{\sqrt{t}}{2})$. For $\gamma = -\frac{\sqrt{t}}{2}$ we have $\lambda_1 = i(1 - \frac{\sqrt{t}}{2})$ and $\lambda_2 = i(1 + \frac{\sqrt{t}}{2})$. In other words, the two eigenvalues have swapped their places without ever actually coalescing.

5.4 Example: The Lateral Pyloric Neuron

Neural models have been studied intensively since the pioneering work of Hodgkin and Huxley [137]. They introduced the clamped Hodgkin–Huxley equations (see §4.6.2) as a model for the electrochemical activity of the giant axon of a squid.

In recent years, more work has been done to obtain models that describe real biological phenomena accurately. The next model was developed at the Center for Applied Mathematics of Cornell University by J. Guckenheimer and coworkers in close collaboration with the department of neurobiology. Apart from its obvious attraction as a model that is based on real experiments, it also presents a large number of interesting mathematical and numerical features.

The lateral pyloric (LP) neuron is one of a group of nerve cells in the stomatogastric ganglion of the crab, *Cancer borealis*. The pyloric sphincter is a circular muscle surrounding the base of the stomach that controls the flow of chyme into the intestines. Golowasch and Marder [105] proposed a 14-dimensional system of ordinary differential equations as a model for the ionic currents in this model. It was later amended by Buchholtz et al. [40]; see also [131]. These models are based on extensive experimental results. They incorporate the effects of many ionic currents with different time scales. Some of their parameters are hard to determine experimentally. One aim of the mathematical modeling is to compare the behavior of the model (studied by numerical methods) with the experimentally observed behavior and to adjust the parameters to make the dynamical behavior of the model match the data.

We study the 13-dimensional model given by

$$
\begin{aligned}
C_m \dot{v} \;=\; & \big(\underbrace{\bar{g}_{Na}\, m_\infty^3 h\, (E_{Na} - v)}_{I_{Na}} + \underbrace{(\bar{g}_{Ca1}\, a_{Ca1}\, b_{Ca1} + \bar{g}_{Ca2}\, a_{Ca2})\, (E_{Ca} - v)}_{I_{Ca}} \\
& + \underbrace{\bar{g}_K\, n^4 (E_K - v)}_{i_{K(V)}} + \underbrace{\bar{g}_{K(Ca)}\, a_{K(Ca)}\, b_{K(Ca)}\, (E_K - v)}_{I_{K(Ca)}} \\
& + \underbrace{(\bar{g}_{Af}\, b_{Af} + \bar{g}_{As}\, b_{As})\, a_A^3 (E_K - v)}_{I_A} + \underbrace{\bar{g}_h\, a_h\, (E_h - v)}_{I_h} \\
& + \underbrace{\bar{g}_l\, (E_l - v)}_{I_l}\big) + I_{ext},
\end{aligned}
$$

$$
I_{Na} : \begin{cases} m_\infty \;=\; \left(1 + 15\, e^{-\frac{v+34}{13}\, \frac{(1-e^{-0.05(v+6)})}{0.11v+0.66}}\right)^{-1}, \\ \dot{h} \;=\; k_h \left(0.08(1-h) e^{-0.125(v+39)} - \frac{h}{1+e^{-(0.2v+8)}}\right), \end{cases}
$$

$$
I_{Ca} : \begin{cases} \dot{Ca} \;=\; -c_{iCa}\, (\bar{g}_{Ca1}\, a_{Ca1}\, b_{Ca1} + \bar{g}_{Ca2}\, a_{Ca2})\, (v - E_{Ca}) + k_{Ca}\, (0.05 - Ca), \\ \dot{a}_{Ca1} \;=\; 50 \left(1 + e^{-\frac{v+11}{7}}\right)^{-1} - 50\, a_{Ca1}, \\ \dot{a}_{Ca2} \;=\; 10 \left(1 + e^{-\frac{v-22}{7}}\right)^{-1} - 10\, a_{Ca2}, \\ \dot{b}_{Ca1} \;=\; 16 \left(1 + e^{0.125(v+50)}\right)^{-1} - 16\, b_{Ca1}, \end{cases}
$$

$$
i_{K(V)} : \left\{ \dot{n} \;=\; \left(282 \left(1 + e^{-\frac{v+25}{17}}\right)^{-1} - 282\, n\right) \left(1 + e^{-\frac{v-10}{22}}\right)^{-1}, \right.
$$

$$
I_{K(Ca)} : \begin{cases} \dot{a}_{K(Ca)} \\ = k_{K(Ca)} \left(\frac{Ca}{(1+e^{-(0.0434783v+0.026087Ca+0.869565)})(1+e^{-(0.166667v+0.1Ca+3.83333)})(2.5+Ca)}\right) \\ \qquad -k_{K(Ca)} a_{K(Ca)}, \\ \dot{b}_{K(Ca)} = \frac{21}{0.6+Ca} - 35\, b_{K(Ca)}, \end{cases}
$$

Table 5.1: State variables and parameters in the model of the LP neuron and their starting values.

	State	Value	Param	Value	Param	Value
1	v	−38.6522	C_m	0.0017	E_{Na}	50.0
2	h	0.119015	k_h	500.0	E_{Ca}	140.0
3	Ca	0.166282	k_{Ca}	360.0	E_K	−86.0
4	a_{Ca1}	0.0188852	$k_{K(Ca)}$	45.0	E_h	−10.0
5	a_{Ca2}	0.00017256	k_{Af}	30.0	E_l	−50.0
6	b_{Ca1}	0.1949	k_{As}	10.0	I_{ext}	0.551636
7	n	0.309369	k_r	0.1	v_r	−110.0
8	$a_{K(Ca)}$	0.00134052	$\bar{g}_{Na}$	2300.0	s_r	12.0
9	$b_{K(Ca)}$	0.783001	$\bar{g}_{Ca1}$	0.21	c_{iCa}	300.0
10	a_A	0.56752	$\bar{g}_{Ca2}$	0.047	v_A	−43.0
11	b_{Af}	0.0200101	$\bar{g}_K$	0.841	v_{kr}	−100.0
12	b_{As}	0.0200101	$\bar{g}_{K(Ca)}$	5.0	s_{kr}	−13.0
13	a_h	0.00261036	$\bar{g}_{Af}$	1.0	g_l	0.1
14			$\bar{g}_{As}$	1.3	v_b	−62
15			$\bar{g}_h$	0.1		

$$I_A : \left\{ \begin{array}{rcl} \dot{a}_A & = & 140\left(\left(1+e^{-\frac{v-v_A}{16}}\right)^{-1} - a_A\right), \\ \dot{b}_{Af} & = & k_{Af}\left(\left(1+e^{\frac{v-v_b}{6}}\right)^{-1} - b_{Af}\right), \\ \dot{b}_{As} & = & k_{As}\left(\left(1+e^{\frac{v-v_b)}{6}}\right)^{-1} - b_{As}\right), \end{array} \right.$$

$$I_h : \left\{ \dot{a}_h \;=\; k_r(1+\exp((v-(v_{kr}))/s_{kr}))\left(\frac{1}{(1+\exp((v-v_r)/s_r))} - a_h\right). \right.$$

Here v is the membrane potential of the neuron and each underbraced term models a particular ionic current through the membrane. For example, the sodium channel is modeled by the I_{Na} term; I_{ext} is an externally applied current and therefore one of the most important parameters. This model contains 29 parameters. Starting values for the state variables and parameters are given in Table 5.1. They correspond to an equilibrium solution where all eigenvalues of the Jacobian matrix have a strictly negative real part.

In this example the dimension of the bialternate product matrix is 78×78 ($= \frac{N(N-1)}{2}$ with $N = 13$). In our computations (using CONTENT) the parameters I_{ext}, $\bar{g}_{Af}$ were first varied; $\bar{g}_{K(Ca)}$ was used as a third parameter when needed. This choice was based on biological considerations of which parameters could be manipulated experimentally.

Starting from the equilibrium point with state and parameter values in Table 5.1 (essentially obtained by biological measurements and experiments) we varied I_{ext} and followed a branch of equilibrium solutions in the direction of increasing values of I_{ext}.

A Hopf point with coordinates given in Table 5.2 was found on this branch. A doubly bordered bialternate product method (see §5.3) was then used to follow a curve of Hopf points by varying I_{ext} and $\bar{g}_{Af}$. In the direction of increasing values of $\bar{g}_{Af}$ on this curve a DH point was detected by a method based on the use of the matrix $S(x, \alpha)$ as defined in (5.54); the parameter values are $I_{ext} = 1.01154$, $\bar{g}_{Af} = 3.65432$ with

Table 5.2: State variables and I_{ext} in the first Hopf point.

	State	Value	Param	Value
1	v	-38.6205	I_{ext}	0.552523
2	h	0.118315		
3	Ca	0.1664109		
4	a_{Ca1}	0.0189691		
5	a_{Ca2}	0.000173341		
6	b_{Ca1}	0.1942803		
7	n	0.3097672		
8	$a_{K(Ca)}$	0.001349397		
9	$b_{K(Ca)}$	0.782869		
10	a_A	0.568005		
11	b_{Af}	0.019906		
12	b_{As}	0.019906		
13	a_h	0.0026035		

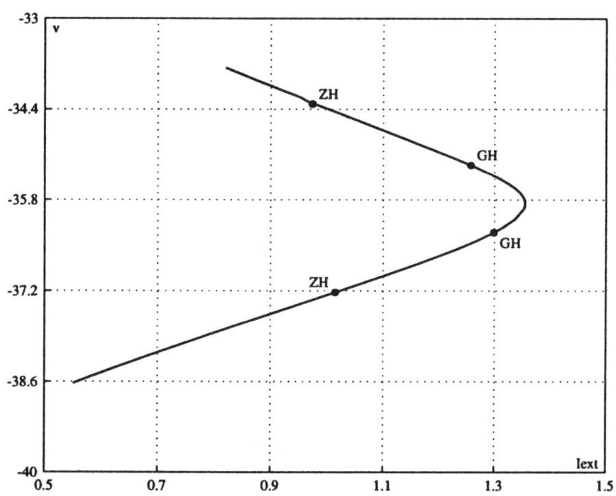

Figure 5.7: The LP neuron: ZH and GH points.

state variables -37.23322, 0.09176, 0.171876, 0.023031, 0.0002113, 0.168562, 0.32748, 0.00179497, 0.77733, 0.5891430, 0.015862, 0.0158627, 0.0023199 in the order of Table 5.1. In this DH point all other eigenvalues have a strictly negative real part. In other words, the point is on the border of the stability region, which makes it particularly interesting from the biological point of view.

Continuing the Hopf curve, we found two ZH points and two GH points; see Figure 5.7. In these points at least one other eigenvalue has a strictly positive real part.

We then varied the parameters I_{ext}, $\bar{g}_{Af}$, and $\bar{g}_{K(Ca)}$ to follow a curve of DH points, using a defining system based on the ideas developed in §5.3.2. On this curve, in the direction of initially increasing values of I_{ext} a BT–Hopf point (BTH in Figure 5.4) was found in the physically meaningful region; the parameter values are 1.821773, 7.77938, 7.37653 and state variable values are -36.23533, 0.076889, 0.175587, 0.026466,

Table 5.3: State variables and I_{ext}, $\bar{g}_{Af}$, $\bar{g}_{K(Ca)}$ at a one-to-one resonant point.

	State	Value	Param	Value
1	v	−62.273	I_{ext}	−2.442
2	h	0.9922	$\bar{g}_{Af}$	−9.2992
3	Ca	0.06922	$\bar{g}_{K(Ca)}$	878.062
4	a_{Ca1}	0.000658		
5	a_{Ca2}	5.9090D-6		
6	b_{Ca1}	0.8226		
7	n	0.10042		
8	$a_{K(Ca)}$	5.351D-6		
9	$b_{K(Ca)}$	0.89656		
10	a_A	0.23066		
11	b_{Af}	0.51137		
12	b_{As}	0.51137		
13	a_h	0.01839		

0.00024369, 0.151796, 0.340537, 0.002193, 0.773607, 0.604151, 0.0134649, 0.0134649, 0.0021352 in the now familiar order. Continuing the curve through the BTH points, one finds Hopf neutral saddle points (HN in Figure 5.4).

In the other direction (initially decreasing values of I_{ext}) a resonant double Hopf point was found in the physically uninteresting region (RDH in Figure 5.4). Continuing the curve, one finds the RT points of Figure 5.4. Since the RDH point is mathematically interesting, we present the coordinates in Table 5.3.

5.5 Notes and Further Reading

A solution curve to (5.1), which is defined for $t \in]-\infty, +\infty[$ and such that the limits

$$x_\infty = \lim_{t\to\infty}, x_{-\infty} = \lim_{t\to-\infty}$$

both exist, is called a *homoclinic orbit* if $x_\infty = x_{-\infty}$ and a *heteroclinic orbit* if $x_\infty \neq x_{-\infty}$. Homoclinic orbits are generic in one-parameter families of dynamical systems, and curves of such orbits arise generically from codimension-2 equilibrium points as discussed in §§5.1, 5.2, and 5.3. They also arise as limit cases of periodic orbits and are associated with the onset of chaotic behavior [31]. Computation of homoclinic and heteroclinic orbits has attracted a lot of attention; we refer to [29], [30], [47], [48], [81], [94], [95].

5.6 Exercises

1. Consider the case $N = 2$ in §5.1. Let

$$A = \begin{pmatrix} a_{11} & a_{12} \\ a_{21} & a_{22} \end{pmatrix}$$

and let $b = (0,1)^T$, $c = (1,0)^T$, $d = 0$. Let M be defined as in (5.5). Let q_0, q_1 be as in (5.15) and (5.17) whenever M is nonsingular. Compute explicitly q_0 and q_1 as functions of $a_{11}, a_{12}, a_{21}, a_{22}$. Prove from this that $q_0 = 0$ if and only if A is singular. Prove also that $q_0 = q_1 = 0$ if and only if A has two zero eigenvalues. Prove that the Jacobian

$$\frac{\partial(q_0, q_1)}{\partial(a_{11}, a_{12}, a_{21}, a_{22})}$$

has full rank in every point where M is nonsingular.

2. Consider the dynamical system

$$\begin{aligned} \dot{x} &= y, \\ \dot{y} &= \mu_1 + \mu_2 y + x^2 + xy, \end{aligned} \tag{5.94}$$

which is an unfolding of the same system with $\mu_1 = \mu_2 = 0$, namely,

$$\begin{aligned} \dot{x} &= y, \\ \dot{y} &= x^2 + xy. \end{aligned} \tag{5.95}$$

The latter one has a BT point in $(0,0)$.

(a) Give an algebraic condition in μ_1, μ_2 that in a neighborhood of $(\mu_1, \mu_2) = (0,0)$ is necessary and sufficient for the system in (5.94) to have an equilibrium solution in which the Jacobian of the system has two equal eigenvalues.

(b) Discuss the stability or instability of the obtained equilibrium solutions in a neighborhood of $(\mu_1, \mu_2) = (0,0)$, the point $(\mu_1, \mu_2) = (0,0)$ itself excluded.

3. Define in a random way some complicated mappings from $\mathbb{R}^2$ into $\mathbb{R}^3$ and look at the two-dimensional surfaces created in this way using mathematical software such as MATHEMATICA, MAPLE, or others. Look for lines of self-intersection and points where they end. Find a few Whitney umbrellas.

4. Prove that if a matrix $A_0 \in \mathcal{M}_n$ has at least one universal unfolding with respect to a group of transformations, then a versal unfolding is universal if and only if it is miniversal.

5. Consider the unfolding

$$\begin{pmatrix} i+\alpha & 1+\beta \\ \gamma & i+\delta \end{pmatrix}$$

of the complex matrix

$$\begin{pmatrix} i & 1 \\ 0 & i \end{pmatrix}$$

by four real parameters $\alpha, \beta, \gamma, \delta$. Prove that it is not a versal unfolding with respect to the group of complex similarity transformations.

6. Let (x^0, α^0) be a regular solution to (5.2) with $m = 2$ and suppose that that this is a nonresonant DH point. Prove that the system consisting of (5.2) and a regular set of two defining equations for DH as obtained in §5.3.2 together form a nonsingular system of $N + 2$ equations in the $N + 2$ unknowns x, α if and only if the real parts of the two Hopf eigenvalues together form a regular parameterization of the equilibrium surface of (5.1) near (x^0, α^0).

7. Prove the remaining cases in Proposition 5.3.4.

8. Prove the remaining cases in Proposition 5.3.6.

9. Let A be a matrix for which 0 is an eigenvalue with geometric multiplicity 1 and algebraic multiplicity 4 and suppose that there are no other zero-sum pairs of eigenvalues. Prove that A is a regular point on the manifold of matrices determined by the equations $S = 0, \det(S_1) = 0$, where S, S_1 are defined as in §5.3.4.

10. Prove that the previous result is no longer true if the algebraic multiplicity of the zero eigenvalue is 5 instead of 4.

11. Prove that the matrices with three different algebraically simple Hopf pairs and no other zero-sum pairs of eigenvalues form a manifold with codimension 3. Give a regular defining system in a neighborhood of a given matrix of this form.

12. Describe all Jordan types of A for which $2A \odot I_n$ has rank defect 2. For which (other than those in Proposition 5.3.4 or 5.3.6) does the defining system (5.64) have rank 2?

13. Describe all Jordan types of A for which $2A \odot I_n$ has rank defect 3. Can you (locally) find regular defining systems for some of these types?

14. Consider the real form

$$A = \begin{pmatrix} a & 1 & 1 & 0 \\ b & a & c & 1 \\ -1 & 0 & a & 1 \\ -c & -1 & b & a \end{pmatrix}$$

of the three-parameter universal unfolding of the resonant DH matrix at $\pm i$. Consider the linear dynamical system $\dot{x} = Ax$, which has an equilibrium at $x = 0$ for all values of a, b, c. Then complete the following steps:

(a) Use a bordering method to give a regular algebraic equation in a, b, c that describes the manifold of Hopf points near $(0, 0, 0)$.

(b) Do the same for DH points with a regular system of two equations.

(c) Use some software to start from an equilibrium point with nonzero parameters, compute a curve of equilibria, detect a Hopf point, compute a curve of Hopf points, detect a DH point, compute a curve of DH points and detect the resonant DH point at $(0, 0, 0)$.

(d) Repeat the same exercise using the matrix

$$A_1 = \begin{pmatrix} 0 & 1 & 1 & 0 \\ b & a & c & 1 \\ -1 & 0 & 0 & 1 \\ -c & -1 & b & a \end{pmatrix},$$

which also unfolds the resonant DH point at $(0,0,0)$. Do you note a different behavior? If so, explain it.

15. Consider the example of the LP neuron in §5.4. Compute Hopf curves in the neighborhood of the one-to-one resonant DH point. Do you have convergence problems? Do you find the self-crossings and sharp turning points predicted in §5.3.5? (See [119]) for a study.)

16. Repeat the previous exercise with the system $\dot{x} = Ax$ from Exercise 14 instead of the LP-neuron.

17. Compute a curve of BT points in the LP neural model.

18. Consider the "new Lorenz" model

$$\begin{cases} \dot{x} &= -y^2 - z^2 - ax + aF, \\ \dot{y} &= xy - bxz - y + G, \\ \dot{z} &= bxy + xz - z. \end{cases}$$

This is a simplified model for atmospheric circulation introduced in [175] and further discussed in [218]. There is a trivial equilibrium at $(0,0,0)$ for the parameter values $a = 0.25, b = 4, F = 0, G = 0$. Compute an equilibrium curve starting at this point with F as a free parameter. Find a Hopf point on this curve. Compute the Hopf curve with F, G free. Find a ZH point and compute a ZH curve with b, F, G free. Find a BT point of order three (triple-point bifurcation) on this curve.

Chapter 6

Singularity Theory

6.1 Contact Equivalence of Nonlinear Mappings

Let $N_1, N_2 \in \mathbb{N}, N_1 \geq N_2$, and U be a nonempty open subset of $\mathbb{R}^{N_1}$. Consider a mapping

$$G : U \to \mathbb{R}^{N_2}. \tag{6.1}$$

For simplicity we will assume that G is a C^∞-(= smooth) mapping. Let $x^\circ \in U$. If $G_x(x^\circ)$ has full rank N_2 then the equation $G(x) = G(x^\circ)$ either has x° as a unique regular solution (this is the case if $N_2 = N_1$) or locally defines a C^∞-manifold with dimension $N_1 - N_2$ through x°. (See §2.1.)

If $G_x(x^\circ)$ does not have full rank, then many questions arise naturally. What does the solution set of $G(x) = G(x^\circ)$ locally look like? If our system contains parameters and we perturb the equation slightly, do we expect to find the same type of solution sets for nearby parameter values? Can we describe qualitatively the solutions for all possible perturbations? These questions belong to the field of singularity theory.

Singularity theory is usually considered to be hard analysis, and it is certainly based on difficult theorems whose meaning and implications are not immediately clear to the nonspecialist. Unfortunately, it is not possible to apply the results without understanding some of the basic elements of the theory. To add to the confusion, there are several approaches to singularity theory. Our aim is to write a systematic description of some results that we consider relevant for numerical work. Mathematical statements will be precise but difficult analytical proofs will be omitted. On the other hand, the numerical implementation will be discussed clearly and we will give some examples.

In this chapter we closely follow the approach in [104], which we consider the basic one. In Chapter 7 we will consider another approach. For a deeper understanding of singularity theory we recommend [64], [86], and [16]. We start with some further motivation.

Suppose that X is a smooth function defined on a neighborhood of $y^\circ \in \mathbb{R}^{N_1}$ with values in $\mathbb{R}^{N_1}$ and such that $X(y^\circ) = x^\circ$ and $X_y(y^\circ)$ is nonsingular. Then if we replace $G(x)$ by $G(X(y))$, the local behavior of the equation $G(x) = 0$ near x° and that of

$G(X(y)) = 0$ near y° are obviously the same in an intuitive sense. Also, if $T(x)$ is a smooth function defined on a neighborhood of $x^\circ \in \mathbb{R}^{N_1}$ and with values in $\mathbb{R}^{N_2 \times N_2}$ such that $T(0)$ is nonsingular, then again $G(x) = 0$ and $T(x)G(x) = 0$ have the same local behavior near x°.

These considerations lead us to the notion of contact equivalence. We first introduce some definitions and notation, mostly following [104].

Definition 6.1.1. A germ of smooth functions (more concisely, a smooth germ or just a germ) from $\mathbb{R}^n$ into $\mathbb{R}^p$ and centered at $x^\circ \in \mathbb{R}^n$ is an equivalence class of smooth functions that are defined in neighborhoods of x° and where two functions are identified if they coincide in a neighborhood of x° (which may be smaller than the intersection of their domains of definition).

According to the above definition, all functions in a germ centered at x° have the same derivatives of all orders in x°. Therefore, it makes sense to talk about the derivatives of the germ itself. We remark that the germ is not completely defined by these derivatives, since there exist smooth functions that do not vanish in any neighborhood of x° but still have zero derivatives of all orders in x°.

Definition 6.1.2. For $n, p \in \mathbb{N}$ we denote by $\mathcal{E}_{n,p}$ and $\mathcal{E}^0_{n,p}$ the set of all smooth germs $f : \mathbf{R}^n \to \mathbf{R}^p$ centered at 0 and its subset for which $f(0) = 0$, respectively; for $p = 1$ we define $\mathcal{E}_n = \mathcal{E}_{n,1}$, $\mathcal{E}^0_n = \mathcal{E}^0_{n,1}$.

Definition 6.1.3. For $n, p \in \mathbb{N}$ let $\mathcal{E}_{n,p,p}$ be the set of all germs centered at 0 from $\mathbf{R}^n$ into the space of all (p,p) matrices. Two germs $f, g \in \mathcal{E}^0_{n,p}$ are said to be $\mathcal{K}$-equivalent (or contact equivalent) if and only if there exist $Z \in \mathcal{E}_{n,p,p}$ and $X \in \mathcal{E}^0_{n,n}$ with $X_x(0)$ and $Z(0)$ nonsingular such that

$$g(x) = Z(x)f(X(x)). \tag{6.2}$$

It is important to establish the link of (6.2) with group theory. Let $\mathcal{G}_1$ be the set of all $Z \in \mathcal{E}_{n,p,p}$ for which $Z(0)$ is nonsingular; it forms a group under matrix multiplication. Also, the set $\{X \in \mathcal{E}^0_{n,n} : X_x(0) \text{ is nonsingular}\}$ is a group under composition. We can define an action of the product group $\mathcal{G}_1 \times \mathcal{G}_2$ on $\mathcal{E}^0_{n,p}$ by $((Z \times X), f)(x) = Z(x)f(X(x))$. In this way two germs $f, g \in \mathcal{E}^0_{n,p}$ are contact equivalent if and only if they belong to the same group orbit of the group $\mathcal{G}_1 \times \mathcal{G}_2$.

In a sense, the most important fact about problems like (6.1) is that they can be reduced to the case where N_1, N_2 are low-dimensional spaces and the Jacobian G_x vanishes. This procedure is called the Lyapunov–Schmidt reduction. In §6.2 we discuss a form of this reduction that is particularly suitable for numerical implementation. The further study of the low-dimensional reduced problems with a vanishing Jacobian is the core of singularity theory; we return to this in §6.3.

6.2 The Numerical Lyapunov–Schmidt Reduction

Consider a smooth function

$$G : U \to \mathbb{R}^{N_2}, \tag{6.3}$$

where U is open in $\mathbb{R}^{N_1}$, $N_1 \geq N_2$, $x^\circ \in U$. Assume that the Jacobian $G_x(x^\circ)$ has rank at least equal to $r < N_2$. By Proposition 3.2.1. there exist smooth matrix functions

$$B : U \to \mathbb{R}^{N_2 \times (N_2 - r)},$$

$$C : U \to \mathbb{R}^{N_1 \times (N_1 - r)},$$

$$D : U \to \mathbb{R}^{(N_1 - r) \times (N_2 - r)}$$

(in fact, B, C, D may be chosen constant) such that

$$M(x) \equiv \begin{pmatrix} G_x(x) & B(x) \\ C^T(x) & D(x) \end{pmatrix} \tag{6.4}$$

is nonsingular in a neighborhood of $x^\circ \in \mathbb{R}^{N_1}$.

For x in a neighborhood of x° and y in a neighborhood of $0 \in \mathbb{R}^{N_1 - r}$ we define $v(x; y) \in \mathbb{R}^{N_1}$ and $g(x; y) \in \mathbb{R}^{N_2 - r}$ by

$$G(x + v(x; y)) - B(x)g(x; y) = G(x), \tag{6.5}$$

$$C^T(x)v(x; y) - D(x)g(x; y) = y. \tag{6.6}$$

We remark that (6.5), (6.6) form a system of $(N_1 + N_2 - r)$ equations in the $(N_1 + N_2 - r)$ unknowns contained in v, g. This system has the zero solution for $y = 0$. By the implicit function theorem and the assumption that $M(x^\circ)$ is nonsingular it has a unique solution for x in a neighborhood of x° and y in a neighborhood of $0 \in \mathbb{R}^{N_1 - r}$. Also, the functions $g(x; y)$ and $v(x; y)$ inherit the differentiability properties of $G(x)$.

We remark that $g(x; y) = 0$ implies $G(x + v(x; y)) = G(x)$. Conversely, if $G(x + z) = G(x)$, then

$$G(x + z) - B(x)0 = G(x),$$

$$C^T(x)z - D(x)0 = C^T(x)z.$$

Hence if $\|z\|$ is sufficiently small, then $v(x; C^T(x)z) = z$ and $g(x; C^T(x)z) = 0$.

Hence there is a one-to-one connection between the solution set of $G(x) = G(x^\circ)$ in $\mathbb{R}^{N_1}$ and the solution set of

$$g(x^\circ; y) = 0 \tag{6.7}$$

in $R^{N_1 - r}$ in appropriate neighborhoods of x° and 0, respectively.

The above constructed connection between solution sets to nonlinear equations in spaces with different dimensions is (a variant of) the Lyapunov–Schmidt reduction. The choice of the matrices $B(x), C(x), D(x)$ of course influences the functions $v(x; y), g(x; y)$. We now proceed to show that the reduced functions $g(x; y)$ for constant x are contact equivalent in the sense of §6.1. Therefore, the choice of B, C, D does not influence the classification in singularity classes that we will discuss in §6.3. Our discussion is based on [148], but we will need some preparation.

Lemma 6.2.1. Let $k(x_1, \ldots, x_l)$ be a smooth function defined in a neighborhood of $(0, \ldots, 0)$. Suppose that there exists an $i \in \{1, \ldots, l\}$ such that $k(x_1, \ldots, x_l) = 0$ if

$x_i = 0$. Then there exists a smooth function $T(x_1, \dots, x_l)$ such that $k(x_1, \dots, x_l) = x_i T(x_1, \dots, x_l)$.

Proof. For ease of notation we assume that $l = 2, i = 1$; the generalization is trivial. By a fundamental analytical result we have

$$k(x_1, x_2) = \int_0^{x_1} k_{x_1}(s, x_2) ds.$$

Substituting $s = s_1 x_1$ we obtain

$$k(x_1, x_2) = x_1 \int_0^1 k_{x_1}(s_1 x_1, x_2) ds_1.$$

The integral in this expression is a smooth function of (x_1, x_2) by classical results on parameterized integrals. □

Lemma 6.2.2. Let $k(x_1, \dots, x_l)$ be a smooth function defined in a neighborhood of $(0, \dots, 0)$. Suppose that there exist $i_1, \dots, i_j \in \{1, \dots, l\}$ such that $k(x_1, \dots, x_l) = 0$ if $x_{i_1} = \dots = x_{i_j} = 0$. Then there exists smooth functions $T_{i_1}(x_1, \dots, x_l), \dots, T_{i_j}(x_1, \dots, x_l)$ such that $k(x_1, \dots, x_l) = x_{i_1} T_{i_1}(x_1, \dots, x_l) + \dots + x_{i_j} T_{i_j}(x_1, \dots, x_l)$.

Proof. For ease of notation we assume that $l = 4$, $j = 3$, $i_1 = 1$, $i_2 = 2$, $i_3 = 3$; the generalization is trivial.

Now we may write

$$k(x_1, x_2, x_3, x_4) = k(x_1, 0, 0, x_4) + (k(x_1, x_2, 0, x_4) - k(x_1, 0, 0, x_4))$$

$$+(k(x_1, x_2, x_3, x_4) - k(x_1, x_2, 0, x_4)).$$

Applying Lemma 6.2.1 three times the result follows. □

Lemma 6.2.3. Let j, l be natural numbers, $j \leq l$. Let $h(x_1, \dots, x_l), k(x_1, \dots, x_l)$ be smooth functions with values in $\mathbb{R}^j$, defined in a neighborhood of $(0, \dots, 0)$ and vanishing in $(0, \dots, 0)$. Suppose that z collects j variables from $(x_1, \dots, x_l)$ and that $\frac{\partial h}{\partial z}$ is nonsingular. If $h(x_1, \dots, x_l) = 0$ implies that $k(x_1, \dots, x_l) = 0$, then there exists a smooth function $T(x_1, \dots, x_l) \in \mathbb{R}^{j \times j}$ such that $k(x_1, \dots, x_l) = T(x_1, \dots, x_l) h(x_1, \dots, x_l)$ in a neighborhood of $(0, \dots, 0)$.

Proof. For ease of notation we decompose $(x_1, \dots, x_l) = (z, z_1)$, where z_1 collects the variables not in z. By the nonsingularity assumption and the implicit function theorem there exists a smooth function $\zeta(y, z_1)$ ($y \in \mathbb{R}^j$, $\zeta(y, z_1) \in \mathbb{R}^j$) such that $z = \zeta(y, z_1)$ is the locally unique solution of the equation $h(z, z_1) = y$. Hence $\zeta(h(z, z_1), z_1) = z$ and $h(\zeta(y, z_1), z_1) = y$ in a neighborhood of $(0, 0)$. Now define $k_1(y, z_1) = k(\zeta(y, z_1), z_1)$. Then $k_1(0, z_1) = 0$ for all z_1 in a neighborhood of 0. By Lemma 6.2.2 there exists a smooth functon $T_1(y, z_1) \in \mathbb{R}^j$ such that $k_1(y, z_1) = T_1(y, z_1) y$. If we now replace y by $h(z, z_1)$ then we obtain $k(z, z_1) = T_1(h(z, z_1)) h(z, z_1)$. □

We now formulate and prove the main result in this section.

Proposition 6.2.4. Let

$$G : U \to \mathbb{R}^{N_2} \tag{6.8}$$

be a smooth function where U is open in $\mathbb{R}^{N_1}$, $N_1 \geq N_2$, $x^\circ \in U$. Suppose that $B_1, B_2 \in C^\infty(U, \mathbb{R}^{N_2 \times (N_2-r)})$, $C_1, C_2 \in C^\infty(U, \mathbb{R}^{N_1 \times (N_1-r)})$, $D_1, D_2 \in C^\infty(U, \mathbb{R}^{(N_1-r)\times(N_2-r)})$ are such that the matrices

$$M_1^\circ \equiv \begin{pmatrix} G_x(x^\circ) & B_1(x^\circ) \\ C_1^T(x^\circ) & D_1(x^\circ) \end{pmatrix}, \tag{6.9}$$

$$M_2^\circ \equiv \begin{pmatrix} G_x(x^\circ) & B_2(x^\circ) \\ C_2^T(x^\circ) & D_2(x^\circ) \end{pmatrix} \tag{6.10}$$

are nonsingular. Let the functions $v_1(y) \in \mathbb{R}^{N_1}$ and $g_1(y) \in \mathbb{R}^{N_2-r}$ be defined for y in a neighborhood of $0 \in \mathbb{R}^{N_1-r}$ by

$$G(x^\circ + v_1(y)) - B_1 g_1(y) = G(x^\circ), \tag{6.11}$$

$$C_1^T v_1(y) - D_1 g_1(y) = y, \tag{6.12}$$

while $v_2(y) \in \mathbb{R}^{N_1}$ and $g_2(y) \in \mathbb{R}^{N_2-r}$ are defined by

$$G(x^\circ + v_2(y)) - B_2 g_2(y) = G(x^\circ), \tag{6.13}$$

$$C_2^T v_2(y) - D_2 g_2(y) = y \tag{6.14}$$

(we suppress the dependence on x in the notation because $x = x^\circ$ throughout the argument). Then $g_1(y), g_2(y)$ are contact equivalent.

Moreover, if $B_1 = B_2$ then we have

$$g_2(y) = g_1(X(y)), v_2(y) = v_1(X(y)) \tag{6.15}$$

for a germ $X(y)$ with values in $\mathbb{R}^{N_1-r}$ and nonsingular Jacobian at $y = 0$.

On the other hand, if $C_1 = C_2$ then then there exists a smooth germ $Z(y)$ with values in $\mathbb{R}^{(N_2-r)\times(N_2-r)}$ such that

$$g_2(y) = Z(y) g_1(y) \tag{6.16}$$

and $Z(0)$ is nonsingular.

Proof. We consider first the case that $B_1 = B_2$. From (6.11) it follows that

$$G(x^\circ + v_1(X(y))) - B_1 g_1(X(y)) = G(x^\circ) \tag{6.17}$$

for every smooth germ $X(y)$. The system

$$C_2^T v_1(X(y)) - D_2 g_1(X(y)) = y \tag{6.18}$$

can be viewed as a system of $N_1 - r$ equations in the $N_1 - r$ components of the not yet known function $X(y)$. The Jacobian matrix of this system is $C_2^T v_{1y} - D_2 g_{1y}$. From (6.11), (6.12) it follows by taking derivatives that

$$\begin{pmatrix} v_{1y} \\ -g_{1y} \end{pmatrix}$$

has full rank $N_1 - r$. If we multiply this from the left with the nonsingular matrix M_2° and take (6.11) into account, as well as the assumption that $B_1 = B_2$, then it follows that $C_2^T v_{1y} - D_2 g_{1y}$ is a nonsingular matrix. Hence (6.18) defines the germ $X(y)$ uniquely; by taking derivatives of the so-obtained identity in y we also find that X_y is nonsingular. Now from a comparison of (6.13), (6.14) with (6.17), (6.18) we obtain (6.15).

Now consider the case where $C_1 = C_2$. We choose any matrices $B_3 \in \mathbb{R}^{N_2 \times (N_2 - r)}$, $D_3 \in \mathbb{R}^{(N_1 - r) \times (N_2 - r)}$ such that

$$M_3^\circ \equiv \begin{pmatrix} G_x(x^\circ) & B_3 \\ C_1^T & D_3 \end{pmatrix} \tag{6.19}$$

is nonsingular.

Now for $y \in \mathbb{R}^{N_1 - r}$, $\epsilon \in \mathbb{R}^{N_2 - r}$ we define in a neighborhood of zero the functions $v_1^*(y, \epsilon) \in \mathbb{R}^{N_1}$, $g_1^*(y, \epsilon) \in \mathbb{R}^{N_2 - r}$, $v_2^*(y, \epsilon) \in \mathbb{R}^{N_1}$, $g_2^*(y, \epsilon) \in \mathbb{R}^{N_2 - r}$ by

$$G(x^\circ + v_1^*(y, \epsilon)) - B_1 g_1^*(y, \epsilon) - B_3 \epsilon = G(x^\circ), \tag{6.20}$$

$$C_1^T v_1^*(y, \epsilon) - D_1 g_1^*(y, \epsilon) - D_3 \epsilon = y, \tag{6.21}$$

$$G(x^\circ + v_2^*(y, \epsilon)) - B_2 g_2^*(y, \epsilon) - B_3 \epsilon = G(x^\circ), \tag{6.22}$$

$$C_1^T v_2^*(y, \epsilon) - D_2 g_2^*(y, \epsilon) - D_3 \epsilon = y. \tag{6.23}$$

By the uniqueness of the solutions to (6.22), (6.23) in a neighborhood of zero it follows that if $g_1^*(y, \epsilon) = 0$, then $g_2^*(y, \epsilon) = 0$ and $v_2^*(y, \epsilon) = v_1^*(y, \epsilon)$.

By taking derivatives of (6.22), (6.23) with respect to ϵ we further obtain

$$\begin{pmatrix} G_u^\circ & B_3 \\ C_1^T & D_3 \end{pmatrix} \begin{pmatrix} v_{1\epsilon}^* \\ -I_{(N_2 - r) \times (N_2 - r)} \end{pmatrix} = \begin{pmatrix} B_2 \\ D_2 \end{pmatrix} g_{2\epsilon}^*. \tag{6.24}$$

Since the left-hand side of (6.24) has full rank, it follows that $g_{2\epsilon}^*$ is nonsingular. By Lemma 6.2.3 it now follows that there exists a smooth function $Z^*(y, \epsilon) \in \mathbb{R}^{(N_2 - r) \times (N_2 - r)}$ such that

$$g_1^*(y, \epsilon) = Z^*(y, \epsilon) g_2^*(y, \epsilon) \tag{6.25}$$

for y, ϵ in a neighborhood of zero.

By a similar argument there exists another smooth function $Z_1^*(y, \epsilon) \in \mathbb{R}^{(N_2 - r) \times (N_2 - r)}$ such that

$$g_2^*(y, \epsilon) = Z_1^*(y, \epsilon) g_1^*(y, \epsilon) \tag{6.26}$$

for y, ϵ in a neighborhood of zero.

Combining (6.25), (6.26) and defining

$$Z_2(y, \epsilon) = Z_1^*(y, \epsilon) Z^*(y, \epsilon)$$

we find that

$$g_2^*(y, \epsilon) = Z_2(y, \epsilon) g_2^*(y, \epsilon).$$

Taking derivatives of this identity with respect to ϵ and evaluating them at zero it follows from the nonsingularity of $g_{2\epsilon}^*$ that at the origin $Z_2(y, \epsilon)$ is the identity $(N_2 - r) \times (N_2 - r)$-matrix. This implies that $Z^*(0, 0)$ is nonsingular.

If we now define $Z(y) = Z_1^{*-1}(y, 0)$ and remark that $g_1^*(y, 0) = g_1(y)$, $g_2^*(y, 0) = g_2(y)$, then (6.25) implies (6.16).

To complete the proof we now consider the general case. It is then possible to find a matrix $D_4 \in \mathbb{R}^{(N_1-r)\times(N_2-r)}$ such that

$$\begin{pmatrix} G_x(x^\circ) & B_1 \\ C_2^T & D_4 \end{pmatrix}$$

is nonsingular. (See Exercise 3.8.3.) Hence it is sufficient to prove the result in the two special cases $B_1 = B_2$ and $C_1 = C_2$ separately. This is precisely what we did. □

In Proposition 6.2.4 the function $z \to G(x^\circ + z) - G(x^\circ)$ represents a germ in $\mathcal{E}^0_{N_1,N_2}$ and g_1, g_2 represent germs in $\mathcal{E}^0_{N_1-r,N_2-r}$. Proposition 6.2.4 is equivalent to the statement that two Lyapunov–Schmidt reduced germs of a germ in $\mathcal{E}^0_{N_1,N_2}$ are $\mathcal{K}$-equivalent if their domain and codomain spaces have the same dimension. This naturally leads us to the following generalization.

Proposition 6.2.5. Let G_1, G_2 be two $\mathcal{K}$-equivalent germs in $\mathcal{E}^0_{N_1,N_2}$. Assume that $N_1 \geq N_2$ and let $r \geq 0$ be an integer that is at most equal to the (common) rank of $G_{1x}(0), G_{2x}(0)$. Suppose that $B_1, B_2 \in \mathbb{R}^{N_2\times(N_2-r)}$, $C_1, C_2 \in \mathbb{R}^{N_1\times(N_1-r)}$, $D_1, D_2 \in \mathbb{R}^{(N_1-r)\times(N_2-r)}$ are such that the matrices

$$M_1 \equiv \begin{pmatrix} G_{1x}(0) & B_1 \\ C_1^T & D_1 \end{pmatrix}, \tag{6.27}$$

$$M_2 \equiv \begin{pmatrix} G_{2x}(0) & B_2 \\ C_2^T & D_2 \end{pmatrix} \tag{6.28}$$

are nonsingular. Let the functions $v_1(y) \in \mathbb{R}^{N_1}$ and $g_1(y) \in \mathbb{R}^{N_2-r}$ be defined for y in a neighborhood of $0 \in \mathbb{R}^{N_1-r}$ by

$$G_1(v_1(y)) - B_1 g_1(y) = 0, \tag{6.29}$$

$$C_1^T v_1(y) - D_1 g_1(y) = y, \tag{6.30}$$

while $v_2(y) \in \mathbb{R}^{N_1}$ and $g_2(y) \in \mathbb{R}^{N_2-r}$ are defined by

$$G_2(v_2(y)) - B_2 g_2(y) = 0, \tag{6.31}$$

$$C_2^T v_2(y) - D_2 g_2(y) = y. \tag{6.32}$$

Then $g_1(y), g_2(y)$ are $\mathcal{K}$-equivalent.

Proof. It is sufficient to prove the result in two special cases, namely, the case where $G_2(x) = G_1(X(x))$ with $X \in \mathcal{E}^0_{N_1,N_1}$ where $X_x(0)$ is nonsingular and the case where $G_2(x) = S(x)G_1(x)$, $S \in \mathcal{E}_{N_1,N_2,N_2}$ and $S(0)$ is nonsingular.

In the first case $G_{2x}(0) = G_{1x}(0)X_x(0)$ and so we may assume by Proposition 6.2.4 that $C_2^T = C_1^T X_x(0)$, $B_2 = B_1$, $D_2 = D_1$. Now $v_1(y), g_1(y)$ are defined by

$$G_1(v_1(y)) - Bg_1(y) = 0 \tag{6.33}$$

and

$$C_1^T v_1(y) - Dg_1(y) = y. \tag{6.34}$$

From (6.33) we infer that

$$G_2(X^{-1}(v_1(y))) - B_2 g_1(y) = 0. \tag{6.35}$$

Now we define a function $\zeta \in \mathcal{E}_{N_1-r,N_1-r}$ by setting

$$C_2^T X_x^{-1}(v_1(y)) - D_2 g_1(y) = \zeta(y). \tag{6.36}$$

From this definition and (6.34) we infer

$$\zeta(y) = y + C_1^T X_x(0) X_x^{-1}(v_1(y)) - C_1^T v_1(y). \tag{6.37}$$

From (6.37) it follows that $\zeta_y(0)$ is the identity matrix and so $\zeta(y)$ is an invertible function near $y = 0$. From (6.35) and (6.36) we infer that $g_2(\zeta(y)) = g_1(y)$; hence g_1 and g_2 are $\mathcal{K}$-equivalent.

Now consider the second case. Since $(S(x)G_1(x))_x(0) = S(0)G_{1x}(0)$ we may assume by Proposition 6.2.4 that $B_2 = S(0)B_1$, $C_2 = C_1$, $D_2 = D_1$. Also, from (6.33), (6.34) we infer that

$$G_2(v_1(y)) - S(v(y))S(0)^{-1}B_2 g_1(y) = 0, \tag{6.38}$$

$$C_2^T v_1(y) - D_2 g_1(y) = y. \tag{6.39}$$

Now we compare (6.38), (6.39) with the equations that define $v_2(y), g_2(y)$. By exactly the same argument as in the proof of Proposition 6.2.4 (case $C_1 = C_2$) we find that there exists a $Z \in \mathcal{E}_{N_2-r,N_2-r,N_2-r}$ that is nonsingular at 0 and such that $g_2(y) = Z(y)g_1(y)$; hence g_1 and g_2 are $\mathcal{K}$-equivalent. □

Definition 6.2.6. Consider a smooth function

$$G : U \to \mathbb{R}^{N_2}, \tag{6.40}$$

where U is open in $\mathbb{R}^{N_1}$, $N_1 \geq N_2$, $x^\circ \in U$. Also assume that $G_x(x^\circ)$ is rank deficient. Then we call a Lyapunov–Schmidt reduction with r equal to the rank of the Jacobian, a maximal Lyapunov–Schmidt reduction, and the resulting function $g(y)$ a maximally Lyapunov–Schmidt reduced function of G at x°.

By Proposition 6.2.4 the maximally reduced function is unique up to $\mathcal{K}$-equivalence; by Proposition 6.2.5 it is even determined up to $\mathcal{K}$-equivalence by the $\mathcal{K}$-equivalence class of the germ $z \to G(x^\circ + z) - G(x^\circ)$.

Proposition 6.2.7. If N_1, N_2, G, U, x° are as in Definition 6.2.6 and $g(y)$ is a maximally Lyapunov–Schmidt reduced function of G, then $g(0) = 0$ and $g_y(0)$ is the $(N_2 - r) \times (N_1 - r)$ zero matrix.

Proof. By construction $g(0) = 0$. From (6.5) and (6.6) we infer that

$$\begin{pmatrix} G_x(x^\circ) & B \\ C^T & D \end{pmatrix} \begin{pmatrix} v_y(0) \\ -g_y(0) \end{pmatrix} = \begin{pmatrix} 0_{N_2 \times (N_1-r)} \\ I_{N_1-r} \end{pmatrix}.$$

From Corollary 3.3.4 it follows that $g_y(0)$ has rank deficiency $N_2 - r$, i.e., that it is a zero matrix.

6.3 Classification of Singularities by Codimension

To introduce the notion of a $\mathcal{K}$-singularity (singularity in a setting with no distinguished bifurcation parameter) one needs a few preparatory definitions and lemmas.

6.3.1 Introduction and Basic Properties

Definition 6.3.1. For $n, k \in \mathbb{N}$ a monomial of order k in the n variables $x_1, \ldots, x_n$ is the germ of a function of the form $x_1^{k_1} x_2^{k_2} \ldots x_n^{k_n}$, where $k_i \geq 0$ for all i and $k_1 + \cdots + k_n = k$.

Obviously $\mathcal{E}_n$ is a ring under the usual pointwise addition and multiplication of functions.

Definition 6.3.2. For $k \geq 1$, $\mathcal{M}_n^k$ is the ideal generated in the ring $\mathcal{E}_n$ by all monomials of order k.

Definition 6.3.3. For any $f \in \mathcal{E}_{n,p}$ and $k \geq 1$ we define its k-jet $j_k(f)$ as the polynomial germ obtained by deleting from the formal Taylor expansion of f all terms with degree higher than k.

Definition 6.3.4. A germ $f \in \mathcal{E}_{n,p}$ is said to be $(\mathcal{K} - k)$-determined if any germ with the same k-jet is $\mathcal{K}$-equivalent to f.

Lemma 6.3.5. If $f \in \mathcal{E}_n$, $k \geq 0$, then $f - j_k(f) \in \mathcal{M}_n^{k+1}$. Furthermore, $f \in \mathcal{M}_n^{k+1}$ if and only if $j_k(f) = 0$.

Proof. We prove the first part by induction. If $k = 0$ then $(f - j_0(f))(0, \ldots, 0) = 0$ and so by Lemma 6.2.2 $f - j_0(f) \in \mathcal{M}_n^1$. Next, assume that the result holds for $k - 1$. So for every n-tuple $(l_1, \ldots, l_n)$ with $l_1 \geq 0$, ..., $l_n \geq 0$, $l_1 + \ldots l_n = k$ there exists a $g_{l_1 \ldots l_n} \in \mathcal{E}_n$ so that

$$f - j_{k-1}(f) = \sum \frac{x_1^{l_1} \ldots x_n^{l_n}}{l_1! \ldots l_n!} g_{l_1 \ldots l_n}, \tag{6.41}$$

where the summation is over the n-tuples of nonnegative indices $l_1, \ldots, l_n$ with sum k. Hence

$$f - j_k(f) = \sum \frac{x_1^{l_1} \ldots x_n^{l_n}}{l_1! \ldots l_n!} \left(g_{l_1 \ldots l_n} - \frac{\partial^k f}{\partial x_1^{l_1} \ldots x_n^{l_n}} \right). \tag{6.42}$$

By taking the derivative of this expression with respect to $x_1^{l_1} \ldots x_n^{l_n}$ and evaluating at zero we find that the functions between brackets in (6.42) all vanish at zero. Hence they are in $\mathcal{M}_n^1$ by Lemma 6.2.2. It follows that $f - j_k(f) \in \mathcal{M}_n^{k+1}$.

The second statement in the lemma follows from the first and the observation that the zero polynomial is the only polynomial of degree less than $k + 1$ in $\mathcal{M}_n^{k+1}$ (take derivatives to prove this). □

Although a group orbit is not a manifold in the sense of Chapter 2, we can associate a so-called tangent space with it. This is a rather abstract notion, which is important mainly because it leads to the notion of the codimension of a singularity.

Definition 6.3.6. If $f \in \mathcal{E}_{n,p}^0$ let J_f be the Jacobian module of f, i.e., the $\mathcal{E}_n$-module generated by the first-order partial derivatives of f.

Definition 6.3.7. The $\mathcal{K}$-tangent space of $f \in \mathcal{E}_{n,p}^0$ is $T_f = \mathcal{E}_{n,p,p}.f + J_f$. The $\mathcal{K}$-codimension of f is the codimension of T_f as a linear subspace of $\mathcal{E}_{n,p}$.

Example 6.3.8. Let $n = 2$, $p = 2$ and consider the germ

$$f_{fh}\begin{pmatrix} x_1 \\ x_2 \end{pmatrix} = \begin{pmatrix} x_1^2 \\ x_2^2 \end{pmatrix}.$$

The first-order partial derivatives (columns of the Jacobian matrix) are

$$\begin{pmatrix} 2x_1 \\ 0 \end{pmatrix}, \begin{pmatrix} 0 \\ 2x_2 \end{pmatrix}.$$

The $\mathcal{K}$-tangent T_f is the collection of all germs in $\mathcal{E}_{n,p}$ of the form

$$T_{f_{fh}} = \begin{pmatrix} x_1 f_1(x,y) + x_1^2 f_2(x,y) + x_2^2 f_3(x,y) \\ x_2 f_4(x,y) + x_1^2 f_5(x,y) + x_2^2 f_6(x,y) \end{pmatrix},$$

where $f_i (i = 1, \dots, 6) \in \mathcal{E}_2$ are arbitrary. It is easily seen that the four germs

$$\begin{pmatrix} 1 \\ 0 \end{pmatrix}, \begin{pmatrix} 0 \\ 1 \end{pmatrix}, \begin{pmatrix} x_2 \\ 0 \end{pmatrix}, \begin{pmatrix} 0 \\ x_1 \end{pmatrix}$$

are linearly independent and span a space that has only the 0 function in common with $T_{f_{fh}}$. Hence the codimension of this germ is at least 4. We will see later that it is precisely 4.

By a Lyapunov–Schmidt reduction one can always reduce the study of the solutions of a singular problem of the form (6.1) to the case where the Jacobian of the system vanishes at the singular point. A $\mathcal{K}$-equivalence class of such germs will be called a $\mathcal{K}$-singularity. It is proved in [104, V, (2.4)] that the $\mathcal{K}$-codimension is invariant under $\mathcal{K}$-equivalence. Hence we may talk about the $\mathcal{K}$-codimension of a $\mathcal{K}$-singularity.

There is an infinite number of $\mathcal{K}$-singularities. Sometimes this class is called a zoo, presumably because we have no global understanding of the whole collection and because one discovers more and more species as the study proceeds. However, some singularities are more common than others. It is natural to start with singularities with low codimension because for them fewer conditions are required to fix the $\mathcal{K}$-tangent spaces.

We start with some preliminary considerations. For practical applications the germs in $\mathcal{E}_{n,n}$ (point singularities) and $\mathcal{E}_{n+1,n}$ (branch singularities) are most important. To limit the scope of our investigation we first prove a lemma on the former case.

Proposition 6.3.9. If $f \in \mathcal{E}_{n,n}^0$ is a germ with vanishing Jacobian, then its $\mathcal{K}$-codimension is at least equal to n^2.

Proof. Denote $f = (f_1, \dots, f_n)^T$. By Lemma 6.3.5 $f_i \in \mathcal{M}_n^2$ for all i and so $\mathcal{E}_{n,n,n} f \subset (\mathcal{M}_n^2)^n$.

The n-vectors with one component either equal to 1 or to x_i for $i \in \{1, \dots, n\}$ and all other components equal to zero span a linear subspace of $\mathcal{E}_{n,n}$ with dimension $n^2 + n$. This subspace has only the zero vector in common with $(\mathcal{M}_n^2)^n$.

By the definition of T_f the result will now follow if we show that J_f is contained in a space of the form $J_f^1 + (\mathcal{M}_n^2)^n$ with J_f^1 a space with dimension at most n.

It is obviously sufficient to fix an index j and to prove that the space $\{f_{x_j} g : g \in \mathcal{E}_n\}$ is contained in the sum of $(\mathcal{M}_n^2)^n$ and a one-dimensional space.

Since $f_i \in \mathcal{M}_n^2$ there exist for all $k(1 \leq k \leq n)$ functions $f_{ijk}^1 \in \mathcal{E}_n$ such that

$$f_{ix_j} = \Sigma_k x_k f_{ijk}^1. \tag{6.43}$$

By Lemma 6.3.5 we can decompose

$$g = g(0) + \Sigma_l x_l g_l$$

with $g_l \in \mathcal{E}_n$. Making a similar decomposition of every function f_{ijk}^1 and inserting these expressions in (6.43) we obtain an expression of the form

$$f_{ix_j} g = \Sigma_k x_k f_{ijk}^1(0) g(0) + h_{ij} \tag{6.44}$$

with $h_{ij} \in \mathcal{M}_n^2$. If we define the vector $v_j = (\Sigma_k x_k f_{ijk}^1(0))_i$, then every vector of the form $f_{x_j} g$ is in the space spanned by $(\mathcal{M}_n^2)^n$ and v_j. This completes the proof. □

The case of germs in $\mathcal{E}_{n+1,n}$ can be handled similarly. In fact, we have the following proposition.

Proposition 6.3.10. If $f \in \mathcal{E}_{n+1,n}^0$ is a germ with vanishing Jacobian, then its $\mathcal{K}$-codimension is at least equal to $n^2 + n - 1$.

Proof. The proof is similar to the proof of Proposition 6.3.9. □

If we limit ourselves to the classification of germs with $\mathcal{K}$-codimension 4 or smaller, then by Propositions 6.3.9, 6.3.10, we need only consider the spaces $\mathcal{E}_{1,1}$, $\mathcal{E}_{2,1}$, and $\mathcal{E}_{2,2}$. We now turn our attention to these cases.

6.3.2 Singularities from $\mathbb{R}$ into $\mathbb{R}$

This case is treated completely in [28]; we summarize the outcome. Let f be any germ in $\mathcal{M}_1^2$. If its derivatives of all orders vanish at zero, then its $\mathcal{K}$-codimension is infinite since the $\mathcal{K}$-tangent space contains only functions with vanishing derivatives of all orders. Otherwise there is an integer $k \geq 0$ such that $g_{x^l}(0) = 0$ for all $l \leq k$ and $g_{x^{k+1}} \neq 0$. Then T_f is the space of all germs whose derivatives of order $k-1$ and less vanish (cf. Lemma 6.3.5). Hence the codimension is equal to k. Since the germ is $(\mathcal{K}-k)$-determined by [104, V, (6.1)] and a nonzero factor can be absorbed in the factor Z in (6.2) it follows that there exists precisely one singularity of codimension k in $\mathcal{E}_1$; it is determined by the defining equations $f = f_x = \cdots = f_{x^k} = 0$ and the nondegenercay condition $f_{x^{k+1}} \neq 0$.

6.3.3 Singularities from $\mathbb{R}^2$ into $\mathbb{R}$

Let $f(x, y)$ be any germ in $\mathcal{M}_2^2$. Following [104] we make a further distinction based on the rank of the Hessian matrix

$$H(f) = \begin{pmatrix} f_{xx} & f_{xy} \\ f_{xy} & f_{yy} \end{pmatrix}. \tag{6.45}$$

It is not difficult to show that the rank of the Hessian matrix and the sign of its determinant (if nonzero) are preserved by $\mathcal{K}$-equivalence; cf. [104], [28]. We will prove it in a somewhat more complicated way, which will provide the key to obtain defining functions and nondegeneracy conditions for several singularities. Let

$$D_2(f) = f_{xx}f_{yy} - f_{xy}^2 \tag{6.46}$$

be the determinant of $H(f)$.

Lemma 6.3.11. Let $f \in \mathcal{E}_2$ and let g be defined by

$$g(x,y) = S(x,y)f(X(x,y),Y(x,y)),$$

where S, X, Y are smooth functions. Then there exist smooth functions $T_{4,4,1}$, $T_{4,4,2}$, $T_{4,4,3}$ such that

$$\begin{pmatrix} g \\ g_x \\ g_y \\ D_2(g) \end{pmatrix} = T_4 \begin{pmatrix} f \\ f_x \\ f_y \\ D_2(f) \end{pmatrix}, \tag{6.47}$$

where the functions f, f_x, f_y, and $D_2(f)$ are evaluated in $(X(x,y), Y(x,y))$ and

$$T_4 = \begin{pmatrix} S & 0 & 0 & 0 \\ S_x & SX_x & SY_x & 0 \\ S_y & SX_y & SY_y & 0 \\ T_{4,4,1} & T_{4,4,2} & T_{4,4,3} & S^2(X_xY_y - X_yY_x)^2 \end{pmatrix}. \tag{6.48}$$

Proof. The first three rows expressed in (6.47) are trivial. By taking derivatives with respect to x, y of the second and third rows of (6.47) we find

$$H(g) = SJ^T H(f)J + R, \tag{6.49}$$

where J is the Jacobian matrix

$$J = \begin{pmatrix} X_x & X_y \\ Y_x & Y_y \end{pmatrix} \tag{6.50}$$

and the entries of the $(2,2)$ matrix R are smooth functions that can be written as sums of product in which each product contains a factor f, f_x, or f_y. So by taking determinants of (6.49) we find an expression of the form as represented in the fourth row of (6.47). □

Lemma 6.3.11 proves that the conditions

$$f = f_x = f_y = 0 \tag{6.51}$$

are preserved under $\mathcal{K}$-equivalence. If they hold, then by (6.49) also the rank of $H(f)$ is invariant.

First assume that the Hessian matrix has corank 0; i.e.,

$$D_2(f) \neq 0. \tag{6.52}$$

Proposition 6.3.12. A germ $f \in \mathcal{E}_2$ is $\mathcal{K}$-equivalent to $x^2 + y^2$ (respectively, $x^2 - y^2$) if and only if (6.51) holds and $D_2(f) > 0$ (respectively, $D_2(f) < 0$).

Proof. By Lemma 6.3.11 these conditions are necessary. Now assume that they hold. By [104, V, (6.2)] it follows that f is $\mathcal{K}$-equivalent to $\pm x^2 \pm y^2$. The first $\pm$ sign can of course be omitted. On the other hand, it follows from Lemma 6.3.11 that the sign of $\det(D_2(f))$ is invariant under $\mathcal{K}$-equivalence. This completes the proof. □

Next consider the case that $H(f)$ has corank 1; i.e.,

$$D_2(f) = 0, \tag{6.53}$$

and at least one of f_{xx}, f_{yy} is nonzero. By [104, V, (6.3)] a singularity of this type has codimension at least 2; it has codimension $k (k \geq 2)$ if and only if it is $\mathcal{K}$-equivalent to $x^{k+1} \pm y^2$.

Consider the case $k = 2$. Obviously here we may omit the $\pm$ sign. Now define $D_3^1(f) = g_{xx}(D_2(f))_y - g_{xy}(D_2(f))_x$, $D_3^2(f) = g_{xy}(D_2(f))_y - g_{yy}(D_2(f))_x$. To obtain a nondegeneracy condition we generalize Lemma 6.3.11.

Lemma 6.3.13. Let the assumptions of Lemma 6.3.11 hold. Then there exists a smooth $(2,4)$ matrix function $T_5^*(x,y)$ such that

$$\begin{pmatrix} g \\ g_x \\ g_y \\ D_2(g) \\ D_3^1(g) \\ D_3^2(g) \end{pmatrix} = T_5 \begin{pmatrix} f \\ f_x \\ f_y \\ D_2(f) \\ D_3^1(f) \\ D_3^2(f) \end{pmatrix}, \tag{6.54}$$

where

$$T_5 = \begin{pmatrix} T_4 & 0_{4,2} \\ T_5^* & S^3(X_xY_y - X_yY_x)^3 J^T \end{pmatrix}. \tag{6.55}$$

Proof. By taking derivatives of the last row in (6.47) we find that

$$\begin{pmatrix} (D_2(g))_x & (D_2(g))_y \end{pmatrix} = S^2(X_xY_y - X_yY_x)^2 \begin{pmatrix} (D_2(f))_x & (D_2(f))_y \end{pmatrix} J + R_1,$$

where the entries of the $(1,2)$ matrix R_1 are smooth functions that can be written as sums of products in which each product contains a factor f, f_x, f_y, or $D_2(f)$. Multiplying this equality from the right with

$$\begin{pmatrix} -g_{xy} & -g_{yy} \\ g_{xx} & g_{xy} \end{pmatrix}$$

and taking (6.49) into account we obtain

$$\begin{pmatrix} D_3^1(g) & D_3^2(g) \end{pmatrix} = S^3(X_xY_y - X_yY_x)^2 \begin{pmatrix} (D_2(f))_x & (D_2(f))_y \end{pmatrix}$$
$$\times J \begin{pmatrix} -X_y & -Y_y \\ X_x & Y_x \end{pmatrix} H(f) J + R_2,$$

where R_2 has properties similar to R_1. Applying the definition of $D_3^1(f), D_3^2(f)$ it follows that

$$\begin{pmatrix} D_3^1(g) & D_3^2(g) \end{pmatrix} = S^3(X_xY_y - X_yY_x)^3 \begin{pmatrix} D_3^1(f) & D_3^2(f) \end{pmatrix} J + R_2.$$

By taking the transpose we can now complete the proof. □

Proposition 6.3.14. A germ $g \in \mathcal{E}_2$ is $\mathcal{K}$-equivalent to $x^3 + y^2$ if and only if (6.51) and (6.53) hold and at least one of $D_3^1(g), D_3^2(g)$ is nonzero.

Proof. The singularity represented by $f(x,y) = x^3 + y^2$ satisfies the mentioned conditions and by Lemma 6.3.13 this carries over to every germ that is $\mathcal{K}$-equivalent to it. Now suppose that g is any germ with these properties. First we remark that we can assume $g_{xx} = 0$ (a linear transformation of (x,y) will do). Second, we can assume that $g_{xxy} = 0$ (a nonlinear transformation of the type $(X(x,y) = x, Y(x,y) = y + \alpha x^2)$ will accomplish this). Then by (6.53) $g_{xy} = 0$ as well and by the conditions on $D_3^1(g), D_3^2(g)$ we must have $g_{yy} \neq 0, g_{xxx} \neq 0$. Now $j_3(g)$ has the form

$$j_3(g) = c_{0,2}y^2 + c_{3,0}x^3 + c_{1,2}xy^2 + c_{0,3}y^3$$

with $c_{0,2} \neq 0, c_{3,0} \neq 0$. Since $(j_3(g))_y$ is the product of y with an invertible germ it follows that $y \in T_{j_3(g)}$. From the form of $(j_3(g))_x$ it now follows that $x^2 \in T_{j_3(g)}$. Then $j_3(g)$ has codimension 2, since its tangent space, together with $1, x$ spans $\mathcal{E}_2$. Also by [104, V, (6.1)] g is $\mathcal{K}$-equivalent to $j_3(g)$ and by [104, V, (2.4)] has the same codimension. By the classification results this implies that g is $\mathcal{K}$-equivalent to $x^3 + y^2$. □

Now consider the case $k = 3$. The two possible singularities with this codimension are $x^4 + y^2$ and $x^4 - y^2$. Both satisfy

$$D_3^1(g) = D_3^2(g) = 0. \tag{6.56}$$

To obtain nondegeneracy conditions we introduce new functions

$$D_4^{j,1}(f) = f_{xx}(D_3^j(f))_y - f_{xy}(D_3^j(f))_x, \tag{6.57}$$

$$D_4^{j,2}(f) = f_{xy}(D_3^j(f))_y - f_{yy}(D_3^j(f))_x \tag{6.58}$$

for $j = 1, 2$. It may be checked by some tedious computations that $D_4^{1,2}(f) = D_4^{2,1}(f)$ for every germ f.

It is easily seen that $D_4^{2,2}(x^4 + y^2) = 1$, $D_4^{2,2}(x^4 - y^2) = -1$, and $D_4^{i,j}(x^4 \pm y^2) = 0$ if $(i,j) \neq (2,2)$.

Lemma 6.3.15. Let the assumptions of Lemma 6.3.11 hold. Then there exists a smooth $(4,6)$ matrix function $T_6^*(x,y)$ such that

$$\begin{pmatrix} g \\ g_x \\ g_y \\ D_2(g) \\ D_3^1(g) \\ D_3^2(g) \\ D_4^{1,1}(g) \\ D_4^{1,2}(g) \\ D_4^{2,1}(g) \\ D_4^{2,2}(g) \end{pmatrix} = T_6 \begin{pmatrix} f \\ f_x \\ f_y \\ D_2(f) \\ D_3^1(f) \\ D_3^2(f) \\ D_4^{1,1}(f) \\ D_4^{1,2}(f) \\ D_4^{2,1}(f) \\ D_4^{2,2}(f) \end{pmatrix}, \tag{6.59}$$

where

$$T_6 = \begin{pmatrix} T_5 & 0_{6,4} \\ T_6^* & S^4(X_xY_y - X_yY_x)^4 J^T \otimes J^T \end{pmatrix}. \tag{6.60}$$

Here $J^T \otimes J^T$ is the $(4,4)$ tensor product matrix defined with respect to lexicographically ordered index pairs by

$$(J^T \otimes J^T)_{2(i-1)+j,2(k-1)+l} = (J^T)_{i,k}(J^T)_{j,l}.$$

This matrix is singular if and only if J is singular.

Proof. By taking derivatives of the last two rows in (6.54) we find

$$\begin{pmatrix} (D_3^1(g))_x & (D_3^1(g))_y \\ (D_3^2(g))_x & (D_3^2(g))_y \end{pmatrix} = S^3(X_xY_y - X_yY_x)^3 J^T \begin{pmatrix} (D_3^1(f))_x & (D_3^1(f))_y \\ (D_3^2(f))_x & (D_3^2(f))_y \end{pmatrix} J + R_3,$$

where the entries of the $(2,2)$ matrix R_3 are smooth functions that can be written as sums of products in which each product contains a factor f, f_x, f_y, $D_2(f)$, $D_3^1(f)$, or $D_3^2(f)$. Multiplying this equality from the right with

$$\begin{pmatrix} -g_{xy} & -g_{yy} \\ g_{xx} & g_{xy} \end{pmatrix}$$

and taking (6.49) into account we obtain

$$\begin{pmatrix} D_4^{1,1}(g) & D_4^{1,2}(g) \\ D_4^{2,1}(g) & D_4^{2,2}(g) \end{pmatrix} = S^4(X_xY_y - X_yY_x)^4 J^T \begin{pmatrix} D_4^{1,1}(f) & D_4^{1,2}(f) \\ D_4^{2,1}(f) & D_4^{2,2}(f) \end{pmatrix} J + R_4$$

by an argument similar to that in Lemma 6.3.13; here R_4 has properties similar to those of R_3. The statement of the lemma is just a reformulation of this result. We remark that if A is a $(2,2)$ matrix, then

$$(J^T \otimes J^T) \begin{pmatrix} A_{1,1} \\ A_{1,2} \\ A_{2,1} \\ A_{2,2} \end{pmatrix} = \begin{pmatrix} (J^TAJ)_{1,1} \\ (J^TAJ)_{1,2} \\ (J^TAJ)_{2,1} \\ (J^TAJ)_{2,2} \end{pmatrix}.$$

This implies that $J^T \otimes J^T$ is singular if and only if J is singular (a more general proof of this statement was given in Proposition 4.4.1(5)). □

Proposition 6.3.16. If a germ $g \in \mathcal{E}_2$ is $\mathcal{K}$-equivalent to $x^4 + y^2$ (respectively, $x^4 - y^2$), then g satisfies (6.51), (6.53), (6.56) and at least one of $D_4^{1,1}(g), D_4^{2,2}(g)$ is strictly positive (respectively, strictly negative). The other one is either zero or has the same sign.

Conversely, if g satisfies (6.51), (6.53), (6.56) and at least one of the four quantities $D_4^{i,j}(g)$ $(i,j = 1,2)$ is nonzero, then g is $\mathcal{K}$-equivalent to either $x^4 + y^2$ or $x^4 - y^2$ (and hence at least one of $D_4^{i,i}, i = 1,2$ is nonzero and its sign determines the precise singularity).

Proof. First assume that g is $\mathcal{K}$-equivalent to $x^4 + y^2$ (the other case being similar). From Lemma 6.3.15 it follows that g satisfies (6.51), (6.53), (6.56), and

$$\begin{pmatrix} D_4^{1,1}(g) & D_4^{1,2}(g) \\ D_4^{2,1}(g) & D_4^{2,2}(g) \end{pmatrix}$$

is a nonvanishing symmetric positive semidefinite matrix so that in particular $D_4^{1,1}(g)$ and $D_4^{2,2}(g)$ are nonnegative and at least one of them is nonzero.

Now let g be a germ that satisfies (6.51), (6.53), (6.56) and for which either $D_4^{1,1}(g)$ or $D_4^{2,2}(g)$ is strictly positive. As in Proposition 6.3.14 we may assume that $g_{xx} = 0, g_{xy} = 0, g_{yy} \neq 0$. Also, by a nonlinear transformation ($X(x,y) = x, Y(x,y) = y + \alpha x^2 + \beta x^3$ for an appropriate choice of α, β) we may assume that $g_{xxy} = g_{xxxy} = 0$. After some tedious computations we find that in the new form $D_4^{i,j} = 0$ if $(i,j) \neq (2,2)$ while $D_4^{2,2} = g_{yy}^3 g_{xxxx}$. By Lemma 6.3.15 this implies that $g_{xxxx} \neq 0$. Now by arguments similar to those in Proposition 6.3.14 we conclude that g has codimension 3 and therefore must be $\mathcal{K}$-equivalent to either x^4+y^2 or x^4-y^2. By the original assumption on $D_4^{1,1}, D_4^{2,2}$ the second case is excluded. □

Now consider the case $k = 4$. The only singularity with this codimension and nonvanishing Hessian is represented by $x^5 + y^2$. It satisfies (6.51), (6.53), (6.56), and

$$D_4^{1,1}(f) = D_4^{2,2}(f) = 0. \tag{6.61}$$

To obtain nondegeneracy conditions we introduce new functions

$$D_5^{i,j,1}(f) = g_{xx}(D_4^{i,j})_y - g_{xy}(D_4^{i,j})_x,$$

$$D_5^{i,j,2}(f) = g_{xy}(D_4^{i,j})_y - g_{yy}(D_4^{i,j})_x$$

for $i, j = 1, 2$.

Remark 6.3.17. The previous case suggests that maybe the values $D_5^{i,j,k}(f)$ are invariant under a permutation of the indices for every germ f. This appears not to be true (it is possible to find a counterexample).

It is easy to check that $D_5^{2,2,2}(x^5+y^2) = 1$ and $D_5^{i,j,k}(x^5+y^2) = 0$ if $(i,j,k) \neq (2,2,2)$. Also Lemma 6.3.15 generalizes to this case. In fact we have the following lemma.

Lemma 6.3.18. Let the assumptions of Lemma 6.3.11 hold. Then there exists a smooth $(8, 10)$ matrix function $T_7^*(x, y)$ such that

$$\begin{pmatrix} g \\ g_x \\ \cdots \\ D_4^{2,2}(g) \\ D_5^{1,1,1}(g) \\ D_5^{1,1,2}(g) \\ D_5^{1,2,1}(g) \\ \cdots \\ D_5^{2,2,2}(g) \end{pmatrix} = T_7 \begin{pmatrix} f \\ f_x \\ \cdots \\ D_4^{2,2}(f) \\ D_5^{1,1,1}(f) \\ D_5^{1,1,2}(f) \\ D_5^{1,2,1}(f) \\ \cdots \\ D_5^{2,2,2} f) \end{pmatrix}, \tag{6.62}$$

where

$$T_7 = \begin{pmatrix} T_6 & 0_{10,8} \\ T_7^* & S^5(X_xY_y - X_yY_x)^5 J^T \otimes J^T \otimes J^T \end{pmatrix}. \tag{6.63}$$

Here $J^T \otimes J^T \otimes J^T$ is the $(8,8)$ tensor product matrix defined with respect to lexicographically ordered index triples by

$$(J^T \otimes J^T \otimes J^T)_{4(i-1)+2(j-1)+k,4(l-1)+2(m-1)+n} = (J^T)_{i,l}(J^T)_{j,m}(J^T)_{k,n}.$$

This matrix is singular if and only if J is singular.

Proof. The proof is a somewhat tedious but straightforward generalization of Lemma 6.3.15. □

Proposition 6.3.19. A germ $g \in \mathcal{E}_2$ is $\mathcal{K}$-equivalent to $x^5 + y^2$ if and only if g satisfies (6.51), (6.53), (6.56), (6.61) and at least one of $D_5^{1,1,1}(g), D_5^{2,2,2}(g)$ is nonzero.

Proof. First let $g \in \mathcal{E}_2$ be $\mathcal{K}$-equivalent to $x^5 + y^2$. Clearly it satisfies (6.51), (6.53), (6.56), (6.61). Also, $D_5^{i,i,i}(g) = S^5(X_xY_y - X_yY_x)^5((J^T)_{i,2})^3$; hence either $D_5^{1,1,1}(g)$ or $D_5^{2,2,2}(g)$ must be nonzero. Now suppose that $g \in \mathcal{E}_2$ satisfies the requirements of the proposition. By Proposition 6.3.16 $D_4^{i,j}(g) = 0$ for all $i, j = 1, 2$. As in Proposition 6.3.16 we can assume that (possibly after a nonlinear transformation) $g_{xx} = g_{xy} = g_{xxy} = g_{xxxy} = g_{xxxxy} = 0$. Also (6.56) and (6.61) imply, respectively, that g_{xxx} and g_{xxxx} are zero. Now $D_5^{1,1,1} = 0$ and hence $D_5^{2,2,2} \neq 0$; this implies that $g_{xxxxx} \neq 0$. By the same arguments as in Proposition 6.3.14 we can prove that g has codimension 4; since its Hessian is nonvanishing it is $\mathcal{K}$-equivalent to $x^5 + y^2$. □

Let us finally consider the case of a vanishing Hessian, i.e.,

$$f = f_x = f_y = f_{xx} = f_{xy} = f_{yy} = 0. \tag{6.64}$$

We define

$$D_\infty(f) = (f_{xxx}f_{yyy} - f_{xxy}f_{xyy})^2 - 4(f_{xxx}f_{xyy} - f_{xxy}^2)(f_{xxy}f_{yyy} - f_{xyy}^2). \tag{6.65}$$

Proposition 6.3.20. Every germ in $\mathcal{M}_2^2$ with vanishing Hessian has $\mathcal{K}$-codimension at least 4. There are precisely two such singularities with codimension 4. One is defined by (6.64) and the nondegeneracy condition

$$D_\infty(f) > 0. \tag{6.66}$$

It may be represented by $x^3 + y^3$. The other one is defined by (6.64) and the condition

$$D_\infty(f) < 0. \tag{6.67}$$

It may be represented by $x^3 - xy^2$.

Proof. The elements of the proof of this result are all in [104], so we arrange them briefly. First consider the homogeneous cubic equation

$$f_{xxx}x^3 + 3f_{xxy}x^2y + 3f_{xyy}xy^2 + f_{yyy}y^3 = 0. \tag{6.68}$$

If all coefficients of this equation vanish, then T_f does not contain any nonzero vector in the span of $(1, x, y, x^2, xy, y^2)$, so f has at least codimension 6. Dismissing this case, we remark as in [104, III] that (6.68) has two coinciding rays of solutions if and only if $D_\infty(f) = 0$. (We note that the determinant of the Hessian matrix of the cubic polynomial in (6.68) is itself a homogeneous quadratic polynomial in x, y and that $D_\infty(f)$ is its discriminant up to a positive factor.) If $D_\infty(f) = 0$, then by a linear transformation of coordinates f is either $\mathcal{K}$-equivalent to a germ with $g_{xxx} = g_{xyy} = g_{yyy} = 0$ or to a germ with $g_{xxy} = g_{xyy} = g_{yyy} = 0$. In the first (respectively, second) case T_f does not contain any nonzero vector in the span of $(1, x, y, xy, y^2)$ (respectively, $(1, x, y, y^2, y^3)$) and so has codimension at least 5.

We note that $x^3 + y^3$ (respectively, $x^3 - xy^2$) satisfies (6.64) and (6.66) ((6.64) and (6.67), respectively). Also, we can check that these properties are preserved under $\mathcal{K}$-equivalence (multiplication with $S(x, y)$ is easy; for right equivalence we remark that the Hessian matrix H of the polynomial in (6.68) is transformed into $J^T HJ$ with J nonsingular). Also, the codimension is 4 since a complementary subspace to T_f is the linear span of $(1, x, y, xy)$ (respectively, $(1, x, y, x^2)$).

Now assume that f satisfies (6.64) and $D_\infty(f) \neq 0$. So either (6.68) has three distinct rays of solutions or one real ray and a conjugate pair of complex rays. In the first case f is equivalent by a linear change of coordinates to a germ with 3-jet $x^3 - xy^2$. By [104] this germ is $(\mathcal{K}-3)$-determined, so f is actually $\mathcal{K}$-equivalent to $x^3 - xy^2$ itself and (6.67) must hold a posteriori. The second case can be handled similarly. $\square$

6.3.4 Singularities from $\mathbb{R}^2$ into $\mathbb{R}^2$

We represent a germ $f \in \mathcal{E}_{2,2}$ by its two coordinate functions

$$f(x, y) = \begin{pmatrix} f_1(x, y) \\ f_2(x, y) \end{pmatrix}.$$

For our classification we need the function

$$D_{2,2}(f) = (f_{1xx}f_{2yy} - f_{2xx}f_{1yy})^2 - 4(f_{1xx}f_{2xy} - f_{2xx}f_{1xy})(f_{1xy}f_{2yy} - f_{2xy}f_{1yy}).$$

Proposition 6.3.21. A germ $f \in \mathcal{E}^0_{2,2}$ with vanishing Jacobian has $\mathcal{K}$-codimension at least 4. There are exactly two such $\mathcal{K}$-singularity classes with codimension 4. One of them is characterized by

$$f_1 = f_2 = f_{1x} = f_{1y} = f_{2x} = f_{2y} = 0 \tag{6.69}$$

and

$$D_{2,2}(f) > 0. \tag{6.70}$$

It is represented by

$$\begin{pmatrix} x^2 + y^2 \\ xy \end{pmatrix}. \tag{6.71}$$

The other one is characterized by (6.69) and

$$D_{2,2}(f) < 0. \tag{6.72}$$

It is represented by

$$\begin{pmatrix} x^2 - y^2 \\ xy \end{pmatrix}. \tag{6.73}$$

Proof. The $\mathcal{K}$-tangent space to a germ in $\mathcal{E}^0_{2,2}$ with vanishing Jacobian intersects the six-dimensional space spanned by

$$\begin{pmatrix} 1 \\ 0 \end{pmatrix}, \begin{pmatrix} 0 \\ 1 \end{pmatrix}, \begin{pmatrix} x \\ 0 \end{pmatrix}, \begin{pmatrix} 0 \\ x \end{pmatrix}, \begin{pmatrix} y \\ 0 \end{pmatrix}, \begin{pmatrix} 0 \\ y \end{pmatrix}$$

in a space with dimension at most 2; therefore, such a germ has codimension at least 4.

We also remark that the conditions on $D_{2,2}(f)$ are really on the geometric properties of the 2-jet of f. It is known (and may be proved by standard arguments) that $D_{2,2}(f) = 0$ if and only if $j_2(f)$ either vanishes in at least one nonzero point or its image is a single ray. Also $D_{2,2}(f) > 0$ if and only if $j_2(f)$ maps no nonzero points onto zero and its image is a proper nontrivial sector of the plane. Finally $D_{2,2}(f) < 0$ if and only if $j_2(f)$ maps no nonzero points onto zero and its image is the whole plane.

All these properties are preserved under $\mathcal{K}$-equivalence. Obviously the germ in (6.71) satisfies (6.69) and (6.70); the germ in (6.73) satisfies (6.69) and (6.72). From [104, (6.1)] it follows that the germs (6.71) and (6.73) are $(\mathcal{K}-2)$-determined and it is easy to check that they have codimension 4.

Now let g be a germ that satisfies (6.69) and (6.70) (the other case being similar). From the proof of [104, (6.7)] it follows that there is a linear change of coordinates that transforms g into a germ whose 2-jet is precisely (6.71). Since (6.71) is $(\mathcal{K}-2)$-determined it follows that g is $\mathcal{K}$-equivalent to (6.71).

From the proof of [104, V, (6.7)] it further follows that a germ that satisfies (6.69) and for which $D_{2,2}(f) = 0$ has codimension at least 5. □

Example 6.3.22. The germ

$$\begin{pmatrix} x^2 \\ y^2 \end{pmatrix}$$

was introduced in Example 6.3.8. For obvious reasons it is sometimes called the folded handkerchief singularity; cf. [104]. We can now see that it is $\mathcal{K}$-equivalent to (6.71). This follows from Proposition 6.3.21, but is also clear from the identity

$$\begin{pmatrix} \frac{1}{2} & \frac{1}{2} \\ \frac{1}{4} & \frac{-1}{4} \end{pmatrix} \begin{pmatrix} (x+y)^2 \\ (x-y)^2 \end{pmatrix} = \begin{pmatrix} x^2 + y^2 \\ xy \end{pmatrix}.$$

6.3.5 A Table of $\mathcal{K}$-Singularities

The classification results obtained in §§6.3.2–6.3.4 for $\mathcal{K}$-singularities without distinguished bifurcation parameter are collected in Table 6.1. This table is meant to be a ready reference for those who apply numerical methods to compute singularities. Representatives of singularities as given in Table 6.1 are often called normal forms; we note that in principle they are not unique although their choice is usually natural in the sense of being "simple."

Table 6.1: Codimension, representative, defining equations and nondegeneracy conditions for all $\mathcal{K}$-singularities without distinguished parameter and codimension at most 4.

Codim.	Representative	Defining equation	Nondeg. cond.
$k \geq 1$	x^{k+1}	$f = f_x = \cdots = f_{x^k} = 0$	$f_{x^{k+1}} \neq 0$
1	$x^2 + y^2$	$f = f_x = f_y = 0$	$D_2(f) > 0$
1	$x^2 - y^2$	$f = f_x = f_y = 0$	$D_2(f) < 0$
2	$x^3 + y^2$	$f = f_x = f_y = 0$ $D_2(f) = 0$	$D_3^1(f) \neq 0$ or $D_3^2(f) \neq 0$
3	$x^4 + y^2$	$f = f_x = f_y = 0$ $D_2(f) = 0$ $D_3^i(f) = 0, i = 1, 2$	$D_4^{1,1}(f) > 0$ or $D_4^{2,2}(f) > 0$
3	$x^4 - y^2$	$f = f_x = f_y = 0$ $D_2(f) = 0$ $D_3^i(f) = 0, i = 1, 2$	$D_4^{1,1}(f) < 0$ or $D_4^{2,2}(f) < 0$
4	$x^5 + y^2$	$f = f_x = f_y = 0$ $D_2(f) = 0$ $D_3^i(f) = 0, i = 1, 2$ $D_4^{i,i}(f) = 0, i = 1, 2$	$D_5^{1,1,1}(f) \neq 0$ or $D_5^{2,2,2}(f) \neq 0$
4	$x^3 + y^3$	$f = f_x = f_y = 0$ $f_{xx} = f_{xy} = f_{yy} = 0$	$D_\infty(f) > 0$
4	$x^3 - xy^2$	$f = f_x = f_y = 0$ $f_{xx} = f_{xy} = f_{yy} = 0$	$D_\infty(f) < 0$
4	$(x^2 + y^2, xy)^T$	$f_1 = f_2 = 0$ $f_{1x} = f_{1y} = f_{2x} = f_{2y} = 0$	$D_{2,2}(f) > 0$
4	$(x^2 - y^2, xy)^T$	$f_1 = f_2 = 0$ $f_{1x} = f_{1y} = f_{2x} = f_{2y} = 0$	$D_{2,2}(f) < 0$

Legend.

$$D_2(f) = f_{xx}f_{yy} - f_{xy}^2,$$

$$D_3^1(f) = f_{xx}(D_2(f))_y - f_{xy}(D_2(f))_x;\ D_3^2(f) = f_{xy}(D_2(f))_y - f_{yy}(D_2(f))_x,$$

$$D_4^{1,1}(f) = f_{xx}(D_3^1(f))_y - f_{xy}(D_3^1(f))_x;\ D_4^{2,2}(f) = f_{xy}(D_3^2(f))_y - f_{yy}(D_3^2(f))_x,$$

$$D_5^{1,1,1}(f) = f_{xx}(D_4^{1,1}(f))_y - f_{xy}(D_4^{1,1}(f))_x;\ D_5^{2,2,2}(f) = f_{xy}(D_4^{2,2}(f))_y - f_{yy}(D_4^{2,2}(f))_x,$$

$$D_\infty(f) = (f_{xxx}f_{yyy} - f_{xxy}f_{xyy})^2 - 4(f_{xxx}f_{xyy} - f_{xxy}^2)(f_{xxy}f_{yyy} - f_{xyy}^2),$$

$$D_{2,2}(f) = (f_{1xx}f_{2yy} - f_{2xx}f_{1yy})^2 - 4(f_{1xx}f_{2xy} - f_{2xx}f_{1xy})(f_{1xy}f_{2yy} - f_{2xy}f_{1yy}).$$

6.3.6 Example: Intersection of a Surface with Its Tangent Plane

Singularities with low codimension are fairly common. As an example, consider a surface in $\mathbb{R}^3$ described by $z = F(x, y)$, where F is a smooth function. Let (x_0, y_0, z_0) be a point on the surface. The tangent plane at this point is described by the equation $(x - x_0)\frac{\partial F}{\partial x_0} + (y - y_0)\frac{\partial F}{\partial y_0} - (z - z_0) = 0$. The intersection of surface and plane satisfies

$$F(x, y) - x\frac{\partial F}{\partial x_0} - y\frac{\partial F}{\partial y_0} = F(x_0, y_0) - x_0\frac{\partial F}{\partial x_0} - y_0\frac{\partial F}{\partial y_0}.$$

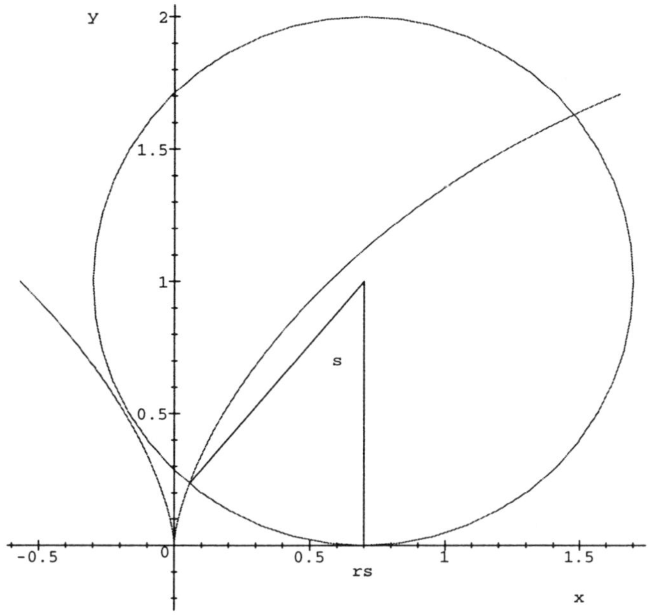

Figure 6.1: A point on a rolling wheel.

So the intersection is an isola center (normal form $x^2 + y^2$) if the surface has a positive curvature at (x_0, y_0, z_0) and a simple bifurcation point (normal form $x^2 - y^2$) if it has a negative curvature. Of course, singularities with higher codimension are found if the surface has zero curvature.

6.3.7 Example: A Point on a Rolling Wheel

The following example is a classic one but illustrates nicely the use of Table 6.1. We consider the movement of a point A on a rolling wheel in a neighborhood of the place where it touches the ground. The movement may be parameterized by the angle s as presented in Figure 6.1.

We easily see that

$$x = r(s - \sin s), \tag{6.74}$$

$$y = r(1 - \cos s), \tag{6.75}$$

where r is the radius of the wheel. Now by elementary calculus (Taylor series of sin and cos) there exist smooth functions f_1, f_2 from $\mathbb{R}$ into $\mathbb{R}$ with $f_1(0) = 0$, $f_2(0) = 0$, $f_1'(0) \neq 0$, $f_2'(0) \neq 0$ such that $1 - \frac{\sin s}{s} = f_1(s^2)$ and $1 - \cos s = f_2(s^2)$. Eliminating s from (6.74) and (6.75) we find $x^2 = r^2 f_2^{-1}(\frac{y}{r})(f_1 f_2^{-1}(\frac{y}{r}))(f_1 f_2^{-1}(\frac{y}{r}))$. Now define

$$H(x, y) \equiv x^2 - r^2 f_2^{-1}\left(\frac{y}{r}\right)\left(f_1 f_2^{-1}\left(\frac{y}{r}\right)\right)\left(f_1 f_2^{-1}\left(\frac{y}{r}\right)\right). \tag{6.76}$$

The equation $H(x, y) = 0$ hence descibes near $(0,0)$ the movement of a point on the wheel in Figure 6.1. The functions $f_2^{-1}(y)$ and $f_1(f_2^{-1}(y))$ vanish at 0 but their derivatives do not vanish. From this it follows that

$$H(0,0) = H_x(0,0) = H_y(0,0) = H_{xy}(0,0) = H_{yy}(0,0) = 0.$$

Hence $D_2(H)(0,0) = 0$. Furthermore, $D_3^1(H)(0,0) = H_{xx}(0,0)H_{yyy}(0,0) \neq 0$. Hence by Table 6.1 the path presents a cusp point at $(0,0)$.

6.4 Unfolding Theory

Intuitively, a singularity represents a point where the solution set to a system of nonlinear equations is unusual, special, and degenerate. If we perturb the equations at such a point, then we expect the behavior to become less degenerate. Unfolding theory makes this precise.

Definition 6.4.1. Let $f(x) \in \mathcal{E}_{n,p}$ represent a singularity. Then $F(x, a) \in \mathcal{E}_{n+k,p}$ is called an unfolding of $f(x)$ if $F(x, 0) = f(x)$ as a germ. F is called a $\mathcal{K}$-transversal unfolding if and only if $F_a(0,0)$ together with T_f spans $\mathcal{E}_{n,p}$.

From Definition 6.3.7 it follows that a $\mathcal{K}$-transversal unfolding can exist only if the codimension of the singularity f is finite; if so, then the number of parameters in a $\mathcal{K}$-transversal unfolding is at least equal to the codimension c of the singularity. On the other hand, if the codimension is finite, then a $\mathcal{K}$-transversal unfolding with c parameters is easily found. One just chooses functions $f_1, \ldots, f_c$, which together with T_f span $\mathcal{E}_{n,p}$; then $F(x, u_1, \ldots, u_c) = f(x) + u_1 f_1 + \cdots + u_c f_c$ is a $\mathcal{K}$-transversal unfolding of $f(x)$.

Definition 6.4.2. Let $F(x, a) \in \mathcal{E}_{n+k,p}$ and $G(x, b) \in \mathcal{E}_{n+l,p}$ be unfoldings of $f(x) \in \mathcal{E}_{n,p}$, respectively. We say that G factors through F if there exist $S(x, b) \in \mathcal{E}_{n+l,p,p}$, $X(x, b) \in \mathcal{E}^0_{n+l,n}$, and $A \in \mathcal{E}^0_{l,k}$ such that $S(x, 0) \equiv I_p$ and $X(x, 0) \equiv x$,

$$G(x, b) = S(x, b)F(X(x, b), A(b)) \tag{6.77}$$

in $\mathcal{E}_{n+l,p}$.

Definition 6.4.3. An unfolding $F(x, a) \in \mathcal{E}_{n+k,p}$ of a singular germ $f \in \mathcal{E}_{n,p}$ is called versal if every unfolding of $f(x)$ factors through $F(x, a)$. It is called universal if it is versal and contains c parameters where c is the codimension of the singularity of f.

The fundamental theorem related to these properties is the following one.

Theorem 6.4.4 (the $\mathcal{K}$-versality theorem). An unfolding $F(x, a) \in \mathcal{E}_{n+k,p}$ of a singular germ $f \in \mathcal{E}_{n,p}$ is $\mathcal{K}$-versal if and only if it is $\mathcal{K}$-transversal.

We will not prove this result. We refer to [104, V, (3.1)] and [180] (but see Exercise 4).

In [28] Beyn remarked that in all the cases that he considered an unfolding $F(x, a)$ of the singular germ with at least c parameters (c being the $\mathcal{K}$-codimension of the germ) is $\mathcal{K}$-transversal if and only if the extended Jacobian matrix of the defining system (i.e., with derivatives with respect to the unfolding parameters added as columns) has full rank. We prove that this actually holds for all singular germs of codimension 4 or less. The idea is that we will show this explicitly for the normal forms and then prove that it carries over to $\mathcal{K}$-equivalent germs.

We first consider the versality.

Proposition 6.4.5. Let $f \in \mathcal{E}_{n,p}$ be a singular germ and let $(S(x), X(x) \in \mathcal{E}_{n,p,p} \times \mathcal{E}^0_{n,n}$ with $X_x(0)$ and $S(0)$ nonsingular. Let $g \in \mathcal{E}_{n,p}$ be the germ determined by

$$g(x) = S(x)f(X(x)). \tag{6.78}$$

For every unfolding $F(x,a)$ of $f(x)$ let $G(x,a)$ be the unfolding of $g(x)$ defined by

$$G(x,a) = S(x)F(X(x),a). \tag{6.79}$$

This defines a one-to-one mapping from the unfoldings of f onto the unfoldings of g. Furthermore, $G(x,a)$ is a versal unfolding of $g(x)$ if and only if $F(x,a)$ is a versal unfolding of $f(x)$.

Proof. We first note that $(S^{-1}(x), X^{-1}(x)) \in \mathcal{E}_{n,p,p} \times \mathcal{E}^0_{n,n}$, that $(X^{-1})_x(0)$ and $S^{-1}(0)$ are nonsingular, and that

$$f(y) = S^{-1}(X^{-1}(y))g(X^{-1}(y)). \tag{6.80}$$

We can therefore map every unfolding $G_1(x,a)$ of $g(x)$ in a natural way to an unfolding $F_1(y,a)$ of $f(x)$ by

$$F_1(y,a) = S^{-1}(X^{-1}(y))G_1(X^{-1}(y),a). \tag{6.81}$$

This mapping is the inverse of the one defined in the statement of the proposition, which therefore has to be one to one and onto.

To complete the proof we show that if $F(x,a)$ is a versal unfolding of $f(x)$, then $G(x,a)$ is a versal unfolding of $g(x)$ (the converse then follows by duality). So let $F(x,a)$ be a versal unfolding of $f(x)$ and let $G_2(x,b)$ be any unfolding of $g(x)$. We define

$$F_2(y,b) = S^{-1}(X^{-1}(y))G_2(X^{-1}(y),b). \tag{6.82}$$

Since $F_2(y,b)$ is an unfolding of $f(y)$ there exist $S_1(y,b) \in \mathcal{E}_{n,p,p}$, $T_1(y,b) \in \mathcal{E}^0_{n,n}$, $A(b)$ with $S_1(y,0) \equiv I_p$, $T_1(y,0) \equiv y$, $A(0) = 0$ such that

$$F_2(y,b) = S_1(y,b)F(T_1(y,b),A(b)). \tag{6.83}$$

From (6.82) and (6.83) we infer

$$G_2(X^{-1}(y),b) = S(X^{-1}(y))S_1(y,b)F(T_1(y,b),A(b)). \tag{6.84}$$

By (6.79) we have

$$G(X^{-1}(T_1(y,b)),A(b)) = S(X^{-1}(T_1(y,b)))F(T_1(y,b),A(b)). \tag{6.85}$$

Combining (6.84) and (6.85) we have

$$G_2(X^{-1}(y),b) = S(X^{-1}(y))S_1(y,b)S^{-1}(X^{-1}(T_1(y,b)))G(X^{-1}(T_1(X(x),b)),A(b)) \tag{6.86}$$

or, equivalently,

$$G_2(X,b) = S(x)S_1(X(x),b)S^{-1}(X^{-1}(T_1(X(x),b)))G(X^{-1}(T_1(X(x),b)),A(b)). \tag{6.87}$$

Now

$$S(x)S_1(X(x),0)S^{-1}(X^{-1}(T_1(X(x),0))) = S(x)S^{-1}(X^{-1}X(x)) \equiv I_p$$

and

$$X^{-1}(T_1(X(x),0)) = X^{-1}X(x) = x.$$

Hence (6.87) proves that $G_2(x,b)$ factorizes through $G(x,a)$. We conclude that $G(x,a)$ is a versal unfolding of $g(x)$. □

Proposition 6.4.6. For all the normal forms in Table 6.1 an unfolding with at least c parameters (with c the $\mathcal{K}$-codimension) is transversal if and only if the extended Jacobian of the defining system has full rank.

Proof. To get the idea, consider the simplest case $f(x) = x^2$ first. Here $c = 1$ and T_f is the $\mathcal{E}_1$-ideal generated by x or, equivalently, by Lemma 6.3.5, the set of germs that vanish at 0. Let $F(x,a)$ be an unfolding of $f(x)$ with $a = (a_1, \ldots, a_l), l \geq 1$. This unfolding is transversal if and only if $F_{a_i}(x,0)$ is not in T_f for at least one index i or, equivalently, if and only if $F_{a_i}(0,0) \neq 0$. On the other hand, the extended Jacobian of the defining system given in Table 6.1 is

$$\begin{pmatrix} F_x(0,0) & F_{a_1}(0,0) & \ldots & F_{a_l}(0,0) \\ F_{xx}(0,0) & F_{xa_1}(0,0) & \ldots & F_{xa_l}(0,0) \end{pmatrix}. \tag{6.88}$$

Now $F_x(0,0) = 0, F_{xx}(0,0) \neq 0$. Hence the matrix in (6.88) is nonsingular if and only if there exists an index i such that $F_{a_i}(0,0) \neq 0$.

The next case is $f(x) = x^3$. Here $c = 2$ and T_f is the $\mathcal{E}_1$-ideal generated by x^2 or, equivalently, by Lemma 6.3.5, the set of germs which, together with their first derivatives, vanish at 0. Let $F(x,a)$ be an unfolding of $f(x)$ with $a = (a_1, \ldots, a_l), l \geq 1$. This unfolding is transversal if and only if there exist two indices i, j such that no nontrivial linear combination of $F_{a_i}(x,0)$ and $F_{a_j}(x,0)$ vanishes at 0 together with its first derivative. This is equivalent to the condition that

$$\begin{pmatrix} F_{a_i}(0,0) & F_{a_j}(0,0) \\ F_{xa_i}(0,0) & F_{xa_j}(0,0) \end{pmatrix} \tag{6.89}$$

is nonsingular.

On the other hand, the extended Jacobian of the defining system given in Table 6.1 is

$$\begin{pmatrix} F_x(0,0) & F_{a_1}(0,0) & \ldots & F_{a_l}(0,0) \\ F_{xx}(0,0) & F_{xa_1}(0,0) & \ldots & F_{xa_l}(0,0) \\ F_{xxx}(0,0) & F_{xxa_1}(0,0) & \ldots & F_{xxa_l}(0,0) \end{pmatrix}. \tag{6.90}$$

Now $F_x(0,0) = 0, F_{xx}(0,0) = 0, F_{xxx}(0,0) \neq 0$. Hence the matrix in (6.90) is nonsingular if and only if there exist indices i and j such that the matrix in (6.89) is nonsingular.

It is now clear that the cases $f(x) = x^k$ for $k \geq 3$ can be handled in the same way. This covers all singularities from $\mathbb{R}$ into $\mathbb{R}$ (cf. Table 6.1).

The next case in Table 6.1 is the singularity from $\mathbb{R}^2$ into $\mathbb{R}$ with normal form $f(x,y) = x^2 + y^2$. From Lemma 6.3.5 if follows that $T_f = \mathcal{E}_2^0 = \{g \in \mathcal{E}_2 | g(0,0) = 0\}$. Let $F(x,a)$ be an unfolding of f, $a = (a_1, \ldots, a_l)$, $l \geq 1$. This unfolding is transversal if and only if there exists an index i such that $F_{a_i}(0,0) \neq 0$. On the other hand, the extended Jacobian of the defining system $f = f_x = f_y = 0$ is

$$\begin{pmatrix} 0 & 0 & F_{a_1}(0,0,0) & \dots & F_{a_l}(0,0,0) \\ 2 & 0 & F_{xa_1}(0,0,0) & \dots & F_{xa_l}(0,0,0) \\ 0 & 2 & F_{ya_1}(0,0,0) & \dots & F_{ya_l}(0,0,0) \end{pmatrix}.$$

Obviously this matrix has full rank if and only if there exists an index i such that $F_{a_i}(0,0) \neq 0$.

The case $f(x,y) = x^2 - y^2$ is similar to the previous one.

Now consider $f(x,y) = x^3 + y^2$. T_f is the ideal generated in $\mathcal{E}_2$ by the germs x^2 and y; hence $T_f \subset \{g \in \mathcal{E}_2 | g(0,0) = g_x(0,0) = 0\}$. On the other hand, if $g \in \mathcal{E}_2$ satisfies the conditions $g(0,0) = g_x(0,0) = 0$, then by Lemma 6.3.5 we have $g - g_y(0,0)y \in \mathcal{M}_2^2 \subset T_f$ so that $g \in T_f$. We conclude that $T_f = \{g \in \mathcal{E}_2 | g(0,0) = g_x(0,0) = 0\}$. Let $F(x,y,a)$ be an unfolding of f, $a = (a_1, \dots, a_l)$, $l \geq 2$. This unfolding is transversal if and only if there exist two indices i, j such that no nontrivial linear combination of $F_{a_i}(x,y,0), F_{a_j}(x,y,0)$ is in T_f, i.e.,

$$\begin{pmatrix} F_{a_i}(0,0,0) & F_{a_j}(0,0,0) \\ F_{xa_i}(0,0,0) & F_{xa_j}(0,0,0) \end{pmatrix} \tag{6.91}$$

is nonsingular. On the other hand, the extended Jacobian of the defining system $f = f_x = f_y = D_2(f) = 0$ is

$$\begin{pmatrix} 0 & 0 & F_{a_1}(0,0,0) & \dots & F_{a_l}(0,0,0) \\ 0 & 0 & F_{xa_1}(0,0,0) & \dots & F_{xa_l}(0,0,0) \\ 0 & 2 & F_{ya_1}(0,0,0) & \dots & F_{ya_l}(0,0,0) \\ 12 & 0 & (D_2(f))_{a_1}(0,0) & \dots & (D_2(f))_{a_l}(0,0) \end{pmatrix}.$$

Obviously, this matrix has full rank if and only if there exist indices i, j such that the matrix in (6.91) is nonsingular.

Next consider $f(x,y) = x^4 + y^2$. T_f is the ideal generated in $\mathcal{E}_2$ by x^3 and y; hence it is included in the set $\{g \in \mathcal{E}_2 | g(0,0) = g_x(0,0) = g_{xx} = 0\}$. Conversely, if g is in this set, then it follows from Lemma 6.3.5 that $h = g - g_y(0,0)y - \frac{1}{2}g_{yy}(0,0)y^2$ is in $\mathcal{M}_2^3 \subset T_f$; hence $g \in T_f$. Let $F(x,a)$ be an unfolding of f, $a = (a_1, \dots, a_l)$, $l \geq 3$. It is transversal if and only if there exist three indices i, j, m such that no nontrivial linear combination of $F_{a_i}(x,y,0), F_{a_j}(x,y,0), F_{a_m}(x,y,0)$ is in T_f, i.e., such that the matrix

$$\begin{pmatrix} F_{a_i}(0,0,0) & F_{a_j}(0,0,0) & F_{a_m}(0,0,0) \\ F_{xa_i}(0,0,0) & F_{xa_j}(0,0,0) & F_{xa_m}(0,0,0) \\ F_{xxa_i}(0,0,0) & F_{xxa_j}(0,0,0) & F_{xxa_m}(0,0,0) \end{pmatrix} \tag{6.92}$$

is nonsingular. The extended Jacobian of the defining system $f = f_x = f_y = D_2(f) = D_3^2(f) = 0$ is

$$\begin{pmatrix} 0 & 0 & F_{a_1}(0,0,0) & \dots & F_{a_l}(0,0,0) \\ 0 & 0 & F_{xa_1}(0,0,0) & \dots & F_{xa_l}(0,0,0) \\ 0 & 2 & F_{ya_1}(0,0,0) & \dots & F_{ya_l}(0,0,0) \\ 0 & 0 & (D_2(F))_{a_1}(0,0,0) & \dots & (D_2(F))_{a_l}(0,0,0) \\ 96 & 0 & (D_3^2(F))_{a_1}(0,0,0) & \dots & (D_3^2(F))_{a_l}(0,0,0) \end{pmatrix}. \tag{6.93}$$

It is checked easily that $(D_2(F))_{a_i} = F_{xxa_i}$ for all indices i. Hence the matrix in (6.93) has full rank if and only if there exist indices i, j, m such that the matrix in (6.92) is nonsingular.

This completes the case $f(x,y) = x^4 + y^2$. The case $f(x,y) = x^4 - y^2$ is similar. The case $f(x,y) = x^5 + y^2$ is an obvious extension, so we leave it as an exercise.

Now consider $f(x,y) = x^3 + y^3$. T_f is the ideal generated in $\mathcal{E}_2$ by the germs x^2 and y^2. By a now familiar argument we can prove that

$$T_f = \{g \in \mathcal{E}_2 | g(0,0) = g_x(0,0) = g_y(0,0) = g_{xy}(0,0) = 0\}.$$

Let $F(x,y,a)$ be an unfolding of f, $a = (a_1, \dots, a_l)$, $l \geq 4$. The extended Jacobian of the defining system $f = f_x = f_y = f_{xx} = f_{xy} = f_{yy} = 0$ is given by

$$\begin{pmatrix} 0 & 0 & F_{a_1}(0,0,0) & \dots & F_{a_l}(0,0,0) \\ 0 & 0 & F_{xa_1}(0,0,0) & \dots & F_{xa_l}(0,0,0) \\ 0 & 0 & F_{ya_1}(0,0,0) & \dots & F_{ya_l}(0,0,0) \\ 6 & 0 & F_{xxa_1}(0,0,0) & \dots & F_{xxa_l}(0,0,0) \\ 0 & 0 & F_{xya_1}(0,0,0) & \dots & F_{xya_l}(0,0,0) \\ 0 & 6 & F_{yya_1}(0,0,0) & \dots & F_{yya_l}(0,0,0) \end{pmatrix} \tag{6.94}$$

and has full rank if and only if there exist indices i, j, m, n such that the submatrix

$$\begin{pmatrix} F_{a_i}(0,0,0) & F_{a_j}(0,0,0) & F_{a_m}(0,0,0) & F_{a_n}(0,0,0) \\ F_{xa_i}(0,0,0) & F_{xa_j}(0,0,0) & F_{xa_m}(0,0,0) & F_{xa_n}(0,0,0) \\ F_{ya_i}(0,0,0) & F_{ya_j}(0,0,0) & F_{ya_m}(0,0,0) & F_{ya_n}(0,0,0) \\ F_{xya_i}(0,0,0) & F_{xya_j}(0,0,0) & F_{xya_m}(0,0,0) & F_{xya_n}(0,0,0) \end{pmatrix}$$

is nonsingular; this is also the transversality condition.

The case $f(x,y) = x^3 - xy^2$ is similar to the previous one.

The first example of a singularity from $\mathbb{R}^2$ into $\mathbb{R}^2$ in Table 6.1 is

$$f(x,y) \equiv \begin{pmatrix} f_1(x,y) \\ f_2(x,y) \end{pmatrix} = \begin{pmatrix} x^2 + y^2 \\ xy \end{pmatrix}.$$

In this case

$$T_f = \left\{ \begin{pmatrix} g_1 \\ g_2 \end{pmatrix} \in (\mathcal{E}_2)^2 \middle| \begin{pmatrix} g_1(0,0) \\ g_2(0,0) \\ g_{1x}(0,0) - 2g_{2y}(0,0) \\ g_{1y}(0,0) - 2g_{2x}(0,0) \end{pmatrix} = 0 \right\} \tag{6.95}$$

since the space defined in (6.95) has codimension 4 in $\mathcal{E}_2^2$ and contains T_f. Let $F(x,y,a)$ be an unfolding of $f(x,y)$, $a = (a_1, \dots, a_l)$, $l \geq 4$. By some easy manipulations on the extended Jacobian of the defining system one shows that it has full rank if and only if there exist indices i, j, m, n such that

$$\begin{pmatrix} F_{1a_i} & F_{1a_j} & F_{1a_m} & F_{1a_n} \\ F_{2a_i} & F_{2a_j} & F_{2a_m} & F_{2a_n} \\ F_{1xa_i} - 2F_{2ya_i} & F_{1xa_j} - 2F_{2ya_j} & F_{1xa_m} - 2F_{2ya_m} & F_{1xa_n} - 2F_{2ya_n} \\ F_{1ya_i} - 2F_{2xa_i} & F_{1ya_j} - 2F_{2xa_j} & F_{1ya_m} - 2F_{2xa_m} & F_{1ya_n} - 2F_{2xa_n} \end{pmatrix}$$

Table 6.2: Universal unfoldings of the singularities in Table 6.1.

Codimension	Representative	Unfolding
$k \geq 1$	x^{k+1}	$x^{k+1} + \alpha_0 + \alpha_1 x + \cdots + \alpha_{k-1}x^{k-1}$
1	$x^2 + y^2$	$x^2 + y^2 + \alpha$
1	$x^2 - y^2$	$x^2 - y^2 + \alpha$
2	$x^3 + y^2$	$x^3 + y^2 + \alpha + \beta x$
3	$x^4 + y^2$	$x^4 + y^2 + \alpha + \beta x + \gamma x^2$
3	$x^4 - y^2$	$x^4 - y^2 + \alpha + \beta x + \gamma x^2$
4	$x^5 + y^2$	$x^5 + y^2 + \alpha + \beta x + \gamma x^2 + \delta x^3$
4	$x^3 + y^3$	$x^3 + y^3 + \alpha + \beta x + \gamma y + \delta xy$
4	$x^3 - xy^2$	$x^3 - xy^2 + \alpha + \beta x + \gamma y + \delta x^2$
4	$(x^2 + y^2, xy)^T$	$(x^2 + y^2 + \alpha + \beta x, xy + \gamma + \delta x)^T$
4	$(x^2 - y^2, xy)^T$	$(x^2 - y^2 + \alpha + \beta x, xy + \gamma + \delta x)^T$

is nonsingular. From the characterization of T_f in (6.95) it follows that this is also the transversality condition.

The other singularity from $\mathbb{R}^2$ into $\mathbb{R}^2$ in Table 6.1

$$f(x,y) \equiv \begin{pmatrix} f_1(x,y) \\ f_2(x,y) \end{pmatrix} = \begin{pmatrix} x^2 + y^2 \\ xy \end{pmatrix}$$

can be handled similarly. □

In Table 6.2 we give universal unfoldings for all $\mathcal{K}$-singularities given in Table 6.1. We leave it as an exercise to prove their correctness (use Proposition 6.4.6). The unfolding of x^k for $k = 1, 2, 3, 4$ is often called a turning point, cusp catastrophe, swallowtail, or butterfly, respectively.

We will now complete the proof that for all singular germs with $\mathcal{K}$-codimension at most 4 an unfolding is versal (equivalently, transversal) if and only if the extended Jacobian of the defining system has full rank.

First consider the cases where we found a defining system with a number of equations equal to the codimension.

Proposition 6.4.7. Let $f(z)$ be any singular germ of codimension 4 or less. Let $F(z, a)$ be an unfolding of $f(z)$ where the number of parameters in a is at least the $\mathcal{K}$-codimension of f. Then F is a $\mathcal{K}$-transversal unfolding (equivalently, a $\mathcal{K}$-versal unfolding) if and only if the extended Jacobian of the defining system in Table 6.1 for this singularity has full rank.

Proof. By Propositions 6.4.5 and 6.4.6 it is sufficient to check that the rank of the extended Jacobian of the defining system is preserved under $\mathcal{K}$-equivalence.

We start with the simplest case $f(x) = x^2$. Let $g(x) = S(x)f(X(x))$ be $\mathcal{K}$-equivalent to $f(x)$, $S(0) \neq 0, X_x(0) \neq 0$. By taking derivatives we find

$$\begin{pmatrix} g(x) \\ g_x(x) \end{pmatrix} = \begin{pmatrix} S(x) & 0 \\ \star & S(x)X_x(x) \end{pmatrix} \begin{pmatrix} f(X(x)) \\ f_x(X(x) \end{pmatrix}, \tag{6.96}$$

where $\star$ denotes a smooth function of x. The matrix in (6.96) is nonsingular at 0 and (6.96) transforms the defining system for f into the defining system for g. If $F(x, a)$ is an unfolding of $f(x)$ and $G(x, a)$ is the corresponding unfolding of $g(x)$ as in (6.79), then we have also

$$\begin{pmatrix} G(x,a) \\ G_x(x,a) \end{pmatrix} = \begin{pmatrix} S(x) & 0 \\ \star & S(x)X_x(x) \end{pmatrix} \begin{pmatrix} F(X(x),a) \\ F_x(X(x),a) \end{pmatrix}. \tag{6.97}$$

If we take derivatives of (6.97) with respect to a_i, where a_i is one of the unfolding parameters in a, and evaluate at $(0,0)$, then we obtain

$$\begin{pmatrix} G_{a_i}(0,0) \\ G_{xa_i}(0,0) \end{pmatrix} = \begin{pmatrix} S(0) & 0 \\ \star & S(0)X_x(0) \end{pmatrix} \begin{pmatrix} F_{a_i}(0,0) \\ F_{xa_i}(0,0) \end{pmatrix}. \tag{6.98}$$

On the other hand,

$$\begin{pmatrix} G_x(0,0) \\ G_{xx}(0,0) \end{pmatrix} = X_x(0) \begin{pmatrix} S(0) & 0 \\ \star & S(0)X_x(0) \end{pmatrix} \begin{pmatrix} F_x(0,0) \\ F_{xx}(0,0) \end{pmatrix}. \tag{6.99}$$

Therefore, the rank of the extended Jacobian of the defining system is invariant under $\mathcal{K}$-equivalence.

The case of $x^k, k \geq 3$ is handled in exactly the same way.

Now consider the normal form $x^3 + y^2$ from $\mathbb{R}^2$ into $\mathbb{R}$. Let $g(x, y) = S(x, y)f(X(x, y), Y(x, y))$ be $\mathcal{K}$-equivalent to it with $S(0,0)$ and $\frac{\partial(X,Y)}{\partial(x,y)}(0,0)$ nonsingular. The basic transformation formula in this case is (6.47). Let $F(x, y, a)$ be an unfolding of f with $a = (a_1, \ldots, a_l)$ and let $G(x, a)$ be the corresponding unfolding of g as in (6.79). Then the relation

$$\begin{pmatrix} G \\ G_x \\ G_y \\ D_2(G) \end{pmatrix} = T_4 \begin{pmatrix} F \\ F_x \\ F_y \\ D_2(F) \end{pmatrix} \tag{6.100}$$

holds with T_4 a matrix that is nonsingular at $(0,0,0)$. By taking derivatives of this relation with respect to x, y and components of a and evaluating them at $(0,0,0)$ one finds that the rank of the extended Jacobian matrix of the defining system of the singularity with normal form $x^3 + y^2$ is invariant under $\mathcal{K}$-equivalence.

The cases $f(x, y) = x^2 + y^2$ and $f(x, y) = x^2 - y^2$ can be handled similarly. For the cases $x^4 + y^2$ and $x^4 - y^2$ one has to start from (6.54) and apply the same type of argument; for $x^5 + y^2$ one has to use (6.59).

Now consider the normal form $x^3 + y^3$ from $\mathbb{R}^2$ into $\mathbb{R}$. Let $g(x, y) = S(x, y)f(X(x, y), Y(x, y))$ be $\mathcal{K}$-equivalent to it with $S(0,0)$ and $\frac{\partial(X,Y)}{\partial(x,y)}(0,0)$ nonsingular. By some easy calculations we find the transformation formulas

$$\begin{pmatrix} g \\ g_x \\ g_y \\ g_{xx} \\ g_{xy} \\ g_{yy} \end{pmatrix} = \begin{pmatrix} S & 0_{1,2} & 0_{1,3} \\ \star_{2,1} & SJ^T & 0_{2,3} \\ \star_{3,1} & \star_{3,2} & SJ_1 \end{pmatrix} \begin{pmatrix} f \\ f_x \\ f_y \\ f_{xx} \\ f_{xy} \\ f_{yy} \end{pmatrix}, \tag{6.101}$$

where the stars denote smooth matrix functions of x, y with the indicated dimensions, J is the nonsingular Jacobian matrix $\frac{\partial(X,Y)}{\partial(x,y)}$, and J_1 is the 3×3 matrix

$$J_1 = \begin{pmatrix} X_x^2 & 2X_xY_x & Y_x^2 \\ X_xX_y & X_xY_y + X_yY_x & Y_xY_y \\ X_y^2 & 2X_yY_y & Y_y^2 \end{pmatrix}. \tag{6.102}$$

Now J_1 is precisely the condensed tensor product $\otimes_2(J^T)$ that was introduced in §4.4.2; in Proposition 4.4.4 we proved that it is nonsingular if J is nonsingular. The rest of the argument is similar to the previous cases.

The case $x^3 - xy^2$ is analogous to the preceding one since the defining system is the same.

Now consider the singularity with normal form

$$f(x,y) \equiv \begin{pmatrix} f_1(x,y) \\ f_2(x,y) \end{pmatrix} = \begin{pmatrix} x^2 + y^2 \\ xy \end{pmatrix}$$

from $\mathbb{R}^2$ into $\mathbb{R}^2$. Let

$$S(x,y) = \begin{pmatrix} S_1(x,y) & S_2(x,y) \\ S_3(x,y) & S_4(x,y) \end{pmatrix}$$

be a smooth matrix function with $S(0,0)$ nonsingular and let $(X(x,y), Y(x,y))^T$ be a transformation whose Jacobian J is nonsingular at $(0,0)$. Then if

$$g(x,y) = S(x,y)f(X(x,y), Y(x,y)),$$

the defining system is transformed into

$$\begin{pmatrix} g_1 \\ \cdots \\ g_{2y} \end{pmatrix} = \begin{pmatrix} S & 0_{2,4} \\ \star_{4,2} & S \otimes J^T \end{pmatrix} \begin{pmatrix} f_1 \\ \cdots \\ f_{2y} \end{pmatrix},$$

where $S \otimes J^T$ is a nonsingular $(4,4)$ tensor product matrix and $\star_{4,2}$ is a smooth 4×2 matrix function of x, y. By taking derivatives of this identity and evaluating them at a point where the defining system is zero, one finds that the rank of the extended Jacobian of the defining system is preserved.

The other singularity from $\mathbb{R}^2$ into $\mathbb{R}^2$ can be handled in the same way since the defining system is the same. □

In a typical application one is mainly interested in the case where the number of unfolding parameters is precisely the codimension of the singularity. In all cases except $x^4 \pm y^2$ and $x^5 + y^2$ the number of defining equations is equal to the sum of the number of state variables and the codimension of the singularity. The unfolding is then universal if and only if the extended Jacobian of the defining system is a nonsingular matrix. The standard Newton method can be used and converges quadratically. We now consider the cases $x^4 \pm y^2$ and $x^5 + y^2$ separately and show that also in this case square systems can be used. To prove this it is convenient to consider a more general setting.

Proposition 6.4.8. Let the definition of $D_2(f), D_3^1(f), \ldots, D_5^{ijk}(f)$ be extended in the natural way to $D_{k+2}^{i_1 i_2 \ldots i_k}$ with $k = 0, 1, 2, \ldots$ and $i_1, \ldots, i_k \in \{1, 2\}$.

If $f_{xx} \neq 0$, then for every $k \geq 0$ and indices $(i_1, \ldots, i_k)$ there exist smooth functions $\phi_{k+2}, \phi_{k+1}, \ldots, \phi_2$ such that

$$D_{k+2}^{i_1 \ldots i_k}(f) = \phi_{k+2} D_{k+2}^{1 \ldots 1}(f) + \phi_{k+1} D_{k+1}^{1 \ldots 1}(f) + \cdots + \phi_2 D_2(f).$$

The functions $\phi_{k+2}, \ldots, \phi_2$ depend also on the upper indices of $D_{k+2}^{i_1 \ldots i_k}$, but for simplicity of notation this dependence is not made explicit.

If $f_{yy} \neq 0$, then for every $k \geq 0$ and indices $(i_1, \ldots, i_k)$ there exist smooth functions $\phi_{k+2}, \phi_{k+1}, \ldots, \phi_2$ such that

$$D_{k+2}^{i_1 \ldots i_k}(f) = \phi_{k+2} D_{k+2}^{2 \ldots 2}(f) + \phi_{k+1} D_{k+1}^{2 \ldots 2}(f) + \cdots + \phi_2 D_2(f).$$

The functions $\phi_{k+2}, \ldots, \phi_2$ depend also on the upper indices of $D_{k+2}^{i_1 \ldots i_k}$, but for simplicity of notation this dependence is not made explicit.

Proof. We will consider the case that $f_{xx} \neq 0$; the other case is similar.

The result is obvious for $k = 0$. We will prove it for other values of k by induction. Assume that the result holds for k. Suppressing the dependence on f in the notation we have

$$D_{k+3}^{i_1 \ldots i_k 1} = f_{xx}(D_{k+2}^{i_1 \ldots i_k})_y - f_{xy}(D_{k+2}^{i_1 \ldots i_k})_x$$

$$= f_{xx}(\phi_{k+2} D_{k+2}^{1 \ldots 1} + \cdots + \phi_2 D_2)_y - f_{xy}(\phi_{k+2} D_{k+2}^{1 \ldots 1} + \cdots + \phi_2 D_2)_x$$

$$= \phi_{k+2} D_{k+3}^{1 \ldots 1} + \phi_{k+1} D_{k+2}^{1 \ldots 1} + \cdots + \phi_2 D_3^1 + \cdots,$$

where the last dots denote a sum of factors each of which contains one of $D_{k+2}^{1 \ldots 1}, \ldots, D_2$. This proves the result if the last upper index is 1.

Next, consider $D_{k+3}^{i_1 \ldots i_k 2}$. From

$$D_{k+3}^{i_1 \ldots i_k 2} = f_{xy}(D_{k+2}^{i_1 \ldots i_k})y - f_{yy}(D_{k+2}^{i_1 \ldots i_k})x$$

and

$$D_{k+3}^{i_1 \ldots i_k 1} = f_{xx}(D_{k+2}^{i_1 \ldots i_k})y - f_{xy}(D_{k+2}^{i_1 \ldots i_k})x$$

we infer that

$$f_{xx} D_{k+3}^{i_1 \ldots i_k 2} - f_{xy} D_{k+3}^{i_1 \ldots i_k 1} = -D_2(f)(D_{k+2}^{i_1 \ldots i_k})_x.$$

Since we assumed that $f_{xx} \neq 0$ this proves the result if the last upper index is 2. $\square$

Proposition 6.4.9. Let $f(z)$ be a singular germ that is $\mathcal{K}$-equivalent to $x^4 \pm y^2$. Let $F(z,a)$ be an unfolding of $f(a)$ where the number of parameters in a is 3, i.e., the codimension of f. Then the following conditions hold:

(a) F is a $\mathcal{K}$-transversal unfolding (equivalently, a $\mathcal{K}$-universal unfolding) if and only if the Jacobian of the defining system

$$(f, f_x, f_y, D_2(f), D_3^1(f), D_3^2(f))^T \tag{6.103}$$

has full rank 5.

(b) If $f_{xx} \neq 0$, then the sixth row of the Jacobian of (6.103) can be dropped so that the condition is that the Jacobian of

$$(f, f_x, f_y, D_2(f), D_3^1(f))^T \tag{6.104}$$

is nonsingular. (Similarly the fifth row in the Jacobian of (6.103) can be dropped if $f_{yy} \neq 0$.)

Proof. In Proposition 6.4.7 (a) was already proved. To prove (b) we note that by Proposition 6.4.8 for $f_{xx} \neq 0$ the sixth row of the Jacobian of (6.103) is a linear combination of the fourth and fifth rows (when evaluated at the singular point that is to be computed). Therefore, it does not contribute to the rank of the Jacobian. The case $f_{yy} \neq 0$ is similar. □

Proposition 6.4.10. Let $f(z)$ be a singular germ that is $\mathcal{K}$-equivalent to $x^5 + y^2$. Let $F(z,a)$ be an unfolding of $f(z)$, where the number of parameters in a is 4, i.e., the codimension of f. Then the following conditions hold:

(a) F is a $\mathcal{K}$-transversal unfolding (equivalently, a $\mathcal{K}$-universal unfolding) if and only if the Jacobian of the defining system

$$(f, f_x, f_y, D_2(f), D_3^1(f), D_3^2(f), D_4^{1,1}(f), D_4^{1,2}(f), D_4^{2,1}(f), D_4^{2,2}(f))^T \tag{6.105}$$

has full rank 6.

(b) If $f_{xx} \neq 0$, then the sixth, eighth, ninth, and tenth rows of the Jacobian of (6.105) can be dropped so that the condition is that the Jacobian of

$$(f, f_x, f_y, D_2(f), D_3^1(f), D_4^{1,1}(f))^T \tag{6.106}$$

is nonsingular. (Similarly the fifth, seventh, eighth, and ninth rows in (6.105) can be dropped if $f_{yy} \neq 0$.)

Proof. In Proposition 6.4.7 (a) was already proved. To prove (b) suppose that $f_{xx} \neq 0$. By Proposition 6.4.8. the sixth, eighth, ninth, and tenth rows are linear combinations of the other rows when evaluated at the singular point. Hence these rows do not contribute to the rank of the Jacobian and may be omitted. □

6.5 Example: The Continuous Flow Stirred Tank Reactor

The continuous flow stirred tank reactor (CSTR) is a model problem from chemical engineering that exhibits multiple solutions. The equilibrium equations can be reduced to the case of one state variable only and so they can be studied without the complications caused by a Lyapunov–Schmidt reduction. They exhibit various singularities that can be studied either analytically or numerically.

The CSTR has been investigated by many authors, starting with [247] and [232]. It is studied in great detail in [109], where it serves as a model example for the use of singularity theory methods. Dynamical aspects are studied extensively in [128], [122], and [2]. We follow [145] to demonstrate the application of numerical methods in singularity theory. Contrary to [109] and [145], we first consider the case where there is no distinguished bifurcation parameter. This will later give us the additional opportunity to point out the differences with the case of a distinguished bifurcation parameter.

6.5.1 Description of the Model

A CSTR is a vessel of unit volume in which a reactant flows continuously at a rate r and undergoes a single, exothermic reaction to form inert products. The reactor is well stirred, so that the concentration c of the reactant and the temperature T are uniform throughout the vessel. The unused reactant and the products leave the vessel at the same rate r as the input. Heat is removed from the reactor by a coolant fluid at temperature T_c. The concentration and temperature in the reactor are modeled by the following pair of coupled ordinary differential equations (ODEs; without the stirring we would have partial differential equations (PDEs) instead of ODEs):

$$\frac{dc}{dt} = r(c_f - c) - Zc\mathcal{A}(T), \tag{6.107}$$

$$\frac{dT}{dt} = r(T_f - T) + k(T_c - T) + hZc\mathcal{A}(T). \tag{6.108}$$

In the first terms on the right of (6.107), (6.108), c_f and T_f denote the concentration and temperature of the incoming reactant. The middle term in (6.108) represents heat removed by the coolant; it contains its own parameter k. The final terms describe the effects of the reaction. Here $\mathcal{A}(T)$ has the Arrhenius form

$$\mathcal{A}(T) = \exp\left\{\frac{T_a}{T_f} - \frac{T_a}{T}\right\}. \tag{6.109}$$

The parameter Z is introduced to allow the scaling $\mathcal{A}(T_f) = 1$. Furthermore T_a is an activation temperature and h is proportional to the heat released by the reaction.

We now restrict our attention to the equilibrium solutions of (6.107), (6.108). Clearly we can solve (6.107) for c and substitute this into (6.108). In order to nondimensionalize this equation we define a normalized temperature $x = (T - T_f)/T_f$. To obtain a simple form we further introduce the parameters $\alpha_1 = hc_f/T_f$, $\alpha_2 = k/Z$, $\alpha_3 = (T_c - T_f)/T_f$, $\alpha_4 = T_a/T_f$, $\lambda = r/k$.

After some trivial computations we find that with these notations the equilibria of the CSTR are described by

$$g(x, \alpha_1, \alpha_2, \alpha_3, \alpha_4, \lambda) = 0, \tag{6.110}$$

where

$$g(x, \alpha_1, \alpha_2, \alpha_3, \alpha_4, \lambda) \equiv \alpha_3 - (1+\lambda)x + \frac{\lambda\alpha_1}{1+\lambda\alpha_2 A(x, \alpha_4)} \tag{6.111}$$

and

$$A(x, \alpha_4) = \exp\{-\alpha_4 x/(1+x)\}. \tag{6.112}$$

We remark that $x > -1$, $\alpha_1 > 0$, $\alpha_2 > 0$, $\alpha_3 > -1$, $\alpha_4 > 0$, $\lambda > 0$ for physically significant solutions to (6.110). Also, in practice $\alpha_4 >> 1$. The main results proved in [109] on the CSTR require $\alpha_4 > \frac{8}{3}$. In our numerical examples we will always have $\alpha_4 = 3$.

6.5.2 Numerical Computation of a Cusp Point

We will now describe some numerical work on (6.110). The bulk of the work actually consists in writing explicit Fortran codes for all the necessary derivatives of $g(x, \alpha_1, \alpha_2, \alpha_3, \alpha_4, \lambda)$. This, of course, is also the part that could be automated most easily. We will not engage in a discussion on the merits of various approaches (symbolic differentiation, finite differences, etc.) although it is clear that in the near future automatic differentiation will make the computation of derivatives a trivial task at least in the case of small dense systems like the CSTR.

For continuation and for computation of special points we used PITCON (§2.4). With this code and our explicit Fortran routines for the derivatives of g the task of computing special points becomes relatively easy; it consists in writing simple driving programs for PITCON, calling the right derivative routines at the right time and place and then examining the outcome.

The actual computations can be organized in many ways. Jepson and Spence [146] studied this case carefully; we choose to reproduce the main line of their results as far as relevant for the case with no distinguished bifurcation parameter. This allows an additional check on the accuracy of our results; also it will allow us later to compare this case with the distinguished bifurcation parameter case.

As in [146] we started by computing a curve in (x, λ)-space with fixed values $\alpha_1 = 10$, $\alpha_2 = 10$, $\alpha_3 = -0.9$, $\alpha_4 = 3$, starting from the trivial solution $(x, \lambda) = (-0.9, 0)$. We found turning points with respect to λ at $(x_1, \lambda_1) = (-0.14356, 1.25066)$ and $(x_2, \lambda_2) = (0.393178, 0.377651)$. There is also a turning point with respect to x at $(x_3, \lambda_3) = (1.33960, 0.961585)$. We further found $g_{xx}(x_1, \lambda_1) = 3.10030$, $g_\lambda(x_1, \lambda_1) = 0.164841$, $g_{xx}(x_2, \lambda_2) = -1.47409$, $g_\lambda(x_2, \lambda_2) = 1.06411$, $g_{\lambda\lambda}(x_3, \lambda_3) = -1.75992$, $g_x(x_3, \lambda_3) = -0.737697$. So the mapping

$$\mathbb{R} \to \mathbb{R}, x \to g(x, \alpha_1, \alpha_2, \alpha_3, \alpha_4, \lambda)$$

has a singularity with codimension 1 at x_1 if $\lambda = \lambda_1$ and at x_2 if $\lambda = \lambda_2$. On the other hand, the mapping

$$\mathbb{R} \to \mathbb{R}, \lambda \to g(x, \alpha_1, \alpha_2, \alpha_3, \alpha_4, \lambda)$$

has a singularity with codimension 1 at λ_3 if $x = x_3$. For the mapping $\mathbb{R} \times \mathbb{R} \to \mathbb{R}, (x, \lambda) \to g(x, \alpha_1, \alpha_2, \alpha_3, \alpha_4, \lambda)$ none of the computed points is a singularity.

Table 6.3: Turning points of (6.113).

	x	λ	β	f_{xx}	f_λ
1	−0.1732895	0.405523	0.608093	0	0.402282
2	0.030654	1.77242	−0.12751	1.62886	0
3	0.25942	1.96889	−0.090723	0	−0.18997

As in [146] we now free a parameter β in such a way that $\alpha_1(\beta) = 10-9\beta$, $\alpha_2 = 10-9\beta$, $\alpha_3 = -0.9 + 0.4\beta$, $\alpha_4 = 3$. The underlying idea is that for $\beta = 0$ this reduces to the previous situation with two singularities in x-variable, while for $\beta = 1$ no such singularities can be found (this can be checked numerically as above). Hence we can expect that something "special" will happen if we free β in the way described above. We continue numerically with PITCON the solution of

$$g = g_x = 0 \tag{6.113}$$

with independent variables x, λ, β and starting from $x = -0.14356, \lambda = 1.25066, \beta = 0$. As a result we find, confirming [146], three limit points with respect to β. Their essential data are given in Table 6.3.

We remark that the determinant of $\frac{\partial(g,g_x)}{\partial(x,\lambda)}$ is given by $g_x g_{x\lambda} - g_\lambda g_{xx}$. Therefore, along a curve with $g = g_x = 0$ limit points with respect to β are precisely points where either $g_\lambda = 0$ or $g_{xx} = 0$. The points in Table 6.3 could therefore be detected and computed by the option in PITCON to compute limit points for the system (6.113).

The first and third points in Table 6.3 are codimension-2 singularities of the mapping

$$\mathbb{R} \to \mathbb{R}, x \to g(x, \alpha_1, \alpha_2, \alpha_3, \alpha_4, \lambda)$$

(we checked that the third derivatives g_{xxx} are nonzero). The second point in Table 6.3 is a codimension-1 singularity of the mapping

$$\mathbb{R} \times \mathbb{R} \to \mathbb{R}, (x, \lambda) \to g(x, \alpha_1, \alpha_2, \alpha_3, \alpha_4, \lambda)$$

(we checked that $D_2(g)$ is nonzero).

We remark that singularities in other plane sections cannot be found simply by looking for limit points with respect to β. For example, the computed solution curves of (6.113) also exhibit several sign changes of g_{α_4}. The points on this curve where α_4 vanishes can also be computed using PITCON.

Points where $g = g_x = g_{\alpha_4} = 0$ are clearly singularities (with codimension at least 1) of the mapping

$$\mathbb{R} \times \mathbb{R} \to \mathbb{R}, (x, \alpha_4) \to g(x, \alpha_1, \alpha_2, \alpha_3, \alpha_4, \lambda).$$

Since λ is a parameter of the problem that in an experimental setting can be changed more easily than α_4 it is a tradition to concentrate on the further study of singularities in (x, λ)-space.

We will now follow a curve of simple bifurcation points in (x, λ)-space. As in [146] we free a new parameter γ so that $\alpha_1(\beta) = 10 - 9\beta + \gamma$, $\alpha_2 = 10 - 9\beta$, $\alpha_3 = -0.9 + 0.4\beta$,

$\alpha_4 = 3$. We follow with PITCON the solution of

$$g = g_x = g_\lambda = 0 \tag{6.114}$$

with independent variables $x, \lambda, \beta, \gamma$ and starting from $x = 0.030654, \lambda = 1.77242, \beta = -0.12751$. From Table 6.1 we know that the solution curve to (6.114) is a curve of simple bifurcation points in (x, λ)-space if $D_2(g) \neq 0$. During the continuation a sign change of $D_2(g)$ was indeed observed. Zooming in at this point in the now familiar way we find a point with coordinates $(x, \alpha_1, \alpha_2, \alpha_3, \alpha_4, \lambda) = (0.173746, 6.44103, 7.34053, -0.781802, 3.0, 1.08077)$, where $g = g_x = g_\lambda = D_2(g) = 0$. Further computations show that $D_3^1(g)$ does not vanish in this point, so it is indeed a cusp point.

6.5.3 The Universal Unfolding of a Cusp Point

To get a priori information on the expected behavior of the solution sets of (6.110) in (x, λ)-space for parameter values $(\alpha_1, \alpha_2, \alpha_3, \alpha_4)$ near cusp point values, we study a universal unfolding of the cusp singularity.

In terms of x, λ the cusp singularity has normal form

$$g(x, \lambda) = x^2 + \lambda^3 \tag{6.115}$$

and is characterized by the defining conditions $g = g_x = g_\lambda = D_2(g) = 0$ and the nondegeneracy condition that either $D_3^1(g) \neq 0$ or $D_3^2(g) \neq 0$; here $D_2(g) = g_{xx}g_{\lambda\lambda} - g_{x\lambda}^2$, $D_3^1(g) = g_{xx}(D_2(g))_\lambda - g_{x\lambda}(D_2(g))_x$, $D_3^2(g) = g_{x\lambda}(D_2(g))_\lambda - g_{\lambda\lambda}(D_2(g))_x$ (cf. Table 6.1).

From Table 6.2 it follows that the two-parameter unfolding

$$g(x, \lambda, \alpha, \beta) = x^2 + \lambda^3 + \alpha + \beta\lambda \tag{6.116}$$

is a universal unfolding of (6.115). Hence by varying the parameters α, β in (6.116) we find (qualitatively) all perturbed behavior that can be expected in a neighborhood of a cusp point in a general nonlinear problem, e.g., in the CSTR. Let us therefore study the zero sets of (6.116) for all possible values of α, β. Clearly they are all symmetric with respect to the λ-axis, but this property is not preserved by contact equivalence. The results of our analysis are illustrated in Figures 6.2 and 6.3.

Figure 6.2 indicates a natural partition of (α, β)-plane and the position of eight typical points, called P1, P2, ..., P8 in this plane. Figure 6.3 contains eight diagrams, each corresponding to one of these points. The diagram is the representation in (x, λ)-space of the zero set of (6.116) for the corresponding values of α, β; λ and x are presented along the horizontal and vertical axes, respectively.

Let us define a function

$$f(\lambda, \alpha, \beta) \equiv -\lambda^3 - \alpha - \beta\lambda \tag{6.117}$$

and its derivative

$$f_\lambda(\lambda, \alpha, \beta) = -3\lambda^2 - \beta. \tag{6.118}$$

Obviously (6.116) is solvable for x only if $f(\lambda, \alpha, \beta) \geq 0$. We also remark that $f(\lambda, \alpha, \beta)$ is always positive for large negative values of λ and negative for large positive values of

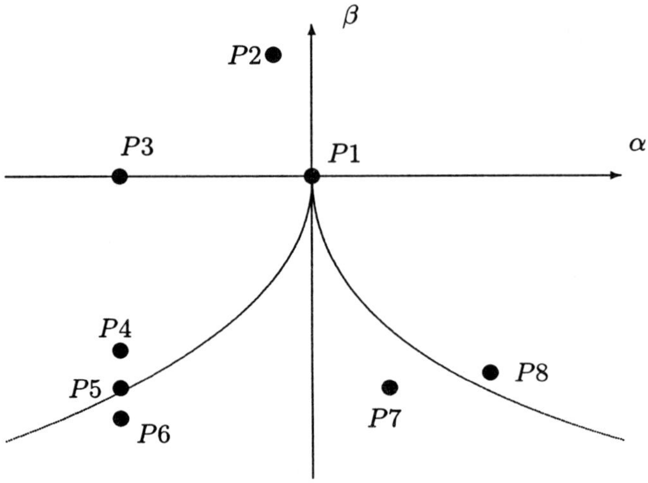

Figure 6.2: Cusp curve $\frac{4}{27}\beta^3 + \alpha^2 = 0$.

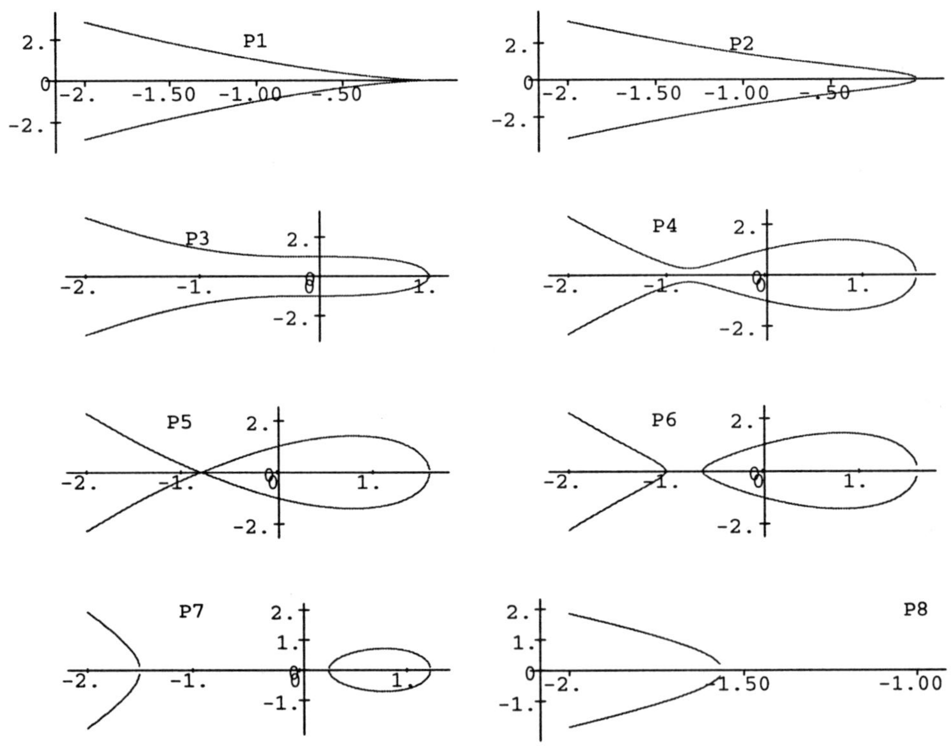

Figure 6.3: Eight diagrams corresponding to points in Figure 6.2.

λ. The diagrams in Figure 6.3 are therefore all bounded at the right side and unbounded at the left side. Again, of course, this property is not preserved by contact equivalence.

For $\alpha = \beta = 0$ (P1 in Figure 6.2) we of course obtain the cusp singularity (6.115) represented at the top left in Figure 6.3.

If $\beta > 0$, then $f(\lambda, \alpha, \beta)$ is a monotone decreasing function of λ. The zero set of (6.116) therefore consists of a single branch with a turning point in a point where $x = 0$ and λ is the unique solution to $f(\lambda, \alpha, \beta) = 0$. An example of this case is P2 with $\alpha = 0, \beta = 1$ (Figures 6.2 and 6.3).

A limit case of the previous one is the case $\beta = 0$. Then $f(\lambda, \alpha, \beta)$ is decreasing everywhere except for $\lambda = 0$ where it is stationary. The zero set of (6.116) therefore has the same properties as in the case $\beta > 0$; if $\alpha < 0$ then for $(x, \lambda) = (\pm\sqrt{-\alpha}, 0)$ it has tangent lines parallel to the λ-axis. An example of this case is P3, where $\alpha = -1, \beta = 0$ (Figures 6.2 and 6.3).

From now on we assume $\beta < 0$. Then f attains a local minimum in the point

$$(\lambda, f(\lambda, \alpha, \beta)) = \left(-\sqrt{\frac{-\beta}{3}}, \frac{2}{3}\beta\sqrt{\frac{-\beta}{3}} - \alpha\right)$$

and a local maximum in the point

$$(\lambda, f(\lambda, \alpha, \beta)) = \left(\sqrt{\frac{-\beta}{3}}, -\frac{2}{3}\beta\sqrt{\frac{-\beta}{3}} - \alpha\right).$$

We now have to distinguish several cases. A first possibility is that the local minimum of f is attained for a positive value of f. This happens precisely if $\alpha < 0, \beta < 0, \frac{4}{27}\beta^3 + \alpha^2 > 0$. Then the zero set of (6.116) is a single branch with a turning point where $x = 0$ and λ is the unique solution to $f(\lambda, \alpha, \beta) = 0$. An example is P4 ($\alpha = -1, \beta = -1.8$). We remark that the form of the corresponding diagram in (x, λ)-space already suggests that we are close to a case with a simple bifurcation point.

A limit case of the preceding one is where the local mimimum is precisely zero. This happens if $\alpha < 0$, $\frac{4}{27}\beta^3 + \alpha^2 = 0$. The zero set of (6.116) is then a curve with a self-crossing in $(x, \lambda) = (-\sqrt{-\beta/3}, 0)$. This is a simple bifurcation point of (6.116). An example is the point P5 ($\alpha = -1, \beta = -3/\sqrt[3]{4}$) (Figures 6.2 and 6.3).

In the remaining cases $\beta < 0$ and the value of the local minimum is negative, i.e.,

$$\frac{2}{3}\beta\sqrt{\frac{-\beta}{3}} - \alpha < 0. \tag{6.119}$$

First assume that the local maximum is positive, i.e.,

$$-\frac{2}{3}\beta\sqrt{\frac{-\beta}{3}} - \alpha > 0. \tag{6.120}$$

The conditions (6.119), (6.120) together are equivalent to

$$\frac{4}{27}\beta^3 + \alpha^2 < 0. \tag{6.121}$$

For such values of α, β the equation $f(\lambda, \alpha, \beta) = 0$ has three distinct real roots, say, w_1, w_2, w_3 with $w_1 < w_2 < w_3$. Then $f(\lambda, \alpha, \beta) \geq 0$ in two disjoint λ-intervals, namely, $[-\infty, w_1]$ and $[w_2, w_3]$. In these intervals (6.116) has zero points. The zero set of (6.116) therefore has two disconnected branches, one of them bounded and the other unbounded. The points P6 ($\alpha = -1, \beta = -2$) and P7 ($\alpha = 0.5, \beta = -1.9$) are examples of this case. We remark that P6 is still close to the case with a self-crossing. In the diagram corresponding to P7 the bounded branch (this is sometimes called an isola) is far from the unbounded one and, in fact, is close to disappearing.

A limit case of the previous one is when the local maximum is exactly zero. This happens if $\alpha > 0$ and

$$\frac{4}{27}\beta^3 + \alpha^2 = 0. \tag{6.122}$$

The isola from the preceding case now reduces to a single point $(x, \lambda) = (0, -\frac{2}{3}\sqrt{-\beta/3} - \alpha)$. In the classification of singularities this is an isola center, i.e., a zero of (6.116) where $g_x = g_\lambda = 0$, $g_{xx}g_{\lambda\lambda} - g_{x\lambda}^2 > 0$.

The remaining case is that the local maximum is attained in a point with negative function value, i.e., $\alpha > 0$ and $\frac{4}{27}\beta^3 + \alpha^2 > 0$. The zero set of (6.116) is then a single branch with a turning point where $x = 0$, λ is the smallest root of the the equation $f(\lambda, \alpha, \beta) = 0$. See point P8 ($\alpha = 1, \beta = -1.8$) in Figures 6.2 and 6.3.

The importance of the solution curve to (6.122) in the parameter plane is now obvious. We remark that it is itself a cusp curve.

6.5.4 Example: Unfolding a Cusp in the CSTR

In §6.5.2 a curve of codimension-1 singularities ($g = g_x = g_\lambda = 0$) was computed numerically, starting with simple bifurcation points ($D_2(g) < 0$), passing through a cusp point ($D_2(g) = 0$), and continuing with isola centers ($D_2(g) > 0$). From the theoretical study in §6.5.3 we know that qualitative changes in the solution pictures to (6.110) in (x, λ)-space are to be expected near this curve, in particular, close to the cusp point itself. We will now show that we indeed find the expected behavior in the CSTR.

To produce pictures of the solution sets of (6.110) in (x, λ)-space for fixed values of α_1, α_2, α_3, α_4 it is not necessary to compute the solution curves by a continuation code. Indeed, for the physically relevant cases ($\lambda > 0, \alpha_2 > 0$), (6.110) is equivalent to the equation

$$-x\alpha_2 A(x, \alpha_4)\lambda^2 + ((\alpha_3 - x)\alpha_2 A(x, \alpha_4) - x + \alpha_1)\lambda + (\alpha_3 - x) = 0, \tag{6.123}$$

which is quadratic in λ. Hence it has at most two solutions in λ if the other variables are known. Pictures of the solution set can be produced easily by any package that handles elementary mathematical operations and produces graphs. We used MAPLE for this purpose.

First, we consider the cusp point $(x, \alpha_1, \alpha_2, \alpha_3, \alpha_4, \lambda) = (0.173746, 6.44103, 7.34053, -0.781802, 3.0, 1.08077)$ computed in §6.5.2. Figure 6.4 presents the solution set to (6.110) (equivalently, (6.123)) in (λ, x)-space near this point. The cusp singularity is clearly visible at the expected place.

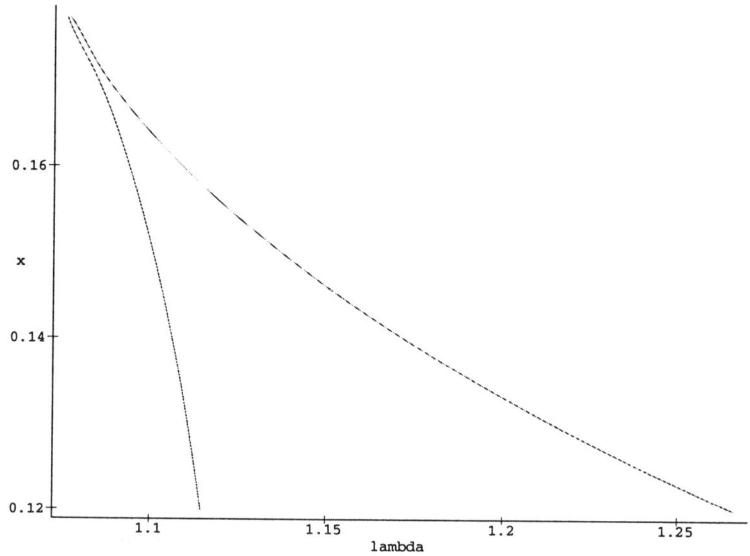

Figure 6.4: A cusp curve in the CSTR model.

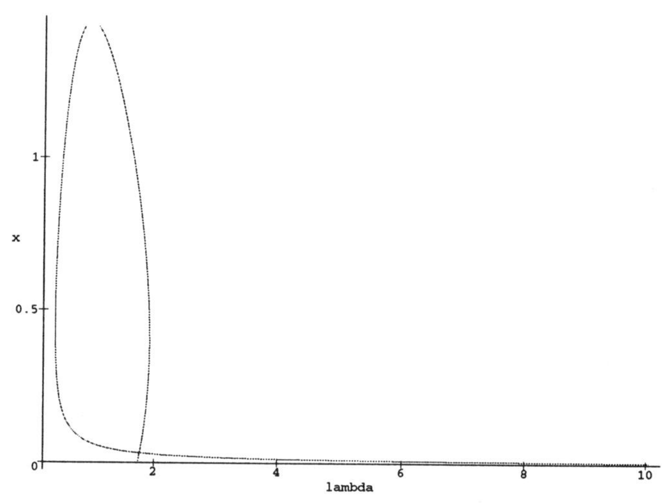

Figure 6.5: Simple bifurcation in the CSTR model.

Next, we choose a simple bifurcation point ($g = g_x = g_\lambda = 0$, $D_2(g) < 0$). One such point computed in §6.5.2 is $(x, \alpha_1, \alpha_2, \alpha_3, \alpha_4, \lambda) = (0.03061799, 11.060448, 11.086435, -0.948286, 3.0, 1.773)$, where $D_2(g) = -0.74906$.

Figure 6.5 presents the solution set in (λ, x)-space close to this point. The simple bifurcation point is clearly visible as well as the closed loop. We have a case as in point P5 in Figure 6.3.

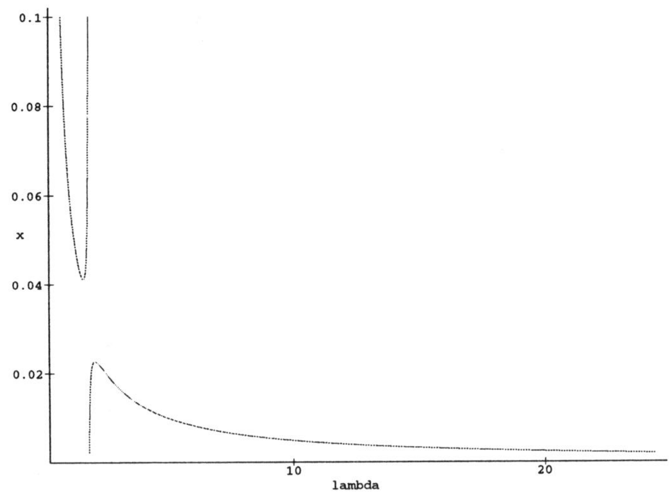

Figure 6.6: Isola formation in the CSTR model.

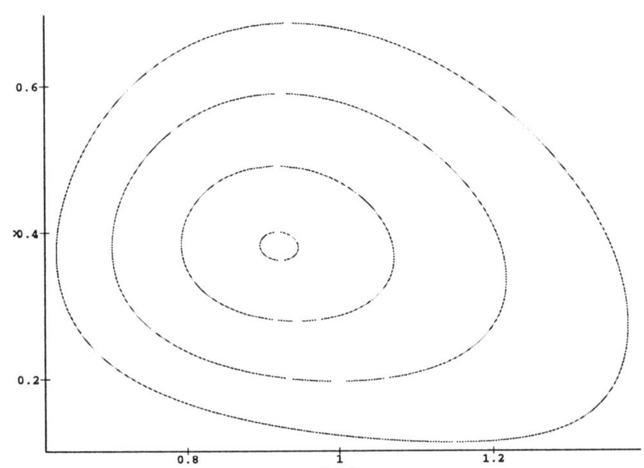

Figure 6.7: Shrinking of an isola in the CSTR model.

Figure 6.6 was obtained by subtracting 0.01 from α_1 with unchanged other parameters. The situation is clearly similar to that in point P6 in Figure 6.3 (isola formation).

Finally, we consider an isola center $g = g_x = g_\lambda = 0$, $D_2(g) > 0$. One such point computed in §6.5.2 is $(x, \alpha_1, \alpha_2, \alpha_3, \alpha_4, \lambda) = (0.326895, 7.650412, 8.65095, -0.840042, 3.0, 0.949178)$, where $D_2(g) = 0.544953$. Figure 6.7 presents four solutions to (6.110) in (λ, x)-space. The outer curve is obtained by adding 0.2 to α_1 and leaving the other parameters unchanged; the inner curves (in that order) are obtained by adding 0.1, 0.03, and 0.01. It is clear that the isola is vanishing as α tends to the value of the isola center.

6.5.5 Pairs of Nondegeneracy Conditions: An Example

Starting from the cusp point computed in §6.5.2 we can compute a curve of cusp points by freeing all variables except α_4. By Table 6.1 a computed cusp point is nondegenerate if at least one of the two quantities $D_3^1(g), D_3^2(g)$ is nonzero where

$$D_3^1(g) = g_{xx}(D_2(g))_\lambda - g_{x\lambda}(D_2(g))_x,$$

$$D_3^2(g) = g_{x\lambda}(D_2(g))_\lambda - g_{\lambda\lambda}(D_2(g))_x.$$

The quantities $D_3^1(g), D_3^2(g)$ were computed along the curve of cusp points. No simultaneous vanishing of both was detected, so the whole curve apparently consists of nondegenerate points. On the other hand, a sign change of $D_3^1(g)$ was detected.

In Tables 6.4 and 6.5 we give the relevant data for four computed points just before the sign change (Table 6.4) and just after the sign change (Table 6.5). In these tables g, g_x, g_λ, and $D_2(g)$ always vanish and $\alpha_4 = 3$.

From the definitions it follows easily that if $D_2(g) = 0$, $D_3^1(g) = 0$, and $D_3^2(g) \neq 0$, then $g_{xx} = g_{x\lambda} = 0$, $g_{\lambda\lambda} \neq 0$, $g_{xxx} \neq 0$. The inverse implication also holds and, in fact, $D_3^2(g) = -g_{\lambda\lambda}^2 g_{xxx}$. These results are largely confirmed by Tables 6.4 and 6.5.

We also remark that in these tables the behavior of g_{xx} and $g_{x\lambda}$ along the curve of cusp points is different: $g_{x\lambda}$ changes sign and g_{xx} does not. To understand why, consider any curve parameterized by s along which $D_2(g)$ vanishes. By taking derivatives with respect to s one obtains $g_{xxs}g_{\lambda\lambda} + g_{xx}g_{\lambda\lambda s} - 2g_{x\lambda}g_{x\lambda s} = 0$. Hence if $g_{xx} = g_{x\lambda} = 0$, $g_{\lambda\lambda} \neq 0$, then necessarily $g_{xxs} = 0$.

It is interesting to note that by taking second derivatives and evaluating in the same point we obtain $g_{xxss}g_{\lambda\lambda} = 2g_{x\lambda s}^2$. Hence g_{xx} attains a regular local extremum with respect to s if and only if the equation $g_{x\lambda} = 0$ provides a regular defining function for the point with $D_3^1(g) = 0$. Since it is easily checked that $(D_3^1(g))_s = -g_{x\lambda s}g_{xxx}g_{\lambda\lambda}$, this happens if and only if $D_3^1(g) = 0$ itself defines the point in a regular way.

6.6 Numerical Methods for $\mathcal{K}$-Singularities

Consider a smooth function

$$G : U \to \mathbb{R}^{N_2}, \tag{6.124}$$

where U is open in $\mathbb{R}^{N_1}$, $N_1 \geq N_2$. We are interested in properties of the solution set $\{x \in U : G(x) = 0\}$ near points where G_x has rank $r < N_2$. In §6.2 we proved that the study of this set can be reduced by a Lyapunov–Schmidt reduction to a case with dimensions $N_1 - r, N_2 - r$, where the Jacobian is the zero matrix. Also, in §6.3 such reduced singular mappings were classified up to $\mathcal{K}$-equivalence. Table 6.1 collects the essential data for $\mathcal{K}$-singularities with codimension 4 or less.

The notion of codimension was defined in a fairly abstract way in Definition 6.3.7 but was related in §6.3 (cf. Table 6.1) to the number of defining equations and in Proposition 6.4.7 to the minimal number of additional parameters that is needed to obtain the singularity as a regular solution to the set of defining equations.

Table 6.4: Four consecutive points on a curve of cusp singularities.

Point	1	2	3	4
x	0.30262	0.30265	0.30269	0.30271
α_1	69.9033	69.9340	69.9648	69.9955
α_2	188.038	188.134	188.229	188.324
α_3	−0.34872	−0.348697	−0.348674	−0.348651
λ	0.151578	0.151538	0.151497	0.151456
g_{xx}	−3.21586E-5	−3.21835E-5	−3.2211E-5	−3.2241E-5
$g_{x\lambda}$	−1.82332E-4	−1.12640E-4	−4.29902E-5	2.66162E-5
$g_{\lambda\lambda}$	−3.73074	−3.73221	−3.73367	−3.73512
g_{xxx}	−1.79985	−1.79961	−1.79936	−1.79912
$D_3^1(g)$	1.42283E-3	9.55148E-4	4.87557E-4	2.00654E-5
$D_3^2(g)$	25.056	25.0709	25.0858	25.1007

Table 6.5: Four more consecutive points on a curve of cusp singularities.

Point	5	6	7	8
x	0.30275	0.30278	0.30282	0.30285
α_1	70.0262	70.057	70.088	70.118
α_2	188.419	188.515	188.610	188.705
α_3	−0.34863	−0.348604	−0.348581	−0.348558
λ	0.151415	0.151374	0.151333	0.151292
g_{xx}	−3.22735E-5	−3.23086E-5	−3.23462E-5	−3.23862E-5
$g_{x\lambda}$	9.61795E-5	1.657E-4	2.35177E-4	3.04612E-4
$g_{\lambda\lambda}$	−3.73658	−3.73804	−3.73950	−3.74095
g_{xxx}	−1.79888	−1.79863	−1.79839	−1.798144
$D_3^1(g)$	−4.47327E-4	−9.14622E-4	−1.38182E-3	−1.84891E-3
$D_3^2(g)$	25.116	25.1305	25.1454	25.1603

So we are led to consider the case where the function G in (6.124) contains a number of additional parameters, say, $\alpha = (\alpha_1, \alpha_2 \ldots)$. These parameters are carried through the Lyapunov–Schmidt reduction (cf. (6.5) and (6.6)) so that the functions $g(x; y), v(x; y)$ are also α-dependent.

We will discuss the numerical detection, computation, and continuation of points in (x, α)-space where the solution manifold of (6.124) presents singularities as in Table 6.1. For simplicity we first consider the easiest case separately.

6.6.1 The Codimension-1 Singularity from $\mathbb{R}$ into $\mathbb{R}$

We now concentrate on the singularity $f(x) = x^2$ from $\mathbb{R}$ into $\mathbb{R}$. From Table 6.1 we know that the codimension is 1. So let α_1 be a selected parameter. Next, we set $N = N_1 = N_2$ and $r = N - 1$ since G_x must have rank defect 1. To perform the maximal Lyapunov–Schmidt reduction we choose $b \in \mathbb{R}^N, c \in \mathbb{R}^N, d \in \mathbb{R}$ such that

$$M(x) \equiv \begin{pmatrix} G_x(x) & -b \\ c^T & -d \end{pmatrix} \tag{6.125}$$

is nonsingular in a neighborhood of the requested point $x \in \mathbb{R}^N$. Let $g(x; y; \alpha_1; t_1) \in \mathbb{R}$, $v(x; y; \alpha_1; t_1) \in \mathbb{R}^N$ be defined by

$$G(x + v(x; y; \alpha_1; t_1); \alpha_1 + t_1) - bg(x; y; \alpha_1; t_1) = G(x; \alpha_1), \tag{6.126}$$

$$c^T v(x; y; \alpha_1; t_1) - dg(x; y; \alpha_1; t_1) = y, \tag{6.127}$$

where $t_1 \in \mathbb{R}$ is the shift of α_1.

Lemma 6.6.1. (a) For all (x, α_1) and $(y, t_1) = (0, 0)$ we have

$$g_x = g_{\alpha_1} = v_x = v_{\alpha_1} = 0. \tag{6.128}$$

(b) For all (x, α_1) and for (y, t_1) in a neighborhood of $(0, 0)$ we have

$$M \begin{pmatrix} v_y \\ g_y \end{pmatrix} = \begin{pmatrix} 0 \\ 1 \end{pmatrix}, \tag{6.129}$$

$$M \begin{pmatrix} v_{t_1} \\ g_{t_1} \end{pmatrix} = \begin{pmatrix} -G_{\alpha_1} \\ 0 \end{pmatrix}, \tag{6.130}$$

$$M \begin{pmatrix} v_{yy} \\ g_{yy} \end{pmatrix} = \begin{pmatrix} -G_{xx} v_y v_y \\ 0 \end{pmatrix}, \tag{6.131}$$

$$M \begin{pmatrix} v_{yt_1} \\ g_{yt_1} \end{pmatrix} = \begin{pmatrix} -G_{xx} v_y v_{t_1} - G_{x\alpha_1} v_y \\ 0 \end{pmatrix}, \tag{6.132}$$

where all derivative functions of G are evaluated at $(x + v(x; y; \alpha_1; t_1), \alpha_1 + t_1)$ and all derivative functions of v, g are evaluated at $(x; y; \alpha_1; t_1)$.

(c) For all (x, α_1) and $(y, t_1) = (0, 0)$ we have

$$M \begin{pmatrix} v_{y\alpha_1} \\ g_{y\alpha_1} \end{pmatrix} = \begin{pmatrix} -G_{x\alpha_1} v_y \\ 0 \end{pmatrix}, \tag{6.133}$$

$$M \begin{pmatrix} v_{yx} \\ g_{yx} \end{pmatrix} = \begin{pmatrix} -G_{xx} v_y \\ 0 \end{pmatrix}. \tag{6.134}$$

In particular, if $X_0 \in \mathbb{R}^N$ is any vector, then

$$M \begin{pmatrix} v_{yx} X_0 \\ g_{yx} X_0 \end{pmatrix} = \begin{pmatrix} -G_{xx} v_y X_0 \\ 0 \end{pmatrix}. \tag{6.135}$$

(d) For all (x, α_1) and $(y, t_1) = (0, 0)$ we have

$$g_{yx} v_y = g_{yy}, \tag{6.136}$$

$$v_{yx} v_y = v_{yy}, \tag{6.137}$$

$$g_{yx} v_{t_1} + g_{y\alpha_1} = g_{yt_1}, \tag{6.138}$$

$$v_{yx} v_{t_1} + v_{y\alpha_1} = v_{yt_1}. \tag{6.139}$$

Proof. (a) follows trivially from the fact that $v(x;0;\alpha_1;0) = g(x;0;\alpha_1;0) = 0$ for all (x,α_1). Next, by taking successive derivatives of (6.126) and (6.127) with respect to y and t_1 we obtain (b). To prove (6.133) we take derivatives of (6.128) with respect to α_1. To prove (6.134) and (6.135) we take derivatives of (6.129) with respect to the components of x.

Now (6.136) and (6.137) both follow from (6.131) and (6.135) with $X_0 = v_y$. Finally, (6.138) and (6.139) both follow from (6.132) and (6.135) with $X_0 = v_{t_1}$. □

Now by fixed $x \in \mathbb{R}^N, \alpha_1 \in \mathbb{R}$, $(y,t_1) = (0,0)$ is a regular solution point to the defining equations

$$g(x;y;\alpha_1;t_1) = g_y(x;y;\alpha_1;t_1) = 0 \tag{6.140}$$

of the singularity if and only if

$$\begin{pmatrix} g_y & g_{t_1} \\ g_{yy} & g_{yt_1} \end{pmatrix} \tag{6.141}$$

is nonsingular. In (x,α_1)-space the singularity is determined by the system

$$G(x;\alpha_1) = g_y(x;0;\alpha_1;0) = 0. \tag{6.142}$$

The point (x,α_1) is a regular solution to this system if

$$\begin{pmatrix} G_x & G_{\alpha_1} \\ g_{yx} & g_{y\alpha_1} \end{pmatrix} \tag{6.143}$$

is nonsingular. We now prove the following.

Proposition 6.6.2. (x,α_1) is a regular solution to the system (6.142) if and only if the reduced unfolding $g(y,t_1)$ is a universal unfolding.

Proof. We will show that (6.143) is nonsingular if and only if (6.141) is nonsingular. It is easy to see that (6.143) is nonsingular if and only if

$$\begin{pmatrix} G_x & -b & 0_{N,1} & G_{\alpha_1} \\ c^T & -d & 1 & 0 \\ 0_{1,N} & 1 & 0 & 0 \\ g_{yx} & 0 & 0 & g_{y\alpha_1} \end{pmatrix} \tag{6.144}$$

is nonsingular. The upper left $(N+1)\times(N+1)$ block in this matrix is precisely $M(x)$, i.e., a nonsingular matrix.

By Lemma 6.6.1(b) (6.144) is nonsingular if and only if

$$\begin{pmatrix} I_{N,N} & 0_{N,1} & v_y & -v_{t_1} \\ 0_{1,N} & 1 & g_y & -g_{t_1} \\ 0_{1,N} & 0 & -g_y & g_{t_1} \\ 0_{1,N} & 0 & -g_{yx}v_y & g_{y\alpha_1} + g_{yx}v_{t_1} \end{pmatrix} \tag{6.145}$$

is nonsingular or, equivalently, if and only if

$$\begin{pmatrix} -g_y & g_{t_1} \\ -g_{yx}v_y & g_{y\alpha_1} + g_{yx}v_{t_1} \end{pmatrix} \tag{6.146}$$

is nonsingular. By Lemma 6.6.1(d) this is equivalent to the condition that (6.141) is nonsingular. □

We now consider the numerical detection, computation, and continuation of the turning point singularity in detail. Consider a smooth curve that passes through such a point in (x, α_1)-space. If $g(x; y; \alpha_1; t_1)$ is a Lyapunov–Schmidt maximally reduced mapping, then $g_y(x; 0; \alpha_1; 0) = 0$, $g_{yy}(x; 0; \alpha_1; 0) \neq 0$ at the singular point (x, α_1). If (generically) $g_{t_1} \neq 0$ then by Proposition 6.6.2 the function $g_y(x; 0; \alpha_1; 0)$ changes sign at (x, α_1). So in principle the singularity can be detected on a curve of, say, equilibrium points by monitoring g_y for a fixed Lyapunov–Schmidt reduction.

For computation we consider the system (6.142). Newton's method requires the solution of linear systems of the form

$$\begin{pmatrix} G_x & G_{\alpha_1} \\ g_{yx} & g_{y\alpha_1} \end{pmatrix} \begin{pmatrix} \Delta x \\ \Delta \alpha_1 \end{pmatrix} = \begin{pmatrix} -G \\ -g_y \end{pmatrix}. \tag{6.147}$$

We present two approaches to solving (6.147). In the first one the components of the matrix in (6.147) are explicitly formed. We still have two possibilities.

Method 6.6.3.a. The matrix elements g_{yx_i} $(1 \leq i \leq N)$ and $g_{y\alpha_1}$ are obtained by solving

$$M \begin{pmatrix} v_{yx_i} \\ g_{yx_i} \end{pmatrix} = \begin{pmatrix} -G_{xx_i}v_y \\ 0 \end{pmatrix}, \tag{6.148}$$

$$M \begin{pmatrix} v_{y\alpha_1} \\ g_{y\alpha_1} \end{pmatrix} = \begin{pmatrix} -G_{x\alpha_1}v_y \\ 0 \end{pmatrix}. \tag{6.149}$$

These formulas are obtained by straightforward derivation of (6.129). We note that (6.149) was already given in (6.133).

Method 6.6.3.b. The matrix elements g_{yx_i} $(1 \leq i \leq N)$ and $g_{y\alpha_1}$ are obtained by solving

$$\begin{pmatrix} w^T & g_y \end{pmatrix} M = \begin{pmatrix} 0_N^T & 1 \end{pmatrix}, \tag{6.150}$$

where $w \in \mathbb{R}^N$, and computing

$$g_{yx} = -w^T G_{xx} v_y, \tag{6.151}$$

$$g_{y\alpha_1} = -w^T G_{x\alpha_1} v_y. \tag{6.152}$$

We note that (6.151) and (6.152) follow from (6.148), (6.149), and (6.150). We note also that Method 6.6.3.b requires one solve only with M^T instead of $N + 1$ solves with M.

In the other approach the solution of (6.147) itself is reduced to the solution of systems with M. The underlying idea is the same as in the proof of Proposition 6.6.2 and is found in, e.g., [126] and [31]. One notes first that (6.147) is equivalent to

$$\begin{pmatrix} G_x & -b & 0_{N,1} & G_{\alpha_1} \\ c^T & -d & 1 & 0 \\ 0_{1,N} & 1 & 0 & 0 \\ g_{yx} & 0 & 0 & g_{y\alpha_1} \end{pmatrix} \begin{pmatrix} \Delta x \\ 0 \\ -c^T \Delta x \\ \Delta \alpha_1 \end{pmatrix} = \begin{pmatrix} -G \\ 0 \\ 0 \\ -g_y \end{pmatrix}. \tag{6.153}$$

After solving the auxiliary problem

$$M \begin{pmatrix} X_0 \\ A_1 \end{pmatrix} = \begin{pmatrix} -G \\ 0 \end{pmatrix} \tag{6.154}$$

and performing an elimination as in the proof of Proposition 6.6.2 we find that (6.153) is equivalent to

$$\begin{pmatrix} I_N & 0_{N,1} & v_y & -v_{t_1} \\ 0_{1,N} & 1 & g_y & -g_{t_1} \\ 0_{1,N} & 0 & -g_y & g_{t_1} \\ 0_{1,N} & 0 & -g_{yy} & g_{yt_1} \end{pmatrix} \begin{pmatrix} \Delta x \\ 0 \\ -c^T \Delta x \\ \Delta \alpha_1 \end{pmatrix} = \begin{pmatrix} X_0 \\ A_1 \\ -A_1 \\ -g_y - g_{yx} X_0 \end{pmatrix}. \tag{6.155}$$

To compute the value of $g_{yx} X_0$ we can solve

$$M \begin{pmatrix} X_1 \\ B_1 \end{pmatrix} = \begin{pmatrix} -G_{xx} v_y X_0 \\ 0 \end{pmatrix}, \tag{6.156}$$

where $X_1 = v_{yx} X_0$, $B_1 = g_{yx} X_0$ by (6.135). Now the solution to (6.155) is simple and we obtain Method 6.6.4.

Method 6.6.4. Solve the systems (6.154), (6.156), and the 2×2 system

$$\begin{pmatrix} -g_y & g_{t_1} \\ -g_{yy} & g_{yt_1} \end{pmatrix} \begin{pmatrix} -c^T \Delta x \\ \Delta \alpha_1 \end{pmatrix} = \begin{pmatrix} -A_1 \\ -g_y - B_1 \end{pmatrix}. \tag{6.157}$$

Then compute $\Delta x = X_0 + v_y (c^T \Delta x) + v_{t_1} \Delta \alpha_1$.

We note that in Methods 6.6.3.b and 6.6.4 the number of solutions of linear systems of order N is independent of N.

For continuation of the turning point singularity we need another free parameter, say, α_2, with shift t_2. The essential linear algebra problem is the solution of systems of the form

$$\begin{pmatrix} G_x & G_{\alpha_1} & G_{\alpha_2} \\ g_{yx} & g_{y\alpha_1} & g_{\alpha_2} \\ a_{31} & a_{32} & a_{33} \end{pmatrix} \begin{pmatrix} \Delta x \\ \Delta \alpha_1 \\ \Delta \alpha_2 \end{pmatrix} = \begin{pmatrix} -G \\ -g_y \\ 0 \end{pmatrix}, \tag{6.158}$$

where $a_{31} \in \mathbb{R}^{1 \times N}$ and $a_{32}, a_{33} \in \mathbb{R}$ are added by the continuation code. The discussion of solution methods for system (6.158) is similar to that for (6.147), and in particular all three Methods 6.6.3.a, 6.6.3.b, and 6.6.4 have an analogue in this case. We note that (6.157) will be replaced by a 3×3 linear system.

During continuation it is advisable to monitor the functions g_{yy} and g_{t_1}. A sign change of g_{yy} indeed indicates that a singularity of higher codimension was encountered; a sign change of g_{t_1} indicates a point where the unfolding in the parameter α_1 is not universal.

6.6.2 Singularities from $\mathbb{R}$ into $\mathbb{R}$ with Codimension Higher than 1

We consider the singularities $f(x) = x^k$ from $\mathbb{R}$ into $\mathbb{R}$ where $k > 2$. To keep the notation simple we assume that $k = 3$ and emphasize the differences with the case $k = 2$ in §6.6.1.

From Table 6.1 we know that the codimension is 2. So let α_1, α_2 be selected parameters. Next, we set $N = N_1 = N_2$ and $r = N - 1$ since G_x must have rank defect 1. To perform the maximal Lyapunov–Schmidt reduction we choose $b \in \mathbb{R}^N, c \in \mathbb{R}^N, d \in \mathbb{R}$ such that

$$M(x) \equiv \begin{pmatrix} G_x(x) & -b \\ c^T & -d \end{pmatrix} \tag{6.159}$$

is nonsingular in a neighborhood of the requested point $x \in \mathbb{R}^N$. Let $g(x; y; \alpha_1, \alpha_2; t_1, t_2) \in \mathbb{R}$, $v(x; y; \alpha_1, \alpha_2; t_1, t_2) \in \mathbb{R}^N$ be defined by

$$G(x+v(x;y;\alpha_1,\alpha_2;t_1,t_2);\alpha_1+t_1,\alpha_2+t_2)-bg(x;y;\alpha_1,\alpha_2;t_1,t_2) = G(x;\alpha_1,\alpha_2), \tag{6.160}$$

$$c^T v(x;y;\alpha_1,\alpha_2;t_1,t_2) - dg(x;y;\alpha_1,\alpha_2;t_1,t_2) = y, \tag{6.161}$$

where $t_1 \in \mathbb{R}$ is the shift of α_1 and $t_2 \in \mathbb{R}$ is the shift of α_2.

Lemma 6.6.5. (a) For all (x, α_1) and for (y, t_1) in a neighborhood of $(0, 0)$ we have

$$M \begin{pmatrix} v_{yyy} \\ g_{yyy} \end{pmatrix} = \begin{pmatrix} -G_{xxx}v_y v_y v_y - 3G_{xx}v_y v_{yy} \\ 0 \end{pmatrix}, \tag{6.162}$$

$$M \begin{pmatrix} v_{yyt_1} \\ g_{yyt_1} \end{pmatrix} = \begin{pmatrix} -G_{xx}v_{t_1}v_{yy} - G_{x\alpha_1}v_{yy} - G_{xxx}v_y v_y v_{t_1} - G_{xx\alpha_1}v_y v_y - 2G_{xx}v_y v_{yt_1} \\ 0 \end{pmatrix}, \tag{6.163}$$

where all derivative functions of G are evaluated at $(x+v(x;y;\alpha_1,\alpha_2;t_1,t_2), \alpha_1+t_1, \alpha_2+t_2)$ and all derivative functions of v, g are evaluated at $(x;y;\alpha_1,\alpha_2;t_1,t_2)$.

(b) For all (x, α_1) and $(y, t_1) = (0, 0)$ we have

$$M \begin{pmatrix} v_{yyx}X_0 \\ g_{yyx}X_0 \end{pmatrix} = \begin{pmatrix} -G_{xx}v_{yy}X_0 - G_{xxx}v_y v_y X_0 - 2G_{xx}v_y v_{yx}X_0 \\ 0 \end{pmatrix}, \tag{6.164}$$

$$\begin{pmatrix} X_0^T v_{yx} & 0 \end{pmatrix} = \left(M^{-T} \begin{pmatrix} X_0^T \\ 0 \end{pmatrix}\right)^T \begin{pmatrix} -G_{xx}v_y \end{pmatrix} \tag{6.165}$$

for any vector $X_0 \in \mathbb{R}^N$:

$$M \begin{pmatrix} v_{yy\alpha_1} \\ g_{yy\alpha_1} \end{pmatrix} = \begin{pmatrix} -G_{xx\alpha_1}v_y v_y - 2G_{xx}v_y v_{y\alpha_1} - G_{x\alpha_1}v_{yy} \\ 0 \end{pmatrix}. \tag{6.166}$$

(c) For all (x, α_1) and $(y, t_1) = (0, 0)$ we have

$$g_{yyx}v_y = g_{yyy}, \tag{6.167}$$

$$v_{yyx}v_y = v_{yyy}, \tag{6.168}$$

$$g_{yyx}v_{t_1} + g_{yy\alpha_1} = g_{yyt_1}, \tag{6.169}$$

$$v_{yyx}v_{t_1} + v_{yy\alpha_1} = v_{yyt_1}. \tag{6.170}$$

Proof. (a) follows by taking derivatives of (6.131) with respect to y and t_1. Next, (b) is obtained by taking derivatives of (6.131) with respect to the components of x, multiplying with the components of X_0, and summing. To prove (6.167) and (6.168) one can either proceed as in Lemma 6.6.1(d) or directly start from (6.131), take derivatives with respect to y and apply the usual rules for derivatives of products, taking (6.137) into account. The proof of (6.169) and (6.170) is similar, taking (6.139) into account. □

Now by fixed $x \in \mathbb{R}^N, \alpha_1, \alpha_2 \in \mathbb{R}$, $(y, t_1, t_2) = (0, 0, 0)$ is a regular solution point to the defining equations

$$g(x; y; \alpha_1, \alpha_2; t_1, t_2) = g_y(x; y; \alpha_1, \alpha_2; t_1, t_2) = g_{yy}(x; y; \alpha_1, \alpha_2; t_1, t_2) = 0 \tag{6.171}$$

of the singularity if and only if

$$\begin{pmatrix} g_y & g_{t_1} & g_{t_2} \\ g_{yy} & g_{yt_1} & g_{yt_2} \\ g_{yyy} & g_{yyt_1} & g_{yyt_2} \end{pmatrix} \tag{6.172}$$

is nonsingular. In (x, α_1, α_2)-space the singularity is determined by the system

$$G(x; \alpha_1, \alpha_2) = g_y(x; 0; \alpha_1, \alpha_2; 0, 0) = g_{yy}(x; 0; \alpha_1, \alpha_2; 0, 0) = 0. \tag{6.173}$$

The point (x, α_1, α_2) is a regular solution to this system if

$$\begin{pmatrix} G_x & G_{\alpha_1} & G_{\alpha_2} \\ g_{yx} & g_{y\alpha_1} & g_{y\alpha_2} \\ g_{yyx} & g_{yy\alpha_1} & g_{yy\alpha_2} \end{pmatrix} \tag{6.174}$$

is nonsingular. We now prove the following.

Proposition 6.6.6. (x, α_1, α_2) is a regular solution to the system (6.173) if and only if the reduced unfolding $g(y, t_1, t_2)$ is a universal unfolding.

Proof. We will show that (6.174) is nonsingular if and only if (6.172) is nonsingular. As in Proposition 6.6.2 we find that (6.174) is nonsingular if and only if

$$\begin{pmatrix} G_x & -b & 0_{N,1} & G_{\alpha_1} & G_{\alpha_2} \\ c^T & -d & 1 & 0 & 0 \\ 0_{1,N} & 1 & 0 & 0 & 0 \\ g_{yx} & 0 & 0 & g_{y\alpha_1} & g_{\alpha_2} \\ g_{yyx} & 0 & 0 & g_{yy\alpha_1} & g_{yy\alpha_2} \end{pmatrix} \tag{6.175}$$

is nonsingular. The upper left $(N+1)\times(N+1)$ block in this matrix is precisely $M(x)$, i.e., a nonsingular matrix.

By Lemma 6.6.1(b) (6.144) is nonsingular if and only if

$$\begin{pmatrix} I_{N,N} & 0_{N,1} & v_y & -v_{t_1} & -v_{t_2} \\ 0_{1,N} & 1 & g_y & -g_{t_1} & -g_{t_2} \\ 0_{1,N} & 0 & -g_y & g_{t_1} & g_{t_2} \\ 0_{1,N} & 0 & -g_{yx}v_y & g_{y\alpha_1}+g_{yx}v_{t_1} & g_{y\alpha_2}+g_{yx}v_{t_2} \\ 0_{1,N} & 0 & -g_{yyx}v_y & g_{yy\alpha_1}+g_{yyx}v_{t_1} & g_{yy\alpha_2}+g_{yyx}v_{t_2} \end{pmatrix} \tag{6.176}$$

is nonsingular or, equivalently, if and only if

$$\begin{pmatrix} -g_y & g_{t_1} & g_{t_2} \\ -g_{yx}v_y & g_{y\alpha_1}+g_{yx}v_{t_1} & g_{y\alpha_2}+g_{yx}v_{t_2} \\ -g_{yyx}v_y & g_{yy\alpha_1}+g_{yyx}v_{t_1} & g_{yy\alpha_2}+g_{yyx}v_{t_2} \end{pmatrix} \tag{6.177}$$

is nonsingular. By Lemma 6.6.1 and Lemma 6.6.5 this is equivalent to the condition that (6.172) is nonsingular. □

The numerical detection, computation, and continuation of the singularity $f(x)=x^3$ is similar to the case $f(x)=x^2$ studied in §6.6.1. We note that in the computation using Method 6.6.3.b one needs the quantities $g_{yy\alpha_1} = -w^T G_{xx\alpha_1}v_y v_y - 2w^T G_{xx}v_y v_{y\alpha_1} - w^T G_{x\alpha_1}v_{yy}$ (6.166) and $g_{yyx} = -w^T G_{xx}v_{yy} - w^T G_{xxx}v_y v_y - 2w^T G_{xx}v_y v_{yx}$ (cf. (6.164)). By using (6.165) we can compute the right-hand side by solving only one linear system with M^T, i.e., without computing v_{yx}. A similar recursion is needed in the application of Method 6.6.4 in the present setting. Indeed, inner products of the form $g_{yyx}X_0$ have to be computed and from (6.164) it is clear that this requires the computation of either products of the form $v_{yx}X_0$ or $Y_0^T v_{yx}$.

During continuation it is advisable to monitor the functions g_{yyy} and $g_{t_1}g_{yt_2}-g_{t_2}g_{yt_1}$. A sign change of g_{yyy} indeed indicates that a singularity of higher codimension was encountered; a sign change of $g_{t_1}g_{yt_2}-g_{t_2}g_{yt_1}$ indicates a point where the unfolding in the parameters α_1,α_2 is not universal.

The further generalization to codimension $3,4,\ldots$ is straightforward, but formulas become more and more complicated and the recursions in the cases of Methods 6.6.3.b and 6.6.4 require more steps. We note, however, that

$$g_{yyyy} = g_{yyyx}v_y. \tag{6.178}$$

This result is a straightforward generalization of (6.136) and (6.167); it will be used in §7.7.6. Some further details are relegated to Exercises 10–13.

We note that in terms of computational work the difference between Method 6.6.3.b and Method 6.6.4 does not seem to be important. As far as we know there is no clear numerical advantage in using one of them either. However, Method 6.6.4 can be applied using only a solver for M, while Method 6.6.3.b always requires solvers for both M and M^T.

A straightforward application of Method 6.6.3.a is conceptually the easiest solution. If the number of state variables is small then there is much to recommend it. Indeed, the derivatives with respect to x and to the free parameters may then be approximated by finite differences, so for the computation of a singularity of codimension k we have to compute derivatives of g up to order k in y, t_1 only.

6.6.3 Singularities from $\mathbb{R}^2$ into $\mathbb{R}$

For simplicity of notation we concentrate on the singularities with codimension 1. We recall from Table 6.1 that there are two such singularities, namely, $f(y_1, y_2) = y_1^2 \pm y_2^2$; both have defining equations $f = f_{y_1} = f_{y_2} = 0$. The nondegeneracy conditions are $\pm D_2(f) > 0$. Since the codimension is 1, we select a free parameter α_1. Next, we need $N_1 = N_2 + 1$ and $r = N_2 - 1$. The Lyapunov–Schmidt reduction requires the choice of $b \in \mathbb{R}^{N_2}$, $C \in \mathbb{R}^{N_1 \times 2}$, $d \in \mathbb{R}^2$ such that

$$M(x) \equiv \begin{pmatrix} G_x(x) & -b \\ C^T & -d \end{pmatrix} \tag{6.179}$$

is nonsingular at the requested point $x \in \mathbb{R}^{N_1}$. For $y_1, y_2, t_1 \in \mathbb{R}$ let $g(x; y_1, y_2; \alpha_1; t_1) \in \mathbb{R}$ and $v(x; y_1, y_2; \alpha_1; t_1) \in \mathbb{R}^{N_1}$ be defined by

$$G(x + v(x; y_1, y_2; \alpha_1; t_1); \alpha_1 + t_1) - bg(x; y_1, y_2; \alpha_1; t_1) = G(x; \alpha_1), \tag{6.180}$$

$$C^T v(x; y_1, y_2; \alpha_1; t_1) - dg(x; y_1, y_2; \alpha_1; t_1) = (y_1, y_2)^T. \tag{6.181}$$

Lemma 6.6.7. (a) For all (x, α_1) and $(y_1, y_2, t_1) = (0, 0, 0)$ we have

$$g_x = g_{\alpha_1} = v_x = v_{\alpha_1} = 0. \tag{6.182}$$

(b) For all (x, α_1) and for (y_1, y_2, t_1) in a neighborhood of $(0, 0, 0)$ we have

$$M \begin{pmatrix} v_{y_1} \\ g_{y_1} \end{pmatrix} = \begin{pmatrix} 0_{N_2} \\ 1 \\ 0 \end{pmatrix}, \tag{6.183}$$

$$M \begin{pmatrix} v_{y_2} \\ g_{y_2} \end{pmatrix} = \begin{pmatrix} 0_{N_2} \\ 0 \\ 1 \end{pmatrix}, \tag{6.184}$$

$$M \begin{pmatrix} v_{t_1} \\ g_{t_1} \end{pmatrix} = \begin{pmatrix} -G_{\alpha_1} \\ 0 \\ 0 \end{pmatrix}. \tag{6.185}$$

For $i, j = 1, 2$ we have

$$M \begin{pmatrix} v_{y_i y_j} \\ g_{y_i y_j} \end{pmatrix} = \begin{pmatrix} -G_{xx} v_{y_i} v_{y_j} \\ 0 \\ 0 \end{pmatrix} \tag{6.186}$$

and

$$M\begin{pmatrix} v_{y_i t_1} \\ g_{y_i t_1} \end{pmatrix} = \begin{pmatrix} -G_{xx}v_{y_i}v_{t_1} - G_{x\alpha_1}v_{y_i} \\ 0 \\ 0 \end{pmatrix}, \tag{6.187}$$

where all derivative functions of G are evaluated at $(x + v(x; y_1, y_2; \alpha_1, t_1), \alpha_1 + t_1)$ and all derivative functions of v, g are evaluated at $(x; y_1, y_2; \alpha_1; t_1)$.

(c) If $X_0 \in \mathbb{R}^{N_1}$ is any vector and $i = 1, 2$, then

$$M\begin{pmatrix} v_{y_i x}X_0 \\ g_{y_i x}X_0 \end{pmatrix} = \begin{pmatrix} -G_{xx}v_{y_i}X_0 \\ 0 \\ 0 \end{pmatrix}. \tag{6.188}$$

(d) For all (x, α_1) and $(y_1, y_2, t_1) = (0, 0, 0)$ we have for $i, j = 1, 2$,

$$g_{y_i x}v_{y_j} = g_{y_i y_j}, \tag{6.189}$$

$$v_{y_i x}v_{y_j} = v_{y_i y_j}, \tag{6.190}$$

$$g_{y_i x}v_{t_1} + g_{y_i \alpha_1} = g_{y_i t_1}, \tag{6.191}$$

$$v_{y_i x}v_{t_1} + v_{y_i \alpha_1} = v_{y_i t_1}. \tag{6.192}$$

Proof. The proof is similar to the proof of Lemma 6.6.1. □

If $x \in \mathbb{R}^N, \alpha_1 \in \mathbb{R}$ are fixed, then $(y_1, y_2, t_1) = (0, 0, 0)$ is a regular solution point to the defining equations

$$g(x; y_1, y_2; \alpha_1; t_1) = g_{y_1}(x; y_1, y_2; \alpha_1; t_1) = g_{y_2}(x; y_1, y_2; \alpha_1; t_1) = 0 \tag{6.193}$$

of the singularity if and only if

$$\begin{pmatrix} g_{y_1} & g_{y_2} & g_{t_1} \\ g_{y_1 y_1} & g_{y_1 y_2} & g_{y_1 t_1} \\ g_{y_2 y_1} & g_{y_2 y_2} & g_{y_2 t_1} \end{pmatrix} \tag{6.194}$$

is nonsingular. In (x, α_1)-space the singularity is determined by the system

$$G(x; \alpha_1) = g_{y_1}(x; 0, 0; \alpha_1; 0) = g_{y_2}(x; 0, 0; \alpha_1; 0) = 0. \tag{6.195}$$

The point (x, α_1) is a regular solution to this system if

$$\begin{pmatrix} G_x & G_{\alpha_1} \\ g_{y_1 x} & g_{y_1 \alpha_1} \\ g_{y_2 x} & g_{y_2 \alpha_1} \end{pmatrix} \tag{6.196}$$

is nonsingular. We now prove the following.

Proposition 6.6.8. (x, α_1) is a regular solution to the system (6.195) if and only if the reduced unfolding $g(y_1, y_2, t_1)$ is a universal unfolding.

Proof. We will show that (6.196) is nonsingular if and only if (6.194) is nonsingular. As in Proposition 6.6.2 we find that (6.196) is nonsingular if and only if

$$\begin{pmatrix} G_x & -b & 0_{N_2,2} & G_{\alpha_1} \\ C^T & -d & I_2 & 0_2 \\ 0_{1,N_1} & 1 & 0_{1,2} & 0 \\ g_{y_1 x} & 0 & 0_{1,2} & g_{y_1 \alpha_1} \\ g_{y_2 x} & 0 & 0_{1,2} & g_{y_2 \alpha_1} \end{pmatrix} \tag{6.197}$$

is nonsingular. The upper left $(N_1+1)\times(N_1+1)$ block in this matrix is precisely $M(x)$, i.e., a nonsingular matrix.

By Lemma 6.6.7(a) (6.197) is nonsingular if and only if

$$\begin{pmatrix} I_{N_1,N_1} & 0_{N_1,1} & v_{y_1} & v_{y_2} & -v_{t_1} \\ 0_{1,N_1} & 1 & g_{y_1} & g_{y_2} & -g_{t_1} \\ 0_{1,N_1} & 0 & -g_{y_1} & -g_{y_2} & g_{t_1} \\ 0_{1,N_1} & 0 & -g_{y_1 x} v_{y_1} & -g_{y_1 x} v_{y_2} & g_{y_1 \alpha_1} + g_{y_1 x} v_{t_1} \\ 0_{1,N_1} & 0 & -g_{y_2 x} v_{y_1} & -g_{y_2 x} v_{y_2} & g_{y_2 \alpha_1} + g_{y_2 x} v_{t_1} \end{pmatrix} \tag{6.198}$$

is nonsingular or, equivalently, by Lemma 6.6.7(d) if and only if

$$\begin{pmatrix} -g_{y_1} & -g_{y_2} & g_{t_1} \\ -g_{y_1 y_1} & -g_{y_1 y_2} & g_{y_1 t_1} \\ -g_{y_2 y_1} & -g_{y_2 y_2} & g_{y_2 t_1} \end{pmatrix} \tag{6.199}$$

is nonsingular. □

Numerical detection, computation, and continuation can now be done as in §6.6.1. An important function to monitor during continuation of curves of codimension-1 singularities from $\mathbb{R}^2$ into $\mathbb{R}$ is $D_2(g) = g_{y_1 y_1} g_{y_2 y_2} - g^2_{y_1 y_2}$. Its sign determines the exact type of the singularity, and sign changes indicate a degeneracy to other singularities. Another interesting function is g_{t_1} since a sign change indicates that a point was encountered where α_1 does not provide a universal unfolding of the singularity.

The generalization of the above study to singularities from $\mathbb{R}^2$ into $\mathbb{R}$ with higher codimension is fairly straightforward. The same numerical methods as in §6.6.2 can be applied. We omit the details.

6.6.4 Singularities from $\mathbb{R}^2$ into $\mathbb{R}^2$

We recall from Table 6.1 that there are no such singularities with codimension less than 4 and just two with codimension 4, namely, $f(y_1, y_2) = (f_1(x,y), f_2(x,y))^T = (y_1^2 \pm y_2^2, y_1 y_2)^T$; both have defining equations $f_1 = f_2 = f_{1y_1} = f_{1y_2} = f_{2y_1} = f_{2y_2} = 0$. The nondegeneracy conditions are $\pm D_{2,2}(f) > 0$. Since the codimension is 4, we select four free parameters and collect them into $\alpha = (\alpha_1, \alpha_2, \alpha_3, \alpha_4)$. Next, we set $N = N_1 = N_2$ and $r = N - 2$. The Lyapunov–Schmidt reduction requires the choice of $B \in \mathbb{R}^{N\times 2}$, $C \in \mathbb{R}^{N\times 2}$, $D \in \mathbb{R}^{2\times 2}$ such that

$$M(x) \equiv \begin{pmatrix} G_x(x) & -B \\ C^T & -D \end{pmatrix} \tag{6.200}$$

is nonsingular at the requested point $x \in \mathbb{R}^N$. For $y_1, y_2 \in \mathbb{R}$ and $t = (t_1, t_2, t_3, t_4) \in \mathbb{R}^4$ let $g(x; y_1, y_2; \alpha; t) \in \mathbb{R}^2$ and $v(x; y_1, y_2; \alpha; t) \in \mathbb{R}^N$ be defined by

$$G(x + v(x; y_1, y_2; \alpha; t); \alpha + t) - bg(x; y_1, y_2; \alpha; t) = G(x; \alpha), \tag{6.201}$$

$$C^T v(x; y_1, y_2; \alpha; t) - dg(x; y_1, y_2; \alpha; t) = (y_1, y_2)^T. \tag{6.202}$$

Lemma 6.6.9. (a) For all (x, α) and $(y_1, y_2, t) = (0, 0, 0)$ we have

$$g_x = g_\alpha = v_x = v_\alpha = 0. \tag{6.203}$$

(b) For all (x, α) and for (y_1, y_2, t) in a neighborhood of $(0, 0, 0)$ we have

$$M \begin{pmatrix} v_{y_1} \\ g_{y_1} \end{pmatrix} = \begin{pmatrix} 0_N \\ 1 \\ 0 \end{pmatrix}, \tag{6.204}$$

$$M \begin{pmatrix} v_{y_2} \\ g_{y_2} \end{pmatrix} = \begin{pmatrix} 0_N \\ 0 \\ 1 \end{pmatrix}, \tag{6.205}$$

$$M \begin{pmatrix} v_t \\ g_t \end{pmatrix} = \begin{pmatrix} -G_\alpha \\ 0 \\ 0 \end{pmatrix}. \tag{6.206}$$

For $i, j = 1, 2$ we have

$$M \begin{pmatrix} v_{y_i y_j} \\ g_{y_i y_j} \end{pmatrix} = \begin{pmatrix} -G_{xx} v_{y_i} v_{y_j} \\ 0 \\ 0 \end{pmatrix} \tag{6.207}$$

and

$$M \begin{pmatrix} v_{y_i t} \\ g_{y_i t} \end{pmatrix} = \begin{pmatrix} -G_{xx} v_{y_i} v_t - G_{x\alpha} v_{y_i} \\ 0 \\ 0 \end{pmatrix}, \tag{6.208}$$

where all derivative functions of G are evaluated at $(x + v(x; y_1, y_2; \alpha; t), \alpha + t)$ and all derivative functions of v, g are evaluated at $(x; y_1, y_2; \alpha; t)$.

(c) If $X_0 \in \mathbb{R}^N$ is any vector and $i = 1, 2$, then

$$M \begin{pmatrix} v_{y_i x} X_0 \\ g_{y_i x} X_0 \end{pmatrix} = \begin{pmatrix} -G_{xx} v_{y_i} X_0 \\ 0 \\ 0 \end{pmatrix}. \tag{6.209}$$

(d) For all (x, α) and $(y_1, y_2, t) = (0, 0, 0)$ we have for $i, j = 1, 2$,

$$g_{y_i x} v_{y_j} = g_{y_i y_j}, \tag{6.210}$$

$$v_{y_i x} v_{y_j} = v_{y_i y_j}, \tag{6.211}$$

$$g_{y_i x} v_t + g_{y_i \alpha} = g_{y_i t}, \tag{6.212}$$

$$v_{y_i x} v_t + v_{y_i \alpha} = v_{y_i t}. \tag{6.213}$$

Proof. The proof is similar to the proof of Lemma 6.6.7. ◻

Now by fixed $x \in \mathbb{R}^N, \alpha \in \mathbb{R}^4$, $(y_1, y_2, t) = (0, 0, 0)$ is a regular solution point to the defining equations

$$g(x; y_1, y_2; \alpha; t) = g_{y_1}(x; y_1, y_2; \alpha; t) = g_{y_2}(x; y_1, y_2; \alpha; t) = 0 \tag{6.214}$$

of the singularity if and only if the 6×6 matrix

$$\begin{pmatrix} g_{y_1} & g_{y_2} & g_t \\ g_{y_1 y_1} & g_{y_1 y_2} & g_{y_1 t} \\ g_{y_2 y_1} & g_{y_2 y_2} & g_{y_2 t} \end{pmatrix} \tag{6.215}$$

is nonsingular. In (x, α)-space the singularity is determined by the system

$$G(x; \alpha) = g_{y_1}(x; 0, 0; \alpha; 0) = g_{y_2}(x; 0, 0; \alpha; 0) = 0. \tag{6.216}$$

The point (x, α) is a regular solution to this system if the $(N + 4) \times (N + 4)$ matrix

$$\begin{pmatrix} G_x & G_\alpha \\ g_{y_1 x} & g_{y_1 \alpha} \\ g_{y_2 x} & g_{y_2 \alpha} \end{pmatrix} \tag{6.217}$$

is nonsingular. We now prove the following.

Proposition 6.6.10. (x, α) is a regular solution to the system (6.216) if and only if the reduced unfolding $g(y_1, y_2, t)$ is a universal unfolding.

Proof. We will show that (6.217) is nonsingular if and only if (6.215) is nonsingular. As in Proposition 6.6.2 we find that (6.217) is nonsingular if and only if

$$\begin{pmatrix} G_x & -B & 0_{N,2} & G_\alpha \\ C^T & -D & I_2 & 0_2 \\ 0_{2,N} & I_2 & 0_{2,2} & 0_2 \\ g_{y_1 x} & 0_{2,2} & 0_{2,2} & g_{y_1 \alpha} \\ g_{y_2 x} & 0_{2,2} & 0_{2,2} & g_{y_2 \alpha} \end{pmatrix} \tag{6.218}$$

is nonsingular. The upper left $(N + 2) \times (N + 2)$ block in this matrix is precisely $M(x)$, i.e., a nonsingular matrix.

By Lemma 6.6.9(a) (6.218) is nonsingular if and only if

$$\begin{pmatrix} I_N & 0_{N,2} & v_{y_1} & v_{y_2} & -v_t \\ 0_{2,N} & I_2 & g_{y_1} & g_{y_2} & -g_t \\ 0_{2,N} & 0_{2,2} & -g_{y_1} & -g_{y_2} & g_t \\ 0_{2,N} & 0_{2,2} & -g_{y_1x}v_{y_1} & -g_{y_1x}v_{y_2} & g_{y_1\alpha} + g_{y_1x}v_t \\ 0_{2,N} & 0_{2,2} & -g_{y_2x}v_{y_1} & -g_{y_2x}v_{y_2} & g_{y_2\alpha} + g_{y_2x}v_t \end{pmatrix} \tag{6.219}$$

is nonsingular or, equivalently, by Lemma 6.6.9(d) if and only if

$$\begin{pmatrix} -g_{y_1} & -g_{y_2} & g_t \\ -g_{y_1y_1} & -g_{y_1y_2} & g_{y_1t} \\ -g_{y_2y_1} & -g_{y_2y_2} & g_{y_2t} \end{pmatrix} \tag{6.220}$$

is nonsingular. □

Numerical detection, computation, and continuation can now be done as in §6.6.1. Of course an important function to monitor during continuation of curves of codimension-4 singularities from $\mathbb{R}^2$ into $\mathbb{R}^2$ is $D_{2,2}(g)$. Its sign determines the exact type of the singularity, and sign changes indicate a degeneracy to other singularities.

6.7 Notes and Further Reading

The geometry of curves and surfaces in low-dimensional Euclidean spaces has fascinated mathematicians for centuries. A highly recommended textbook is [39]. The notes in [15] are a challenge to further study.

6.8 Exercises

1. Prove that two germs $f, g \in \mathcal{E}^0_{n,p}$ are $\mathcal{K}$-equivalent if and only if there exist $X \in \mathcal{E}^0_{n,n}$ and $H \in \mathcal{E}^0_{n+p,p}$ (written as $H(x,y)$ with $x \in \mathbb{R}^n, y \in \mathbb{R}^p$) such that

$$g(x) = H(x, f(X(x)))$$

and

(a) $X_x(0)$ and $H_y(0,0)$ are nonsingular,

(b) $H(x,0) = 0$ for all $x \in \mathbb{R}^n$

(in [64] this is the fundamental formulation).

(*Hint:* Use Lemma 6.2.3.)

2. Consider the germs f_1, f_2 from $\mathbb{R}^2$ into $\mathbb{R}$ where $f_1(x_1,x_2) = x_1^2 - x_2^2$ and $f_2(x_1,x_2) = x_1^2 - x_2^2 + 0.1x_1x_2$. Prove explicitly (i.e., by giving the transformations) that they are $\mathcal{K}$-equivalent.

3. Suppose that in Proposition 6.2.4 we have $B_1 = B_2$ and $C_1 = C_2$. Does this imply that the germs $g_1(y)$ and $g_2(y)$ are the same?

4. Use Proposition 6.2.5 to solve Exercise 3.8.4.

5. Prove that a versal unfolding of a singular germ is transversal (this is the easy part of the $\mathcal{K}$-versality theorem 6.4.4).

   (*Hint:* Start with any given other germ $g(x)$ and consider the unfolding $f(x)+vg(x)$ where v is an unfolding parameter.)

6. Consider the germ F from $\mathbb{R}^2$ into $\mathbb{R}$ obtained in §5.3.5 (section of the Whitney umbrella) defined by $F(x,y) = y^2 - rx^3 - syx^2$, where r, s are parameters. Prove that it is $\mathcal{K}$-equivalent to $x^3 + y^2$ if $r \neq 0$ and to $x^4 - y^2$ if $r = 0, s \neq 0$. What is the codimension if $r = s = 0$?

7. Prove that an unfolding of $f(x,y) = x^5 + y^2$ is transversal if and only if the extended Jacobian of the defining system in Table 6.1 for this singularity has full rank. Can you generalize this result to higher powers of x (not in the table, but with the obvious extension for the defining system)?

8. Repeat the experiment on the CSTR described in §6.5.2 up to the continuation of a branch of points with $g = g_x = 0$. Detect and compute a point where g_{α_4} vanishes. Then compute a branch of simple bifurcation points in (x, α_4)-space (i.e., $g = g_x = g_{\alpha_4} = 0$) by freeing $(x, \alpha_1, \alpha_4, \lambda)$ and freezing (α_2, α_3). Detect and compute a pitchfork bifurcation point in (x, α_4)-space. Can you also detect a cusp point in (x, α_4)-space?

9. Prove that the CSTR model in §6.5.1 does not allow cusp points in (x, α_4)-space for admissible values of the variables ($x > -1$, $\alpha_1 > 0$, $\alpha_2 > 0$, $\alpha_3 > -1$, $\alpha_4 > 0$, $\lambda > 0$).

10. In the case of singularities from $\mathbb{R}$ into $\mathbb{R}$ prove the following formulas, which allow one to compute the cases with codimension 3 and 4 with Method 6.6.3.a if a finite difference method is used to compute the derivatives with respect to state variables and parameters (the second formula only to monitor the nondegeneracy):

$$M \begin{pmatrix} v_{yyyy} \\ g_{yyyy} \end{pmatrix} = \begin{pmatrix} T_3 \\ 0 \end{pmatrix},$$

$$M \begin{pmatrix} v_{yyyyy} \\ g_{yyyyy} \end{pmatrix} = \begin{pmatrix} T_4 \\ 0 \end{pmatrix},$$

where

$$T_3 = -G_{xxxx}v_yv_yv_yv_y - 6G_{xxx}v_yv_yv_{yy} - 4G_{xx}v_yv_{yyy} - 3G_{xx}v_{yy}v_{yy},$$

$$T_4 = -G_{xxxxx}v_yv_yv_yv_yv_y - 10G_{xxxx}v_yv_yv_yv_{yy} - 10G_{xxx}v_yv_yv_{yyy} - 15G_{xxx}v_yv_{yy}v_{yy}$$
$$-4G_{xx}v_yv_{yyyy} - 10G_{xx}v_{yy}v_{yyy} - G_{xx}v_yv_{yyyy}.$$

11. In the case of singularities from $\mathbb{R}$ into $\mathbb{R}$ prove the following formulas, which are needed in both Methods 6.6.3.a and 6.6.4 for the recursion steps

$$M\begin{pmatrix} v_{yyyx} \\ g_{yyyx} \end{pmatrix} = \begin{pmatrix} T_{31} \\ 0 \end{pmatrix},$$

$$M\begin{pmatrix} v_{yyyyx} \\ g_{yyyyx} \end{pmatrix} = \begin{pmatrix} T_{41} \\ 0 \end{pmatrix},$$

where

$$T_{31} = -G_{xxxx}v_yv_yv_y - 3G_{xxx}v_yv_{yy} - 3G_{xxx}v_yv_yv_{yx} - 3G_{xx}v_yv_{yyx} - 3G_{xx}v_{yy}v_{yx},$$

$$\begin{aligned} T_{41} = &-G_{xxxxx}v_yv_yv_yv_y - 4G_{xxxx}v_yv_yv_yv_{yx} - 6G_{xxxx}v_yv_yv_{yy} \\ &-12G_{xxx}v_yv_{yy}v_{yx} - 6G_{xxx}v_yv_yv_{yyx} - 4G_{xxx}v_yv_{yyy} - 4G_{xx}v_yv_{yyyx} \\ &-4G_{xx}v_{yyy}v_{yx} - 3G_{xxx}v_{yy}v_{yy} - 6G_{xx}v_{yy}v_{yyx} - G_{xx}v_{yyyy}. \end{aligned}$$

12. In the case of singularities from $\mathbb{R}$ into $\mathbb{R}$ prove the following formulas, which (together with those in Exercises 10 and 11) allow one to compute the cases with codimension 3 and 4 with Method 6.6.3.a:

$$M\begin{pmatrix} v_{yyy\alpha_1} \\ g_{yyy\alpha_1} \end{pmatrix} = \begin{pmatrix} T_{32} \\ 0 \end{pmatrix},$$

$$M\begin{pmatrix} v_{yyyy\alpha_1} \\ g_{yyyy\alpha_1} \end{pmatrix} = \begin{pmatrix} T_{42} \\ 0 \end{pmatrix},$$

where

$$\begin{aligned} T_{32} = &-G_{xxx\alpha_1}v_yv_yv_y - 3G_{xxx}v_yv_yv_{y\alpha_1} - 3G_{xx\alpha_1}v_yv_{yy} - 3G_{xx}v_yv_{yy\alpha_1} \\ &-3G_{xx}v_{yy}v_{y\alpha_1} - G_{x\alpha_1}v_{yyy}, \end{aligned}$$

$$\begin{aligned} T_{42} = &-G_{xxxx\alpha_1}v_yv_yv_yv_y - 4G_{xxxx}v_yv_yv_yv_{y\alpha_1} - 6G_{xxx\alpha_1}v_yv_yv_{yy} \\ &-12G_{xxx}v_yv_{yy}v_{y\alpha_1} - 6G_{xxx}v_yv_yv_{yy\alpha_1} - 4G_{xx\alpha_1}v_yv_{yyy} - 4G_{xx}v_yv_{yyy\alpha_1} \\ &-4G_{xx}v_{yyy}v_{y\alpha_1} - 3G_{xx\alpha_1}v_{yy}v_{yy} - 6G_{xx}v_{yy}v_{yy\alpha_1} - G_{x\alpha_1}v_{yyyy}. \end{aligned}$$

13. In the case of singularities from $\mathbb{R}$ into $\mathbb{R}$ prove the following formulas, which (together with those in Exercises 10 and 11) allow one to compute the cases with codimension 3 and 4 with Method 6.6.4:

$$M\begin{pmatrix} v_{yyyt_1} \\ g_{yyyt_1} \end{pmatrix} = \begin{pmatrix} T_{33} \\ 0 \end{pmatrix},$$

$$M\begin{pmatrix} v_{yyyyt_1} \\ g_{yyyyt_1} \end{pmatrix} = \begin{pmatrix} T_{43} \\ 0 \end{pmatrix},$$

where

$$T_{33} = -G_{xxxx}v_yv_yv_yv_{t_1} - G_{xxx\alpha_1}v_yv_yv_y - 3G_{xxx}v_yv_yv_{yt_1} - 3G_{xxx}v_yv_{yy}v_{t_1}$$
$$-3G_{xx\alpha_1}v_yv_{yy} - 3G_{xx}v_yv_{yyt_1} - 3G_{xx}v_{yy}v_{yt_1} - G_{xx}v_{t_1}v_{yyy} - G_{x\alpha_1}v_{yyy},$$
$$T_{43} = -G_{xxxxx}v_yv_yv_yv_yv_{t_1} - G_{xxxx\alpha_1}v_yv_yv_yv_y - 4G_{xxxx}v_yv_yv_yv_{yt_1} - 6G_{xxxx}v_yv_yv_{yy}v_{t_1}$$
$$-6G_{xxx\alpha_1}v_yv_yv_{yy} - 12G_{xxx}v_yv_{yy}v_{yt_1} - 6G_{xxx}v_yv_yv_{yyt_1} - 4G_{xxx}v_yv_{yyy}v_{t_1}$$
$$-4G_{xx\alpha_1}v_yv_{yyy} - 4G_{xx}v_yv_{yyyt_1} - 4G_{xx}v_{yyy}v_{yt_1} - 3G_{xxx}v_{yy}v_{yy}v_{t_1}$$
$$-3G_{xx\alpha_1}v_{yy}v_{yy} - 6G_{xx}v_{yy}v_{yyt_1} - G_{xx}v_{yyyy}v_{t_1} - G_{x\alpha_1}v_{yyyy}.$$

Chapter 7

Singularity Theory with a Distinguished Bifurcation Parameter

As in §6.1 we consider $N_1, N_2 \in \mathbb{N}, N_1 \geq N_2$ and a mapping

$$G : U \to \mathbb{R}^{N_2} \tag{7.1}$$

with U a nonempty open subset of $\mathbb{R}^{N_1}$. We assume that G is a C^∞-mapping and $x^\circ \in U$. We are again interested in the singular case, i.e., the case where $G_x(x^\circ)$ does not have full rank. However, we now assume that problem (7.1) contains a distinguished parameter λ for which the behavior of the solutions to (7.1) in (x, λ) space (with fixed other parameters if there are any) deserves special attention. The resulting approach is called λ-singularity theory because λ is the standard name for the distinguished bifurcation parameter.

The popularity of λ-singularities may be partly due to numerical methods. Numerical continuation is usually done by freeing one parameter of a problem, which then obviously plays a special role in the description of the phenomena that one observes. The very notion of a turning point implicitly refers to the choice of a distinguished parameter. We refer to [150], [68], and [153] among many references. Furthermore, this approach was made extremely popular by the standard textbooks [109] and [110].

In [161] augmented systems for the computation of singularities are derived from the classification results in [109]. Our approach is related but somewhat simpler in the sense that we follow [109] more closely. The Lyapunov–Schmidt reduction (an analytical tool) is replaced by the numerical Lyapunov–Schmidt reduction as described in §6.2. The solution of the *recognition problem* (i.e., how to recognize the λ-singularity to which a given germ is equivalent) is reformulated as a family of defining systems for λ-singularities, which can be used in a numerical implementation.

We remark that in problems with several parameters any parameter may be taken as the "distinguished" one. It seems natural to study singular points first in a setting with no distinguished bifurcation parameters (cf. Chapter 6).

7.1 Singularities with a Distinguished Bifurcation Parameter

Definition 7.1.1. Let $n, p \in \mathbb{N}$ and let (x, λ) denote a generic point in $\mathbb{R}^{n+1}$ where $\lambda \in \mathbb{R}$ is the distinguished parameter. Two germs $f, g \in \mathcal{E}^0_{n+1,p}$ are said to be $(\lambda - \mathcal{K})$-equivalent if and only if there exist $Z \in \mathcal{E}_{n+1,p,p}$, $X \in \mathcal{E}^0_{n+1,n}$, and $\Lambda \in \mathcal{E}^0_1$ with $X_x(0,0)$ and $Z(0,0)$ nonsingular, $\Lambda_\lambda(0) \neq 0$ such that

$$g(x, \lambda) = Z(x, \lambda) f(X(x, \lambda), \Lambda(\lambda)). \tag{7.2}$$

The germs f, g are said to be strongly $(\lambda - \mathcal{K})$-equivalent if in addition $\Lambda(\lambda) \equiv \lambda$.

As in Chapter 6, the most important fact about problems like (7.1) is that they can be reduced to the case where N_1, N_2 are low-dimensional spaces and the Jacobian G_x vanishes. The Lyapunov–Schmidt reduction was studied extensively in §6.2 and carries over immediately to the present setting, preserving strong $(\lambda - \mathcal{K})$-equivalence for any parameter that one wants to consider as the distinguished one.

The further study of the low-dimensional reduced problems where the Jacobian with respect to the variables in x vanishes is the core of $(\lambda - \mathcal{K})$-singularity theory. This is developed extensively in [109] and [110]. We will recall the main results with some additions that are of interest for numerical implementations.

The notions of tangent space, codimension, unfolding, universal unfolding, etc., all have their λ-pendants. We will discuss the classification up to codimension 3, i.e., the cases where three additional parameters are needed for a universal unfolding. Since the distinguished bifurcation parameter is also present, this is roughly comparable with the codimension-4 case considered in Chapter 6. In §7.2 we will discuss $(\lambda - \mathcal{K})$-singularities from $\mathbb{R}$ into $\mathbb{R}$, in §7.3 from $\mathbb{R}^2$ into $\mathbb{R}^2$. We will not consider $(\lambda - \mathcal{K})$-singularities from $\mathbb{R}^2$ into $\mathbb{R}$.

7.2 Classification of $(\lambda - \mathcal{K})$-Singularities from $\mathbb{R}$ into $\mathbb{R}$

The classification given in [109] for this case is slightly finer than the one derived from Definition 7.1.1. In fact, for this case Definition 7.1.1 is replaced in [109] by Definition 7.2.1.

Definition 7.2.1. Two germs $f, g \in \mathcal{E}^0_{2,1}$ are said to be $(\lambda - \mathcal{K})$-equivalent if and only if there exist $Z \in \mathcal{E}_2$, $X \in \mathcal{E}^0_2$, and $\Lambda \in \mathcal{E}^0_1$ with $X_x(0,0) > 0$, $Z(0,0) > 0$, $\Lambda_\lambda(0) > 0$ such that

$$g(x, \lambda) = Z(x, \lambda) f(X(x, \lambda), \Lambda(\lambda)). \tag{7.3}$$

The germs f, g are said to be strongly $(\lambda - \mathcal{K})$-equivalent if in addition $\Lambda(\lambda) \equiv \lambda$.

The finer distinction made in Definition 7.2.1 with regard to the signs of Z, X_x, and Λ_λ is intended to preserve certain orientation and stability properties. For a geometric study of the solutions to (7.1) it is irrelevant. In the classification of singularities it leads to a splitting of the singularity classes according to certain signs. Except for this splitting

Table 7.1: Singularities from $\mathbb{R}$ into $\mathbb{R}$ with codimension ≤ 3.

Normal form		Codimension	Name
(1)	$\varepsilon x^2 + \delta\lambda$	0	Turning point
(2)	$\varepsilon(x^2 - \lambda^2)$	1	Transcritical bifurcation
(3)	$\varepsilon(x^2 + \lambda^2)$	1	Isola center
(4)	$\varepsilon x^3 + \delta\lambda$	1	Hysteresis point
(5)	$\varepsilon x^2 + \delta\lambda^3$	2	Asymmetric cusp
(6)	$\varepsilon x^3 + \delta\lambda x$	2	Pitchfork
(7)	$\varepsilon x^4 + \delta\lambda$	2	Quartic turning point
(8)	$\varepsilon x^2 + \delta\lambda^4$	3	
(9)	$\varepsilon x^3 + \delta\lambda^2$	3	Winged cusp
(10)	$\varepsilon x^4 + \delta\lambda x$	3	Degenerate pitchfork
(11)	$\varepsilon x^5 + \delta\lambda$	3	

there are precisely 11 singularities of codimension 3 or less. They are collected in Table 7.1, which we took from [109, IV, Table 2.1].

In Table 7.1 ε, δ can take the values ± 1 only. So (1) contains four singularities, namely, $x^2 + \lambda$, $x^2 - \lambda$, $-x^2 + \lambda$, and $-x^2 - \lambda$. We remark that the bifurcation diagrams for the first and fourth are the same, and similarly for the second and third. Also, the remaining two cases differ only by a reflection that leaves the x-axis fixed.

The 11 types in Table 7.1 can be organized naturally in four families, namely,

$$\begin{array}{lll} \text{(I)} & \varepsilon x^k + \delta\lambda & (k \geq 2)\text{: numbers 1, 4, 7, 11;} \\ \text{(II)} & \varepsilon x^k + \delta\lambda x & (k \geq 3)\text{: numbers 6, 10;} \\ \text{(III)} & \varepsilon x^2 + \delta\lambda^k & (k \geq 2)\text{: numbers 2, 3, 5, 8;} \\ \text{(IV)} & \varepsilon x^3 + \delta\lambda^2 & \text{number 9.} \end{array}$$

Each singularity in Table 7.1 can be characterized by a set of defining equations (the number of equations is equal to the codimension) and nondegeneracy conditions, i.e., inequality relations that make a further distinction between singularities with the same defining equations. They are collected in Table 7.2.

We make the following remarks:

1. The conditions $g = g_x = 0$ are not mentioned in the lists of defining equations but are always implicitly assumed.

2. The expressions $\delta = \text{sgn}\ (*)$ or $\varepsilon = \text{sgn}\ (*)$ imply that $(*)$ is nonzero and has the sign of δ (respectively, ε).

Table 7.2: Defining equations and nondegeneracy conditions.

Normal form	Defining equations	Nondegeneracy conditions
$\varepsilon x^k + \delta\lambda \ (k \geq 2)$	$g_{xx} = \cdots = \dfrac{\partial^{k-1} g}{\partial x^{k-1}} = 0$	$\varepsilon = \text{sgn}\left(\dfrac{\partial^k g}{\partial x^k}\right), \delta = \text{sgn}\ (g_\lambda)$
$\varepsilon x^k + \delta\lambda x \ (k \geq 3)$	$g_{xx} = \cdots = \dfrac{\partial^{k-1} g}{\partial x^{k-1}} = g_\lambda = 0$	$\varepsilon = \text{sgn}\left(\dfrac{\partial^k g}{\partial x^k}\right), \delta = \text{sgn}\ (g_{x\lambda})$
$\varepsilon x^2 + \delta\lambda^k (k \geq 2)$	$g_\lambda = D_2(g) = \cdots = D_{k-1}(g) = 0$	$\varepsilon = \text{sgn}(g_{xx}), \delta = \text{sgn}(g_{xx})^{k+1}\text{sgn}(D_k(g))$
$\varepsilon x^3 + \delta\lambda^2$	$g_\lambda = g_{xx} = g_{x\lambda} = 0$	$\varepsilon = \text{sgn}\ (g_{xxx}), \delta = \text{sgn}\ (g_{\lambda\lambda})$

Table 7.3: Universal unfoldings of singularities from $\mathbb{R}$ into $\mathbb{R}$.

(1)	$\varepsilon x^2 + \delta\lambda$
(2,3)	$\varepsilon(x^2 + \delta\lambda^2 + \alpha)$
(4)	$\varepsilon x^3 + \delta\lambda + \alpha x$
(5)	$\varepsilon x^2 + \delta\lambda^3 + \alpha + \beta\lambda$
(6)	$\varepsilon x^3 + \delta\lambda x + \alpha + \beta x^2$
(7)	$\varepsilon x^4 + \delta\lambda + \alpha x + \beta x^2$
(8)	$\varepsilon x^2 + \delta\lambda^4 + \alpha + \beta\lambda + \gamma\lambda^2$
(9)	$\varepsilon x^3 + \delta\lambda^2 + \alpha + \beta x + \gamma\lambda x$
(10)	$\varepsilon x^4 + \delta\lambda x + \alpha + \beta\lambda + \gamma x^2$
(11)	$\varepsilon x^5 + \delta\lambda + \alpha x + \beta x^2 + \gamma x^3$

3. The expressions $D_k(g)$ are determined inductively by $D_2(g) = g_{xx}g_{\lambda\lambda} - g_{x\lambda}^2$, $D_{k+1}(g) = g_{xx}(D_k(g))_\lambda - g_{x\lambda}(D_k(g))_x$ for $k \geq 2$.

The results in Table 7.2 for the first, second, and fourth families are taken from [109]. For the third family we used the defining functions obtained in [118].

The sets of defining functions in Table 7.2 have the important property that an unfolding of a singularity is universal if and only if the extended Jacobian (i.e., with respect to x, λ and the unfolding parameters) is a square nonsingular matrix. For a proof of this we refer again to [109] and [118].

It is therefore easy to give universal unfoldings of the normal forms in Table 7.1. These are, of course, not unique, but in all cases there are universal unfoldings that look particularly simple. We collect them in Table 7.3.

7.3 Classification of $(\lambda - \mathcal{K})$-Singularities from $\mathbb{R}^2$ into $\mathbb{R}^2$

Let $f(x, y, \lambda) = (f_1(x, y, \lambda), f_2(x, y, \lambda))^T \in \mathcal{E}^0_{3,2}$ with distinguished bifurcation parameter λ. It is proved in [109, IX] that in the classification with respect to Definition 7.1.1 there are no singularities with codimension less than 3 and precisely four with codimension 3. They can be represented by

$$\left(\; x^2 - y^2 + \lambda, \quad 2xy \; \right)^T, \tag{7.4}$$

$$\left(\; x^2 + \lambda, \quad y^2 + \lambda \; \right)^T, \tag{7.5}$$

$$\left(\; x^2 + \lambda, \quad y^2 - \lambda \; \right)^T, \tag{7.6}$$

$$\left(\; x^2 - \lambda, \quad y^2 - \lambda \; \right)^T. \tag{7.7}$$

Only in the case of (7.4) is a system of defining equations and nondegeneracy conditions explicitly given in [109]. It uses the function $D_{2,2}(f)$ introduced already in Chapter 6.

$$D_{2,2}(f) = (f_{1xx}f_{2yy} - f_{2xx}f_{1yy})^2 - 4(f_{1xx}f_{2xy} - f_{2xx}f_{1xy})(f_{1xy}f_{2yy} - f_{2xy}f_{1yy}).$$

Proposition 7.3.1. The germ $f(x_1, x_2, \lambda)$ is $(\lambda - \mathcal{K})$-equivalent to (7.4) if and only if

$$f_1 = f_2 = f_{1x} = f_{1y} = f_{2x} = f_{2y} = 0, \tag{7.8}$$

$$D_{2,2}(f) < 0, \tag{7.9}$$

and at least one of $f_{1\lambda}, f_{2\lambda}$ is nonzero.

Proof. See [109, IX, Theorem 2.2]. □

For (7.5), (7.6), (7.7) the condition

$$D_{2,2}(f) > 0 \tag{7.10}$$

is a necessary condition. Further classification is done in [109] by geometric considerations. Assume that (7.8) and (7.10) hold. Consider the 2-jet of f:

$$j_2(f) = \begin{pmatrix} (j_2(f))_1 \\ (j_2(f))_2 \end{pmatrix} = \begin{pmatrix} p_1x^2 + p_2xy + p_3y^2 \\ q_1x^2 + q_2xy + q_3y^2 \end{pmatrix},$$

where

$$p_1 = \frac{1}{2}f_{1xx}(0,0,0), p_2 = f_{1xy}(0,0,0), p_3 = \frac{1}{2}f_{1yy}(0,0,0), \tag{7.11}$$

$$q_1 = \frac{1}{2}f_{2xx}(0,0,0), q_2 = f_{2xy}(0,0,0), q_3 = \frac{1}{2}f_{2yy}(0,0,0). \tag{7.12}$$

It is easy to see that the determinant of the Jacobian of $j_2(f)$ is given by

$$\det(J_{x,y}(j_2(f))) = ax^2 + bxy + cy^2, \tag{7.13}$$

where

$$a = 2(p_1q_2 - p_2q_1), b = 4(p_1q_3 - q_1p_3), c = 2(p_2q_3 - q_2p_3). \tag{7.14}$$

Also $b^2 - 4ac = D_{2,2}(f) > 0$; hence $j_2(f)$ is a singular transformation precisely along the two distinct real lines determined by

$$ax^2 + 2bxy + cy^2 = 0. \tag{7.15}$$

Let these lines be spanned by vectors w_1, w_2, respectively. Then the image of $j_2(f)$ is the wedge W (closed sector comprising less than a half-plane) bounded by the rays that contain $j_2(f)(w_1), j_2(f)(w_2)$. This can be proved by elementary algebraic and topological arguments. Furthermore, it is proved in [109, IX, Theorem 3.2] that f is $(\lambda - \mathcal{K})$-equivalent to (7.5) if and only if

$$f_\lambda(0,0,0) = \begin{pmatrix} f_{1\lambda}(0,0,0) \\ f_{2\lambda}(0,0,0) \end{pmatrix}$$

is in the interior of W. It is $(\lambda - \mathcal{K})$-equivalent to (7.7) if and only if $f_\lambda(0,0,0)$ is in the interior of $-W$. It is $(\lambda - \mathcal{K})$-equivalent to (7.6) if and only if $f_\lambda(0,0,0)$ is in the interior of one of the two remaining sectors.

We will express these geometric conditions in terms of the derivatives of f at $(0,0,0)$. First remark that a vector $(X,Y)^T$ is a multiple of either $j_2(f)(w_1)$ or $j_2(f)(w_2)$ if and only if the system consisting of (7.15) and

$$X(q_1x^2 + q_2xy + q_3y^2) - Y(p_1x^2 + p_2xy + p_3y^2) = 0$$

admits a nonzero solution in (x,y). By standard algebra results this is equivalent to

$$\det \begin{pmatrix} a & b & c & 0 \\ 0 & a & b & c \\ q_1X - p_1Y & q_2X - p_2Y & q_3X - p_3Y & 0 \\ 0 & q_1X - p_1Y & q_2X - p_2Y & q_3X - p_3Y \end{pmatrix} = 0$$

or

$$rX^2 + sXY + tY^2 = 0, \tag{7.16}$$

where

$$r = \det \begin{pmatrix} a & b & c & 0 \\ 0 & a & b & c \\ q_1 & q_2 & q_3 & 0 \\ 0 & q_1 & q_2 & q_3 \end{pmatrix}, \tag{7.17}$$

$$s = -\det \begin{pmatrix} a & b & c & 0 \\ 0 & a & b & c \\ q_1 & q_2 & q_3 & 0 \\ 0 & p_1 & p_2 & p_3 \end{pmatrix} - \det \begin{pmatrix} a & b & c & 0 \\ 0 & a & b & c \\ p_1 & p_2 & p_3 & 0 \\ 0 & q_1 & q_2 & q_3 \end{pmatrix}, \tag{7.18}$$

$$t = \det \begin{pmatrix} a & b & c & 0 \\ 0 & a & b & c \\ p_1 & p_2 & p_3 & 0 \\ 0 & p_1 & p_2 & p_3 \end{pmatrix}. \tag{7.19}$$

It follows that (7.16) defines the two distinct lines that separate the wedges discussed in [109, IX, Theorem 3.2]. We shall prove that W contains at least one point where $rX^2 + sXY + tY^2 < 0$. If $a \neq 0$, then obviously $(p_1, q_1)^T \in W$ and it follows by routine computations that $rp_1^2 + sp_1q_1 + tq_1^2 = -\frac{a^2}{16}(b^2 - 4ac) < 0$. If $c \neq 0$ then $(p_3, q_3)^T \in W$ and

$rp_3^2 + sp_3q_3 + tq_3^2 = -\frac{c^2}{16}(b^2 - 4ac) < 0$. If $a = c = 0$ then necessarily $p_2 = q_2 = 0, b \neq 0$, $(p_{13}, q_{13})^T = (\frac{1}{2}(p_1 + p_3), \frac{1}{2}(q_1 + q_3))^T \in W$ and $rp_{13}^2 + sp_{13}q_{13} + tq_{13}^2 = -\frac{1}{4}b^4 < 0$.

Hence $rX^2 + sXY + tY^2$ is strictly negative in the interior of W, and $-W$ is strictly positive in the interior of the other sectors. In particular we obtain Proposition 7.3.2.

Proposition 7.3.2. The germ $f(x, y, \lambda)$ is $(\lambda - \mathcal{K})$-equivalent to (7.6) if and only if (7.8), (7.10), and

$$rf_{1\lambda}^2 + sf_{1\lambda}f_{2\lambda} + tf_{2\lambda}^2 > 0 \tag{7.20}$$

hold. Here r, s, t are defined by (7.17), (7.18), and (7.19). □

To distinguish further between (7.5) and (7.7) we consider the bilinear form

$$Q((X_1, Y_1)^T, (X_2, Y_2)^T) = rX_1X_2 + \frac{1}{2}s(X_1Y_2 + X_2Y_1) + tY_1Y_2. \tag{7.21}$$

By standard results in elementary projective geometry $Q((X_1, Y_1)^T, (X_2, Y_2)^T) = 0$ if and only if the lines spanned by $(X_1, Y_1)^T$ and $(X_2, Y_2)^T$ are a harmonically conjugate pair to the pair of solutions to (7.16). Therefore, the form (7.21) is nonpositive if $(X_1, Y_1)^T, (X_2, Y_2)^T$ are both in W and nonnegative if one of them is in W and the other is in $-W$. Since $(p_1, q_1)^T$ is nonzero, in W we obtain the following proposition.

Proposition 7.3.3. The germ $f(x, y, \lambda)$ is $(\lambda - \mathcal{K})$-equivalent to (7.5) if and only if (7.8), (7.10), and

$$rf_{1\lambda}^2 + sf_{1\lambda}f_{2\lambda} + tf_{2\lambda}^2 < 0, \tag{7.22}$$

$$rp_1f_{1\lambda} + \frac{1}{2}s(q_1f_{1\lambda} + p_1f_{2\lambda}) + tq_1f_{2\lambda} < 0 \tag{7.23}$$

hold. Here p_1, q_1, r, s, t are defined by (7.11), (7.12), (7.14), (7.17), (7.18), and (7.19). The germ $f(x, y, \lambda)$ is $(\lambda - \mathcal{K})$-equivalent to (7.7) if and only if (7.8), (7.10), (7.22), and

$$rp_1f_{1\lambda} + \frac{1}{2}s(q_1f_{1\lambda} + p_1f_{2\lambda}) + tq_1f_{2\lambda} > 0 \tag{7.24}$$

hold. □

7.4 Numerical Methods for $(\lambda - \mathcal{K})$-Singularities

Consider a smooth function

$$G : U \to \mathbb{R}^N, \tag{7.25}$$

where U is open in $\mathbb{R}^{N+1+k}$. We decompose $z \in \mathbb{R}^{N+1+k}$ as $z = (x, \alpha) = (x, \alpha_1, \ldots, \alpha_{k+1})$ and call x the state variable, α_1 the distinguished or bifurcation parameter, and $\alpha_2, \ldots, \alpha_{k+1}$ the unfolding parameters. The choice of α_1 among the parameters is arbitrary but has a strong influence on our perception of the problem since we are mainly interested in the behavior of the solution set of (7.25) in the (x, α_1)-plane for constant values of $\alpha_2, \ldots, \alpha_{k+1}$ near points where G_x has rank $r < N$ (the case $r = N$ is covered by the implicit function theorem). If we restrict to codimension 3 or less, then only the cases $r = N - 1$ and $r = N - 2$ are possible. We start with the case $r = N - 1$, which is by far the most important one.

7.4.1 Numerical Methods for $(\lambda-\mathcal{K})$-Singularities with Corank 1

To perform the maximal Lyapunov–Schmidt reduction in this case we choose $b \in \mathbb{R}^N, c \in \mathbb{R}^N, d \in \mathbb{R}$ such that

$$M(x,\alpha) \equiv \begin{pmatrix} G_x(x,\alpha) & -b \\ c^T & -d \end{pmatrix} \tag{7.26}$$

is nonsingular in a neighborhood of $(x,\alpha) \in \mathbb{R}^{N+1+k}$. Then $g(x;y;\alpha;t) \in \mathbb{R}$, $v(x;y;\alpha;t) \in \mathbb{R}^N$ are defined for $y \in \mathbb{R}$, $t \in \mathbb{R}^{1+k}$, (y,t) in a neighborhood of $(0,0)$, by

$$G(x+v(x;y;\alpha;t);\alpha+t) - bg(x;y;\alpha;t) = G(x;\alpha), \tag{7.27}$$

$$c^T v(x;y;\alpha;t) - dg(x;y;\alpha;t) = y, \tag{7.28}$$

where t is the shift of α. The basic results on the computation of derivatives of v and g are essentially the same as in §6.6. However, since the classification is now with respect to both x and α_1, the shift t_1 of α_1 is also distinguished and we may need derivatives of g_{t_1} as well (cf. Table 7.2).

Lemma 7.4.1. (a) For all (x,α) and $(y,t) = (0,0)$, $1 \le i \le k+1$ we have

$$g_x = g_{\alpha_i} = v_x = v_{\alpha_i} = 0. \tag{7.29}$$

(b) For all (x,α) and for (y,t) in a neighborhood of $(0,0)$, $1 \le i,j \le k+1$ we have

$$M\begin{pmatrix} v_y \\ g_y \end{pmatrix} = \begin{pmatrix} 0 \\ 1 \end{pmatrix}, \tag{7.30}$$

$$M\begin{pmatrix} v_{t_i} \\ g_{t_i} \end{pmatrix} = \begin{pmatrix} -G_{\alpha_i} \\ 0 \end{pmatrix}, \tag{7.31}$$

$$M\begin{pmatrix} v_{yy} \\ g_{yy} \end{pmatrix} = \begin{pmatrix} -G_{xx}v_y v_y \\ 0 \end{pmatrix}, \tag{7.32}$$

$$M\begin{pmatrix} v_{yt_i} \\ g_{yt_i} \end{pmatrix} = \begin{pmatrix} -G_{xx}v_y v_{t_i} - G_{x\alpha_i}v_y \\ 0 \end{pmatrix}, \tag{7.33}$$

$$M\begin{pmatrix} v_{t_it_j} \\ g_{t_it_j} \end{pmatrix} = \begin{pmatrix} -G_{xx}v_{t_i}v_{t_j} - G_{x\alpha_i}v_{t_j} - G_{x\alpha_j}v_{t_i} - G_{\alpha_i\alpha_j} \\ 0 \end{pmatrix}, \tag{7.34}$$

$$M\begin{pmatrix} v_{t_it_i} \\ g_{t_1t_i} \end{pmatrix} = \begin{pmatrix} -G_{xx}v_{t_i}v_{t_i} - 2G_{x\alpha_i}v_{t_i} - G_{\alpha_i\alpha_i} \\ 0 \end{pmatrix}, \tag{7.35}$$

$$M\begin{pmatrix} v_{yyy} \\ g_{yyy} \end{pmatrix} = \begin{pmatrix} -G_{xxx}v_y v_y v_y - 3G_{xx}v_y v_{yy} \\ 0 \end{pmatrix}, \tag{7.36}$$

$$M\begin{pmatrix} v_{yyt_i} \\ g_{yyt_i} \end{pmatrix} = \begin{pmatrix} -G_{xxx}v_y v_y v_{t_i} - G_{xx\alpha_i}v_y v_y - 2G_{xx}v_y v_{yt_i} - G_{xx}v_{t_i}v_{yy} - G_{x\alpha_i}v_{yy} \\ 0 \end{pmatrix}, \tag{7.37}$$

$$M\begin{pmatrix} v_{yt_it_i} \\ g_{yt_it_i} \end{pmatrix} = \begin{pmatrix} T_8 \\ 0 \end{pmatrix}, \tag{7.38}$$

$$M\begin{pmatrix} v_{t_it_it_i} \\ g_{t_it_it_i} \end{pmatrix} = \begin{pmatrix} T_9 \\ 0 \end{pmatrix}, \tag{7.39}$$

where T_8 and T_9 denote

$$-G_{xxx}v_yv_{t_i}v_{t_i} - 2G_{xx\alpha_i}v_yv_{t_i} - 2G_{xx}v_{yt_i}v_{t_i} - G_{xx}v_yv_{t_it_i} - G_{x\alpha_i\alpha_i}v_y - 2G_{x\alpha_i}v_{yt_i}$$

and

$$-G_{xxx}v_{t_i}v_{t_i}v_{t_i} - 3G_{xx\alpha_i}v_{t_i}v_{t_i} - 3G_{xx}v_{t_i}v_{t_it_i} - 3G_{x\alpha_i\alpha_i}v_{t_i} - 3G_{x\alpha_i}v_{t_it_i} - G_{\alpha_i\alpha_i\alpha_i},$$

respectively, and all derivative functions of G are evaluated at $(x + v(x;y;\alpha;t), \alpha + t)$ and all derivative functions of v, g are evaluated at $(x;y;\alpha;t)$.

(c) For all (x,α) and $(y,t) = (0,0)$, $1 \le i \le k+1$, we have

$$M\begin{pmatrix} v_{y\alpha_i} \\ g_{y\alpha_i} \end{pmatrix} = \begin{pmatrix} -G_{x\alpha_i}v_y \\ 0 \end{pmatrix}, \tag{7.40}$$

$$M\begin{pmatrix} v_{yx} \\ g_{yx} \end{pmatrix} = \begin{pmatrix} -G_{xx}v_y \\ 0 \end{pmatrix}, \tag{7.41}$$

$$M\begin{pmatrix} v_{t_1\alpha_i} \\ g_{t_1\alpha_i} \end{pmatrix} = \begin{pmatrix} -G_{x\alpha_i}v_{t_1} - G_{\alpha_1\alpha_i} \\ 0 \end{pmatrix}, \tag{7.42}$$

$$M\begin{pmatrix} v_{t_1x} \\ g_{t_1x} \end{pmatrix} = \begin{pmatrix} -G_{xx}v_{t_1} - G_{\alpha_1x} \\ 0 \end{pmatrix}. \tag{7.43}$$

In particular, if $X_0 \in \mathbb{R}^N$ is any vector, then

$$M\begin{pmatrix} v_{yx}X_0 \\ g_{yx}X_0 \end{pmatrix} = \begin{pmatrix} -G_{xx}v_yX_0 \\ 0 \end{pmatrix}, \tag{7.44}$$

$$M\begin{pmatrix} v_{t_1x}X_0 \\ g_{t_1x}X_0 \end{pmatrix} = \begin{pmatrix} -G_{xx}v_{t_1}X_0 - G_{x\alpha_1}X_0 \\ 0 \end{pmatrix}. \tag{7.45}$$

(d) For all (x,α) and $(y,t) = (0,0)$, $1 \le i \le k+1$ we have

$$g_{yx}v_y = g_{yy}, \tag{7.46}$$

$$v_{yx}v_y = v_{yy}, \tag{7.47}$$

$$g_{yx}v_{t_i} + g_{y\alpha_i} = g_{yt_i}, \tag{7.48}$$

$$v_{yx}v_{t_i} + v_{y\alpha_i} = v_{yt_i}, \tag{7.49}$$

$$g_{t_1x}v_{t_i} + g_{t_1\alpha_i} = g_{t_1t_i}, \tag{7.50}$$

$$v_{t_1x}v_{t_i} + v_{t_1\alpha_i} = v_{t_1t_i}. \tag{7.51}$$

Proof. The lemma is proved by straightforward derivation of (7.27) and (7.28) (see also Lemma 6.6.1). □

Proposition 6.6.2 also carries over to the case of $(\lambda - \mathcal{K})$-singularities, in fact to all singularities with codimension 3 or less with the defining functions in Table 7.2. So in each case the singular point is in (x, α)-space the regular solution to its system of defining equations, if and only if the Lyapunov–Schmidt maximally reduced function is universally unfolded by the same (shifted) parameters.

Details on the numerical detection, computation, and continuation of $\lambda - \mathcal{K}$ are similar to the case of $\mathcal{K}$-singularities although we have a larger number of singularities to consider. In particular, Methods 6.6.3.a, 6.6.3.b, and 6.6.4 can all be applied in this case. In §7.7 we will discuss a numerical example in detail.

7.4.2 Numerical Methods for $(\lambda - \mathcal{K})$-Singularities with Corank 2

To perform the maximal Lyapunov–Schmidt reduction in this case we choose $B \in \mathbb{R}^{N \times 2}$, $C \in \mathbb{R}^{N \times 2}$, $D \in \mathbb{R}^{2 \times 2}$ such that

$$M(x, \alpha) \equiv \begin{pmatrix} G_x(x, \alpha) & -B \\ C^T & -D \end{pmatrix} \tag{7.52}$$

is nonsingular in a neighborhood of $(x, \alpha) \in \mathbb{R}^{N+1+k}$. Then $g(x; y; \alpha; t) \in \mathbb{R}^2$, $v(x; y; \alpha; t) \in \mathbb{R}^N$ are defined for $y \in \mathbb{R}^2$, $t \in \mathbb{R}^{1+k}$, (y, t) in a neighborhood of $(0, 0)$ by

$$G(x + v(x; y; \alpha; t); \alpha + t) - Bg(x; y; \alpha; t) = G(x; \alpha), \tag{7.53}$$

$$C^T v(x; y; \alpha; t) - Dg(x; y; \alpha; t) = y, \tag{7.54}$$

where t is the shift of α. The basic results on the computation of derivatives of v and g are essentially the same as in §7.4.1. The only formal difference is that y now has two components. In particular, Lemma 7.4.1 generalizes easily to this case and provides all the formulas necessary for a numerical study of this case since all singularities (with codimension 3 or less!) have the same set of defining equations, namely, $g = g_{y_1} = g_{y_2} = 0$. The second derivatives are needed for the nondegeneracy conditions, i.e., further classification with respect to $(\lambda - \mathcal{K})$-equivalence (cf. §7.3). They can, of course, also be used in the Newton steps.

7.5 Interpretation of Simple Singularities with Corank 1

Let G be as in (7.25), $G(x, \alpha) = 0$ and suppose that $G_x(x, \alpha)$ has rank $r \geq N - 1$. Let a Lyapunov–Schmidt reduction be described by (7.27), (7.28). We will now relate some of the more common singularities in the maximally reduced functions to properties in the original (x, α)-setting.

Proposition 7.5.1. $g_y(x; 0; \alpha; 0) = 0$ if and only if $G_x(x, \alpha)$ is singular. If in this case ϕ is a right singular vector of G_x, then $v_y = \frac{1}{c^T \phi} \phi$.

Proof. The proof is immediate from (7.30). □

Proposition 7.5.2. Let $g_y(x; 0; \alpha; 0) = 0$. Then the following hold:

1. $g_{t_1} = 0$ if and only if G_{α_1} is in the linear space spanned by the columns of G_x.

2. If this is the case, then $v_{t_1} = \theta - \frac{c^T\theta}{c^T\phi}\phi$ with ϕ as in Proposition 7.5.1 and θ is the unique solution to the system

$$\begin{cases} G_x\theta + G_{\alpha_1} = 0, \\ \phi^T\theta = 1. \end{cases}$$

Proof. If $g_{t_i} = 0$, then by (7.31) G_{α_1} is in the column span of G_x. Conversely, suppose that G_{α_1} is in this space. Then the system in θ has a unique solution. Furthermore, for each $r \in \mathbb{R}$ we have

$$\begin{bmatrix} G_x & -b \\ c^T & -d \end{bmatrix} \begin{bmatrix} \theta + r\phi \\ 0 \end{bmatrix} + \begin{bmatrix} G_{\alpha_1} \\ 0 \end{bmatrix} = \begin{bmatrix} 0 \\ (c^T\theta) + r(c^T\phi) \end{bmatrix}. \tag{7.55}$$

Here $c^T\phi \neq 0$ by Proposition 7.5.1; if we choose $r = -\frac{c^T\theta}{c^T\phi}$, then by (7.31) and (7.55) it follows that $g_{t_1} = 0$ and $v_{t_1} = \theta - \frac{c^T\theta}{c^T\phi}\phi$. □

Corollary 7.5.3. If $g_y = g_{t_1} = 0$, then $[G_x, G_{\alpha_1}]$ has a two-dimensional null space.

Proposition 7.5.4. Let $g = g_x = 0$. Then

1. $g_{yy} = 0$ if and only if $G_{xx}\phi\phi \in \mathcal{R}(G_x)$.

2. If this is the case, then

$$v_{yy} = \frac{\eta}{(c^T\phi)^2} - \frac{(c^T\eta)}{(c^T\phi)^3}\phi$$

with ϕ as in Proposition 7.5.1 and η uniquely determined by

$$\begin{cases} G_x\eta + G_{xx}\phi\phi = 0, \\ \phi^T\eta = 0. \end{cases}$$

Proof. If $g_{yy} = 0$, then $G_{xx}\phi\phi \in \mathcal{R}(G_x)$ by (7.32). Since $v_y = \frac{1}{c^T\phi}\phi$ (Proposition 7.5.1) $G_{xx}\phi\phi \in \mathcal{R}(G_x)$. Conversely, if $G_{xx}\phi\phi \in \mathcal{R}(G_x)$, then the system in η has a unique solution. For every $r \in \mathbb{R}$ we have

$$\begin{bmatrix} G_x & -b \\ c^T & -d \end{bmatrix} \begin{bmatrix} \frac{\eta}{(c^T\phi)^2} + r\phi \\ 0 \end{bmatrix} + \begin{bmatrix} G_{xx}v_yv_y \\ 0 \end{bmatrix} = \begin{bmatrix} 0 \\ \frac{c^T\eta}{(c^T\phi)^2} + r(c^T\phi) \end{bmatrix}.$$

If in particular we choose $r = -\frac{c^T\eta}{(c^T\phi)^3}$, then from a comparison with (7.32) it follows that $g_{yy} = 0$ and

$$v_{yy} = \frac{\eta}{(c^T\phi)^2} - \frac{c^T\eta}{(c^T\phi)^3}\phi.$$

□

Proposition 7.5.5. Let $g = g_y = g_{yy} = 0$. Then we have the following conditions:

1. $g_{yyy} = 0$ if and only if $3G_{xx}\phi\eta + G_{xxx}\phi\phi\phi \in \mathcal{R}(G_x)$.

2. If this is the case, then

$$v_{yyy} = \frac{1}{(c^T\phi)^3}\zeta - 3\frac{c^T\eta}{(c^T\phi)^4}\eta - \frac{(c^T\zeta)(c^T\phi) - 3(c^T\eta)^2}{(c^T\phi)^5}\phi,$$

where ϕ, η are defined in Propositions 7.5.1 and 7.5.4 and ζ is the unique solution to the system

$$\begin{cases} G_x\zeta + 3G_{xx}\phi\eta + G_{xxx}\phi\phi\phi = 0, \\ \phi^T\zeta = 0. \end{cases}$$

Proof. Taking derivatives from (7.32) we obtain

$$\begin{bmatrix} G_x & -b \\ c^T & -d \end{bmatrix} \begin{bmatrix} v_{yyy} \\ g_{yyy} \end{bmatrix} + \begin{bmatrix} 3G_{xx}v_y v_{yy} + G_{xxx}v_y v_y v_y \\ 0 \end{bmatrix} = \begin{bmatrix} 0 \\ 0 \end{bmatrix}. \tag{7.56}$$

Now

$$3G_{xx}v_y v_{yy} + G_{xxx}v_y v_y v_y = 3G_{xx}\frac{\phi}{c^T\phi}\left(\frac{\eta}{(c^T\phi)^2} - \frac{c^T\eta}{(c^T\phi)^3}\phi\right) + \frac{1}{(c^T\phi)^3}G_{yyy}\phi\phi\phi$$

$$= \frac{1}{(c^T\phi)^3}\{3G_{xx}\phi\eta + G_{xxx}\phi\phi\phi\} - 3\frac{c^T\eta}{(c^T\phi)^4}G_{xx}\phi\phi. \tag{7.57}$$

If $g_{yyy} = 0$, then $3G_{xx}\phi\eta + G_{xxx}\phi\phi\phi \in \mathcal{R}(G_x)$ (since $G_{xx}\phi\phi \in \mathcal{R}(G_x)$ by Proposition 7.5.4).

Conversely, suppose that $3G_{xx}\phi\eta + G_{xxx}\phi\phi\phi \in \mathcal{R}(G_x)$. Then the system in ζ has a unique solution. Furthermore,

$$\begin{bmatrix} G_x & -b \\ c^T & -d \end{bmatrix} \begin{bmatrix} \zeta \\ 0 \end{bmatrix} + \begin{bmatrix} 3G_{xx}\phi\eta + G_{xxx}\phi\phi\phi \\ -c^T\zeta \end{bmatrix} = \begin{bmatrix} 0 \\ 0 \end{bmatrix}, \tag{7.58}$$

$$\begin{bmatrix} G_x & -b \\ c^T & -d \end{bmatrix} \begin{bmatrix} \eta \\ 0 \end{bmatrix} + \begin{bmatrix} G_{xx}\phi\phi \\ -c^T\eta \end{bmatrix} = \begin{bmatrix} 0 \\ 0 \end{bmatrix}, \tag{7.59}$$

$$\begin{bmatrix} G_x & -b \\ c^T & -d \end{bmatrix} \begin{bmatrix} r\phi \\ 0 \end{bmatrix} + r\begin{bmatrix} 0 \\ -c^T\phi \end{bmatrix} = \begin{bmatrix} 0 \\ 0 \end{bmatrix} \tag{7.60}$$

for all $r \in \mathbb{R}$. From (7.58)–(7.60) it follows that

$$\begin{bmatrix} G_x & -b \\ c^T & -d \end{bmatrix} \begin{bmatrix} \frac{1}{(c^T\phi)^3}\zeta - 3\frac{c^T\eta}{(c^T\phi)^4}\eta + r\phi \\ 0 \end{bmatrix} + \begin{bmatrix} 3G_{xx}v_y v_{yy} + G_{xxx}v_y v_y v_y \\ -\frac{c^T\zeta}{(c^T\phi)^3} - 3\frac{c^T\eta}{(c^T\phi)^4}c^T\eta - rc^T\phi \end{bmatrix} = \begin{bmatrix} 0 \\ 0 \end{bmatrix}.$$

If in particular we choose $r = -\frac{c^T\zeta}{(c^T\phi)^4} - 3\frac{(c^T\eta)^2}{(c^T\phi)^5}$, then from a comparison with (7.56) it follows that $g_{yyy} = 0$ and that v_{yyy} is given as in the statement of Proposition 7.5.5. ◻

7.6 Examples in Low-Dimensional Spaces

We will first deal with the instructive case of low-dimensional spaces. In §7.6.1 we return to the CSTR model defined in §6.5. This is a particularly clear case since it has only one state variable and therefore does not require a Lyapunov–Schmidt reduction. In §7.6.2 we consider a case with three state variables. It requires a Lyapunov–Schmidt reduction, but we will be able to do most of the analysis by hand in the unreduced space and also to perform the reduction explicitly.

7.6.1 Winged Cusps in the CSTR

Winged cusps in the CSTR are studied analytically in [109], especially with respect to their universal unfolding, which serves as a motivation example for the use of singularity theory methods in general. Numerically they are treated in [146].

We will rather stress a special feature. Namely, the winged cusp case provides an interesting link between singularity theory with and singularity theory without a distinguished bifurcation parameter. In §6.5.5 a curve of cusp points ($g = g_x = g_\lambda = D_2(g) = 0$) was computed and a point with ($D_3^1(g) = 0, D_3^2(g) \neq 0$) was detected on this curve. We noted that this is, in fact, equivalent to ($g_{xx} = g_{x\lambda} = 0, g_{\lambda\lambda} \neq 0, g_{xxx} \neq 0$). In the approach with distinguished bifurcation parameter λ this means precisely that we are dealing with a winged cusp; see Table 7.2.

We now prove the relation between the two phenomena in a precise and generally useful way.

Proposition 7.6.1. Let $g(x, \lambda, \alpha)$ be a smooth function of two single variables x, λ and a multidimensional parameter α. Consider the set of defining conditions

$$g = g_x = g_\lambda = g_{xx} = g_{x\lambda} = 0, \tag{7.61}$$

nondegeneracy conditions

$$g_{xxx} \neq 0, \qquad g_{\lambda\lambda} \neq 0, \tag{7.62}$$

defining conditions

$$g = g_x = g_\lambda = D_2(g) = D_3^1(g) = 0, \tag{7.63}$$

and, finally, nondegeneracy condition

$$D_3^2(g) \neq 0. \tag{7.64}$$

Then the two sets (7.61), (7.62) and (7.63), (7.64) are equivalent. Furthermore, if they are satisfied at a point, then there exists a nonsingular matrix $M \in \mathbb{R}^{5\times 5}$ such that

$$\begin{pmatrix} g_z \\ g_{xz} \\ g_{\lambda z} \\ (D_2(g))_z \\ (D_3^1(g))_z \end{pmatrix} = M \begin{pmatrix} g_z \\ g_{xz} \\ g_{\lambda z} \\ g_{xxz} \\ g_{x\lambda z} \end{pmatrix} \tag{7.65}$$

for every variable z in (x, λ, α). In particular, all rank conditions on the Jacobian of (7.61) are equivalent to rank conditions on the Jacobian of (7.63).

Proof. Obviously (7.61) implies (7.63), and it is easy to check that then $D_3^2(g) = -g_{\lambda\lambda}^2 g_{xxx}$. Hence (7.61), (7.62) together imply (7.63), (7.64).

Now assume that, conversely, (7.63), (7.64) hold. Since $D_3^2(g) \neq 0$, $g_{x\lambda}, g_{\lambda\lambda}$ cannot both be zero. Since $D_2(g) = 0$ and $D_3^1(g) = 0$ this implies that $g_{xx} = g_{x\lambda} = 0$; i.e., (7.61) holds. But then $D_3^2(g) = -g_{\lambda\lambda}^2 g_{xxx}$; i.e., (7.62) also holds.

Now let z be any variable in (x, λ, α). At a point where the conditions (7.61), (7.62) are satisfied, we have $(D_2(g))_z = g_{xxz} g_{\lambda\lambda}$, $(D_3^1(g))_z = g_{xxz}(D_2(g))_\lambda - g_{x\lambda z} g_{xxx} g\lambda\lambda$. If we define

$$M_2 = \begin{pmatrix} g_{\lambda\lambda} & 0 \\ (D_2(g))_\lambda & -g_{xxx} g_{\lambda\lambda} \end{pmatrix},$$

then M_2 is a nonsingular matrix and we have proved that

$$\begin{pmatrix} (D_2(g))_z \\ (D_3^1(g))_z \end{pmatrix} = M_2 \begin{pmatrix} g_{xxz} \\ g_{x\lambda z} \end{pmatrix}. \tag{7.66}$$

Now let M be the matrix obtained by replacing the bottom right 2×2-block in the 5×5 identity matrix by M_2. Then (7.66) implies (7.65). □

7.6.2 An Eutrophication Model

We consider the eutrophication model developed in [10] and studied numerically in [237], especially with respect to Hopf bifurcations. It was already mentioned briefly in Exercise 2.5.14. The dynamical equations for this model are

$$\begin{aligned} \dot{x}_1 &= x_1(0.2(\lambda_1 - x_1 - x_2) - 0.445 x_3 - 4), \\ \dot{x}_2 &= -0.0455 x_2 x_3 + 4 x_1, \\ \dot{x}_3 &= \lambda_2(10 - x_3) - 2.67 x_3(0.445 x_1 + 0.0455 x_2). \end{aligned} \tag{7.67}$$

We will fix $\lambda_2 = 0.7$ and consider the starting point $(x_1^o = 0, x_2^o = 0, x_3^o = 10, \lambda_1^o = 42.25)$ in $(x_1, x_2, x_3, \lambda_1)$-space. The aim of this study is to describe all equilibrium solutions to (7.67) in a neighborhood of the starting point by using a Lyapunov–Schmidt reduction.

First we remark that these solutions can be found directly because of the special form of (7.67). There is a trivial solution

$$(x_1 = 0, x_2 = 0, x_3 = 10) \tag{7.68}$$

for every value of λ_1. The tangent vector to this curve is

$$v^{(1)} = (0, 0, 0, 1)^T. \tag{7.69}$$

It is easily seen that the hyperplane $x_1 = 0$ in $(x_1, x_2, x_3, \lambda_1)$-space contains no other equilibrium solutions to (7.67). Therefore, the equilibria of (7.67) different from those given by (7.68) satisfy

$$\begin{aligned} &0.2(\lambda_1 - x_1 - x_2) - 0.445 x_3 - 4 = 0, \\ &-0.0455 x_2 x_3 + 4 x_1 = 0, \\ &\lambda_2(10 - x_3) - 2.67 x_3(0.445 x_1 + 0.0455 x_2) = 0. \end{aligned} \tag{7.70}$$

The $(x_1, x_2, x_3, \lambda_1)$-Jacobian of this system in the starting point is

$$\begin{pmatrix} -0.2 & -0.2 & -0.445 & 0.2 \\ 4 & -0.455 & 0 & 0 \\ -2.67 \times 4.45 & -2.67 \times 0.455 & -0.7 & 0 \end{pmatrix}. \tag{7.71}$$

Since it has full rank, system (7.70) determines a curve through the starting point. The right singular vector $v^{(2)}$ of (7.71) is a tangent vector to this curve. One obtains (up to a scalar multiple)

$$v^{(2)} = \begin{pmatrix} -0.0161493 & -0.141972 & 0.520504 & 1 \end{pmatrix}^T. \tag{7.72}$$

Obviously $v^{(1)}, v^{(2)}$ are not parallel. Hence (7.67) has two equilibrium branches through the starting point with different tangent vectors. This suggests that the starting point is a transcritical bifurcation point (not a pitchfork bifurcation since both tangent vectors have nonzero λ_1-components).

Now we consider the equilibrium equations

$$\begin{aligned} G_1(x, \lambda_1) &\equiv x_1(0.2(\lambda_1 - x_1 - x_2) - 0.445x_3 - 4) = 0, \\ G_2(x, \lambda_1) &\equiv -0.0455x_2x_3 + 4x_1 = 0, \\ G_3(x, \lambda_1) &\equiv \lambda_2(10 - x_3) - 2.67x_3(0.445x_1 + 0.0455x_2) = 0, \end{aligned} \tag{7.73}$$

where $x = (x_1, x_2, x_3)$. For this system

$$G_x^o = \begin{pmatrix} 0 & 0 & 0 \\ 4 & -0.455 & 0 \\ -2.67 \times 4.45 & -2.67 \times 0.455 & -0.7 \end{pmatrix} \tag{7.74}$$

is singular with rank deficiency 1. For a maximal Lyapunov–Schmidt reduction we must choose $b \in \mathbb{R}^3, c \in \mathbb{R}^3, d \in \mathbb{R}$ such that

$$M = \begin{pmatrix} G_x^o & b \\ c^T & d \end{pmatrix} \tag{7.75}$$

is nonsingular. As usual, this is easy, and an obvious choice is

$$b = c = \begin{pmatrix} 1 \\ 0 \\ 0 \end{pmatrix}, \qquad d = 0. \tag{7.76}$$

We recall that the essential results of the analysis are independent of this choice (proved in §6.2).

Let us denote the Lyapunov–Schmidt reduced variables by x (state) and t (parameter, shifted from λ_1), respectively. The vector v_x and scalar g_x are now obtained by solving the system

$$\begin{pmatrix} 0 & 0 & 0 & 1 \\ 4 & -0.455 & 0 & 0 \\ -2.67 \times 4.45 & -2.67 \times 0.455 & -0.7 & 0 \\ 1 & 0 & 0 & 0 \end{pmatrix} \begin{pmatrix} v_{x1} \\ v_{x2} \\ v_{x3} \\ g_x \end{pmatrix} = \begin{pmatrix} 0 \\ 0 \\ 0 \\ 1 \end{pmatrix}. \tag{7.77}$$

Clearly $g_x = 0$ (as expected) and $v_{x1} = 1$. Substituting these values in (7.77), we obtain a 2×2 system of linear equations in v_{x2}, v_{x3} from which we compute $v_{x2} = 8.79121$, $v_{x3} = -32.2307$.

The system dual to (7.77) is

$$\begin{pmatrix} w_1 & w_2 & w_3 & g_x \end{pmatrix} \begin{pmatrix} 0 & 0 & 0 & 1 \\ 4 & -0.455 & 0 & 0 \\ -2.67 \times 4.45 & -2.67 \times 0.455 & -0.7 & 0 \\ 1 & 0 & 0 & 0 \end{pmatrix}$$

$$= \begin{pmatrix} 0 & 0 & 0 & 1 \end{pmatrix}. \tag{7.78}$$

The solution to this system is $w_1 = 1$, $w_2 = 0$, $w_3 = 0$, $g_x = 0$.

The vector v_t and scalar g_t are obtained by solving the system

$$M \begin{pmatrix} v_t \\ g_t \end{pmatrix} + \begin{pmatrix} G^o_{\lambda_1} \\ 0 \end{pmatrix} = 0. \tag{7.79}$$

From (7.73) it follows that $G^o_{\lambda_1} = 0$. Hence $v_t = 0, g_t = 0$.

To determine the singularity type we compute the second-order derivatives of g. From the theory we know that

$$g_{xx} = -w^T G_{xx} v_x v_x. \tag{7.80}$$

Substituting the values of the components of w we find $g_{xx} = -G_{1xx} v_x v_x$. Further,

$$G_{1,11} = -0.4, G_{1,12} = -0.2, G_{1,13} = -0.445, G_{1,22} = 0, G_{1,23} = 0, G_{1,33} = 0.$$

Hence

$$g_{xx} = 0.4 v_{x1} v_{x1} + 0.2 v_{x1} v_{x2} + 0.445 v_{x1} v_{x3} + 0.2 v_{x2} v_{x1} + 0.445 v_{x3} v x1$$

$$= -24.7689 < 0.$$

Again, from the theory

$$g_{xt} = -w^T (G_{xx} v_x v_t + G_{x\lambda_1} v_x). \tag{7.81}$$

We already know that $v_t = 0$. From (7.73) it follows that the 3×3 matrix $G_{x\lambda_1}$ has only one nonzero component, namely, the $(1,1)$ entry with value 0.2. Hence

$$g_{xt} = -1 \times 0.2 \times 1 = -0.2.$$

Finally, the formula for g_{tt} is

$$g_{tt} = -w^T (G_{xx} v_t v_t + 2 G_{x\lambda_1} v_t + G_{\lambda_1 \lambda_1}). \tag{7.82}$$

Since $v_t = 0$ and $G^o_{\lambda_1 \lambda_1} = 0$ we conclude that $g_{tt} = 0$.

Summarizing the above results we have $g_{xx} < 0$, $D_2(g) = g_{xx} g_{tt} - g_{xt}^2 < 0$. So the Lyapunov–Schmidt reduced form is λ-contact equivalent with the normal form $-x^2 + t^2$ of the transcritical bifurcation.

The equation $-x^2 + t^2 = 0$ has two solution branches through $(0,0)$ with different tangent vectors, none of which is orthogonal to the t-axis. This property is obviously preserved under λ-contact equivalence. So let $(x(s), t(s))$ be a parameterization of a branch of $g(x,t) = 0$ with $x^o = x(0) = 0$, $t^0 = t(0) = 0$, $(x_s^o, t_s^o) \neq (0,0)$. By taking second derivatives of the identity $g(x(s), t(s)) \equiv 0$ and evaluating them in $(0,0)$ we obtain

$$g_{xx}^o (x_s^o)^2 + 2 g_{xt}^o x_s^o t_s^o = 0$$

(where we have taken into account that $g_x^o = g_t^o = g_{tt}^o = 0$). So the tangent vectors (x_s^o, t_s^o) are $(0,1)^T$ and $(2g_{xt}^o, -g_{xx}^o)^T$ (up to a scalar multiple).

The Lyapunov–Schmidt reduction relates the branches in (x,t)-space to branches in (x, λ_1)-space. The shift $(v(x,t), t)$ from (x, λ_1) is parameterized by $(v(x(s), t(s)), t(s))$ with tangent vector $(v_x^{oT} x_s^o + v_t^{oT} t_s^o, t_s^0)^T$. Since $v_t^o = 0$ the two possible tangent vectors are $(0,0,0,1)^T$ and $(2g_{xt}^o v_x^{oT}, -g_{xx}^o)^T$. A trivial calculation shows that these vectors are parallel to the vectors $v^{(1)}, v^{(2)}$ in (7.69) and (7.72), respectively.

7.7 Example: The One-Dimensional Brusselator

We will now describe the computational study of some singularities in the continuous Brusselator model introduced in §4.2.1. The equilibrium equations in this model are (equivalent to, modulo some scaling)

$$F(u, D_X, D_Y, A_0, L, D_A, B) = 0, \tag{7.83}$$

where u is the state vector with $2n$ components (n is routinely set to 42) and D_X, D_Y, A_0, L, D_A, B are six parameters. F has $2n$ components, explicitly given in (4.33), (4.34) and the formulas on which these depend.

In subsequent computations L was chosen as the distinguished bifurcation parameter; its shift is further denoted by t. A subroutine DERGSUB was written to compute the quantities g_y, g_t, g_{yy}, g_{yt}, g_{tt}, g_{yyy}, g_{yyt}, g_{ytt}, and g_{ttt} by the formulas obtained in Lemma 7.4.1(b), given as input the values of u, the parameters, and the bordering vectors $b, c \in \mathbb{R}^n$ (d is routinely set to zero). Most of the programming work (done in Fortran double precision) consists in writing the code for the computation of the right-hand side expressions in the formulas in Lemma 7.4.1(b), exploiting the sparsity of the problem.

DERGSUB can also return the solutions to (6.129) and (6.150). If $g_y = 0$, i.e., F_u is singular, then the first $2n$ components of these solutions are the right and left singular vectors of F_u, respectively.

7.7.1 Computational Study of a Curve of Equilibria

As in §4.2.2 we start with parameter values $D_X = 0.0016$, $D_Y = 0.008$, $A_0 = 2.0$, $L = 0.0$, $D_A = 0.5$, and $B = 4.6$. The state values are $u_i = A_0$ for i odd, $u_i = B/A_0$ for i even. Since only values $L > 0$ have a physical meaning, we start by freeing L and following a curve of equilibrium solutions in the direction of increasing L.

By a simple continuation code (Chapter 2) we compute 500 points along this curve. To perform a Lyapunov–Schmidt reduction in every computed point one has to choose b, c and a scalar d such that the bordered matrix

$$M \equiv \begin{pmatrix} F_u & -b \\ c^T & -d \end{pmatrix}$$

is nonsingular.

In §4.1.3 we discussed the choice of b, c, d and their adaptation along the curve. However, in a given problem (as opposed to writing software), it is usually not difficult to deal with the case of a singular bordered extension. To show this we choose in the present example $c_i = b_i = i/\sqrt{\sum_{j=1}^{2n} j^2}$, $d = 0$. This choice is completely arbitrary, so we are not guaranteed that the bordered matrix is nonsingular.

For the points 1 to 173 we always have $g_y > 0$ indicating that nothing of interest happens to the solution curve of (7.83). However, g_y changes sign between the points 173 and 174 and again between 178 and 179. We provide some data.

We first note that g_y, g_t, and g_{yy} all change sign between the points 173 and 174 and the evolution of their values from points 172 to 175 is monotonous. This suggests that they are all zero in an equilibrium point between the computed points 173 and 174. Also, g_{yt} and g_{yyy} do not change sign between 173 and 174 although their evolution is also monotonous. From Table 7.2 we conclude that there is probably a pitchfork bifurcation point between the computed points 173 and 174 for which $g_{yt} < 0, g_{yyy} < 0$. In other words, the new branch should be found for values of L smaller than that of the bifurcation point.

Two remarks are in order. First, we have only made a guess from limited information. But we can easily get more convincing evidence by repeating the continuation with smaller stepsize. Also, we can set up a Newton system to compute an equilibrium point with L as the only free parameter and $g_y = 0$ as an additional equation. The resulting system is not regular (we will treat this issue in §8.1.5) but nevertheless converges to a point for which $g_t = 0$, $g_{yy} = 0$, $g_{yt} < 0$, $g_{yyy} < 0$.

Second, we know from Table 7.1 that a pitchfork is a singularity with codimension 2. So it may seem strange that we find it in a problem with only one (the distinguished) free parameter. The reason is that the Brusselator model is not quite generic: It has a $\mathbb{Z}_2$-symmetry induced by the reflection $z \to 1 - z$; cf. §1.4. The computed branch of equilibrium solutions is fully in the symmetric space, i.e., $u_{2i-1} = u_{2n-2i+1}$ and $u_{2i} = u_{2n+2-2i}$ $(1 \leq i \leq n)$, while the bifurcating branch is in a nonsymmetric direction. We will discuss this further in §8.1.

Now we consider again Tables 7.4 and 7.5. Between the points 178 and 179 all singularity data change sign. However, the evolution of the values is not monotonous. For example, g_y is decreasing from points 172 to 178 and again from 179 to 180. But between 178 and 179 it jumps from a negative value to a positive value. This suggests that there is an intermediate point where g_y does not vanish but becomes infinite. The same remark applies to the other singularity data. The phenomenon is easily understood if we assume that between points 178 and 179 the bordered matrix M as used in Lemma 7.4.1 becomes singular. Since the phenomenon is merely a consequence of the choice of

Table 7.4: Singularity data for nine consecutive points on an equilibrium curve.

Counter	L	g_y	g_t
172	0.256867	3.929318D-03	−9.491181D-02
173	0.257365	1.575605D-03	−3.818792D-02
174	0.257638	−7.021743D-04	1.705852D-02
175	0.257793	−2.663405D-03	6.480366D-02
176	0.257970	−6.082842D-03	0.148294
177	0.258173	−1.340559D-02	0.327663
178	0.258400	−3.962518D-02	0.971869
179	0.258652	0.170482	−4.200970
180	0.258928	3.658241D-02	−0.907361

Table 7.5: Some more singularity data for nine consecutive points on an equilibrium curve.

Counter	g_{yy}	g_{yt}	g_{yyy}
172	−0.144964	8.175237D-02	1.131025D-02
173	−0.111003	−3.864631	−0.696984
174	8.028245D-02	−12.604828	−40.545827
175	0.429516	−25.491676	−176.462754
176	1.603080	−63.690298	−1010.193924
177	7.649723	−240.354493	−10989.037305
178	105.418069	−2835.164169	−774851.015630
179	−5033.730359	121252.66577	456552734.97699
180	−32.661419	724.851452	99023.162912

the bordering vectors b, c we can check it by repeating the experiment with other choices for b, c.

The next change of sign of g_y occurs between the points 192 and 193 (we are not interested in the zeros of other singularity data if $g_y \neq 0$). Again, we present some singularity data.

The evolution of the singularity data in Tables 7.6 and 7.7 is quite monotonous, and we infer that between the points 192 and 193 there is a point with $g_y = 0$, $g_t < 0$, $g_{yy} < 0$. From Table 7.2 we conclude that this is a turning point. This is the only codimension-0 singularity in Table 7.1 and hence the only that we can expect generically. We note that in Table 7.6 the values of L are increasing before the suspected turning point and decreasing after it. This is of course what we expect.

The computed curve of equilibria further contains a turning point between the points 290 and 291 ($g_t < 0$, $g_{yy} > 0$), a pitchfork between 341 and 342 ($g_{yt} > 0$, $g_{yyy} < 0$), another pitchfork between 434 and 435 ($g_{yt} > 0$, $g_{yyy} > 0$), and another turning point between 457 and 458 ($g_t > 0$, $g_{yy} > 0$).

7.7.2 Computational Study of a Curve of Turning Points

In §7.7.1 we computed a curve of equilibria and found reasonable evidence to suspect a turning point in (u, L)-space between the computed points 192 and 193. The next step is

Table 7.6: Singularity data for eight consecutive points on an equilibrium curve.

Counter	L	g_y	g_t
189	0.260590	8.289490D-03	−0.271538
190	0.260666	6.758256D-03	−0.253130
191	0.260725	4.499694D-03	−0.228811
192	0.260752	7.144452D-04	−0.192164
193	0.260749	−2.547407D-03	−0.162689
194	0.260726	−8.290850D-03	−0.113032
195	0.260675	−2.086268D-02	−8.478206D-03
196	0.260581	−6.865790D-02	0.377538

Table 7.7: Some more singularity data for eight consecutive points on an equilibrium curve.

Counter	g_{yy}	g_{yt}	g_{yyy}
189	−0.786730	9.939521	−62.882583
190	−1.006741	11.518895	−153.569673
191	−1.520379	15.620631	−403.815685
192	−2.954467	27.375722	−1427.339752
193	−4.876651	43.114935	−3442.917988
194	−10.181612	86.284114	−12075.742743
195	−33.258352	272.540272	−87818.480843
196	−377.587444	3029.790325	−5019127.847
197	5835.990161	−46551.45004	474508771.14

to locate this point more accurately. To this end we set up a Newton system consisting of the equations in (7.83) and the additional equation $g_y = 0$; the unknowns are the components of u and the parameter L. From §7.2 we know that this is a regular system in the turning point. As the starting point for the Newton iteration we take the point 192 computed in §7.7.1. The bordering vectors in the Lyapunov–Schmidt reduction are those from §7.7.1. In the Newton iteration the derivatives of g_y with respect to u and L are simply computed by finite differences (this is Method 6.6.3.a from §6.6).

The method appears to converge quite well; the norms of the successive Newton corrections are 1.182797E-02, 4.133588E-04, 6.542175E-07, 2.137159E-12. The value of g_y tends to zero while g_t and g_{yy} tend to nonzero values; see Table 7.8.

Let the computed point be denoted T_1. Its computed value of L is 0.26075331599577; T_1 is in the symmetric space, and its 42 first coordinates are given in Table 7.9.

Starting from T_1 we now compute a curve of turning points. The parameters L and A_0 are now free, and the equation $g_y = 0$ is added to (7.83) to form a set of $2n+1$ equations in $2n+2$ variables. The derivatives of g_y with respect to u, L, and A_0 are all computed by finite differences. We choose the direction of initially increasing values of A_0 (it happens that L is initially increasing). Along the path of turning points the Lyapunov–Schmidt reduction was performed with vectors b, c chosen as in §7.7.1. The most important singularity data are now g_t and g_{yy}, since nothing interesting happens if

Table 7.8: Values of g_y, g_t, and g_{yy} before, between, and after Newton corrections.

g_y	g_t	g_{yy}
7.1444526989679$E-04$	−0.19216460976738	−2.9544670651958
−2.8837604239740E-05	−0.18532604048285	−3.3322353139454
−4.5175972247626E-08	−0.18558818080633	−3.3166343604266
−1.4504994404110E-13	−0.18558859246620	−3.3166099761107
−5.3342582566039E-15	−0.18558859246747	−3.3166099760336

Table 7.9: u-coordinates of the turning point T_1.

1.9257148565779	2.3873496394386	1.8499648908861	2.4746642058879
1.7718930421225	2.5617799549844	1.6913062572866	2.6483892483784
1.6086213230212	2.7340504801707	1.5247536341193	2.8182121972717
1.4409635608554	2.9002479755865	1.3586834258828	2.9794967525101
1.2793507476107	3.0553026988663	1.2042698800857	3.1270494462563
1.1345161133551	3.1941852264291	1.0708865685397	3.2562376166779
1.0138936702461	3.3128185074105	0.96379141587518	3.3636212142282
0.92062253597274	3.4084122318755	0.88427532505967	3.4470200866472
0.85454128576807	3.4793233165234	0.83116765397497	3.5052390281858
0.81390155668183	3.5247129207553	0.80252457982588	3.5377112233212
0.79687778374385	3.5442146976244		

they are both nonzero. It turns out that both g_t and g_{yy} vary in a monotonous way and both stay negative in the first 116 computed points. However, g_{yy} changes sign between the points 116 and 117. We present some relevant data in Tables 7.10 and 7.11.

The data in Tables 7.10 and 7.11 strongly suggest that between points 116 and 117 there is a point with $g_y = g_{yy} = 0$, $g_t \neq 0$, $g_{yyy} \neq 0$. By the classification results in Table 7.2 this represents a hysteresis point ($\epsilon x^3 + \delta\lambda = 0$) in λ-singularity theory with L as the distinguished bifurcation parameter. We note that in the sense of $\mathcal{K}$-singularity theory we deal with a singularity from $\mathbb{R}$ into $\mathbb{R}$ which is $\mathcal{K}$-equivalent to x^3 (see §6.3.2).

Table 7.10 suggests that both A_0 and L attain extremal values in the hysteresis point. To explain this by singularity theory, we use the unfolding results in §6.4. It is shown there that the $\mathcal{K}$-singularity represented by x^3 has the universal unfolding $x^3 + \alpha x + \beta$. If we let α, β vary, then the singular points of the two-parameter problem are defined by the two equations $x^3 + \alpha x + \beta = 0$, $3x^2 + \alpha = 0$. In (x, α, β)-space they form a smooth curve that can be smoothly parameterized by the variable x itself, and we easily find that $\alpha = -3x^2$, $\beta = 2x^3$. Hence the derivatives of both α and β with respect to arclength of that curve vanish at the hysteresis point.

The projection of the curve of singular points onto (α, β)-space is obtained by elimination of x; we easily obtain the equation $4\alpha^2 + 27\beta^3 = 0$. In other words, this projection is a cusp curve. This can be roughly checked by plotting the values of L, A_0 in Table 7.10.

Now we look at the values of g_t. It turns out that g_t is negative in the first 251 points and changes sign in point 252. We present the relevant data in Tables 7.12 and 7.13.

Table 7.10: Singularity data for six consecutive points on a curve of turning points.

Counter	L	A_0	g_y
114	0.14295587833097	2.5767720226383	3.8616193075828D-08
115	0.14258097347766	2.5778343549669	5.3686487252509D-08
116	0.14236320935075	2.5784576200251	8.5595105897013D-08
117	0.14238885829761	2.5783792364485	2.4704134197519D-08
118	0.14278424308635	2.5771874069439	7.4636546848250D-08
119	0.14405949175490	2.5730888237358	6.6774022950047D-08

Table 7.11: Some more singularity data for six consecutive points on a curve of turning points.

Counter	g_t	g_{yy}	g_{yyy}
114	−1.6487509776572D-02	−3.1290241421297D-04	6.0926777138435D-03
115	−1.5945080853886D-02	−1.9399010859465D-04	5.8150234741239D-03
116	−1.5537521102208D-02	−6.5242373775603D-05	5.5458253605939D-03
117	−1.5312453600796D-02	8.2184467842062D-05	5.2822470456775D-03
118	−1.5331730476806D-02	2.6165435404682D-04	5.0159453936188D-03
119	−1.5855754696751D-02	5.2829885601286D-04	4.7235132911713D-03

Tables 7.12 and 7.13 strongly suggest that between points 251 and 252 there is a point with $g_y = g_t = 0$, $g_{yy} < 0$, $D_2(g) = g_{yy}g_{tt} - g_{yt}^2 < 0$. By the classification results in Table 7.2 this represents a transcritical bifurcation point $(-x^2 + \lambda^2 = 0)$ with λ as the distinguished bifurcation parameter.

7.7.3 Computational Study of a Curve of Hysteresis Points

In §7.7.2 we found reasonable evidence to suspect a hysteresis point in (u, L)-space between the computed points 116 and 117. We will now compute this point more accurately and then use it as a starting point for continuation of a curve of hysteresis points.

To compute the hysteresis point we add the equations $g_y = 0$, $g_{yy} = 0$ to (7.83) to form a system of $2n + 2$ equations in the $2n$ state variables and L, A_0. The derivatives of g_y, g_{yy} are computed by finite differences. As starting point of the Newton iteration we use the computed point 116 of the curve of turning points. Let the computed hysteresis point be called H_1. Its parameter values are $L = 0.14234756258276$, $A_0 = 2.5784996009488$. The point is in the symmetric space with the first 42 components given in Table 7.14.

Starting from H_1 we now compute a curve of hysteresis points in (u, L)-space. The parameters L, A_0, D_A are free and the equations $g_y = 0$, $g_{yy} = 0$ are added to (7.83) to form a set of $2n+2$ equations in $2n+3$ variables. The derivatives of g_y, g_{yy} are computed by finite differences. Along the path of hysteresis points the Lyapunov–Schmidt reduction is performed with fixed vectors b, c chosen as in §7.7.1. The important singularity data to monitor are now g_{yyy} and g_t, since a degeneracy can occur only if one of them vanishes.

The chosen direction of continuation is with initially increasing values of D_A; it turns

Table 7.12: Singularity data for six consecutive points on a curve of turning points.

Counter	L	A_0	g_y
249	0.21673975201882	1.8715972551873	3.2453344942375D-08
250	0.21683283885693	1.8716434722980	3.6365184240755D-08
251	0.21693916283608	1.8716694152090	3.9966812848077D-08
252	0.21705646000402	1.8716677420835	4.3004710656045D-08
253	0.21718255269296	1.8716325136044	4.5099546597646D-08
254	0.21731528069361	1.8715591661264	4.5672319205528D-08

Table 7.13: Some more singularity data for six consecutive points on a curve of turning points.

Counter	g_t	g_{yy}	$D_2(g)$
249	−1.3799415309754D-02	−2.5801048405106D-03	−0.19886458971121
250	−7.9512868505643D-03	−3.1032182232441D-03	−0.18603065775567
251	−2.4353339778566D-03	−3.5708663971932D-03	−0.17370880874212
252	2.7810886607745D-03	−3.9775952414482D-03	−0.16183792850979
253	7.7267841054216D-03	−4.3162165758346D-03	−0.15036489465565
254	1.2427254338127D-02	−4.5774979351065D-03	−0.13924395724130

Table 7.14: u-coordinates of the hysteresis point H_1.

2.5443160463982	1.7994335223238	2.5102847410377	1.8148081876291
2.4765644434248	1.8300303992082	2.4433197244510	1.8450225353307
2.4107193578647	1.8597067083309	2.3789345529236	1.8740053491006
2.3481370896474	1.8878418143860	2.3184974164162	1.9011410023903
2.2901827661043	1.9138299628464	2.2633553408987	1.9258384889536
2.2381706079322	1.9370996802712	2.2147757384833	1.9475504676868
2.1933082134379	1.9571320937870	2.1738946076580	1.9657905441890
2.1566495565029	1.9734769275231	2.1416748995472	1.9801478036373
2.1290589899554	1.9857654611613	2.1188761533028	1.9902981467204
2.1111862770050	1.9937202488268	2.1060345109260	1.9960124397669
2.1034510610471	1.9971617786979		

out that L is initially decreasing and A_0 is initially increasing. Furthermore, it turns out that $g_t < 0$ and $g_{yyy} > 0$ in the first 32 computed points and that g_t changes sign in the thirty-third point. We present some relevant singularity data in Tables 7.15, 7.16, and 7.17.

Tables 7.15, 7.16, and 7.17 suggest that between the computed points 32 and 33 there is a point where $g_y = g_t = g_{yy} = 0$, $g_{yt} < 0$, $g_{yyy} > 0$. By the classification results in Table 7.2 this indicates a pitchfork bifurcation in (u, L)-space. We note that such a point is special only with respect to the choice of L as a distinguished bifurcation parameter. In the setting of $\mathcal{K}$-singularity theory it is not distinguished among other hysteresis points.

Table 7.15: Data for six consecutive points on a curve of hysteresis points.

Counter	L	A_0	D_A
30	0.12326928524989	2.5765907521002	3.2938969585393
31	0.12253944324383	2.5756854978557	3.7858330648367
32	0.12183036359671	2.5747410589955	4.3772152705928
33	0.12114649424129	2.5737689098667	5.0877362732414
34	0.12049099254938	2.5727800038225	5.9410502625583
35	0.11987976954582	2.5718070512091	6.9401205465577

Table 7.16: Singularity data for six consecutive points on a curve of hysteresis points.

Counter	g_y	g_t	g_{yy}
30	−8.5498608796581D-11	−6.7538494360761D-04	8.2094585165173D-11
31	−9.6285713417404D-11	−3.3902536888094D-04	9.0979014114631D-11
32	−1.0446281095622D-10	−3.7623226293806D-05	9.7149923512846D-11
33	−1.0963830121536D-10	2.2814637598016D-04	1.0040380346386D-10
34	−1.1176060085589D-10	4.5880231930698D-04	1.0084949858731D-10
35	−9.6984462858400D-11	6.5165078611233D-04	8.6316114438862D-11

Table 7.17: More singularity data for six consecutive points on a curve of hysteresis points.

Counter	g_{yt}	g_{tt}	g_{yyy}
30	−6.8543562618369D-03	−0.67703425722211	2.7346732113433D-03
31	−7.4032017002651D-03	−0.64871793581977	2.6670303137137D-03
32	−7.9328343206711D-03	−0.62028767986055	2.6030946385972D-03
33	−8.4403024835011D-03	−0.59198499402144	2.5430457534567D-03
34	−8.9236670096673D-03	−0.56401968477544	2.4869312106714D-03
35	−9.3717014285121D-03	−0.53718612413738	2.4358445419410D-03

7.7.4 Computational Study of a Curve of Transcritical Bifurcation Points

In §7.7.2 we found reasonable evidence to suspect a transcritical bifurcation point in (u, L)-space between the computed points 251 and 252. We will now compute this point more accurately and then use it as a starting point for continuation of a curve of transcritical bifurcation points.

To compute the transcritical bifurcation point we add the equations $g_y = 0$, $g_t = 0$ to (7.83) to form a system of $2n + 2$ equations in the $2n$ state variables and L, A_0. The derivatives of g_y, g_t are computed by finite differences. As a starting point of the Newton iteration we use the computed point 251 of the curve of turning points. Let the computed transcritical bifurcation point be called S_1. Its parameter values are $L = 0.21699159478803$, $A_0 = 1.8716693460333$. The point is in the symmetric space

Table 7.18: u-coordinates of the transcritical bifurcation point S_1.

1.9683797827132	2.4107281957191	2.0621100736219	2.3646667073422
2.1498835979462	2.3204135107417	2.2289102402836	2.2788110019368
2.2967849466730	2.2405965215285	2.3516607658045	2.2063601369548
2.3923750806501	2.1765135507618	2.4185154898209	2.1512732412537
2.4304213149933	2.1306588488953	2.4291257089456	2.1145058196791
2.4162501285492	2.1024897714271	2.3938667548154	2.0941591348325
2.3643453585332	2.0889723359696	2.3301997063224	2.0863360178610
2.2939456402560	2.0856413887737	2.2579791808196	2.0862965777514
2.2244790288164	2.0877537354022	2.1953341904421	2.0895304249370
2.1720945150583	2.0912255145867	2.1559399780385	2.0925302463689
2.1476636740961	2.0932353890117		

with the first 42 components given in Table 7.18.

Starting from S_1 we now compute a curve of transcritical bifurcation points in (u, L)-space. The parameters L, A_0, D_A are free, and the equations $g_y = 0$, $g_t = 0$ are added to (7.83) to form a set of $2n + 2$ equations in $2n + 3$ variables. The derivatives of g_y, g_t are computed by finite differences. Along the path of hysteresis points the Lyapunov–Schmidt reduction is performed with fixed vectors b, c chosen as in §7.7.1. The important singularity data to monitor are now g_{yy} and $D_2(g)$, since a degeneracy can occur only if one of them vanishes.

The chosen direction of continuation is with initially decreasing values of D_A; it turns out that L and A_0 are initially decreasing. It turns out that g_{yy} and $D_2(g)$ are both negative in the first computed 22 points; in the twenty-third point g_{yy} is positive. We present the important data in Tables 7.19, 7.20, and 7.21.

Tables 7.19, 7.20, and 7.21 suggest that between the computed points 22 and 23 there is a point with $g_y = g_t = g_{xx} = 0$, $g_{yt} < 0$, $g_{yyy} > 0$, i.e., a pitchfork bifurcation in (u, L)-space by the classification results in Table 7.2.

7.7.5 A Winged Cusp on a Curve of Pitchfork Bifurcations

In §7.7.3 we found reasonable evidence to suspect a pitchfork bifurcation point in (u, L)-space between the computed points 32 and 33 of the curve of hysteresis points. To compute this point more accurately we add the equations $g_y = 0$, $g_t = 0$, $g_{yy} = 0$ to (7.83) to form a system of $2n + 3$ equations in the $2n$ state variables and L, A_0, D_A. Derivatives of g_y, g_t, and g_{yy} are computed by finite differences. As the starting point for the Newton iteration we used point 32 of the curve of hysteresis points. Let the computed pitchfork bifurcation point be called P_1. Its parameter values are $L = 0.12173732624441$, $A_0 = 2.5746123604809$, $D_A = 4.4649043872268$. The point is in the symmetric space with first 42 components given in Table 7.22.

Starting from P_1 we now compute a curve of pitchfork bifurcation points in (u, L)-space. The parameters L, A_0, D_A, B are free, and the equations $g_y = 0$, $g_t = 0$, and $g_{yy} = 0$ are added to (7.83) to form a set of $2n + 3$ equations in $2n + 4$ variables. The derivatives of g_y, g_t, g_{yy} are computed by finite differences. Along the path of hysteresis points the Lyapunov–Schmidt reduction is performed with fixed vectors b, c chosen as in

Table 7.19: Data for six consecutive points on a curve of transcritical bifurcation points.

Counter	L	A_0	D_A
20	0.21413027274225	1.8556105477205	0.29728324425771
21	0.21390715795656	1.8542639883958	0.28611089747583
22	0.21363888686953	1.8526267192158	0.27333360960622
23	0.21347770151757	1.8516334402503	0.26598338465848
24	0.21328409829529	1.8504308945357	0.25746305643177
25	0.21305156936512	1.8489728493074	0.24765393583841

Table 7.20: Singularity data for six consecutive points on a curve of transcritical bifurcation points.

Counter	g_y	g_t	g_{yy}
20	−3.6001335840887D-13	4.7625316353755D-11	−6.0068573285283D-04
21	−9.9166113684286D-13	7.5272361964362D-10	−3.8441269277924D-04
22	−2.7477099837659D-12	7.9891938900890D-10	−1.2973824082863D-04
23	2.1973202247885D-12	−4.9432809564720D-10	2.0503336901704D-05
24	−2.9118891347153D-13	3.2146356092307D-10	1.9825093707651D-04
25	−7.6599352260583D-13	1.1490645038797D-10	4.0788190920194D-04

Table 7.21: More singularity data for six consecutive points on a curve of transcritical bifurcation points.

Counter	g_{yt}	$D_2(g)$	g_{yyy}
20	−0.44112503581854	−0.17938736345251	2.5710254787360D-02
21	−0.43583382272077	−0.18016491116352	2.6024516661547D-02
22	−0.42942760397322	−0.18108270215350	2.6324421603748D-02
23	−0.42555562104471	−0.18162522792345	2.6465802836714D-02
24	−0.42088236275509	−0.18226810981567	2.6599038192906D-02
25	−0.41523726704185	−0.18302770442680	2.6708800238089D-02

Table 7.22: u-coordinates of the pitchfork bifurcation point P_1.

2.5597270021498	1.7924359310895	2.5449195554636	1.7981647370463
2.5302678895997	1.8038328402920	2.5158494192158	1.8094101729347
2.5017406509934	1.8148670776082	2.4880167297236	1.8201744748689
2.4747509887728	1.8253040283518	2.4620145095764	1.8302283063075
2.4498756945213	1.8349209382059	2.4383998571800	1.8393567651870
2.4276488334024	1.8435119832472	2.4176806162581	1.8473642781737
2.4085490172936	1.8508929513669	2.4003033560437	1.8540790358264
2.3929881792360	1.8569054016990	2.3866430106764	1.8593568509118
2.3813021324156	1.8614202005197	2.3769943974777	1.8630843544902
2.3737430741985	1.8643403637312	2.3715657220620	1.8651814742252
2.3704740988421	1.8656031631861		

Table 7.23: u-coordinates of the winged cusp bifurcation point W_1.

4.4185838410822	1.7166251083831	4.2918010162071	1.7678240768051
4.1635037412018	1.8191140088368	4.0334866676873	1.8704237908616
3.9019060321544	1.9216053060640	3.7692536952839	1.9724388384093
3.6363193117995	2.0226412639882	3.5041414041338	2.0718768282817
3.3739496256917	2.1197699918001	3.2471017096470	2.1659195737336
3.1250193208393	2.2099132724050	3.0091271501161	2.2513416156209
2.9007991077069	2.2898104945304	2.8013144898588	2.3249516397704
2.7118256992968	2.3564306696348	2.6333377206708	2.3839526300581
2.5666982996088	2.4072652111516	2.5125968082830	2.4261600313354
2.4715691891116	2.4404725090696	2.4440061677226	2.4500808897997
2.4301620782259	2.4549049694101		

§7.7.1. The important singularity data to monitor are now g_{yyy} and g_{yt} since a degeneracy can occur only if one of them vanishes.

The chosen direction of continuation is with initially decreasing values of L; it turns out that A_0 and B are initially increasing; D_A is initially decreasing.

It also turns out that g_{yyy} is positive and g_{yt} is negative in the first 174 computed points. In point 175 g_{yt} and g_{yyy} are positive. So we suspect that between the points 174 and 175 there is a point with $g_y = g_t = g_{yy} = g_{yt} = 0$ and $g_{yyy} > 0$. Furthermore, $g_{tt} < 0$. By the results in Table 7.2 this indicates a winged cusp.

We then set up a Newton system to compute the point more accurately. We add the equations $g_y = 0$, $g_t = 0$, $g_{yy} = 0$, $g_{yt} = 0$ to (7.83) to form a system of $2n+4$ equations in the $2n$ state variables and L, A_0, D_A, B. Derivatives of g_y, g_t, g_{yy}, and g_{yt} are computed by finite differences. As the starting point for the Newton iteration we used point 175 of the curve of pitchfork bifurcation points. Let the computed winged cusp point be called W_1. Its parameter values are $L = 0.13539057957389$, $A_0 = 4.5444340712390$, $D_A = 2.0970914551929D-02$, $B = 7.5687934071543$. The point is in the symmetric space with the first 42 components given in Table 7.23.

7.7.6 A Degenerate Pitchfork on a Curve of Pitchfork Bifurcations

In §7.7.4 we found reasonable evidence to suspect a pitchfork bifurcation point in (u, L)-space between the computed points 22 and 23 of the curve of transcritical bifurcation points. To compute this point more accurately we add the equations $g_y = 0$, $g_t = 0$, $g_{yy} = 0$ to (7.83) to form a system of $2n+3$ equations in the $2n$ state variables and L, A_0, D_A. Derivatives of g_y, g_t, and g_{yy} are computed by finite differences. As the starting point for the Newton iteration we used point 23 of the curve of transcritical bifurcation points. Let the computed pitchfork bifurcation point be called P_2. Its parameter values are $L = 0.21349982935047$, $A_0 = 1.8517702247076$, $D_A = 0.26697837716870$. The point is in the symmetric space with the first 42 components given in Table 7.24.

Starting from P_2 we now compute a curve of pitchfork bifurcation points in (u, L)-space. The parameters L, A_0, D_A, B are free, and the equations $g_y = 0$, $g_t = 0$, and $g_{yy} = 0$ are added to (7.83) to form a set of $2n+3$ equations in $2n+4$ variables. The

Table 7.24: u-coordinates of the pitchfork bifurcation point P_2.

1.9742893367231	2.4260577339689	2.0929721429512	2.3691606585935
2.2039320637318	2.3145706950771	2.3034984844712	2.2633728075026
2.3885111320831	2.2165143891064	2.4565726919138	2.1747450661096
2.5062259330825	2.1385739315219	2.5370362501672	2.1082485682461
2.5495773067795	2.0837565781393	2.5453319614631	2.0648471398282
2.5265301439781	2.0510679619332	2.4959492182157	2.0418120314148
2.4567015589372	2.0363686027677	2.4120301625218	2.0339736147760
2.3651276077950	2.0338558457191	2.3189877041159	2.0352763841066
2.2762934713097	2.0375602298102	2.2393402152388	2.0401199273215
2.2099887693569	2.0424719787542	2.1896417028431	2.0442473305366
2.1792345228973	2.0451974529118		

derivatives of g_y, g_t, g_{yy} are computed by finite differences. Along the path of hysteresis points the Lyapunov–Schmidt reduction is performed with fixed vectors b, c chosen as in §7.7.1. The important singularity data to monitor are now g_{yyy} and g_{yt}, since a degeneracy can occur only if one of them vanishes.

The chosen direction of continuation is with initially decreasing values of L; it turns out that A_0, D_A, and B are initially increasing.

It also turns out that g_{yyy} is positive and g_{yt} is negative in the first 35 computed points. In point 36 g_{yt} and g_{yyy} are both negative. So we suspect that between the points 35 and 36 there is a point with $g_y = g_t = g_{yy} = g_{yyy} = 0$ and $g_{yt} < 0$. By the classification results in Table 7.2 this indicates a degenerate pitchfork bifurcation ($\epsilon x^4 + \delta \lambda x$), at least if $g_{yyyy} \neq 0$. Now g_{yyyy} is not in the list of singularity data that we compute routinely. However, we can obtain it from (6.178), since both v_y and (a finite difference approximation to) g_{yyyx} are computed anyway. So g_{yyyy} is available at practically no extra cost. We found $g_{yyyy} > 0$.

Again, we can set up a Newton system to compute the point more accurately. We add the equations $g_y = 0$, $g_t = 0$, $g_{yy} = 0$, $g_{yyy} = 0$ to (7.83) to form a system of $2n + 4$ equations in the $2n$ state variables and L, A_0, D_A, B. Derivatives of g_y, g_t, g_{yy}, and g_{yyy} are computed by finite differences. As the starting point for the Newton iteration we used point 36 of the curve of pitchfork bifurcation points. Let the computed degenerate pitchfork point be called DP_1. Its parameter values are $L = 0.20266719518791$, $A_0 = 2.2649986147020$, $D_A = 0.62134499890079$, $B = 5.5742627948698$. The point is in the symmetric space with the first 42 components given in Table 7.25.

7.7.7 Computation of Branches of Cusp Points and Quartic Turning Points

A winged cusp like W_1 satisfies the cusp equations ($g = g_x = g_t = D_2(g) = 0$) and therefore can be used to start the computation of a curve of cusp points. Similarly, a degenerate pitchfork like DP_1 satisfies the equations for a quartic turning point ($g = g_x = g_{xx} = g_{xxx} = 0$) and therefore can be used to start the computation of a curve of quartic turning points. This was done in the Brusselator example; we omit the straightforward details.

Table 7.25: u-coordinates of the degenerate pitchfork bifurcation point DP_1.

2.3402891546959	2.4273176284379	2.4131972067471	2.3942802628139
2.4814026077224	2.3626030004393	2.5427733768539	2.3329053812399
2.5954922505150	2.3057248719544	2.6381652171478	2.2814901382832
2.6699011162824	2.2605012727889	2.6903550363203	2.2429186509612
2.6997327518186	2.2287610871661	2.6987578932329	2.2179129646011
2.6886072281296	2.2101391735368	2.6708219380237	2.2051061031278
2.6472039798158	2.2024066247398	2.6197066238892	2.2015869627013
2.5903272875630	2.2021735266707	2.5610091176591	2.2036981187833
2.5335557268685	2.2057203623769	2.5095613633135	2.2078466601877
2.4903568765613	2.2097454155454	2.4769703698432	2.2111585883205
2.4701005494595	2.2119098737070		

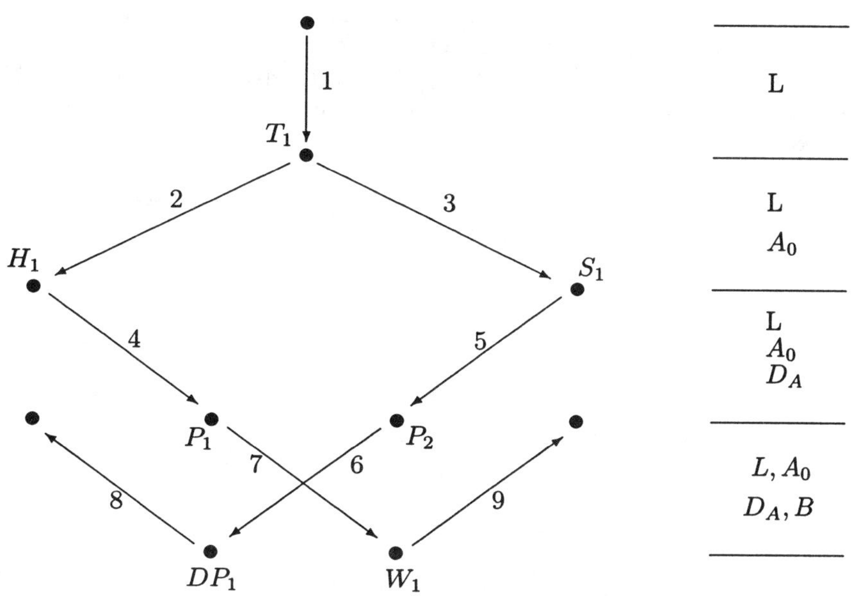

Figure 7.1: Scheme of the singularities computed in the Brusselator example, ordered vertically by codimension. Free parameters are at the right.

A scheme of the computed branches and points is given in Figure 7.1. There are nine numbered arrows, each corresponding to a curve of special points; the curves are discussed in §§7.7.1, 7.7.2, 7.7.2, 7.7.3, 7.7.4, 7.7.5, 7.7.6, 7.7.7, and 7.7.7, respectively. The special points T_1, H_1, S_1, P_1, W_1, P_2, and DP_1 are computed in §§7.7.2, 7.7.3, 7.7.4, 7.7.5, 7.7.5, 7.7.6, and 7.7.6, respectively.

7.8 Numerical Branching

Let $N \in \mathbb{N}$ and U be a nonempty open subset of $\mathbb{R}^{N+1}$. Consider a C^∞ mapping

$$G : U \to \mathbb{R}^N. \tag{7.84}$$

Let $x^\circ \in U$, $G(x^\circ) = 0$. If $G_x(x^\circ)$ has full rank N, then the equation $G(x) = 0$ locally defines a C^∞-curve through x° (see §2.1). If $G_x(x^\circ)$ is rank deficient, then the situation may be more complicated. There may be more than one branch or no branch at all; also, the solution branches may be nonsmooth in x°.

The computation of such branches is a natural precursor of singularity theory and can be developed to a considerable degree of sophistication without it; see, e.g., [68], [150], [153]. However, singularity theory is the proper mathematical setting to get an overview of the many possible cases and to classifiy them in a logical way. Also, it tells us how likely a particular case is in a particular situation. In combination with the Lyapunov–Schmidt reduction it also leads to reliable numerical methods, which combine a solid mathematical base with numerical efficiency.

It is natural to start with the situation where there is no distinguished bifurcation parameter in (7.84). By the results in Chapter 6 (see in particular Table 6.1) all such cases with codimension 4 or less (i.e., that can be expected in problems with no more than four free parameters) have a Jacobian with rank $N - 1$ and can be reduced to singularities from $\mathbb{R}^2$ into $\mathbb{R}$. The Lyapunov–Schmidt reduction requires the choice of vectors $B \in \mathbb{R}^N$, $D \in \mathbb{R}^2$, and a matrix $C \in \mathbb{R}^{N\times 2}$ such that

$$M(x) \equiv \begin{pmatrix} G_x(x) & B \\ C^T & D \end{pmatrix} \tag{7.85}$$

is nonsingular in a neighborhood of $x^\circ \in \mathbb{R}^{N+1}$.

For x in a neighborhood of x° and y in a neighborhood of $0 \in \mathbb{R}^2$ we define $v(x; y) \in \mathbb{R}^N$ and $g(x; y) \in \mathbb{R}$ by

$$G(x + v(x; y)) - Bg(x; y) = G(x), \tag{7.86}$$

$$C^T v(x; y) - Dg(x; y) = y. \tag{7.87}$$

Equations (7.86), (7.87) form a system of $N+1$ equations in the $N+1$ unknowns contained in v, g. The problem of computing solution branches of

$$G(x) = 0 \tag{7.88}$$

in a neighborhood of x° is now equivalent to solving

$$g(x^\circ; y) = 0 \tag{7.89}$$

for y in a neighborhood of $0 \in \mathbb{R}^2$. The equivalence has the following precise sense. If $y(t)$ is a solution to (7.89) for t in a neighborhood of 0 and with $y(0) = 0$, then $x^\circ + v(x^\circ; y(t))$ is a solution to (7.88). Conversely, if $x(t)$ is a solution to (7.88) for t in a neighborhood of 0 and with $x(0) = x^\circ$, then $C^T(x(t) - x^\circ)$ is a solution to (7.89). Since v is a smooth function of both x and y, all smoothness properties of solution branches of (7.88) or (7.89) carry over to the other situation.

We now consider the most important cases from Table 6.1. For simplicity we denote

$$y = \begin{pmatrix} y_1 \\ y_2 \end{pmatrix}.$$

Afterward, we will specialize to the case of $(\lambda - \mathcal{K})$-singularities.

7.8.1 Simple Bifurcation Point and Isola Center

In these cases $g = g_{y_1} = g_{y_2} = 0$ and $D_2(g) = g_{y_1y_1}g_{y_2y_2} - g_{y_1y_2}^2 \neq 0$. If $D_2(g) > 0$ (isola center) then there are no solution branches, so we assume $D_2(g) < 0$ (simple bifurcation point). Since (7.89) is $\mathcal{K}$-equivalent to the normal form $y_1^2 - y_2^2 = 0$ there must be two smooth solution branches with different tangent directions. For practical purposes, it is sufficient to compute the tangents to the corresponding curves in (7.88) since continuation methods can do the rest. So we first compute the tangents to the solution curves of (7.89).

Let $(y_1(t), y_2(t))$ be a parameterization of a solution curve of (7.89), i.e.,

$$g(y_1(t), y_2(t)) = 0, \tag{7.90}$$

where we have omitted the dependence on x and where t is in a neighborhood of 0, $y_1(0) = 0$, $y_2(0) = 0$. Taking derivatives of (7.90) we obtain

$$g_{y_1}y_{1t} + g_{y_2}y_{2t} = 0 \tag{7.91}$$

and

$$g_{y_1y_1}y_{1t}^2 + 2g_{y_1y_2}y_{1t}y_{2t} + g_{y_2y_2}y_{2t}^2 + g_{y_1}y_{1tt} + g_{y_2}y_{2tt} = 0. \tag{7.92}$$

Evaluating (7.92) at $t = 0$ we find

$$g_{y_1y_1}y_{1t}^2 + 2g_{y_1y_2}y_{1t}y_{2t} + g_{y_2y_2}y_{2t}^2 = 0. \tag{7.93}$$

Since $D_2(g) < 0$ this defines two different tangent directions for the curve $(y_1(t), y_2(t))$. If $g_{y_1y_1} \neq 0$ then they are given (up to scaling) by $(-g_{y_1y_2} \pm \sqrt{-D_2(g)}, g_{y_1y_1})$. If $g_{y_1y_1} = 0$, then they are given (again up to scaling) by $(0, 1)$ and $(g_{y_2y_2}, -2g_{y_1y_2})$. To each of them a solution curve $x^\circ + v(x^\circ; y(t))$ corresponds with tangent direction $v_{y_1}y_{1t} + v_{y_2}y_{2t}$. The vectors v_{y_1}, v_{y_2} can be obtained from (6.183) and (6.184).

7.8.2 Cusp Points in $\mathcal{K}$-Singularity Theory

In this case $g = g_{y_1} = g_{y_2} = 0$,

$$D_2(g) = g_{y_1y_1}g_{y_2y_2} - g_{y_1y_2}^2 = 0, \tag{7.94}$$

and either

$$D_3^1(g) = g_{y_1y_1}(D_2(f))_{y_2} - g_{y_1y_2}(D_2(f))_{y_1} \tag{7.95}$$

or

$$D_3^2(g) = g_{y_1y_2}(D_2(f))_{y_2} - g_{y_2y_2}(D_2(f))_{y_1} \tag{7.96}$$

are nonzero. Since (7.89) is $\mathcal{K}$-equivalent to the normal form $y_1^3 + y_2^2 = 0$ (Table 6.1), there are two nonsmooth solution branches with the same tangent direction. So it is not sufficient to compute a tangent direction. But the normal form allows parameterizations $y_1(t) = -t^2, y_2(t) = \pm t^3$. So the solution curves to (7.89) may be parameterized by $(y_1(t), y_2(t))$ where the derivatives of second order are not both zero. We will compute enough derivatives of $y_1(t), y_2(t)$ to distinguish between the two branches. Of course, (7.90), (7.91), (7.92) still hold. Taking more derivatives we find

$$\begin{gathered} g_{y_1y_1y_1}y_{1t}^3 + 3g_{y_1y_1y_2}y_{1t}^2y_{2t} + 3g_{y_1y_2y_2}y_{1t}y_{2t}^2 + g_{y_2y_2y_2}y_{2t}^3 \\ + 3g_{y_1y_1}y_{1t}y_{1tt} + 3g_{y_1y_2}y_{1tt}y_{2t} + 3g_{y_1y_2}y_{1t}y_{2tt} + 3g_{y_2y_2}y_{2t}y_{2tt} \\ + g_{y_1}y_{1ttt} + g_{y_2}y_{2ttt} = 0, \end{gathered} \tag{7.97}$$

$$\begin{gathered} g_{y_1y_1y_1y_1}y_{1t}^4 + 4g_{y_1y_1y_1y_2}y_{1t}^3y_{2t} + 6g_{y_1y_1y_2y_2}y_{1t}^2y_{2t}^2 + 4g_{y_1y_2y_2y_2}y_{1t}y_{2t}^3 + g_{y_2y_2y_2y_2}y_{2t}^4 \\ + 6g_{y_1y_1y_1}y_{1t}^2y_{1tt} + 12g_{y_1y_1y_2}y_{1t}y_{1tt}y_{2t} + +6g_{y_1y_1y_2}y_{1t}^2y_{2tt} \\ + 6g_{y_1y_2y_2}y_{1tt}^2y_{2t} + 12g_{y_1y_2y_2}y_{1t}y_{2t}y_{2tt} + 6g_{y_2y_2y_2}y_{2t}^2y_{2tt} \\ + 3g_{y_1y_1}y_{1tt}^2 + 4g_{y_1y_1}y_{1t}y_{1ttt} + 4g_{y_1y_2}y_{1ttt}y_{2t} + 6g_{y_1y_2}y_{1tt}y_{2tt} \\ + 4g_{y_1y_2}y_{1t}y_{2ttt} + 4g_{y_2y_2}y_{2t}y_{2ttt} + 3g_{y_2y_2}y_{2tt}^2 + g_{y_1}y_{1tttt} + g_{y_2}y_{2tttt} = 0. \end{gathered} \tag{7.98}$$

Evaluating (7.92) at $t = 0$ and taking (7.94) into account we find

$$g_{y_1y_1}y_{1t} + g_{y_1y_2}y_{2t} = 0, \tag{7.99}$$

$$g_{y_1y_2}y_{1t} + g_{y_2y_2}y_{2t} = 0. \tag{7.100}$$

Evaluating (7.97) at $t = 0$ and taking (7.99) and (7.100) into account we find

$$g_{y_1y_1y_1}y_{1t}^3 + 3g_{y_1y_1y_2}y_{1t}^2y_{2t} + 3g_{y_1y_2y_2}y_{1t}y_{2t}^2 + g_{y_2y_2y_2}y_{2t}^3 = 0. \tag{7.101}$$

We note that (7.93) can be written symbolically as

$$\left(y_{1t}\frac{\partial}{\partial y_1} + y_{2t}\frac{\partial}{\partial y_2}\right)^2 g = 0 \tag{7.102}$$

and that (7.101) may be written similarly as

$$\left(y_{1t}\frac{\partial}{\partial y_1} + y_{2t}\frac{\partial}{\partial y_2}\right)^3 g = 0. \tag{7.103}$$

Further, by some trivial manipulations it follows from (7.94), (7.99), and (7.100) that

$$g_{y_1y_2}g_{y_2y_2}\left(y_{1t}\frac{\partial}{\partial y_1} + y_{2t}\frac{\partial}{\partial y_2}\right)^3 g = -y_{1t}^3 D_3^1(g), \tag{7.104}$$

$$g_{y_1y_2}g_{y_1y_1}\left(y_{1t}\frac{\partial}{\partial y_1} + y_{2t}\frac{\partial}{\partial y_2}\right)^3 g = y_{2t}^3 D_3^2(g). \tag{7.105}$$

From (7.99), (7.100), (7.101), (7.104), (7.105), and the fact that at least one of $D_3^1(g), D_3^2(g)$ is nonzero, it follows that $y_{1t} = y_{2t} = 0$. Now evaluating (7.98) at $t = 0$ we find

$$\left(y_{1tt}\frac{\partial}{\partial y_1} + y_{2tt}\frac{\partial}{\partial y_2}\right)^2 g = 0 \tag{7.106}$$

from which we infer

$$g_{y_1y_1}y_{1tt} + g_{y_1y_2}y_{2tt} = 0, \tag{7.107}$$

$$g_{y_1y_2}y_{1tt} + g_{y_2y_2}y_{2tt} = 0. \tag{7.108}$$

This information is not sufficient to distinguish between the two nonsmooth branches of the cusp, but it fixes the direction of the common tangent vector $(y_{1tt}, y_{2tt})^T$. Indeed, by (7.95) and (7.96) it implies that $(y_{1tt}, y_{2tt})^T$ is a (nonzero) scalar multiple of the nonzero vector $(D_3^2(g), -D_3^1(g))^T$.

We will consider two further derivatives of (7.98) and evaluate them at $t = 0$. Since $y_{1t} = y_{2t} = 0$ it will not be necessary to take all the terms of (7.98) into account but only those that may lead to a nonzero first or second derivative at $t = 0$. For the first derivative of (7.98) we are so reduced to

$$\begin{gathered} g_{y_1y_1y_1}(15y_{1t}y_{1tt}^2 + 10y_{1t}^2y_{1ttt}) \\ + g_{y_1y_1y_2}(15y_{1tt}^2y_{2t} + 30y_{1t}y_{1tt}y_{2tt} + 20y_{1t}y_{2t}y_{1ttt} + 10y_{1t}^2y_{2ttt}) \\ + g_{y_1y_2y_2}(15y_{2tt}^2y_{1t} + 30y_{2t}y_{2tt}y_{1tt} + 20y_{2t}y_{1t}y_{2ttt} + 10y_{2t}^2y_{1ttt}) \\ + g_{y_2y_2y_2}(15y_{2t}y_{2tt}^2 + 10y_{2t}^2y_{2ttt}) + g_{y_1y_1}(10y_{1tt}y_{1ttt} + 5y_{1t}y_{1tttt}) \\ + g_{y_1y_2}(10y_{1ttt}y_{2tt} + 5y_{1tttt}y_{2t} + 10y_{1tt}y_{2ttt} + 5y_{1t}y_{2tttt}) \\ + g_{y_2y_2}(10y_{2tt}y_{2ttt} + 5y_{2t}y_{2tttt}) + g_{y_1}y_{1ttttt} + g_{y_2}y_{2ttttt} = 0. \end{gathered} \tag{7.109}$$

Evaluating at $t = 0$ we find

$$g_{y_1y_1}y_{1tt}y_{1ttt} + g_{y_1y_2}(y_{1ttt}y_{2tt} + y_{1tt}y_{2ttt}) + g_{y_2y_2}y_{2tt}y_{2ttt} = 0. \tag{7.110}$$

This provides no new information since it already follows from (7.107) and (7.108). Taking derivatives of (7.109) and evaluating immediately at $t = 0$ we obtain

$$\begin{gathered} 15g_{y_1y_1y_1}y_{1tt}^3 + 45g_{y_1y_1y_2}y_{1tt}^2y_{2tt} + 45g_{y_1y_2y_2}y_{1tt}y_{2tt}^2 + 15g_{y_2y_2y_2}y_{2tt}^3 \\ + g_{y_1y_1}(15y_{1tt}y_{1tttt} + 10y_{1ttt}^2) \\ + g_{y_1y_2}(15y_{1tttt}y_{2tt} + 20y_{1ttt}y_{2ttt} + 15y_{1tt}y_{2tttt}) \\ + g_{y_2y_2}(15y_{2tt}y_{2tttt} + 10y_{2ttt}^2) = 0. \end{gathered} \tag{7.111}$$

By (7.107) and (7.108) this implies

$$3\left(y_{1tt}\frac{\partial}{\partial y_1} + y_{2tt}\frac{\partial}{\partial y_2}\right)^3 g + 2\left(y_{1ttt}\frac{\partial}{\partial y_1} + y_{2ttt}\frac{\partial}{\partial y_2}\right)^2 g = 0. \tag{7.112}$$

We note that the first term in (7.112) cannot vanish because as in the case of y_{1t}, y_{2t} this would imply that $y_{1tt} = y_{2tt} = 0$, a situation that was excluded a priori. Further, at least one of $g_{y_1y_1}, g_{y_2y_2}$ must be nonzero; if they are both nonzero, they must have the same sign. Let us assume that $g_{y_iy_i} \neq 0$. By (7.112) it follows that

$$g_{y_iy_i}\left(y_{1tt}\frac{\partial}{\partial y_1}+y_{2tt}\frac{\partial}{\partial y_2}\right)^3 g>0. \tag{7.113}$$

This fixes $(y_{1tt}, y_{2tt})^T$ up to a positive multiple.

The parameterization $(y_1(t), y_2(t))$ is not unique. Let us consider the effect of a smooth reparameterization $t(s)$ with $t_s(0) \neq 0$. We find

$$y_{is} = y_{it}t_s, \tag{7.114}$$

$$y_{iss} = y_{itt}t_s^2, \tag{7.115}$$

$$y_{isss} = y_{ittt}t_s^3 + 3y_{itt}t_st_{ss}, \tag{7.116}$$

where $i \in \{1,2\}$ and all functions are evaluated at zero. So by a positive rescaling of t we may assume that

$$y_{1tt} = \epsilon D_3^2(g), y_{2tt} = -\epsilon D_3^1(g), \tag{7.117}$$

where

$$\epsilon = \text{sign}\left((g_{y_1y_1}+g_{y_2y_2})\left(D_3^2(g)\frac{\partial}{\partial y_1}-D_3^1(g)\frac{\partial}{\partial y_2}\right)^3 g\right). \tag{7.118}$$

Also, we find that

$$y_{1sss}y_{1ss}+y_{2sss}y_{2ss} = t_s^3((y_{1ttt}y_{1tt}+y_{2ttt}y_{2tt})t_s^2+3(y_{1tt}^2+y_{2tt}^2)t_{ss}).$$

We may choose t_{ss} so that this expression vanishes. Hence we may assume that

$$. \ y_{1ttt} = \alpha D_3^1(g), \qquad y_{2ttt} = \alpha D_3^2(g), \tag{7.119}$$

where α is determined via (7.112) by

$$3\epsilon\left(D_3^2(g)\frac{\partial}{\partial y_1}-D_3^1(g)\frac{\partial}{\partial y_2}\right)^3 g+2\alpha^2\left(D_3^1(g)\frac{\partial}{\partial y_1}+D_3^2(g)\frac{\partial}{\partial y_2}\right)^2 g=0 \tag{7.120}$$

or, equivalently,

$$3\epsilon\left(D_3^2(g)\frac{\partial}{\partial y_1}-D_3^1(g)\frac{\partial}{\partial y_2}\right)^3 g+2\alpha^2(g_{y_1y_2}+g_{y_2y_2})((D_3^1(g))^2+(D_3^2(g))^2)=0. \tag{7.121}$$

Multiplying with $g_{y_1y_2}g_{y_2y_2}$ we get

$$-3\epsilon(D_3^2(g))^3D_3^1(g)+2\alpha^2g_{y_1y_2}g_{y_2y_2}(g_{y_1y_2}+g_{y_2y_2})((D_3^1(g))^2+(D_3^2(g))^2)=0. \tag{7.122}$$

Multiplying with $g_{y_1y_2}g_{y_1y_1}$ we get

$$-3\epsilon(D_3^1(g))^3D_3^2(g)+2\alpha^2g_{y_1y_2}g_{y_1y_1}(g_{y_1y_2}+g_{y_2y_2})((D_3^1(g))^2+(D_3^2(g))^2)=0. \tag{7.123}$$

Adding (7.122) and (7.123) we obtain

$$2\alpha^2 g_{y_1y_2}(g_{y_1y_1}+g_{y_2y_2})^2 = 3\epsilon D_3^1(g)D_3^2(g). \tag{7.124}$$

This allows us to compute α if $g_{y_1y_2}\neq 0$. If $g_{y_1y_2}=0$ then from (7.120) we obtain

$$g_{y_2y_2}=0 \Rightarrow 2\alpha^2 = 3\epsilon g_{y_1y_1}g^2_{y_2y_2y_2}, \tag{7.125}$$

$$g_{y_1y_1}=0 \Rightarrow 2\alpha^2 = 3\epsilon g_{y_2y_2}g^2_{y_1y_1y_1}. \tag{7.126}$$

To summarize the above results, we note that the formulas (7.124), (7.125), (7.126) allow us to compute both ϵ and two opposite values for α; it is not necessary to use (7.118). Then (7.117) and (7.119) allow parameter predictions in a parameter t for the two branches of the cusp that are accurate up to order 3 in t.

If we denote by α the positive root of (7.124), (7.125), or (7.126), then the two corresponding branches in x-spaces are parameterized as $x(t)=x^\circ+v(x^\circ;y(t))$, where

$$v(x^\circ;y(t)) = \frac{1}{2}v_1t^2+\frac{1}{6}\delta v_2t^3+O(t^4) \tag{7.127}$$

and $\delta=\pm1$,

$$v_1 = v_{y_1}y_{1tt}+v_{y_2}y_{2tt}, \qquad v_2 = v_{y_1}y_{1ttt}+v_{y_2}y_{2ttt}. \tag{7.128}$$

The vectors v_{y_1}, v_{y_2} can be obtained from (6.183) and (6.184). Omitting the $O(t^4)$-term in (7.127) we obtain predictions for the two branches of the cusp. These are analogous to the simpler tangent predictions that are ubiquitous in numerical continuation (see Chapter 2). For computation of the branch a corrector step is also needed. Consider Figure 7.2 and let A be the predicted point. For simplicity we may do the correction step in a hyperplane through A. At least by a geometric intuition the hyperplane orthogonal to the predicted vector step (OA in Figure 7.2) does not seem to be recommendable, since it intersects the other branch as well. A more natural choice is the hyperplane with normal vector $(v_2^Tv_2)v_1-(v_2^Tv_1)v_2$. This hyperplane is indeed parallel to v_2 and orthogonal to the (v_1,v_2)-plane.

7.8.3 Transcritical and Pitchfork Bifurcations in $(\lambda-\mathcal{K})$-Singularity Theory

These two cases correspond to the simple bifurcation point in $\mathcal{K}$-singularity theory and can be treated similarly (§7.8.1). Instead of two reduced state variables y_1,y_2 we now have a state variable y and a (distinguished) parameter t_i, which is the shift of a parameter α_i in the original, unreduced problem. In both cases we have $g_y=g_{t_i}=0$ and $D_2(g)=g_{yy}g_{t_it_i}-g^2_{yt_i}<0$. The tangents to the bifurcating branches have the form

$$v_y\dot{y}+v_{t_i}\dot{t}_i,$$

where v_y, v_{t_i} are determined by (7.30) and (7.31), respectively, and $\dot{y},\dot{t}_i$ satisfy the quadratic equation

$$g_{yy}\dot{y}^2+2g_{yt_i}\dot{y}\dot{t}_i+g_{t_it_i}\dot{t}_i^2=0.$$

In the pitchfork case we have additionally $g_{yy}=0$, so that there is one branch where $\dot{t}_i=0$; i.e., the tangent direction is v_y.

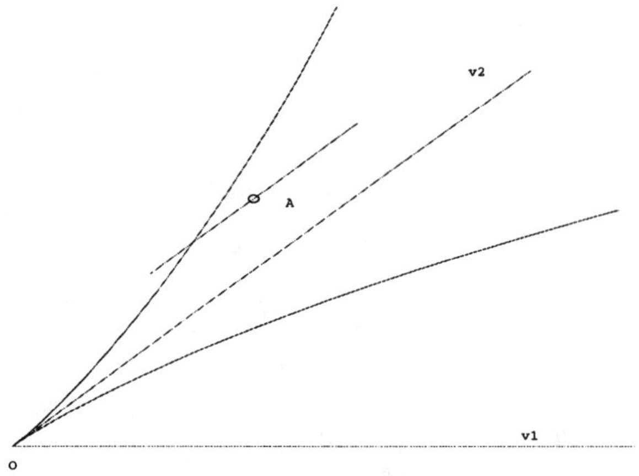

Figure 7.2: Finding a branch of a cusp curve.

7.8.4 Branching Point on a Curve of Equilibria

When a simple bifurcation point appears on a curve of equilibria in a nonlinear system

$$G(z) = 0 \tag{7.129}$$

with $z \in \mathbb{R}^{N+1}, G(z) \in \mathbb{R}^N$, then it is called a *branching point* (BP). We note that this phenomenon is ungeneric; cf. Table 7.1. However, it occurs rather frequently in model examples with symmetries or invariant subspaces. Computation of branching points is therefore routinely provided in several software packages. In a typical application one follows a branch of equilibria, detects a branching point, and switches to the other branch (or continues the original one). So there are three tasks: detecting the branching point, computing it, and computing the tangent vector to the new branch.

It is convenient to start with the last task. In the branching point the right singular space of G_z is spanned by two linearly independent vectors $\phi_1, \phi_2 \in \mathbb{R}^{N+1}$. The left singular space is spanned by a vector $\psi \in \mathbb{R}^N$. If $z(t)$ is a parameterization of a branch with the branching point at $t = 0$, then by a routine argument we have

$$\psi^T G_{zz}\dot{z}(0)\dot{z}(0) = 0. \tag{7.130}$$

Now $\dot{z}(0) = \alpha\phi_1 + \beta\phi_2$ for some $\alpha, \beta \in \mathbb{R}$. Substituting this into (7.130) we obtain the so-called algebraic bifurcation equations

$$a_{11}\alpha^2 + 2a_{12}\alpha\beta + a_{22}\beta^2 = 0 \tag{7.131}$$

with $a_{11} = \psi^T G_{zz}\phi_1\phi_1$, $a_{12} = \psi^T G_{zz}\phi_1\phi_2$, $a_{22} = \psi^T G_{zz}\phi_2\phi_2$, and $a_{12}^2 - a_{11}a_{22} > 0$. So the tangent directions are determined by (7.131). In particular, if ϕ_1 is the tangent vector to an already computed branch, then $a_{11} = 0$, $a_{12} \neq 0$, and $a_{22}\phi_1 - 2a_{12}\phi_2$ is a tangent vector to the new branch.

For the detection we note that the $(N+1) \times (N+1)$ matrix

$$G_z^e = \begin{pmatrix} G_z \\ \phi_1^T \end{pmatrix},$$

defined along the already computed curve with ϕ_1 a tangent vector, is singular at the branching point. This suggests that we use $\det(G_z^e)$ as a test function for branching. An alternative is to border G_z^e with an additional row and column, to form a nonsingular matrix G_z^{ee}, to solve

$$G_z^{ee} \begin{pmatrix} r_b \\ s_b \end{pmatrix} = \begin{pmatrix} 0_{N+1} \\ 1 \end{pmatrix},$$

where $r_b \in \mathbb{R}^{N+1}$, $s_b \in \mathbb{R}$, and to use s_b as a test function. To prove that $\det(G_z^e)$ and s_b have a regular zero at the branching point we first note that the left singular vector of G_z^e is $(\psi^T, 0)^T$ and the right singular vector is a vector ϕ_2 in the right singular space of G_z and orthogonal to ϕ_1. If σ denotes arclength along the already computed equilibrium curve, then by the now familiar arguments and an appropriate choice of ϕ_1, ϕ_2 we have

$$(s_b)_\sigma = -(\psi^T, 0)(G_z^e)_z \phi_1 \phi_2 = -\psi^T G_{zz} \phi_1 \phi_2 = -a_{12} \neq 0$$

in the branching point. Hence s_b has a regular zero at the branching point and by the argument developed in §3.1 this carries over to $\det(G_z^e)$.

In [187] and [184] a regular system is obtained for the computation of the branching point. It uses the left singular vector p and an artificial unfolding parameter γ as additional unknowns. The system is

$$\begin{array}{lcl} G(z) + \gamma p & = & 0, \\ p^T G_z(z) & = 0, & \\ p^T p - 1 & = & 0. \end{array} \tag{7.132}$$

This is a system of $2N+2$ equations with $2N+2$ unknowns; in the branching point $\gamma = 0$. The proof of the regularity is easy and left to the exercises. Alternatively, one can set up a system of $N+2$ equations in $N+2$ unknowns by using the first set of equations in (7.132) and further expressing that G_z has rank defect 1 by the bordered matrix method in §3.4 (this gives two scalar conditions). The regularity of the system can be proved by the now familiar methods.

7.9 Exercises

1. Consider the example in §1.4. Prove that the singular point found in $(3, 3, \frac{3}{h^2 e^3})$ is indeed a pitchfork bifurcation point.

2. Repeat the computations done in the study of the eutrophication model (§7.6.2) for another choice of b, c, d. Show that the conclusions for the (u, λ_1)-space are the same, in spite of having a different (but λ-contact equivalent) function $g(x, t)$ in (x, t)-space.

3. Consider the eutrophication model (§7.6.2) in $u = (x_1, x_2, x_3)$ state and $\lambda = \lambda_1$ parameter space for fixed value $\lambda_2 = 0$. Take the same starting point as in §7.6.2. Prove by an explicit Lyapunov–Schmidt reduction that this is a singularity with codimension higher than 3 in λ-singularity theory. Then compute directly all equilibrium solutions of the original system in (u, λ_1)-space.

4. Consider the following functions:

   (a) $g(x, \lambda) = \lambda$,
   (b) $g(x, \lambda) = \lambda^2$,
   (c) $g(x, \lambda) = x$,
   (d) $g(x, \lambda) = \lambda x$,
   (e) $g(x, \lambda) = x^2 - \lambda^2 + 3x^2\lambda$,
   (f) $g(x, \lambda) = x^2 + \lambda^3 + x^2\lambda$,
   (g) $g(x, \lambda) = x^2 - x^2 \cos x + 5\lambda \sin x + 6x^2\lambda^2$,
   (h) $g(x, \lambda) = x^3 - \lambda x + \lambda^2$,
   (i) $g(x, \lambda) = \cos^2 \lambda - \cos^2 x$.

   Which of these functions represent λ-singularities in $(x, \lambda) = (0, 0)$? If they do, is their type in the list of singularities with codimension 3 or less?

5. Show explicitly how the unfolding $H(x, \lambda, \beta) = \varepsilon x^2 + \delta\lambda + \beta x$ of the turning point $g(x, \lambda) = \varepsilon x^2 + \delta\lambda$ factors through the 0-parameter unfolding in the sense of λ-singularity theory.

6. Show explicitly (in λ-singularity theory) how the unfolding $H(x, \lambda, \beta_1, \beta_2) = \varepsilon x^3 + \delta\lambda + \beta_1 x + \beta_2 x^2$ of the hysteresis point $g(x, \lambda) = \varepsilon x^3 + \delta\lambda$ factors through the universal unfolding $G(x, \lambda, \alpha) = \varepsilon x^3 + \delta\lambda + \alpha x$.

7. Prove that the unfolding $H(x, \lambda, \beta) = \varepsilon x^3 + \delta\lambda + \beta x^2$ of the hysteresis point is not universal.

8. Consider the equation

$$F(x, y) \equiv x^4 + 64y + 32xy + y^2 + 2yx^2 = 0. \tag{7.133}$$

   (a) Compute (using a continuation code, e.g., PITCON) a branch of solutions of (7.133) that passes through all solution points on the y-axis.
   (b) Compute all turning points on this branch, both with respect to x and y.
   (c) Draw a sketch of the computed branch and indicate the limit points and intersections with the y-axis.
   (d) Find the above turning points by analytical means and classify them with respect to λ-singularity theory, where λ plays the role of either x or y.
   (e) Equation (7.133) also has an isola center solution in the sense of $\mathcal{K}$-singularity theory. Find it and prove that it is an isola center.

9. Consider the equation

$$F(x,\lambda) \equiv ax^2 + bx\lambda^2 + c\lambda^4 = 0 \tag{7.134}$$

with $a, b, c \in \mathbb{R}, a \neq 0, b \neq 0, c \neq 0$.

(a) Classify the singularity of (7.134) at $(x, \lambda) = (0,0)$ in the list of λ-singularities given in §7.2 in the case that $4ac - b^2 \neq 0$ and give the normal form.

(b) Prove that the singularity is not in the list if $4ac - b^2 = 0$.

10. Consider the system

$$\begin{cases} 0.5x^2 + y - \lambda &= 0, \\ xy - \lambda &= 0, \end{cases}$$

with state variables x, y and distinguished parameter λ. Classify the point $(0,0,0)$ in $(\lambda - \mathcal{K})$-singularity theory.

11. Let $f \in \mathcal{E}^0_{3,2}$ be defined by $f(x,y,\lambda) = (x^2 - \lambda, y^2 + \lambda)^T$, where λ is the distinguished bifurcation parameter. Prove explicitly that f is $(\lambda - \mathcal{K})$-equivalent to (7.6).

12. Plot a picture in (L, A_0)-space of the six points given in Table 7.10. Indicate also the projection of the point H_1 (computed in §7.7.3) on this space. Does this look like a collection of points on a cusp curve with H_1 as the cusp point?

13. In §7.7.6 the formula (6.178) for g_{yyyy} proved to be useful. Give similar formulas for g_{yyyt}, g_{yytt}, g_{yttt}, and g_{tttt} that allow us to compute these quantities from the computations described in §7.7 with practically no extra cost.

14. Prove the regularity of the system (7.132).

(*Hint:* Assume that the linearized system has a right singular vector. Then proceed in a way similar to the proofs of Propositions 4.1.1, 4.3.3, and 4.3.4.)

15. Provide the details of the bordered matrix alternative to the system in (7.132) to compute a branching point.

16. Repeat the computations done in §7.7.1 and compute at each equilibrium point all the eigenvalues of the Jacobian matrix of the discretized system of (7.83). Use some good software for this purpose, e.g., LAPACK [11]. In the interesting region there is a large number of eigenvalues close to zero. Do you find eigenvalue crossings at the turning points and pitchfork bifurcations of the equilibrium curve? Would you recommend detecting these phenomena by eigenvalue computations?

17. Consider the problem in Exercise 4.8.22. Apply the methods used in §7.7 to show that the points computed in Exercise 22(c), §4.8. (L-values close to L_1 and L_2) are transcritical bifurcation points.

Chapter 8

Symmetry-Breaking Bifurcations

We consider once more the dynamical system

$$\dot{x} = G(x, \alpha), (x \in \mathbb{R}^N, \alpha \in \mathbb{R}^m, G(x, \alpha) \in \mathbb{R}^N) \tag{8.1}$$

and its equilibrium solutions for which

$$G(x, \alpha) = 0. \tag{8.2}$$

The bifurcation behavior of the solutions to (8.2) often changes dramatically if G satisfies equivariance relations with respect to a group of transformations, i.e.,

$$G(Sx, \alpha) = SG(x, \alpha) \tag{8.3}$$

for all $S \in \mathcal{S}$, where $\mathcal{S}$ is a subgroup of the group $\mathbf{GL}(N)$ of nonsingular $N \times N$ matrices. The most striking effect is that certain bifurcations that are very ungeneric in general (high codimension) become common. In §§8.1–8.3 we consider in detail the simplest case, where $\mathcal{S}$ is a group of matrices isomorphic to the additive group $\mathbb{Z}_2$. This situation already presents many interesting numerical features that can be related directly to fairly complicated mathematical results. For the theory we will refer to [109, VI] (corank 1) and [110, XIX] (corank 2). We note that the case of corank 1 was discussed from a numerical point of view in [149].

The computed phenomena are in a sense quite common, since they appear generically in problems with $\mathbb{Z}_2$ symmetry and only one (corank 1) or two (corank 2) free parameters. Also, they are related to important dynamical features of (8.1) such as the interaction of steady state and Hopf bifurcations, invariant tori.

For a recent comprehensive study of bifurcation behavior in the case of general groups we refer to [92]. In §8.4 we restrict our attention to the simplest case, i.e., finite groups, and the branches whose existence is guaranteed by the equivariant branching lemma; this is a numerically well-understood situation for which robust methods can be given.

8.1 The $\mathbb{Z}_2$-Case: Corank 1 and Symmetry Breaking

8.1.1 Basic Results on $\mathbb{Z}_2$-Equivariance

We follow partially the notation of [238]. Let $S \in \mathbb{R}^{N\times N}$ represent an involution operator in $\mathbf{GL}(N)$, i.e.,

$$S \neq I, \qquad S = S^{-1}. \tag{8.4}$$

So $\mathcal{S} = \{I, S\}$ is a group of matrices under matrix multiplication. We assume that $G(x,\alpha)$ is $\mathcal{S}$-equivariant; i.e., (8.3) holds for all $x \in \mathbb{R}^N, \alpha \in R^m$. A simple example with $N = 2$,

$$S = \begin{pmatrix} 0 & 1 \\ 1 & 0 \end{pmatrix} \tag{8.5}$$

was encountered in §1.4. The combustion example in §1.3 is also of this type with S the N-dimensional analogue of (8.5). The Brusselator in §4.2 also exhibits a $\mathbb{Z}_2$-symmetry with

$$S = \begin{pmatrix} 0 & 0 & . & . & 1 & 0 \\ 0 & 0 & . & . & 0 & 1 \\ . & . & . & . & . & . \\ . & . & . & . & . & . \\ 1 & 0 & . & . & 0 & 0 \\ 0 & 1 & . & . & 0 & 0 \end{pmatrix}. \tag{8.6}$$

In general, we define the symmetric and antisymmetric subspaces $X_s(S)$ and $X_a(S)$ of $\mathbb{R}^N$ by

$$X_s(S) = \{x \in \mathbb{R}^N : Sx = x\}, \qquad X_a(S) = \{x \in \mathbb{R}^N : Sx = -x\}. \tag{8.7}$$

Obviously, every $x \in \mathbb{R}^N$ can be written uniquely as $x = x_s + x_a$ with $x_s \in X_s(S), x_a \in X_a(S)$ (in fact, $x_s = (x + Sx)/2$, $x_a = (x - Sx)/2$). The dependence on S will usually be omitted if confusion is unlikely.

In the above examples S is also orthogonal; hence $S = S^{-1} = S^T$. That is, S is symmetric as well. This case is important in applications because symmetries in a problem are often induced by symmetries in a domain of definition. If S is orthogonal, then for $x \in X_s$, $y \in X_a$ we have

$$x^T y = (Sx)^T y = x^T S^T y = x^T S y = -x^T y;$$

hence $x^T y = 0$. In other words, $\mathbb{R}^N = X_s \oplus X_a$ is an orthogonal direct decomposition of $\mathbb{R}^N$. For the general case we have the following proposition.

Proposition 8.1.1. Let $S \in \mathbb{R}^{n\times n}$. Then

1. $S^2 = I$ if and only if there exist a diagonal matrix J with $J_{ii} = \pm 1$ for all i and a nonsingular $T \in \mathbb{R}^{n\times n}$ such that $S = T^{-1}JT$.

2. $X_s(S)$ and $X_s(S^T)$ have the same dimension, namely, the number of positive diagonal elements of J.

3. $X_a(S)$ and $X_a(S^T)$ also have the same dimension, namely, the number of negative diagonal elements of J.

4. $\mathbb{R}^N = X_s(S) \oplus X_a(S^T)$ and $\mathbb{R}^N = X_s(S^T) \oplus X_a(S)$ are orthogonal direct decompositions.

Proof. The condition in (1) is obviously sufficient. Conversely, if $S^2 = I$ then we can write S in the complex Jordan normal form $T^{-1}JT$, where T is a complex matrix. However, $J^2 = I$ implies that J must be a diagonal matrix with diagonal elements equal to ± 1. So all eigenvalues of S are real and there exists a base of eigenvectors that spans $\mathbb{C}^n$. Hence there exists also a real Jordan decomposition $T^{-1}JT$. It is now easily seen that $x \in X_s(S)$ if and only if $Tx \in X_s(J)$ and $x \in X_s(S^T)$ if and only if $T^{-T}x \in X_s(J)$. This implies (2) since the dimension of $X_s(J)$ is the number of positive diagonal elements of J. Finally (3) is proved in the same way and (4) is then obvious. □

We note that $X_a(S) \neq \{0\}$ by the assumption $S \neq I$. However, $X_s(S) = \{0\}$ is possible if $S = -I$.

In general, symmetry matters only in points where it is broken. Typical results are therefore proven in points (x, α) where $x \in X_s(S)$. We first formulate Proposition 8.1.2.

Proposition 8.1.2. Let (x, α) be a solution of (8.2), $x \in X_s(S), \alpha \in \mathbb{R}^m$. Then the following conditions hold:

1. $SG_x = G_xS$.

2. $SG_{xx} = G_{xx}SS$; i.e., $SG_{xx}v_1v_2 = G_{xx}(Sv_1)(Sv_2)$ for all $v_1, v_2 \in \mathbb{R}^N$.

3. $SG_{\alpha_i} = G_{\alpha_i}$.

4. If $v_s, v_s^1, v_s^2 \in X_s(S)$, $v_a, v_a^1, v_a^2 \in X_a(S)$, then $G_xv_s \in X_s(S)$, $G_xv_a \in X_a(S)$, $G_{xx}v_s^1v_s^2 \in X_s(S)$, $G_{xx}v_sv_a \in X_a(S)$, $G_{xx}v_av_a \in X_s(S)$.

Proof. (1), (2), and (3) follow from (8.3) by straightforward arguments; (1) and (2) imply (4). □

Let $B_s(S)$, $B_a(S)$, $B_s(S^T)$, and $B_a(S^T)$ be column bases for $X_s(S)$, $X_a(S)$, $X_s(S^T)$, and $X_a(S^T)$, respectively. We then have $B_s^T(S^T)B_a(S) = 0$ and $B_a^T(S^T)B_s(S) = 0$. By rearranging $B_s(S^T)$ and $B_a(S^T)$ if necessary we may assume that $B_s^T(S^T)B_s(S) = I$, $B_a^T(S^T)B_a(S) = I$. (If S is symmetric, then we can choose B_s and B_a as orthogonal bases for X_s and X_a, respectively.)

Proposition 8.1.3. Let (x, α) be a solution of (8.2), $x \in X_s, \alpha \in \mathbb{R}^m$. Then the following conditions hold:

1. $X_s(S)$ and $X_a(S)$ are invariant subspaces of G_x.

2. Define $G_x^s = B_s^T(S^T)G_xB_s(S)$, $G_x^a = B_a^T(S^T)G_xB_a(S)$. Then with respect to the basis $(B_s(S), B_a(S))$, G_x has the block diagonal form

$$\begin{aligned} G_x^{(s,a)} &= \begin{pmatrix} B_s(S) & B_a(S) \end{pmatrix}^{-1} G_x \begin{pmatrix} B_s(S) & B_a(S) \end{pmatrix} \\ &= \begin{pmatrix} B_s^T(S^T) \\ B_a^T(S^T) \end{pmatrix} G_x \begin{pmatrix} B_s(S) & B_a(S) \end{pmatrix} = \begin{pmatrix} G_x^s & 0 \\ 0 & G_x^a \end{pmatrix}. \end{aligned} \tag{8.8}$$

3. If G_x has rank defect 1, then either G_x^s or G_x^a are singular but not both. In the former case the left and right singular vectors of G_x are in $X_s(S^T)$ and $X_s(S)$, respectively; in the latter case they are in $X_a(S^T)$ and $X_a(S)$, respectively.

Proof. (1) follows from Proposition 8.1.2(4) and (2) is proved by trivial computations. Finally, (2) implies (3). □

8.1.2 Symmetry Breaking on a Branch of Equilibria: Generic Scenario

We temporarily restrict our attention to smooth branches of solutions to (8.2) through a point in the symmetric space $X_s(S)$. With a small abuse of notation the parameter is denoted by α.

In a point where G_x is nonsingular, the tangent vector to the curve may be represented as $t = \begin{pmatrix} t_1^T & t_2 \end{pmatrix}^T$, where $t_1 \in \mathbb{R}^N$, $t_2 \in \mathbb{R}$, and t satisfies $G_x t_1 + G_\alpha t_2 = 0$, $t_2 \neq 0$. By Proposition 8.1.2 we have also that $G_x(St_1) + G_\alpha t_2 = 0$. Hence t_1 is in the symmetric space. So the branch cannot leave the symmetric space if no point with singular G_x is encountered.

Now consider a point where G_x is singular. We restrict our attention to the most common case where G_x has rank defect 1. By Proposition 8.1.3(3) there are two possibilities: either G_x^s is singular or G_x^a is singular.

First assume that G_x^s is singular. As in the previous case we find $G_x(St_1 - t_1) = 0$. Hence by Proposition 8.1.3(3) $St_1 - t_1 \in X_s(S)$, i.e., $t_1 \in X_s(S)$. By Proposition 8.1.3(3) the left singular vector of G_x is in $X_s(S^T)$. If G_α is not orthogonal to it (this is the generic case), then the point is a regular solution point of (8.2) and so we have a smooth branch that is completely in the symmetric space. The generic situation is hence a quadratic turning point in the symmetric space. If G_α is orthogonal to the left singular vector of G_x, then branching behavior is possible but only in the symmetric space.

Now we consider the case that G_x^a is singular. Since $x \in X_s(S)$ there exists a $x_{r0} \in \mathbb{R}^{dim(X_s)}$ such that $x = B_s(S)x_{r0}$. Now

$$B_s^T(S^T)G(B_s(S)x_r, \alpha) = 0 \tag{8.9}$$

is a system of $\dim(X_s)$ equations in the $(\dim(X_s)+1)$ unknowns x_r, α. Its Jacobian with respect to x_r is the nonsingular matrix G_x^s. So (8.2) admits a regular solution branch that is completely in the symmetric space.

By Proposition 8.1.3(3) the left singular vector of G_x is in $X_a(S^T)$, and hence it is orthogonal to $G_\alpha \in X_s$. So G_α is in the range space of G_x and the point under consideration is not an ordinary solution point of (8.2). If g is a Lyapunov–Schmidt reduced function of G as in Chapter 7, then $g_y = g_t = 0$ by Propositions 7.5.1 and 7.5.2. Furthermore, by Propositions 8.1.2(2) and 7.5.4 we have also $g_{yy} = 0$. So the reduced function satisfies the defining equations of a pitchfork bifurcation (Table 7.2). By the results in §7.8.3 the right singular vector of G_x is the tangent vector to the branch orthogonal to the α-axis. This tangent vector is in X_a. Hence the bifurcation is really symmetry breaking: there is a branch that stays in the symmetric space and another one that points in an antisymmetric direction.

8.1.3 The Lyapunov–Schmidt Reduction with Symmetry-Adapted Bordering

Let (x, α) be a solution point of (8.2) with $x \in X_s(S)$ and assume that G_x has rank defect at most 1. For a systematic study of the possible bifurcation behavior we perform a maximal Lyapunov–Schmidt reduction as in §7.4.1. We choose $b \in \mathbb{R}^N, c \in \mathbb{R}^N, d \in \mathbb{R}$ such that

$$M(x,\alpha) \equiv \begin{pmatrix} G_x(x,\alpha) & -b \\ c^T & -d \end{pmatrix} \tag{8.10}$$

is nonsingular in a neighborhood of $(x, \alpha) \in \mathbb{R}^{N+m}$. Then $g(x; y; \alpha; t) \in \mathbb{R}$, $v(x; y; \alpha; t) \in \mathbb{R}^N$ are defined for $y \in \mathbb{R}$, $t \in \mathbb{R}^m$, (y, t) in a neighborhood of $(0, 0)$, by

$$G(x + v(x; y; \alpha; t); \alpha + t) - bg(x; y; \alpha; t) = G(x; \alpha), \tag{8.11}$$

$$c^T v(x; y; \alpha; t) - dg(x; y; \alpha; t) = y, \tag{8.12}$$

where t is the shift of α.

We know that the choice of b, c, d does not really matter from a numerical point of view. However, it turns out that g also has equivariance properties if we choose b, c in a special way; we will then be able to use the classification results on $\mathbb{Z}_2$-equivariant germs obtained in [109] and so get the mathematical picture "correct."

To be precise, we choose $b \in X_a(S), c \in X_a(S^T)$. We note that this is always possible in the case that really matters, i.e., near points where G_x^a is singular. However, it is not allowed if G_x^s is singular, because then $M(x, \alpha)$ would be singular.

By multiplying (8.11) with S and some cosmetic changes we obtain

$$G(x + Sv(x; y; \alpha; t); \alpha + t) - b(-g(x; y; \alpha; t)) = G(x; \alpha), \tag{8.13}$$

$$c^T(Sv(x; y; \alpha; t)) - d(-g(x; y; \alpha; t)) = -y. \tag{8.14}$$

It follows that

$$g(x; -y; \alpha; t) = -g(x; y; \alpha; t), \tag{8.15}$$

$$v(x; -y; \alpha; t) = Sv(x; y; \alpha; t). \tag{8.16}$$

Suppressing the dependence on (x, α) we infer that the smooth germ $g(y, t) \in \mathcal{E}^0_{1+m}$ exhibits the $\mathbb{Z}_2$-symmetry

$$g(-y, t) = -g(y, t). \tag{8.17}$$

If $r(z, t) \in \mathcal{E}_{1+m}$ is any smooth germ, then clearly the germ

$$g(y, t) = yr(y^2, t) \tag{8.18}$$

has the same equivariance. It can be proved that the converse holds; i.e., (8.17) implies the existence of a smooth germ $r(z, t)$ such that (8.18) holds. See [109, VI, Corollary 2.2]. The classification of $\mathbb{Z}_2$-equivariant germs in $\mathcal{E}_2$ with distinguished parameter α_i is based on the derivatives of r; so it is necessary to relate these to the derivatives of g, which we can compute by the methods described in §7.4.1. Taking derivatives of (8.18) we obtain

$$g_y = r + 2y^2 r_z, \qquad g_{t_i} = yr_{t_i}, \tag{8.19}$$

$$g_{yy} = 6yr_z + 4y^3 r_{zz}, \quad g_{yt_i} = r_{t_i} + 2y^2 r_{zt_i}, \quad g_{t_it_i} = yr_{t_it_i}, \tag{8.20}$$

$$g_{yyy} = 6r_z + 24y^2 r_{zz} + 8y^4 r_{zzz}, \quad g_{yyt_i} = 6yr_{zt_i} + 4y^3 r_{zzt_i}, \tag{8.21}$$

$$g_{yt_it_i} = r_{t_it_i} + 2y^2 r_{zt_it_i}, \quad g_{t_it_it_i} = yr_{t_it_it_i}, \tag{8.22}$$

$$g_{yyyy} = 60yr_{zz} + 80y^3 r_{zzz} + 16y^5 r_{zzzz}, \quad g_{yyyt_i} = 6r_{zt_i} + 24y^2 r_{zzt_i} + 8y^4 r_{zzzt_i}, \tag{8.23}$$

$$g_{yyt_it_i} = 6yr_{zt_it_i} + 4y^3 r_{zzt_it_i}, \quad g_{yt_it_it_i} = r_{t_it_it_i} + 2y^2 r_{zt_it_it_i}, \quad g_{t_it_it_it_i} = yr_{t_it_it_it_i}. \tag{8.24}$$

Evaluating at $y = 0, t = 0$ we find

$$g_y = r, \quad g_{yt_i} = r_{t_i}, \quad g_{yyy} = 6r_z, \quad g_{yt_it_i} = r_{t_it_i}, \tag{8.25}$$

$$g_{yyyt_i} = 6r_{zt_i}, \ g_{yt_it_it_i} = r_{t_it_it_i}, \ g_{yyyyy} = 60r_{zz}, \ g_{yyyt_it_i} = 6r_{zt_it_i}, \ g_{yt_it_it_it_i} = r_{t_it_it_it_i}, \tag{8.26}$$

and all other derivatives of order 5 or less are zero.

It is clear that in general one needs derivatives of order $2k + 1$ of g to compute derivatives of order k of r.

8.1.4 The Classification of $\mathbb{Z}_2$-Equivariant Germs

In [109, VI, Definition 2.5] $\mathbb{Z}_2$-equivariant germs are classified by $\mathbb{Z}_2$-equivalence, i.e., equivalence that respects the $\mathbb{Z}_2$-equivariance. For the sake of completeness we recall the definition.

Definition 8.1.4. Let $g(y,t)$ and $h(y,t)$ be $\mathbb{Z}_2$-equivariant germs. We say that they are $\mathbb{Z}_2$-equivalent if

$$h(y,t) = S(y,t)g(X(y,t), T(t)), \tag{8.27}$$

where S, X, T define a $(\lambda - \mathcal{K})$-equivalence (cf. Definition 7.1.1), and X is odd in y, and S is even in y.

In [109, VI] $\mathbb{Z}_2$-equivariant germs are classified by $\mathbb{Z}_2$-equivalence and of course this leads to a natural definition of $\mathbb{Z}_2$-codimension. A list of normal forms up to $\mathbb{Z}_2$-codimension 3 is given in [109, Table 5.1]. We repeat it in Table 8.1.

In Table 8.1 $\varepsilon, \delta, \phi, \sigma$ can take the values ± 1 only.

We note that (8) is a whole family of singularities because it contains a parameter m. Such a family is called a moduli family. For a study of this family and its relation to the singularities (9), (10), (11) we refer to [109, VI].

For numerical purposes it is important to know the defining equations and nondegeneracy conditions. We collect them in Table 8.2. It is convenient to formulate them in terms of the representation $g(y,t) = yr(y^2, t)$, i.e., using the derivatives of r.

We make the following remarks:

1. The condition $r = 0$ is not mentioned in the lists of defining equations but always is implicitly assumed.

2. The use of the expression sgn $(*)$ implicitly implies that $(*)$ is nonzero.

3. The expressions $D_k(r)$ are determined inductively by $D_2(r) = r_{yy}r_{tt} - r_{yt}^2$, $D_{k+1}(r) = r_{yy}(D_k(r))_t - r_{yt}(D_k(r))_y$ for $k \geq 2$.

Table 8.1: Normal forms for singularities of $\mathbb{Z}_2$-codimension ≤ 3.

Normal form		$\mathbb{Z}_2$-codimension	Codimension
(1)	$\varepsilon y^3 + \delta t y$	0	2
(2)	$\varepsilon y^3 + \delta t^2 y$	1	5
(3)	$\varepsilon y^5 + \delta t y$	1	4
(4)	$\varepsilon y^3 + \delta t^3 y$	2	8
(5)	$\varepsilon y^7 + \delta t y$	2	6
(6)	$\varepsilon y^3 + \delta t^4 y$	3	11
(7)	$\varepsilon y^9 + \delta t y$	3	8
(8)	$\varepsilon y^5 + 2mty^3 + \delta t^2 y$	3	9
	$m^2 \neq \varepsilon\delta$		
(9)	$\phi y^7 + \varepsilon y^5 + 2\sigma t y^3 + \varepsilon t^2 y$	3	9
(10)	$\varepsilon y^5 + \sigma t y^3 + \phi t^3 y$	3	11
(11)	$\phi y^7 + \sigma t y^3 + \varepsilon t^2 y$	3	11

Table 8.2: Defining equations and nondegeneracy conditions.

Normal form	Defining equations	Nondegeneracy conditions
$y(\varepsilon z^k + \delta t)$ $(k \geq 2)$	$r_z = \cdots = r_{z^{k-1}} = 0$	$\varepsilon = \text{sgn}\ (r_{z^k}), \delta = \text{sgn}\ (r_t)$
$y(\varepsilon z + \delta t^k)$ $(k \geq 2)$	$r_t = \cdots = r_{t^{k-1}} = 0$	$\varepsilon = \text{sgn}\ (r_z), \delta = \text{sgn}\ (r_{t^k})$
$y(\varepsilon z^2 + 2mtz + \delta t^2)$	$r_z = r_t = 0$	$\varepsilon = \text{sgn}(r_{zz}), \delta = \text{sgn}(r_{tt}), m^2 \neq \varepsilon\delta$
$y(\phi z^3 + \varepsilon z^2 + 2\delta t z + \varepsilon t^2)$	$r_z = r_t = D_2(r) = 0$	$\varepsilon = \text{sgn}\ (r_{zz}), \delta = \text{sgn}\ (r_{zt})$
		$\phi = -\text{sgn}(r_{zt} r_{zz} D_3(r))$
$y(\varepsilon z^2 + 2\delta t z + \phi t^3)$	$r_z = r_t = r_{tt} = 0$	$\varepsilon = \text{sgn}(r_{zz}), \delta = \text{sgn}(r_{zt})$
		$\phi = \text{sgn}(r_{ttt})$
$y(\varepsilon z^3 + \delta z t + \phi t^2)$	$r_z = r_{zz} = r_t = 0$	$\varepsilon = \text{sgn}(r_{zzz}), \delta = \text{sgn}(r_{zt})$
		$\phi = \text{sgn}(r_{tt})$

Table 8.2 is copied from [109, VI, Table 5.3] except for the fourth row. In [109] the singular vector v of the Hessian of $r(z,t)$ is used. This is numerically awkward and leads to a certain ambiguity, because v is only defined up to a scalar multiple, including the choice of a sign. The proposed change can be justified by the argument used in [118, Proposition 2.3].

We note that the condition $m^2 \neq \varepsilon\delta$ in the moduli family is equivalent to $D_2(g) \neq 0$. The nondegeneracy conditions in the last two singularities of Table 8.2 also imply $D_2(g) \neq 0$.

8.1.5 Numerical Detection, Computation, and Continuation

To simplify notation we will in the sequel assume that S is orthogonal, and hence is symmetric. This is the case in most interesting examples anyway, and the changes to recover the general case are obvious (see also Proposition 8.4.1).

On a branch of equilibrium solutions to a $\mathbb{Z}_2$-equivariant problem (8.2) two phenomena are equally likely: a simple nonsymmetry breaking turning point and a symmetry-breaking pitchfork bifurcation. Both singularities were found in the Brusselator example in §7.7.1. Both can be detected by a sign change in the state derivative g_y of the Lyapunov–Schmidt reduced function. In §7.7.1 the two situations were further distinguished by the fact that in the pitchfork bifurcation case two other derivatives (g_t and g_{yy}) also change sign.

It is possible to distinguish between the two situations without recourse to other derivatives if the symmetry in the problem is known a priori. It may not seem obvious that this is always the case. However, in the applications that we are aware of symmetry is usually induced in an obvious way by physical symmetries in the underlying problem (as is the case in the Brusselator example). So suppose that the orthogonal involution matrix S is known a priori. By Proposition 8.1.3 and the analysis in §8.1.3 we know that in the turning point case v_y is in X_s and so $v_y^T S v_y > 0$; in the symmetry-breaking case v_y is in X_a and so $v_y^T S v_y < 0$. Now if g_y is computed by (7.30) along the curve, then v_y is available for free and the scalar function $v_y^T S v_y$ can be monitored easily.

We check this numerically in the Brusselator example by repeating the computations in §7.7.1, now computing $v_x^T S v_x$ in every computed point. By the simple structure of S in (8.6) we have

$$v_x^T S v_x = \Sigma_{i=1}^n ((v_x)_{2i-1}(v_x)_{2n+1-2i} + (v_x)_{2i}(v_x)_{2n+2-2i}).$$

In Table 8.3 we give the computed values in a number of points. As expected, $v_x^T S v_x$ is positive near the three quadratic turning points (192–193, 290–291, and 457–458) and negative near the three symmetry breaking pitchfork bifurcations (173–174, 341–342, and 434–435).

Naturally, the next problem is the exact computation of symmetry-breaking bifurcation points. The system

$$\begin{aligned} G(x,\alpha) &= 0, \\ g_y(x,\alpha) &= 0 \end{aligned} \tag{8.28}$$

of $N+1$ equations in $N+1$ variables (x, α with $\alpha = L$ in the Brusselator problem) determines the requested point but does not constitute a regular system. Therefore, it is

Table 8.3: Recognition of three turning points and three pitchfork bifurcations on a curve of symmetric equilibria.

Counter	$v_x^T S v_x$	Counter	$v_x^T S v_x$
172	−8.256942	340	−54.992472
173	−22.377409	341	−42.206454
174	−36.531073	342	−33.293473
175	−48.069262	343	−29.800404
191	1443.7817	433	−677.78021
192	2401.9724	434	−124.35859
193	3450.1428	435	−90.946141
194	5785.5098	436	−75.975531
289	76.265645	456	10.084259
290	73.630410	457	25.450203
291	71.123520	458	85.887678
292	68.742905	459	447.65321

not very suitable for computation or continuation. However, it can be replaced easily by a regular system. To this end, we choose a bordering of G_x by vectors b, c and a scalar d such that the bordered extension (8.10) is nonsingular. Then we consider the system

$$\begin{aligned} G(x,\alpha) + rb &= 0, \\ g_y(x,\alpha) &= 0, \\ c^T x &= 0 \end{aligned} \tag{8.29}$$

of $N+2$ equations in $N+2$ variables x, α, r. This system has the requested point and $r = 0$ as a solution. To prove that it is a regular system, we note that its Jacobian matrix is nonsingular if and only if the matrix

$$\begin{pmatrix} G_x & -b & 0 & G_\alpha & -b \\ c^T & -d & 1 & 0 & -d \\ 0 & 1 & 0 & 0 & 0 \\ g_{yx} & 0 & 0 & g_{y\alpha} & 0 \\ c^T & -d & 0 & 0 & 0 \end{pmatrix}$$

is nonsingular. By an elimination procedure as in Propositions 6.6.2 and 6.6.6 it follows that system (8.29) is regular if and only if the matrix

$$\begin{pmatrix} -g_y & g_t & -1 \\ -g_{yy} & g_{yt} & 0 \\ -1 & 0 & d \end{pmatrix} \tag{8.30}$$

is nonsingular.

Since $g_y = g_t = g_{yy} = 0$, $g_{yt} \neq 0$ at the symmetry-breaking point, the result follows.

We now apply the above method to the computation of the first symmetry-breaking bifurcation point on the curve of equilibria computed in §7.7.1, taking as a starting point the 173rd computed point. In Table 8.4 we give the essential data at each Newton step

Table 8.4: Computation of a symmetry-breaking bifurcation point by Newton's method.

r	g_y	g_t	g_{yy}
0.	8.97687243726D-04	−5.45314948551D-10	3.81471974726D-11
−1.13770542D-13	−1.49056224D-05	−5.59611202D-11	−1.40796339D-11
3.05766612D-16	−4.66861433D-09	−5.62890112D-10	9.63041495D-14
−1.20397796D-18	−6.33645520D-16	−4.28995452D-10	−7.82330506D-15

Table 8.5: State coordinates of a symmetry-breaking bifurcation point.

1.9002777914382	2.3990123903918	1.8001975486039	2.4976623198457
1.6999984543576	2.5954659123723	1.6004488025810	2.6918324926887
1.5026765511685	2.7861018588985	1.4079836952867	2.8775863980886
1.3176706346813	2.9656119873516	1.2328927973414	3.0495524483901
1.1545631988255	3.1288541819239	1.0833044335000	3.2030498469427
1.0194448011398	3.2717619268186	0.96304772744820	3.3346983291434
0.91396173955470	3.3916426942672	0.87187927953660	3.4424419756860
0.83639534872975	3.4869933541322	0.80706019243264	3.5252319043008
0.78342313498806	3.5571198274174	0.76506685206746	3.5826375886181
0.75163273978402	3.6017769811669	0.74283871854733	3.6145359642096
0.73849097487920	3.6209150538301		

in the case where $b = c$ are first chosen as antisymmetric vectors with $b_i = c_i = i$ for $1 \le i \le n$ and then normalized; we set $d = 0$.

To conclude this discussion we give the exact coordinates of the first symmetry-breaking bifurcation point on the curve computed in §7.7.1. We have $L = 0.25756746437116$; the first 42 components of u are given in Table 8.5.

The above method for computation of a symmetry-breaking bifurcation point can also be used for the continuation of such points if another parameter is freed. We omit the obvious details.

8.1.6 Branching and Numerical Study of a Nonsymmetric Branch

If a symmetry-breaking pitchfork bifurcation is discovered, one may want to compute numerically the nonsymmetric branch. It is not hard to start the continuation since the tangent vector to the nonsymmetric branch has v_y as state component and no component in the parameter direction; see §7.8.3. The numerical continuation of the nonsymmetric branch is to be done by general methods; i.e., symmetry does not play a role. However, symmetry still influences the bifurcation behavior because the nonsymmetric branch may intersect the symmetric space again, causing the appearance of pitchfork bifurcations. Since they are now found on the "wrong" (i.e., nonsymmetric) branch, they appear as degenerate turning points where a new (two-sided) branch originates. Of course we have again $g_y = g_t = g_{yy} = 0$, $g_{yt} \neq 0$, $g_{yyy} \neq 0$.

As an example we followed the nonsymmetric branch in the point with u-coordinates

Table 8.6: State coordinates of a symmetry-breaking bifurcation point.

2.0023962277707	2.2984345601407	2.0047820038271	2.2968772489707
2.0071465212815	2.2953361627244	2.0094786367602	2.2938193352472
2.0117669369768	2.2923347056993	2.0139998160841	2.2908900836409
2.0161655627336	2.2894931118464	2.0182524555569	2.2881512272520
2.0202488655888	2.2868716204963	2.0221433639873	2.2856611945583
2.0239248332658	2.2845265230320	2.0255825801553	2.2834738086053
2.0271064481453	2.2825088423249	2.0284869277257	2.2816369642357
2.0297152623666	2.2808630259746	2.0307835483249	2.2801913558822
2.0316848264577	2.2796257271667	2.0324131643577	2.2791693296129
2.0329637272876	2.2788247452824	2.0333328365938	2.2785939285884
2.0335180145065	2.2784781910643		

in Table 8.5 in the direction of decreasing values of L. On this branch a new pitchfork bifurcation (symmetric) point was found for $L = 0.15897726187161$; the first 42 components of u are given in Table 8.6.

We note that the points in Tables 8.5 and 8.6 are both in the symmetric space, but they are connected by a branch in the nonsymmetric space. Of course, there is also a symmetric branch through the point in Table 8.6. Its tangent direction is $v_y\dot{y}+v_t\dot{t}$, where $\dot{y}, \dot{t}$ are obtained up to a scale factor by the equation

$$2g_{yt}\dot{y} + g_{tt}\dot{t} = 0$$

(see §7.8.3). So it is possible to find new symmetric branches by following a nonsymmetric branch, just as stable equilibria are often found by following unstable ones.

8.2 The $\mathbb{Z}_2$-Case: Corank 2 and Mode Interaction

We noted in Chapter 6 (see Table 6.1) that a square Jacobian G_x with corank 2 in a problem like (8.1) can in general be expected only in problems with at least four free parameters. However, a special situation arises for equilibria in the symmetric space of a $\mathbb{Z}_2$-equivariant problem. Proposition 8.1.3 shows that then there is a canonical base in $\mathbb{R}^n$ (canonical here means "dependent on the orthogonal involution matrix S but independent of the function G") with respect to which the Jacobian matrix has the block form

$$G_x^{(s,a)} = \begin{pmatrix} B_s & B_a \end{pmatrix}^{-1} G_x \begin{pmatrix} B_s & B_a \end{pmatrix} = \begin{pmatrix} G_x^s & 0 \\ 0 & G_x^a \end{pmatrix}, \qquad (8.31)$$

where G_x^s and G_x^a are square matrices whose dimensions are those of X_s and X_a, respectively. Now if G_x^s and G_x^a are both singular, then G_x has rank defect 2. So this situation can be expected generically in a two-parameter problem if we follow either a curve of turning points (G_x^s is singular) or a curve of symmetry-breaking bifurcation points (G_x^a is singular).

We note that the computation of derivative information for the classification of singularities as performed in, e.g., §7.7, implicitly assumes that G_x has rank defect 1. It is not obvious whether this information has any value near points where G_x has rank defect

Table 8.7: Norms of five vectors obtained as a byproduct of singularity computations.

0.205802	8.578	202.8	20.50	1571	87850
0.204458	8.351	242.7	27.65	2169	187175
0.203109	8.133	321.2	42.46	3372	554386
0.201758	7.925	535.7	82.78	6451	3378414
0.200404	7.727	3049	550.8	19715	841123095
0.199048	7.533	636.9	132.6	707245	10598821
0.197691	7.352	256.4	61.44	33973	975332
0.196334	7.178	148.4	40.54	6109	269757
0.194978	7.011	98.68	30.37	2260	112879
0.193623	6.852	70.92	24.26	1090	58862

2. Also, the defining system for a turning point ($G = 0, g_y = 0$) is itself undefined. This does not necessarily prevent continuation through the point, but it can hardly be recommended as a general strategy. This is not only a matter of shaky theoretical base; there are indeed cases where it practically always leads to a breakdown of the continuation. So it is worse than the situation of a transcritical or pitchfork bifurcation on a curve of equilibria where the defining system is defined at the singular point but is merely singular. In §8.2.2 we will describe a robust approach to this situation, but we first consider a numerical experiment in the Brusselator example where, indeed, continuation through the corank-2 point is possible.

8.2.1 Numerical Example: A Corank-2 Point on a Curve of Turning Points

We repeat the experiment described in §7.7.2. Between the sixtieth and sixty-ninth points g_t is negative and increases montonously, so no particular bifurcation behavior is suspected if we assume that F_u has rank defect 1 throughout. Now we compute also the norms of the vectors in the left-hand sides of (7.30), (7.31), (7.32), (7.33), (7.34). These vectors are computed during the continuation anyway, so the overhead is small. We provide the results in Table 8.7 in the given order; the left column is the value of L.

It is striking that the norms of all computed vectors attain a maximum in one of the two middle points, except for the vector

$$\begin{pmatrix} v_y \\ g_y \end{pmatrix} \tag{8.32}$$

whose norm is in the second column of Table 8.7. The fact that the vector norms in the third, fourth, fifth, and sixth columns attain a maximum near the same point strongly suggests that the matrix

$$M \equiv \begin{pmatrix} G_x & -b \\ c^T & -d \end{pmatrix} \tag{8.33}$$

is singular in a nearby point. Now it is a well-known fact that solving a nearly singular linear system by a backward stable method leads to an explosive growth of the computed solution except when the right-hand side vector is in the range of the nearby singular matrix. So this suggests that the right-hand side vector in the system

$$M\begin{pmatrix} v_y \\ g_y \end{pmatrix} = \begin{pmatrix} 0_N \\ 1 \end{pmatrix} \tag{8.34}$$

is in the range of M in the point where M becomes singular. Since $g_y = 0$ along the computed curve, this is equivalent to saying that G_x has a singular vector ϕ for which $c^T\phi = 1$ and this is indeed always possible if G_x has rank defect 2.

To summarize the above discussion, if we compute curves of corank-1 points by the methods applied in §7.7 (i.e., using an extension of G_x with one bordering row and column), then a corank-2 point can usually be detected either by monitoring the norms of some computed vectors or simply by the breakdown of continuation.

8.2.2 Continuation of Turning Points by Double Bordering

We consider the previous problem again. If a point with corank 2 is suspected on the curve, then it is natural to border G_x with two rows and columns. So M now has the form

$$M \equiv \begin{pmatrix} G_x & -B \\ C^T & -D \end{pmatrix}, \tag{8.35}$$

where $B, C \in \mathbb{R}^{N\times 2}$, $D \in \mathbb{R}^{2\times 2}$. With this bordering the Lyapunov–Schmidt reduced function is a mapping $g : \mathbb{R}^2 \to \mathbb{R}^2$ that consists of two scalar functions $g_1(y_1, y_2), g_2(y_1, y_2)$ (see §6.2 for the general results on Lyapunov–Schmidt reductions). The first derivatives of g_1, g_2 are obtained from

$$M\begin{pmatrix} v_{y_1} & v_{y_2} \\ g_{1y_1} & g_{1y_2} \\ g_{2y_1} & g_{2y_2} \end{pmatrix} = \begin{pmatrix} 0_N & 0_N \\ 1 & 0 \\ 0 & 1 \end{pmatrix}. \tag{8.36}$$

This Lyapunov–Schmidt reduction is not maximal in the points where G_x has corank smaller than 2. The curve of turning points is now defined by the system $G = 0, \det(g_y) = g_{1y_1}g_{2y_2} - g_{2y_1}g_{1y_2} = 0$. This system is somewhat more involved than the one used in §8.2.1, but the increase in computational work is marginal. On the other hand, it presents the great advantage that it is perfectly well defined in the corank-2 point (under the generic assumption that M is nonsingular). Moreover, it characterizes the corank-2 point by the fact that $g_{1y_1} = g_{1y_2} = g_{2y_1} = g_{2y_2} = 0$ in that point.

We now repeat the computation in §8.2.1 with a simple choice for the double borders. We first set $B(i,1) = C(i,1) = i$, $B(i,2) = C(i,2) = 2n - i$ and then rescale so that all columns of B, C have Euclidean norm 1; D is simply set to 0. In Table 8.8 we give the values of g_{1y_1}, g_{1y_2}, g_{2y_1}, g_{2y_2}, and $\det(g_y)$ in a number of points along the same curve as in Table 8.7. The points are not exactly the same because different defining systems lead to different continuation points. We give the values of L to make a comparison possible.

We note that between the fourth and fifth point the first-order derivatives g_{1y_1}, g_{1y_2}, g_{2y_1}, g_{2y_2} all change sign. This suggests that they all vanish at an intermediate point

Table 8.8: Detection of a corank-2 point on a curve of turning points.

L	g_{1y_1}	g_{1y_2}	g_{2y_1}	g_{2y_2}	$\det(g_y)$
0.200257	−3.3770D-05	2.9678D-05	3.43417D-05	−3.0179D-05	−1.2394D-14
0.200221	−2.2522D-05	1.9795D-05	2.29041D-05	−2.0129D-05	−2.1694D-14
0.200177	−9.0308D-06	7.9394D-06	9.18519D-06	−8.0721D-06	−2.7154D-14
0.200151	−9.3303D-07	8.2334D-07	9.52222D-07	−8.3391D-07	−5.9412D-15
0.200147	3.3317D-07	−2.8276D-07	−3.27946D-07	2.9822D-07	6.6303D-15
0.200146	6.1428D-07	−5.4045D-07	−6.25178D-07	5.4907D-07	−5.8770D-16
0.200145	1.0555D-06	−9.2775D-07	−1.07325D-06	9.4360D-07	3.1543D-16
0.200143	1.5779D-06	−1.3871D-06	−1.60468D-06	1.4105D-06	−4.3669D-17

where then by (8.36) and Corollary 3.3.4 G_x has rank defect 2. This corank-2 point appears in a two-parameter problem, and so the situation is similar to the computation of a double Hopf point by a doubly bordered biproduct matrix as in §5.3.2. We will now specialize the choice of the bordering matrices B, C, D in such a way that two equations are sufficient. For a good understanding we need some further theory.

8.2.3 The $\mathbb{Z}_2$-Equivariant Reduction by a Symmetry-Adapted Double Bordering

We consider the special choice $B = (b_1\ b_2)$, $C = (c_1\ c_2)$, where $b_1, c_1 \in X_a$, $b_2, c_2 \in X_s$. This is expressed more elegantly by the relations

$$SB = BT, \qquad SC = CT, \tag{8.37}$$

where

$$T = \begin{pmatrix} -1 & 0 \\ 0 & 1 \end{pmatrix}. \tag{8.38}$$

We note that T is itself an orthogonal involution operator. Next, we require that D commutes with T, i.e.,

$$TD = DT. \tag{8.39}$$

It is easily seen that this is equivalent to the requirement that D is a diagonal matrix. Now by multiplying the first N equations in the Lyapunov–Schmidt equations from the left with S and the last two equations with T and by using the invariance property (8.3) and (8.37), (8.39), we find the fundamental invariance properties

$$g_1(-y_1, y_2) = -g_1(y_1, y_2), \qquad g_2(-y_1, y_2) = g_2(y_1, y_2) \tag{8.40}$$

or, in a more elegant form,

$$g(T(y)) = T(g(y)) \tag{8.41}$$

in every symmetric point $x \in X_s$. It now follows immediately that in every $x \in X_s$ we have for $(y_1, y_2) = (0, 0)$ that

$$g_1 = 0, \quad g_{1y_2} = 0, \quad g_{1y_1y_1} = 0, \quad g_{1y_2y_2} = 0, \tag{8.42}$$

$$g_{2y_1} = 0, \qquad g_{2y_1y_2} = 0, \tag{8.43}$$

while for every shifted parameter t,

$$g_{1t} = 0, \quad g_{1y_2t} = 0, \quad g_{2y_1t} = 0, \quad g_{1tt} = 0. \tag{8.44}$$

With this $\mathbb{Z}_2$-equivariant reduction the corank-2 point is characterized by the system

$$\begin{aligned} G(x, \alpha_1, \alpha_2) &= 0, \\ g_{1y_1} &= 0, \\ g_{2y_2} &= 0 \end{aligned} \tag{8.45}$$

if (α_1, α_2) are the two free parameters (e.g., L, A_0 in §8.2). Unfortunately, the corank-2 point is not a regular solution to this system of equations. We will replace it by the following system:

$$\begin{aligned} G(x, \alpha_1, \alpha_2) + rb_1 &= 0, \\ g_{1y_1} &= 0, \\ g_{2y_2} &= 0, \\ c_1^T x &= 0 \end{aligned} \tag{8.46}$$

of $N+3$ equations in the $N+3$ unknowns x, α_1, α_2, r. The corank-2 point with $r = 0$ is a solution to this system. We will now prove that (8.46) is generically (i.e., under appropriate conditions on the unfolding by the parameters α_1, α_2) a regular system.

Proposition 8.2.1. The following are equivalent:

1. System (8.46) is regular at the corank-2 point.

2. The system of three equations

$$g_2 = g_{1y_1} = g_{2y_2} = 0 \tag{8.47}$$

in the three variables y_2, t_1, t_2 has full rank in the origin, i.e., the matrix

$$\begin{pmatrix} 0 & g_{2t_1} & g_{2t_2} \\ g_{1y_1y_1} & g_{1y_1t_1} & g_{1y_1t_2} \\ g_{2y_1y_2} & g_{2y_2t_1} & g_{2y_2t_2} \end{pmatrix} \tag{8.48}$$

is nonsingular.

Proof. System (8.46) is regular if and only if the matrix

$$\begin{pmatrix} G_x & -B & 0 & G_{\alpha_1} & G_{\alpha_2} & -b_1 \\ C^T & -D & I_2 & 0 & 0 & -d_l \\ 0 & I_2 & 0 & 0 & 0 & 0 \\ g_{1y_1x} & 0 & 0 & g_{1y_1\alpha_1} & g_{1y_1\alpha_2} & 0 \\ g_{2y_2x} & 0 & 0 & g_{2y_2\alpha_1} & g_{2y_2\alpha_2} & 0 \\ c_1^T & -d_t & 0 & 0 & 0 & 0 \end{pmatrix} \tag{8.49}$$

is nonsingular, where d_l is the first column of D and d_t is its first row. By the elimination technique used already in Propositions 6.6.2 and 6.6.6 this is equivalent to the nonsingularity of the matrix

$$\begin{pmatrix} -g_{1y_1} & -g_{1y_2} & g_{1t_1} & g_{1t_2} & -1 \\ -g_{2y_1} & -g_{2y_2} & g_{2t_1} & g_{2t_2} & 0 \\ -g_{1y_1y_1} & -g_{1y_1y_2} & g_{1y_1t_1} & g_{1y_1t_2} & 0 \\ -g_{2y_1y_2} & -g_{2y_2y_2} & g_{2y_2t_1} & g_{2y_2t_2} & 0 \\ -1 & 0 & 0 & 0 & d_{11} \end{pmatrix}. \tag{8.50}$$

Now at the corank-2 point the first row of this matrix contains only a nonzero in the last column. If we delete the first row and last column, then the remaining last row has only a nonzero in the first column. Deleting now the last row and first column we are left with (8.48), except for a sign change in the first column. □

We note that the proof of the preceding proposition has a numerical significance: it shows how to solve the Newton systems derived from (8.46) by solving only linear systems with M and with the 3×3 matrix in (8.48). Details are as in §6.6.

We further note that the condition in Proposition 8.2.1(b) is very similar to the full rank conditions for universal unfoldings obtained in Proposition 6.4.7 in the case of no equivariance. It can indeed be proved that it represents the condition for a universal unfolding in the case of $\mathbb{Z}_2$-equivariance. We will consider this in some more detail in §8.2.6 for the case with a distinguished bifurcation parameter.

In §8.2.4 we will use the system (8.46) to compute numerically the corank-2 point that was discovered in §8.2.2. At present we note that the symmetry-adapted bordering has also interesting implications for the continuation of turning points as performed in §8.2.2. Namely, in every point of the curve F_u^s is singular. By the choice of the borders, this implies that the matrices

$$\begin{pmatrix} G_x & b_1 \\ c_1^T & d_{11} \end{pmatrix}, \quad \begin{pmatrix} G_x & b_2 \\ c_1^T & d_{12} \end{pmatrix}, \quad \begin{pmatrix} G_x & b_1 \\ c_2^T & d_{21} \end{pmatrix} \tag{8.51}$$

are all singular. By Corollary 3.3.4 this implies that $g_{1y_2}, g_{2y_1}, g_{2y_2}$ are all zero. Also, we can now replace the condition $\det(g_y) = 0$ by the somewhat simpler condition $g_{2y_2} = 0$. We now repeat the computations in §8.2.2 choosing $b_1(i) = c_1(i) = i$, $b_2(i) = c_2(i) = 2n - i$ for $i = 1, \ldots, n$ and extending the definition to $n+1 \le i \le 2n$ so that b_1, c_1 are in X_a and b_2, c_2 in X_s. Afterward, b_1, b_2, c_1, and c_2 are normalized and D is set to zero.

In Table 8.9 we give the essential data for two points near the starting point of the curve and six points near the corank-2 point.

8.2.4 Computation of a Corank-2 Point

To compute the corank-2 point we use the system (8.46) with the symmetry-adapted choices for the borders used also in §8.2.3. The convergence of the Newton method is clear from Table 8.10.

For the corank-2 point we obtained $L = 0.20014865528426$, $A_0 = 2.3301815004061$. Other parameters were fixed and the 42 first state variables are given in Table 8.11.

Table 8.9: Continuation of turning points by symmetry-adapted double bordering.

L	g_{1y_1}	g_{1y_2}	g_{2y_1}	g_{2y_2}
0.260753	−5.5353D-02	1.3279D-16	2.0493D-17	2.8269D-16
0.260651	−5.5178D-02	1.3375D-15	3.6258D-16	6.5234D-13
...	...	...	...	...
0.201975	−7.2629D-04	−1.0323D-14	−3.6683D-15	6.5755D-13
0.201298	−4.5294D-04	−2.0666D-15	−7.8000D-16	2.4816D-14
0.200485	−1.3136D-04	7.8698D-12	2.8841D-12	7.3694D-14
0.199998	5.8259D-05	$-4.8112D-12$	−1.7763D-12	3.5982D-13
0.199705	1.7085D-04	1.1922D-10	4.4213D-11	2.8020D-14
0.199353	3.0481D-04	3.5522D-12	1.3243D-12	6.9818D-14

Table 8.10: Essential data during computation of a corank-2 point by Newton's method.

L	g_{1y_1}	g_{1y_2}	g_{2y_1}	g_{2y_2}
0.19999806325	5.825D-05	−4.811D-12	−1.776D-12	3.594D-13
0.20014896727165	−5.5580D-08	1.002D-15	2.462D-15	3.634D-08
0.20014865528479	2.189D-13	6.436D-17	5.418D-17	2.654D-13
0.20014865528426	−7.767D-17	4.420D-17	−1.931D-17	−5.445D-16

Table 8.11: State coordinates of a corank-2 point.

2.2649428598040	2.0226717063539	2.1994393804475	2.0711301723529
2.1335658856898	2.1193195948366	2.0673864372289	2.1670539355236
2.0011211395760	2.2141148814674	1.9351252353704	2.2602566175714
1.8698621056544	2.3052121827034	1.8058723830202	2.3487008952053
1.7437417066220	2.3904362589237	1.6840696473032	2.4301337573556
1.6274420199751	2.4675180088254	1.5744082457784	2.5023288755808
1.5254647322179	2.5343262720783	1.4810445204217	2.5632935776764
1.4415128064657	2.5890397042213	1.4071674489518	2.6113999837583
1.3782432606175	2.6302361179174	1.3549187458618	2.6454354679153
1.3373239619428	2.6569099675718	1.3255483098433	2.6645949191611
1.3196472606494	2.6684478918039		

8.2.5 Analysis and Computation of the Singularity Properties of a Corank-2 Point

Singularities from $\mathbb{R}^2$ into $\mathbb{R}^2$ were discussed in §6.3.4 (no distinguished bifurcation parameter) and §7.3 (distinguished bifurcation parameter). We will now use the obtained results to classify the corank-2 point computed in §8.2.4.

First, classification according to §6.3.4 is based on the function

$$\begin{aligned} D_{2,2}(g) = {} & (g_{1y_1y_1}g_{2y_2y_2} - g_{2y_1y_1}g_{1y_2y_2})^2 \\ & - 4(g_{1y_1y_1}g_{2y_1y_2} - g_{2y_1y_1}g_{1y_1y})(g_{1y_1y}g_{2y_2y_2} - g_{2y_1y_2}g_{1y_2y_2}) \end{aligned} \tag{8.52}$$

(see Table 6.1) and thus requires the computation of six second derivatives $g_{iy_jy_k}$, i.e., the solution of three linear systems of the form

$$M \begin{pmatrix} v_{y_i y_j} \\ g_{y_i y_j} \end{pmatrix} = \begin{pmatrix} -G_{xx} v_{y_i} v_{y_j} \\ 0 \\ 0 \end{pmatrix}; \tag{8.53}$$

see (6.207). Of course, the six second derivatives depend on the choice of the borders in M and so does $D_{2,2}(g)$, but the sign of $D_{2,2}(g)$ does not. We did the computations with the borders described in §8.2.2 (not symmetry adapted) and found $g_{1y_1y_1} = -3.3298286063346D{-}02$, $g_{1y_1y_2} = 7.6453386727344D{-}04$, $g_{1y_2y_2} = 1.7321250558094D{-}02$, $g_{2y_1y_1} = 2.0372656670567D - 02$, $g_{2y_1y_2} = 3.4915014422862D - 03$, $g_{2y_2y_2} = -2.8820536266628D{-}02$; hence $D_{2,2}(g) = 3.2468753598552D{-}07$. We note that $D_{2,2}(g)$ is fairly small in absolute value, but this is merely due to scaling (see the size of the second derivatives and the way in which $D_{2,2}(g)$ is formed) and does not cast doubt upon the conclusion that $D_{2,2}(g) > 0$. By the results in Table 6.1 it follows that the corank-2 point has the singularity of the germ

$$\begin{pmatrix} x^2 + y^2 \\ xy \end{pmatrix} \tag{8.54}$$

from $\mathbb{R}^2$ into $\mathbb{R}^2$, i.e., the so-called folded handkerchief (cf. Example 6.3.22).

We can now adopt, e.g., L as the distinguished bifurcation parameter and classify further by the formulas derived in §7.3. This requires the computation of the first-order derivatives g_{1t}, g_{2t}, where t is the shift of L, i.e., the solution of

$$M \begin{pmatrix} v_t \\ g_t \end{pmatrix} = \begin{pmatrix} -G_L \\ 0 \\ 0 \end{pmatrix}. \tag{8.55}$$

We find $g_{1t} = -4.9382359483528D - 02$, $g_{2t} = -4.8508341894692D - 02$. The conditions of Proposition 7.3.1 are not satisfied, so we now have to compute the quantity

$$r g_{1t}^2 + s g_{1t} g_{2t} + t g_{2t}^2, \tag{8.56}$$

where r, s, t are defined in (7.17), (7.18), and (7.19), respectively, from information that is now available since it involves only the second derivatives of g with respect to y_1, y_2. We find $r = -4.8649676499723D - 11$, $s = -1.0610915382311D - 10$, $t = -4.6864787565027D - 11$, and so

$$r g_{1t}^2 + s g_{1t} g_{2t} + t g_{2t}^2 = -4.8309341931191D - 13$$

with the same remark about scaling as above.

The conditions of Proposition 7.3.2 are not satisfied, so we have to continue and compute the quantity

$$r p_1 f_{1\alpha} + \frac{1}{2} s (q_1 f_{1\alpha} + p_1 f_{2\alpha}) + t q_1 f_{2\alpha}, \tag{8.57}$$

where $p_1 = \frac{1}{2}g_{1y_1y_1}$, $q_1 = \frac{1}{2}g_{2y_1y_1}$. We find the value $-3.3001865634287D-14$. From Proposition 7.3.3 it now follows that in the point with corank 2 the germ with distinguished bifurction parameter L is $(\lambda - \mathcal{K})$-equivalent to

$$\begin{pmatrix} x^2+\lambda \\ y^2+\lambda \end{pmatrix}. \tag{8.58}$$

As we previously remarked, the details of the above computations depend strongly on the choice of the borders but the final conclusions do not. It is therefore worthwhile to repeat the computations for several choices if only to get a feeling of confidence about the outcome. We will not describe such further tests but rather return to the special case where the borders are symmetry adapted.

As in §8.2.3 we consider $B = (b_1\ b_2)$, $C = (c_1\ c_2)$, where $b_1, c_1 \in X_a$, $b_2, c_2 \in X_s$ and D is a diagonal matrix. We assume that the matrix M in (8.35) is nonsingular; this amounts to saying that b_1, b_2 are not in the column range of G_x and c_1, c_2 not in the row range of G_x at the corank-2 point (we are now taking a formal mathematical viewpoint; a numerical test will follow).

By the results in (8.42) and (8.43) the expression for $D_{2,2}(g)$ now simplifies to

$$D_{2,2}(g) = 4g^2_{1y_1y_2}g_{2y_1y_1}g_{2y_2y_2}. \tag{8.59}$$

The expressions for the quantities a, b, c, r, s, t in §7.3 also simplify. We obtain

$$a = -g_{1y_1y_2}g_{2y_1y_1}, \quad b = 0, \quad c = g_{1y_1y_2}g_{2y_2y_2}, \tag{8.60}$$

$$r = g^2_{1y_1y_2}g^2_{2y_1y_1}g^2_{2y_2y_2}, \quad s = 0, \quad t = -g^4_{1y_1y_2}g_{2y_1y_1}g_{2y_2y_2}. \tag{8.61}$$

By (8.44) we have $g_{1t} = 0$. So

$$rg^2_{1t} + sg_{1t}g_{2t} + tg^2_{2t}$$

reduces to $-g^4_{1y_1y_2}g_{2y_1y_1}g_{2y_2y_2}g^2_{2t}$. We note that the sign of this quantity is always opposite to that of $D_{2,2}(g)$ (provided that G_L is not the nullvector).

Next, $rp_1f_{1\alpha} + \frac{1}{2}s(q_1f_{1\alpha} + p_1f_{2\alpha}) + tq_1f_{2\alpha}$ reduces to $-\frac{1}{2}g^4_{1y_1y_2}g^2_{2y_1y_1}g_{2y_2y_2}g_{2t}$.

Summarizing the above results, we conclude that the corank-2 point is as follows:

1. $\mathcal{K}$-equivalent to $(x^2-y^2, xy)^T$ if and only if $g_{2y_1y_1}g_{2y_2y_2} < 0$, $g_{1y_1y_2} \neq 0$.
2. $\mathcal{K}$-equivalent to $(x^2+y^2, xy)^T$ if and only if $g_{2y_1y_1}g_{2y_2y_2} > 0$, $g_{1y_1y_2} \neq 0$.
3. $(\lambda-\mathcal{K})$-equivalent to $(x^2-y^2+\lambda, 2xy)^T$ if and only if $g_{2y_1y_1}g_{2y_2y_2} < 0$, $g_{1y_1y_2} \neq 0$, $g_{2t} \neq 0$.
4. $(\lambda-\mathcal{K})$-equivalent to $(x^2+\lambda, y^2+\lambda)^T$ if and only if $g_{2y_1y_1}g_{2y_2y_2} > 0$, $g_{2y_2y_2}g_{2t} > 0$, $g_{1y_1y_2} \neq 0$.
5. $(\lambda-\mathcal{K})$-equivalent to $(x^2-\lambda, y^2-\lambda)^T$ if and only if $g_{2y_1y_1}g_{2y_2y_2} > 0$, $g_{2y_2y_2}g_{2t} < 0$, $g_{1y_1y_2} \neq 0$.

As an interesting conclusion we find that the singularity $(x^2+\lambda, y^2-\lambda)$ is not possible in the present situation. Also, if a symmetry-adapted double bordering is used, then the classification follows from an inspection of the signs of $g_{2y_1y_1}$, $g_{2y_2y_2}$, and g_{2t} and a check that $g_{1y_1y_2}$ is nonzero.

Numerically we cannot expect to find $g_{1y_1y_1}$, $g_{1y_2y_2}$, $g_{2y_1y_2}$, and g_{1t} to be exactly zero. But they should at least be small compared with the other quantities and the signs of these should determine the singularity type. We repeat the previous numerical computations, now using the symmetry-adapted double bordering described in §8.2.3 and find

$$g_{1y_1y_1} = 1.4989475386236D-11, \qquad g_{1y_1y_2} = -2.9600821200839D-02,$$

$$g_{1y_2y_2} = -4.3466417001217D-11, \qquad g_{2y_1y_1} = -1.0903608371643D-02,$$

$$g_{2y_1y_2} = -6.9606215293581D-12, \qquad g_{2y_2y_2} = -1.2159946927583D-02,$$

$$g_{1t} = 4.8659450723979D-10, \qquad g_{2t} = -9.3380767716168D-02.$$

Further computations give $D_{2,2}(g) = 4.6469653531342D-07$ and the expressions in (8.56) and (8.57) evaluate to $-8.8762973550842D-13$ and $-5.1822057430639D-14$, respectively. These data largely support our analysis.

8.2.6 The $\mathbb{Z}_2$-Equivariant Classification of Corank-2 Points with Distinguished Bifurcation Parameter

We consider again the $\mathbb{Z}_2$-equivariant reduction in §8.2.3. As in §8.1 the equivariance relations (8.40) imply the existence of smooth germs $p(u,v)$, $q(u,v)$ such that

$$g_1(y_1,y_2) = y_1 q(u,v), \quad g_2(y_1,y_2) = p(u,v), \quad u = y_2, \quad v = y_1^2. \tag{8.62}$$

If a distinguished bifurcation parameter α_1 is selected, then its shift t_1 filters through the decomposition in (8.62) and p, q are routinely denoted as $p(u,v,t_1), q(u,v,t_1)$. In [110, XIX, §2] a classification of the corank-2 germs with distinguished parameter is given, based on the values of p, q at the origin. For the sake of completeness and to show the relation with our previous computations we copy the classification results. First, Table 8.12 gives the normal forms of $\mathbb{Z}_2$-equivariant germs with topological $\mathbb{Z}_2$-codimension ≤ 2. For a discussion of topological codimension we refer to [110]; we note, however, that Table 8.12 contains all normal forms with $\mathbb{Z}_2$-codimension ≤ 2 but not all with $\mathbb{Z}_2$-codimension 3.

In Table 8.12 ε_j can take the values ±1 only. We note that (4) and (5) are families of singularities since they contain a parameter m. These are again examples of moduli families.

For numerical purposes the defining conditions and nondegeneracy conditions are as indispensable as ever. They are given in [110, XIX, Table 2.2] and reproduced here as Table 8.13.

The conditions $p = 0, q = 0, p_u = 0$ are not mentioned in Table 8.13 but are always implicitly part of the defining conditions.

Table 8.12: Normal forms for singularities with corank 2 and topological $\mathbb{Z}_2$-codimension ≤ 2.

	Normal form $[p, q]$	$\mathbb{Z}_2$-codimension
(1)	$[\varepsilon_1 u^2 + \varepsilon_2 v + \varepsilon_3 t_1,\ \varepsilon_4 u]$	1
(2)	$[\varepsilon_1 u^3 + \varepsilon_2 v + \varepsilon_3 t_1,\ \varepsilon_4 \text{ u}]$	2
(3)	$[\varepsilon_1 u^2 + \varepsilon_2 v^2 + \varepsilon_3 t_1,\ \varepsilon_4 \text{ u}]$	2
(4)	$[\varepsilon_1 u^2 + \varepsilon_2 v + m t_1^2,\ \varepsilon_3 u + \varepsilon_4 t_1]$	3
	$m \neq 0, -\varepsilon_1$	
(5)	$[\varepsilon_1 u^2 + \varepsilon_2 v + \varepsilon_3 t_1,\ m u^2 + \varepsilon_4 v]$	3
	$m \neq 0, \varepsilon_1\varepsilon_2\varepsilon_4$	

Table 8.13: Defining and nondegeneracy conditions for Table 8.12.

	Defining conditions $[p, q]$	Nondegeneracy conditions
(1)	-	$\varepsilon_1 = \text{sgn} p_{uu}, \varepsilon_2 = \text{sgn} p_v$
		$\varepsilon_3 = \text{sgn} p_{t_1}, \varepsilon_4 = \text{sgn} q_u$
(2)	$p_{uu} = 0$	$\varepsilon_1 = \text{sgn} p_{uuu}, \varepsilon_2 = \text{sgn} p_v$
		$\varepsilon_3 = \text{sgn} p_{t_1}, \varepsilon_4 = \text{sgn} q_u$
(3)	$p_v = 0$	$\varepsilon_1 = \text{sgn} p_{uuu}, \varepsilon_2 = \text{sgn}(p_{vv} q_u^2 - 2p_{uv} q_u + p_{uu} q_v^2)$
		$\varepsilon_3 = \text{sgn} p_{t_1}, \varepsilon_4 = \text{sgn} q_u$
(4)	$p_{t_1} = 0$	$\varepsilon_1 = \text{sgn} p_{uu}, \varepsilon_2 = \text{sgn} p_v$
	$m = \varepsilon_1 \frac{[p_{uu} p_{t_1 t_1} - p_{ut_1}^2] q_u^2}{[q_{t_1} p_{uu} - q_u p_{ut_1}]^2}$	$m \neq 0, -\varepsilon_1, \varepsilon_3 = \text{sgn} q_u, \varepsilon_4 = \varepsilon_1 \text{sgn}(q_{t_1} p_{uu} - q_u p_{ut_1})$
(5)	$q_u = 0$	$\varepsilon_1 = \text{sgn} p_{uu}, \varepsilon_2 = \text{sgn} p_v$
	$m = \varepsilon_3 \frac{\lvert p_v \rvert [q_{uu} p_{t_1} - q_{t_1} p_{uu}]}{\lvert p_{uu} \rvert \lvert q_v p_{t_1} - q_{t_1} p_v \rvert}$	$m \neq 0, \varepsilon_1\varepsilon_2\varepsilon_4, \varepsilon_3 = \text{sgn} p_{t_1}, \varepsilon_4 = \varepsilon_3 \text{sgn}(q_v p_{t_1} - q_{t_1} p_v)$

Numerically we can only compute derivatives of g_1 and g_2. The formulas to relate these to the derivatives of p, q are obtained by taking derivatives of the relations in (8.62) and evaluating them at the origin. By straightforward computations we get

$$\begin{aligned} g_{1y_1} &= q, \quad g_{1y_1y_2} = q_u, \quad g_{1y_1y_1y_1} = 6q_v, \\ g_{1y_1y_2y_2} &= q_{uu}, \quad g_{1y_1y_1y_1y_2} = 6q_{uv}, \quad g_{1y_1y_2y_2y_2} = q_{uuu}, \end{aligned} \tag{8.63}$$

$$\begin{aligned} g_2 = p, g_{2y_2} &= p_u, \quad g_{2y_1y_1} = 2p_v, \quad g_{2y_2y_2} = p_{uu}, \\ g_{2y_1y_1y_2} &= 2p_{uv}, \quad g_{2y_2y_2y_2} = p_{uuu}, \quad g_{2y_1y_1y_1y_1} = 12p_{vv}. \end{aligned} \tag{8.64}$$

In the formulas in (8.63) and (8.64) one can formally take derivatives with respect to all shifted parameters in any order, distinguished or not. The collection of formulas

Table 8.14: Universal $\mathbb{Z}_2$-unfoldings for the $\mathbb{Z}_2$-singularities in Table 8.12.

(1)	$[\varepsilon_1 u^2 + \varepsilon_2 v + \varepsilon_3 t_1, \varepsilon_4 u + \alpha]$
(2)	$[\varepsilon_1 u^3 + \varepsilon_2 v + \varepsilon_3 t_1 + \beta u, \varepsilon_4 u + \alpha]$
(3)	$[\varepsilon_1 u^2 + \varepsilon_2 v^2 + \varepsilon_3 t_1 + \beta v, \varepsilon_4 u + \alpha]$
(4)	$[\varepsilon_1 u^2 + \varepsilon_2 v + \mu t_1^2 + \beta, \varepsilon_3 u + \varepsilon_4 t_1 + \alpha]$
(5)	$[\varepsilon_1 u^2 + \varepsilon_2 v + \varepsilon_3 t_1, \mu u^2 + \beta u + \varepsilon_4 v + \alpha]$

obtained in this way is sufficient for computation and continuation of all singularities in Table 8.12.

We can now classify the corank-2 point computed in §8.2.4 with respect to $\mathbb{Z}_2$-equivalence with distinguished bifurcation parameter L. We found $p = g_2 = 0$, $q = g_{1y_1} = 0$, $p_u = g_{2y_2} = 0$; i.e., the defining conditions of (1) in Table 8.13 are satisfied. Furthermore, $p_{uu} = g_{2y_2y_2} = -1.21599469275830D-02 < 0$, $p_v = \frac{1}{2}g_{2y_1y_1} = -1.0903608371643D-02/2 < 0$, $p_{t_1} = g_{2t_1} = -9.3380767716168D-02 < 0$, and $q_u = g_{1y_1y_2} = -2.9600821200839D-02 < 0$.

So we conclude that our corank-2 point has the normal form (1) in Table 8.12 with precise form

$$[p, q] = [-u^2 - v - t_1, -u];$$

i.e.,

$$(g_1, g_2) = (-y_1y_2, -y_1^2 - y_2^2 - t_1).$$

The universal $\mathbb{Z}_2$-unfoldings of the $\mathbb{Z}_2$-bifurcation problems in Table 8.12 are given in [110, XIX, Table 2.3]. We copy them here since we know that universal unfoldings are intimately related to the regularity of defining systems and hence are of critical numerical importance (see §6.4 for a study of the case with no equivariance and no distinguished parameter). In Table 8.14 α, β are near zero and μ is near m (the modal parameter).

It is checked easily that in the case of (1) the Jacobian matrix

$$\frac{\partial(p, q, p_u)}{\partial(u, t_1, \alpha)}$$

is nonsingular. In terms of g_1, g_2 this is precisely the matrix

$$\frac{\partial(g_2, g_{1y_1}, g_{2y_2})}{\partial(y_2, t_1, t_2)}$$

encountered in Proposition 8.2.1, where α plays the role of the additional parameter t_2. This confirms that there is, again, a relation between universality of unfoldings and regularity of defining systems. We note that also in the other cases in Table 8.12 the defining systems are regular with respect to the variables u, t_1 and the unfolding parameters.

8.3 Rank Drop on a Curve of Singular Points

In Chapters 6 and 7 and in §8.1 singularities were computed in a maximally reduced setting where the Jacobian is zero. For a theoretical study this is the only natural setting. Numerically it requires the a priori knowledge of the rank drop. Since rank drop higher than 1 is rare in generic situations (requires four free parameters), this is usually not a big obstacle. However, in §8.2 we found a situation with group invariance where rank drop 2 is generic on a curve of turning points, i.e., in a problem with two free parameters only.

Such rank drop is also generic on other singularity curves. If this possibility is ignored, then one of two things happens. One situation is that the point with rank drop 2 just remains undetected; then we miss interesting information (cf. §8.2.1). The other situation is that the rank drop 2 point causes the continuation code to break down. This is quite possible since the defining system of the computed curve is undefined at the corank-2 point; in a neighborhood of the corank-2 point the evaluation of the defining functions is itself an ill-conditioned problem (this has to be distinguished from the more common situation where the defining system is well defined but not regular).

In §8.2.2 we solved this problem by a double bordering in the case of a curve of turning points. We now discuss it in a more general setting. It will serve as a proof of the power of bordered matrix methods: We will detect and compute in a numerically stable way a $\mathbb{Z}_2$-equivariant germ with $\mathbb{Z}_2$-codimension higher than 2 on a nontrivial branch of cusp points.

8.3.1 Corank-1 Singularities in Two State Variables

Consider a smooth function $F : \mathbb{R}^3 \to \mathbb{R}^2$, where

$$F(x_1, x_2, \alpha) = \begin{pmatrix} F_1(x_1, x_2, \alpha) \\ F_2(x_1, x_2, \alpha) \end{pmatrix} \tag{8.65}$$

and suppose that (x_1, x_2, α) is a root of $F = 0$. We consider x_1, x_2 as the state variables and α as a parameter. We assume temporarily that F_x has rank defect at most 1; the aim is to obtain defining equations for singularities that are defined even in points with rank defect 2.

Choose $r, s \in \{1, 2\}$ such that $F_{rx_s} \neq 0$ (numerically we will take the entry of F_x with largest absolute value). Let $i, j \in \{1, 2\}, i \neq r, j \neq s$. The maximal Lyapunov–Schmidt reduction is described in §7.4.1. In the present case we can choose

$$M(x, \alpha) \equiv \begin{pmatrix} F_{1x_1} & F_{1x_2} & \delta_{1i} \\ F_{2x_1} & F_{2x_2} & \delta_{2i} \\ \delta_{1j} & \delta_{2j} & 0 \end{pmatrix} \tag{8.66}$$

with the usual convention that $\delta_{p,q} = 1$ if $p = q$, $\delta_{p,q} = 0$ if $p \neq q$. Obviously, M is nonsingular in a neighborhood of (x, α). Let y be the Lyapunov–Schmidt reduced variable and t the shift of α as in §7.4.1.

Now $v_y \in \mathbb{R}^2, g_y \in \mathbb{R}$ are obtained from (7.30) and we find

$$v_{yj} = 1, \quad v_{ys} = -\frac{F_{rx_j}}{F_{rx_s}}, \quad g_y = \frac{F_{ix_s}F_{rx_j} - F_{rx_s}F_{ix_j}}{F_{rx_s}}. \tag{8.67}$$

The defining equation for turning points is hence

$$F_{1x_1}F_{2x_2} - F_{2x_1}F_{1x_2} = 0 \tag{8.68}$$

and so it is independent of the actual choice of r, s as we expected.

Next, $v_t \in \mathbb{R}^2, g_t \in \mathbb{R}$ are obtained from (7.31) and we find

$$v_{tj} = 0, \quad v_{ts} = -\frac{F_{r\alpha}}{F_{rx_s}}, \quad g_t = \frac{F_{ix_s}F_{r\alpha} - F_{rx_s}F_{i\alpha}}{F_{rx_s}}. \tag{8.69}$$

The defining equations for a simple bifurcation point are now

$$F_{1x_1}F_{2x_2} - F_{2x_1}F_{1x_2} = F_{ix_s}F_{r\alpha} - F_{rx_s}F_{i\alpha} = 0. \tag{8.70}$$

From (7.32), (7.33), and (7.34) we obtain

$$g_{yy} = \frac{F_{ix_s}F_{rxx}v_yv_y - F_{rx_s}F_{ixx}v_yv_y}{F_{rx_s}}, \tag{8.71}$$

$$g_{yt} = \frac{F_{ix_s}(F_{rxx}v_yv_t + F_{rx\alpha}v_y) - F_{rx_s}(F_{ixx}v_yv_t + F_{ix\alpha}v_y)}{F_{rx_s}}, \tag{8.72}$$

$$g_{tt} = \frac{F_{ix_s}(F_{rxx}v_tv_t + 2F_{rx\alpha}v_t + F_{r\alpha\alpha}) - F_{rx_s}(F_{ixx}v_tv_t + 2F_{ix\alpha}v_t + F_{i\alpha\alpha})}{F_{rx_s}}. \tag{8.73}$$

To work this out we need also

$$F_{xx}v_yv_y = \frac{1}{F_{rx_s}^2}\begin{pmatrix} F_{1x_jx_j}F_{rx_s}^2 - 2F_{1x_jx_s}F_{rx_j}F_{rx_s} + F_{1x_sx_s}F_{rx_j}^2 \\ F_{2x_jx_j}F_{rx_s}^2 - 2F_{2x_jx_s}F_{rx_j}F_{rx_s} + F_{2x_sx_s}F_{rx_j}^2 \end{pmatrix}, \tag{8.74}$$

$$F_{xx}v_yv_t = \frac{F_{r\alpha}}{F_{rx_s}^2}\begin{pmatrix} -F_{1x_jx_s}F_{rx_s} + F_{1x_sx_s}F_{rx_j} \\ -F_{2x_jx_s}F_{rx_s} + F_{2x_sx_s}F_{rx_j} \end{pmatrix}, \tag{8.75}$$

$$F_{xx}v_tv_t = \frac{F_{r\alpha}^2}{F_{rx_s}^2}\begin{pmatrix} F_{1x_sx_s} \\ F_{2x_sx_s} \end{pmatrix}, \tag{8.76}$$

$$F_{x\alpha}v_y = \frac{1}{F_{rx_s}}\begin{pmatrix} F_{rx_s}F_{1x_j\alpha} - F_{rx_j}F_{1x_s\alpha} \\ F_{rx_s}F_{2x_j\alpha} - F_{rx_j}F_{2x_s\alpha} \end{pmatrix}, \tag{8.77}$$

$$F_{x\alpha}v_t = \frac{-F_{r\alpha}}{F_{rx_s}}\begin{pmatrix} F_{1x_s\alpha} \\ F_{2x_s\alpha} \end{pmatrix}. \tag{8.78}$$

The defining equations for a cusp are now those in (8.70) together with

$$g_{yy}g_{tt} - g_{yt}^2 = 0. \tag{8.79}$$

For a winged cusp the equations are those in (8.70) together with

$$g_{yy} = g_{yt} = 0. \tag{8.80}$$

Using (8.71)–(8.78) one can express (8.79) and (8.80) in terms of the derivatives of F only. We omit the details. Other singularities (hysteresis, pitchfork, etc.) can be computed similarly.

8.3.2 The Case of a Symmetry-Adapted Bordering

We apply the preceding formulas to the special case of a $\mathbb{Z}_2$-equivariant system where F is the Lyapunov–Schmidt reduced function by a symmetry-adapted double bordering. To avoid confusion (we are dealing with two Lyapunov–Schmidt reductions) we keep the notation of §8.3.1. In a singular point in the symmetric space we have $F_{1x_2} = F_{2x_1} = F_{2x_2} = 0$. Hence $r = s = 1, i = j = 2$. The equation for a turning point is $F_{2x_2} = 0$; for a simple bifurcation point we have $F_{2x_2} = F_{2\alpha} = 0$.

Next, we obtain

$$F_{xx}v_yv_y = \begin{pmatrix} 0 \\ F_{2x_2x_2} \end{pmatrix},$$

$$F_{xx}v_yv_t = F_{xx}v_tv_t = \begin{pmatrix} 0 \\ 0 \end{pmatrix},$$

$$F_{x\alpha}v_y = \begin{pmatrix} 0 \\ F_{2x_2\alpha} \end{pmatrix},$$

$$F_{x\alpha}v_t = 0,$$

$$g_{yy} = -F_{2x_2x_2}, \quad g_{yt} = -F_{2x_2\alpha}, \quad g_{tt} = -F_{2\alpha\alpha}.$$

Hence the defining equations for the cusp singularity are

$$F_{2x_2} = F_{2\alpha} = F_{2x_2x_2}F_{2\alpha\alpha} - F^2_{2x_2\alpha} = 0. \tag{8.81}$$

For the winged cusp they are

$$F_{2x_2} = F_{2\alpha} = F_{2x_2x_2} = F_{2x_2\alpha} = 0. \tag{8.82}$$

A corank-2 point on a branch of singular points is characterized by $F_{1x_1} = 0$. It is fairly easy to establish the link with the classification in §8.2.6. If we actually deal with a branch of simple bifurcation points, then the condition $F_{2\alpha} = 0$ translates as $p_t = 0$, i.e., the computed corank-2 point has (generically) type (4) in Table 8.12 ($\mathbb{Z}_2$-codimension 3). If we deal with a branch of cusp points, then the condition $F_{2x_2x_2}F_{2\alpha\alpha} - F^2_{2x_2\alpha} = 0$ translates as $p_{uu}p_{t_1t_1} - p^2_{ut_1} = 0$. In other words, the corank-2 point has topological $\mathbb{Z}_2$-codimension higher than 2 and does not appear in Table 8.12.

A further classification up to $\mathbb{Z}_2$-codimension 4 is given in [65]. One of the normal forms obtained in that paper is denoted $(8)_{131}$ and has the form

$$[p, q] = [p(u, v, t), q(u, v, t)] = [u^2 + \epsilon_1 t^3 + \epsilon_2 v, u + \epsilon_3 t] \tag{8.83}$$

and $\mathbb{Z}_2$-codimension 3. The defining conditions are

$$p_u = 0, \qquad q = 0 \tag{8.84}$$

and

$$p_t = 0, \qquad D_{6,1} \equiv \frac{1}{4}(p_{uu}p_{tt} - p^2_{ut}) = 0 \tag{8.85}$$

and nondegeneracy conditions

$$q_u \neq 0, \quad p_v \neq 0, \quad p_{uu} \neq 0, \quad D_{8,1} \equiv q_t - \alpha q_u \neq 0, \tag{8.86}$$

$$D_{8,2} \equiv \frac{1}{6}p_{uuu}\alpha^3 - \frac{1}{2}p_{uut}\alpha^2 + \frac{1}{2}p_{utt}\alpha - \frac{1}{6}p_{ttt} \neq 0, \tag{8.87}$$

where $\alpha = \frac{p_{ut}}{p_{uu}}$. The values ϵ_i are given by

$$\epsilon_1 = \text{sign}(p_{uu}D_{8,2}), \quad \epsilon_2 = \text{sign}(p_{uu}p_v), \quad \epsilon_3 = \text{sign}(q_u D_{8,1}). \tag{8.88}$$

A corank-2 point on a branch of cusp points necessarily satisfies the defining conditions (8.84) and (8.85) and the Z_2-codimension can be expected to be 3. The nondegeneracy conditions (8.86) and (8.87) of course have to be checked numerically and $\epsilon_1, \epsilon_2, \epsilon_3$ have to be computed. We now turn to an example.

8.3.3 Numerical Example: A Corank-2 Point on a Curve of Cusps

In §7.7.5 a winged cusp bifurcation point W_1 was detected on a curve of pitchfork bifurcations in the Brusselator example and its coordinates were computed (Table 7.23). With four free parameters L, A_0, D_A, B we can compute a curve of cusp points through W_1. If one chooses the direction of initially increasing values of L, then codes based on single bordering (as applied in Chapter 7) tend to have convergence problems near values $L \simeq 0.1270$ where L is already decreasing. This is due to the appearance of a corank-2 point on this curve.

To make this fully clear we repeat the computation of this curve of cusp points using the symmetry-adapted double bordering described in §8.2.3 and the defining function in (8.81). In the starting point W_1 we find numerically that

$$\frac{\partial(g_1, g_2)}{\partial(y_1, y_2, t)} = \begin{pmatrix} 5.1541570923D-03 & -1.3904618706D-17 & -1.0244074198D-18 \\ -7.0535748456D-17 & -1.7848727134D-09 & 4.9830417563D-15 \end{pmatrix}.$$

We note that $g_{1y_2} = g_{2y_1} = g_{1t}$ have to be zero by the symmetry-adapted construction while $g_{2y_2} = g_{2t}$ are zero because we are in a cusp point, i.e., by the first two equations in (8.81). These quantities have to vanish in all points of the cusp curve (numerically only up to the tolerance in the continuation code). On the other hand, g_{1y_1} is only zero in a corank-2 point, so this is our detection function.

The Hessian matrix of G with respect to the state variables in W_1 is

$$H_s = \begin{pmatrix} 2.7492908831797D-17 & -1.5340022004744D-03 & 3.0252420945498D-17 \\ -1.4800189759526D-03 & 7.4951129608848D-18 & 6.7967235574923D-17 \end{pmatrix}.$$

We note that $g_{1y_1y_1} = g_{1y_2y_2} = g_{2y_1y_2} = 0$ by the symmetry-adapted construction. On the other hand, $g_{2y_2y_2}$ vanishes only because we are in a winged cusp point; see (8.82). This quantity can indeed be used to detect winged cusp points on a curve of cusp points.

The mixed state-parameter derivatives in W_1 are

$$\frac{\partial(g_{1t}, g_{2t})}{\partial(y_1, y_2)} = \begin{pmatrix} -0.26711401261450 & 1.9224787226549D-10 \\ 2.9046723957557D-11 & 4.8177056096454D-15 \end{pmatrix}.$$

We note that $g_{2y_1t} = g_{1y_2t} = 0$ by the symmetry-adapted construction and that g_{2y_2t} vanishes because the point is a winged cusp; cf. (8.82).

In the point preceding a sign change in g_{1y_1} we find

$$\frac{\partial(g_1, g_2)}{\partial(y_1, y_2, t)} = \begin{pmatrix} 2.5360831183D-05 & 1.0788470530D-13 & 1.5106091310D-14 \\ -3.4425573901D-13 & -3.4794573464D-10 & -3.3768582216702D-08 \end{pmatrix}.$$

The Hessian matrix of G with respect to the state variables is

$$H_s = \begin{pmatrix} -4.1235955003459D-13 & -9.0498893860793D-05 & -2.7468691299150D-13 \\ -6.4121515933838D-04 & -8.3499889181402D-14 & -3.2499495934694D-04 \end{pmatrix}.$$

We note that $g_{2y_2y_2}$ no longer vanishes. The mixed state-parameter derivatives are

$$\frac{\partial(g_{1t}, g_{2t})}{\partial(y_1, y_2)} = \begin{pmatrix} -0.26691444356699 & 3.1137259697382D-11 \\ 5.0895860783701D-12 & 4.6894458666183D-02 \end{pmatrix}.$$

We note that g_{2y_2t} no longer vanishes.

In the point just after a sign change in g_{1y_1} we find

$$\frac{\partial(g_1, g_2)}{\partial(y_1, y_2, t)} = \begin{pmatrix} -6.0727629802D-05 & -5.2157494285D-10 & -6.3315590356D-11 \\ -4.8949739271D-12 & -2.2748410444D-07 & -7.7748308024D-08 \end{pmatrix}.$$

The Hessian matrix of G with respect to the state variables is

$$H_s = \begin{pmatrix} 1.9986824156326D-09 & -7.1558268005582D-05 & 1.3217183562136D-09 \\ -6.0691206366987D-04 & 3.9852606017913D-10 & -3.2594666508791D-04 \end{pmatrix}.$$

The mixed state-parameter derivatives are

$$\frac{\partial(g_{1t}, g_{2t})}{\partial(y_1, y_2)} = \begin{pmatrix} -0.26367969692468 & 2.9396296775649D-08 \\ 2.3247842131914D-08 & 4.8230187196134D-02 \end{pmatrix}.$$

Finally, we locate the corank-2 point accurately by setting up the system

$$\begin{aligned} F(u, D_X, D_Y, A_0, L, D_A, B) + rb_1 &= 0, \\ g_{2y_2} &= 0, \\ g_{2t} &= 0, \\ g_{2y_2y_2}g_{2tt} - g_{2y_2t}^2 &= 0, \\ g_{1y_1} &= 0, \\ c_1^T u &= 0, \end{aligned} \tag{8.89}$$

which is a natural extension of (8.46). This system has $2n+5$ equations and $2n+5$ unknowns, i.e., u, A_0, L, D_A, B, r where r is a dummy unknown introduced to make the system (8.89) regular; as in (8.46) b_1 and c_1 are the first columns of B, C, respectively. In our test the Newton iteration with the system (8.89) converged satisfactorily and we found $A_0 = 6.5977343845795$, $L = 0.12869163312226$, $D_A = 9.1859393992030D-03$, $B = 11.143305066117$. The first 42 state coordinates are given in Table 8.15.

Table 8.15: State coordinates of a corank-2 point with $\mathbb{Z}_2$-equivariant normal form $[p, q] = [u^2 - t^3 + v, u + t]$.

6.4306619818080	1.7592365584193	6.2573028929472	1.8307144495253
6.0724999956823	1.9043543505868	5.8724409805245	1.9808290014986
5.6548023319576	2.0604896013371	5.4188736049615	2.1433384711317
5.1656164331275	2.2290165637040	4.8976192039366	2.3168136959786
4.6189234886811	2.4057062333863	4.3347215720357	2.4944221630491
4.0509514277354	2.5815279503395	3.7738390170461	2.6655267547284
3.5094504743721	2.7449549855025	3.2633143481682	2.8184646680086
3.0401575205770	2.8848824769181	2.8437734056555	2.9432414226517
2.6770152838677	2.9927864712406	2.5418877738905	3.0329594874697
2.4396987985655	3.0633711288738	2.3712327755109	3.0837677109923
2.3369106504396	3.0940001109967		

To check the nondegeneracy conditions (8.86) and (8.87) we compute the quantities $q_u = g_{1y_1y_2}$, $p_v = \frac{1}{2}g_{2y_1y_1}$, $p_{uu} = g_{2y_2y_2}$, $\alpha = g_{2y_2t}/p_{uu}$, $D_{8,1} = g_{1y_1t} - \alpha q_u$, and

$$D_{8,2} = \frac{1}{6}g_{2y_2y_2y_2}\alpha^3 - \frac{1}{2}g_{2y_2y_2t}\alpha^2 + \frac{1}{2}g_{2y_2tt}\alpha - \frac{1}{6}g_{2ttt}$$

at every Newton step during the computation of the corank-2 point. The convergence looks quite robust and we find as limit values $q_u = -8.48489D-05$, $p_v = -3.15663D-04$, $p_{uu} = -3.25412D - 04$, $\alpha = -145.31275$, $D_{8,1} = -0.27834$, $D_{8,2} = 7145.219758$. As a consequence, $\epsilon_1 < 0$, $\epsilon_2 > 0$, and $\epsilon_3 > 0$.

So the normal form in the classification in [65] is

$$[p, q] = [u^2 - t^3 + v, u + t].$$

Remark. In Figure 8.1 we present graphically the X and Y components of the corank-2 point from the data provided in Table 8.15. We note that these functions are quite smooth. This helps us to understand why fixed equidistant meshes work so well in this example.

8.4 Other Symmetry Groups

8.4.1 Symmetry-Adapted Bases

We consider again the dynamical system

$$\dot{x} = G(x, \alpha), \qquad (x \in \mathbb{R}^N, \alpha \in \mathbb{R}^m, G(x, \alpha) \in \mathbb{R}^N) \tag{8.90}$$

and its equilibrium solutions for which

$$G(x, \alpha) = 0. \tag{8.91}$$

We assume that G is equivariant with respect to a group of transformations, i.e.,

$$G(Sx, \alpha) = SG(x, \alpha) \tag{8.92}$$

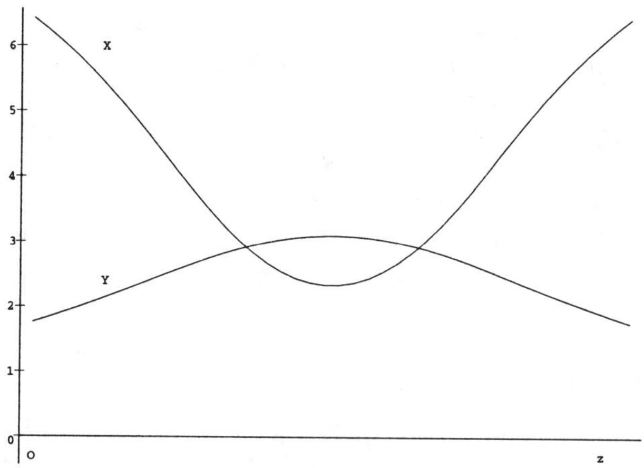

Figure 8.1: X- and Y-components of the corank-2 point in Table 8.15.

for all $S \in \mathcal{S}$ where $\mathcal{S}$ is a compact subgroup of the group $\mathbf{GL}(N)$ of nonsingular $N \times N$ matrices.

A group that can be represented in this way is called a compact Lie group. We will rely on [110, Chapter XII] for mathematical background, but it is useful to reformulate some of the basic results in terms that are related more directly to the numerical methods. The first states that we may essentially assume that $\mathcal{S}$ is a subgroup of the group $\mathcal{O}(N)$ of orthogonal matrices.

Proposition 8.4.1.

1. There exists a nonsingular matrix $P \in \mathbb{R}^{N \times N}$ such that the conjugate group

$$P\mathcal{S}P^{-1} = \{PSP^{-1} : S \in \mathcal{S}\}$$

is a group of orthogonal matrices.

2. If we define $z = Px \in \mathbb{R}^N$ then z satisfies

$$\dot{z} = H(z, \alpha), \tag{8.93}$$

where the nonlinear function $H(z, \alpha) = PG(P^{-1}z, \alpha)$ is equivariant with respect to $P\mathcal{S}P^{-1}$.

Proof. The first statement is a matrix-oriented reformulation of Proposition 1.3 in [110, Chapter XII]. The second statement is obvious. □

In typical examples, $\mathcal{S}$ is a group of orthogonal matrices. From now on we shall make this assumption, thus avoiding some technicalities in the exposition that are irrelevant in most applications.

A representation of $\mathcal{S}$ is a map $\mathcal{S} \to \mathrm{GL}(k)$, which preserves the group operations. A representation is irreducible if the only invariant subspaces of $\mathbb{R}^k$ are the trivial one and

$\mathbb{R}^k$ itself. Two representations are called $\mathcal{S}$-isomorphic if they are induced in the natural way by an isomorphism of the underlying vector spaces.

In particular, every invariant subspace V of $\mathbb{R}^N$ defines a representation of $\mathcal{S}$ by restriction. If an orthogonal base of V is chosen, this induces a representation ϑ of $\mathcal{S}$ as orthogonal matrices for which $SV = V\vartheta(S)$; changing the orthogonal base amounts to replacing the matrix group by a conjugate group. Such a representation is irreducible if and only if the invariant subspace is minimal under all invariant subspaces of $\mathbb{R}^N$, i.e., if it contains no proper nontrivial invariant subspace. We have Proposition 8.4.2.

Proposition 8.4.2.

1. Up to $\mathcal{S}$-isomorphism, there exists a finite number of distinct $\mathcal{S}$-irreducible subspaces of $\mathbb{R}^N$. Call them $U_1, \ldots, U_t$.

2. For $k = 1, \ldots, t$ let the columns of $W_k \in \mathbb{R}^{N \times d_k}$ form an orthogonal base for the span of all $\mathcal{S}$-irreducible subspaces of $\mathbb{R}^N$ that are isomorphic to U_k and let $\tilde{\vartheta}_k$ be the corresponding matrix representation. Then $W = [W_1, \ldots, W_t]$ is in $\mathcal{O}(N)$ and
$$SW = WD_S, \qquad D_S = \mathrm{Diag}(\tilde{\vartheta}_1(S), \ldots, \tilde{\vartheta}_t(S)) \tag{8.94}$$
for all $S \in \mathcal{S}$.

3. Let ϑ_k be the irreducible representation of $\mathcal{S}$ on W_k and let n_k be the dimension of this representation. Then by changing if necessary the orthogonal base W_k may be written further as $W_k = [W_k^1, \ldots, W_k^{m_k}]$, where each W_k^i has dimension n_k (so $n_k m_k = d_k$) and
$$SW = WD_S, \qquad D_S = \mathrm{Diag}(\vartheta_1(S), \ldots, \vartheta_1(S), \vartheta_2(S), \ldots, \vartheta_t(S)), \tag{8.95}$$
where each $\vartheta_k(S)$ appears precisely m_k times.

Proof. See Proposition 2.5 in [110, Chapter XII] and subsequent results. □

The spaces W_k are called the isotypic components of $\mathbb{R}^N$ because the actions of $\mathcal{S}$ on all irreducible subspaces of W_k are $\mathcal{S}$-isomorphic. We note that the decomposition of $\mathbb{R}^N$ into the subspaces W_k is essentially unique, while the further decomposition of W_k in the third statement of Proposition 8.4.2 is not.

We now return to (8.90). Let $\mathrm{Fix}(\mathcal{S}) \equiv \{x \in \mathbb{R}^N : Sx = x \text{ for all } S \in \mathcal{S}\}$ be the set of fixed points of $\mathcal{S}$, i.e., the set of all $x \in \mathbb{R}^N$ that exhibit the full symmetry of $\mathcal{S}$.

Let $(x^*, \alpha^*) \in \mathrm{Fix}(\mathcal{S}) \times \mathbb{R}^m$. If we set $A = G_x(x^*, \alpha^*)$ then we have
$$AS = SA \tag{8.96}$$
for all $S \in \mathcal{S}$. Multiplying with the matrix W defined in (8.95) we find $AWD_S = WD_SW^{-1}AW$ or, equivalently,
$$(W^{-1}AW)D_S = D_S(W^{-1}AW). \tag{8.97}$$
If we decompose $W^{-1}AW$ blockwise as
$$W^{-1}AW = \begin{pmatrix} B_{11} & \ldots & B_{1t} \\ \vdots & & \vdots \\ B_{t1} & \ldots & B_{tt} \end{pmatrix}, \tag{8.98}$$

where B_{ij} has dimension $d_i \times d_j$, then (8.97) implies in the setting of (8.94) that

$$\tilde{\vartheta}_i(S)B_{ij} = B_{ij}\tilde{\vartheta}_j(S) \tag{8.99}$$

for all $i, j \in \{1, \ldots, t\}$. It can be proved that for $i \neq j$ this implies $B_{ij} = 0$. Now consider the case $i = j$. We further decompose B_{kk} blockwise as

$$\begin{pmatrix} (B_{kk})_{11} & \ldots & (B_{kk})_{1m_k} \\ \vdots & & \vdots \\ (B_{kk})_{m_k 1} & \ldots & (B_{kk})_{m_k m_k} \end{pmatrix}, \tag{8.100}$$

where $(B_{kk})_{ij} \in \mathbb{R}^{n_k \times n_k}$ for all i, j. In the setting of (8.95) we then have

$$\vartheta_k(S)(B_k)_{ij} = (B_k)_{ij}\vartheta_k(S) \tag{8.101}$$

for all $S \in \mathcal{S}$; i.e., every $(B_{kk})_{ij}$ commutes with $\vartheta_k(S)$.

A representation ϑ is called *absolutely irreducible* if the multiples of the identity matrix are the only matrices that commute with all $\vartheta(S)$. According to [110, Proposition 3.2, p. 82] an irreducible representation is generically absolutely irreducible. We shall make this assumption, leaving the discussion to [110]. Then by (8.100) every $(B_k)_{ij}$ has the form

$$(B_{kk})_{ij} = b_{ij} I_{n_k}. \tag{8.102}$$

We conclude that

$$AW = W\text{Diag}(B_{11}, \ldots, B_{tt}). \tag{8.103}$$

Since W is orthogonal this implies

$$A^T W = W\text{Diag}(B_{11}^T, \ldots, B_{tt}^T) \tag{8.104}$$

and every B_{tt} has the form described in (8.102).

We say that the columns of W form a *symmetry-adapted base* for $\mathbb{R}^N$. This generalizes the decomposition into symmetric and antisymmetric subspaces of the state space discussed in §§8.1.1–8.1.2. If we introduce $z = W^{-1}x$ as a new state variable, then we have $\dot{z} = H(z, \alpha)$ with $H(z, \alpha) = W^{-1}G(Wz, \alpha)$ and $H_z = W^{-1}G_x W$ is precisely the above obtained matrix $\text{Diag}(B_{11}, \ldots, B_{tt})$.

8.4.2 The Equivariant Branching Lemma

Let $(x^*, \alpha^*) \in \text{Fix}(\mathcal{S}) \times \mathbb{R}^m$ be a singular point of G with corank $k \geq 1$, i.e.,

$$G(x^*, \alpha^*) = 0, \qquad \dim \text{Ker } G_x(x^*, \alpha^*) = k. \tag{8.105}$$

We recall briefly the analysis of the generic *spontaneous symmetry-breaking bifurcation* (cf. [110, Chapter XIII]):

The solution set of $G(x, \alpha) = 0$ is locally identified with the solution set of a *bifurcation equation* $g(y, t) = 0$, where $g : \mathbb{R}^k \times \mathbb{R}^m \to \mathbb{R}^k$ is the reduced version of G obtained via the Lyapunov–Schmidt reduction (see §8.4.4 for an explicit form). The space $\mathbb{R}^k$ of the *reduced state space* variables y is identified with Ker $G_x(x^*, \alpha^*)$ by the choice of a basis in Ker $G_x(x^*, \alpha^*)$. The parameter t is the shift of α, namely, $t = \alpha - \alpha^*$. The isomorphism between the solution sets of G and g relates the singular point $(x^*, \alpha^*) \in \mathbb{R}^N \times \mathbb{R}^m$ to the origin of $\mathbb{R}^k \times \mathbb{R}^m$.

The reduced operator g shares the invariance property of G: the k-dimensional representation of $\mathcal{S}$ on Ker $G_x(x^*, \alpha^*)$, obtained by restricting every $S \in \mathcal{S}$ to Ker $G_x(x^*, \alpha^*)$, is identified with a homomorphism $\vartheta : \mathcal{S} \to \mathbf{GL}(k)$, such that $g(\vartheta(S)y, t) = \vartheta(S)g(y, t)$ for each $(y, t) \in \mathbb{R}^k \times \mathbb{R}^m$ and each group element $S \in \mathcal{S}$.

We assume that ϑ is an absolutely irreducible representation. Since $g_y(\vartheta(S)y, t)\vartheta(S) = \vartheta(S)g_y(y, t)$ for all $S \in \mathcal{S}$, $y \in \mathbb{R}$, $t \in \mathbb{R}^m$, it follows that $g(0, t)$ is a multiple of the identity for all t in a neighborhood of 0. Since $g_y(0, 0) = 0$ this implies that a rank drop k occurs in (x^*, α^*).

We say that the representation is trivial if $\vartheta(S)$ is the identity matrix for all $S \in \mathcal{S}$. Under the assumption of absolute irreducibility this happens if and only if $k = 1$ and Ker $G_x\,(x^*, \alpha^*)$ is in Fix($\mathcal{S}$). In this case all solution points of (8.91) near (x^*, α^*) are in Fix($\mathcal{S}$) and from the point of view of symmetry nothing special happens (in the $\mathbb{Z}_2$-setting this corresponds to the case where G_x^s is singular in §8.1.2).

From now on we assume that ϑ is a *nontrivial representation*; i.e., $\text{Fix}(\vartheta(\mathcal{S})) \equiv \{y \in \mathbb{R}^k : \vartheta(S)y = y \text{ for all } S \in \mathcal{S}\}$ is trivial. In other words, Ker $G_x\,(x^*, \alpha^*) \cap \text{Fix}(\mathcal{S}) = \{0\}$. Consequently, the solution sets of $G\,(x, \alpha) = 0$ and $g\,(y, t) = 0$, restricted to the symmetric states $x \in \text{Fix}(\mathcal{S})$ and $y \in \text{Fix}(\vartheta(\mathcal{S}))$, respectively, consist (locally) of just one branch that can be parametrised by α and t, respectively. The Lyapunov–Schmidt reduction links the symmetric branch of $G\,(x, \alpha) = 0$ with the trivial branch of $g\,(y, t) = 0$. The other branches, if they exist, do not possess the full symmetry of $\mathcal{S}$.

Let us consider a subgroup Σ of $\mathcal{S}$, of the form

$$\Sigma = \{S \in \mathcal{S} : \vartheta(S)y_0 = y_0\}$$

for a given fixed $y_0 \in \mathbb{R}^k$. We assume

$$\dim \text{Fix}(\vartheta(\Sigma)) = 1, \quad \text{Fix}(\vartheta(\Sigma)) \equiv \{y \in \mathbb{R}^k : \vartheta(S)y = y \text{ for all } S \in \Sigma\}; \tag{8.106}$$

i.e., $\dim(\text{Ker } G_x\,(x^*, \alpha^*) \cap \text{Fix}(\Sigma)) = 1$. The equivariant branching lemma, first proved in [233] (see [110, Theorem 3.3, p. 82] for details), states that the operator g, restricted as $g : \text{Fix}(\vartheta(\Sigma)) \times \mathbb{R}^1 \to \text{Fix}(\vartheta(\Sigma))$, generically has a nontrivial solution branch emanating from the origin; here one parameter in α is singled out as the distinguished parameter. By the isomorphism of the solution sets of $G\,(x, \alpha) = 0$ and $g\,(y, t) = 0$, there is a nonsymmetric solution branch of $G : \text{Fix}(\Sigma) \times \mathbb{R}^1 \to \text{Fix}(\Sigma)$ emanating from (x^*, α^*).

Let us choose a vector $\mathbf{d} \in \mathbb{R}^k$ that spans Fix $(\vartheta(\Sigma))$; we shall call $\mathbf{d}$ an *isotropy vector*.

The symmetry group of the operator $G : \text{Fix}\,(\Sigma) \times \mathbb{R}^m \to \text{Fix}\,(\Sigma)$ is the normalizer $\mathcal{N}_{\mathcal{S}}(\Sigma)$ of Σ on $\mathcal{S}$, i.e., the largest subgroup of $\mathcal{S}$ that leaves Fix $(\vartheta(\Sigma))$ invariant as a set. This is also called the *isotropy subgroup* of Fix $(\vartheta(\Sigma))$. By [110, Exercise 2.2, p. 79], $\mathcal{N}_{\mathcal{S}}(\Sigma)$ acts on Fix $(\vartheta(\Sigma))$ either as the identity or as a $\mathbb{Z}_2$-group. Choosing a particular isotropy direction with normalized $\mathbf{d}$, we define $h : \mathbb{R}^m \times \mathbb{R}^m \to \mathbb{R}^1$ by setting

$$h(\xi, t) \equiv \mathbf{d}^\top g(\xi\,\mathbf{d}, t). \tag{8.107}$$

By the isotropy assumption we have $g(\xi\,\mathbf{d}, t) = h(\xi, t)\mathbf{d}$. The solution set of $h(\xi, t) = 0$ is constrained to have a trivial branch, namely, $h(0, t) \equiv 0$. If the normalizer acts as the identity, then $h(\cdot, t) : \mathbb{R}^1 \to \mathbb{R}^1$ exhibits no symmetry. If the normalizer acts as $\mathbb{Z}_2$, then $h(\cdot, t) : \mathbb{R}^1 \to \mathbb{R}^1$ is $\mathbb{Z}_2$-equivariant for each t. Hence, $h(-\xi, t) = -h(\xi, t)$ on a neighborhood of the origin.

A singular point (x^*, α^*) that satisfies the above resumed assumptions is called an *$\mathcal{S}$-symmetry-breaking bifurcation point.*

We note that k and ϑ are group theoretic data that characterize a particular $\mathcal{S}$-symmetry-breaking bifurcation point. Moreover, nonsymmetric branches are determined by an isotropy direction $\mathbf{d}$. Often there are several branches. In particular, if $\mathbf{d}$ is an isotropy direction and $S \in \mathcal{S}$, then $\vartheta(S)\mathbf{d}$ also determines an isotropy direction, namely, the one associated with the subgroup $S\Sigma S^{-1}$ conjugate to Σ. Therefore, the two isotropy directions are also called conjugate. But there also may be nonconjugate isotropy directions.

We refer to [110, pp. 132–137] for a discussion of the genericity of the sketched scenario. The above assumptions are generically satisfied in the case that $\mathcal{S}$ is a dihedral group $\mathbf{D}_n$; see [110, Ch. XIII, §5]. We recall that the dihedral group $\mathbf{D}_n$ is the symmetry group of the regular n-gon in the plane. In §8.4.2 we will consider $\mathbf{D}_4$ in detail.

In the m-parameter setting one may expect degeneracies of the above sketched scenario of symmetry breaking. For example, the assumption concerning the irreducibility of the representation on Ker $G_x(x^*, \alpha^*)$ can be violated. This leads to the scenario of *mode interaction*, which we met already in §8.2 in the case of $\mathbb{Z}_2$-symmetry. See [102] for a study of the case of $\mathbf{D}_6$. Also, the dimension k of Ker $G_x(x^*, \alpha^*)$ might be different from the dimension of Ker $(G_x(x^*, \alpha^*))^2$, as one generically expects for an operator $G : \mathbb{R}^N \to \mathbb{R}^N$. This leads to a *Takens–Bogdanov bifurcation* scenario. Let us call all degeneracies affected by the properties of the Jacobian $G_x(x^*, \alpha^*)$ *linear degeneracies.* On the other hand, a degeneracy may be caused by nongeneric properties of the Taylor expansion of $G(x, \alpha)$ at (x^*, α^*). We refer to [110, pp. 218–223] for an example of a generic nonlinear degeneracy of a $\mathbf{D}_3$-symmetry-breaking bifurcation with $m = 2$, $\mathcal{S} = \mathbf{D}_3$, and $k = 2$. The representation does not have to be specified, since there is only one two-dimensional irreducible representation of $\mathbf{D}_3$. Let us call this kind of degeneracy a *nonlinear degeneracy.*

The nonlinear degeneracy of a particular $\mathcal{S}$-symmetry-breaking bifurcation point can be classified partly by the properties of the bifurcation diagram of $g(y, t, 0) = 0$, where all parameters but one are fixed and y is restricted to the one-dimensional state space spanned by the isotropy direction $\mathbf{d}$. If the normalizer $\mathcal{N}_{\mathcal{S}}(\Sigma)$ acts on Fix($\vartheta(\Sigma)$) as the identity, then, generically, the bifurcation diagram of h consists of the trivial branch and a nontrivial transcritical branch. If the normalizer acts as $\mathbb{Z}_2$, then h has generically a (symmetric) pitchfork bifurcation diagram.

The study of irreducible representations of a given compact Lie group and of the isotropy directions is a group theoretic task that can be performed independently of any applications in bifurcation theory and may require considerable sophistication. We refer to [214], [90], and, of course, [110].

Figure 8.2: 4-box coupled oscillator.

8.4.3 Example: A System with $\mathbf{D}_4$-Symmetry

Coupled oscillators with dihedral symmetry are among the most popular models for illustrating symmetry-breaking phenomena. Figure 8.2 illustrates the case of four boxes coupled with a $\mathbf{D}_4$-symmetry. Suppose that the state in the box B_i is described by the two variables x_{2i-1} and x_{2i}. The full dynamical system is described by eight equations of the form

$$\begin{aligned} \dot{x}_{2i-1} &= f_1(x_{2i-1}, x_{2i}, x_{2i-3}, x_{2i-2}, x_{2i+1}, x_{2i+2}, \alpha), \\ \dot{x}_{2i} &= f_2(x_{2i-1}, x_{2i}, x_{2i-3}, x_{2i-2}, x_{2i+1}, x_{2i+2}, \alpha) \end{aligned} \tag{8.108}$$

for $i = 1, 2, 3, 4$; it is understood that $x_9 = x_1$, $x_{10} = x_2$, $x_0 = x_8$, $x_{-1} = x_7$, and f_1, f_2 are invariant under the simultaneous interchange of the third and fourth arguments with the fifth and sixth arguments, respectively. This type of equations is called a reaction-diffusion system if f_1, f_2 have the form

$$\begin{aligned} f_1(x_{2i-1}, x_{2i}, x_{2i-3}, x_{2i-2}, x_{2i+1}, x_{2i+2}, \alpha) &= d_1(x_{2i-3} - 2x_{2i-1} + x_{2i+1}) + f_1^1(x_{2i-1}, x_{2i}, \alpha_1), \\ f_2(x_{2i-1}, x_{2i}, x_{2i-3}, x_{2i-2}, x_{2i+1}, x_{2i+2}, \alpha) &= d_2(x_{2i-2} - 2x_{2i} + x_{2i+2}) + f_2^1(x_{2i-1}, x_{2i}, \alpha_1), \end{aligned} \tag{8.109}$$

where d_1, d_2, α_1 form a splitting of α.

It is easily seen (and understood from the geometry of the problem) that system (8.108) is equivariant with respect to multiplication with

$$\kappa^{(8)} = \begin{pmatrix} & & & & & & 1 & \\ & & & & & & & 1 \\ & & & & 1 & & & \\ & & & & & 1 & & \\ & & 1 & & & & & \\ & & & 1 & & & & \\ 1 & & & & & & & \\ & 1 & & & & & & \end{pmatrix}$$

and

$$\rho^{(8)} = \begin{pmatrix} & & 1 & & & & & \\ & & & 1 & & & & \\ & & & & 1 & & & \\ & & & & & 1 & & \\ & & & & & & 1 & \\ & & & & & & & 1 \\ 1 & & & & & & & \\ & 1 & & & & & & \end{pmatrix}.$$

Hence it is also equivariant with respect to the group of matrices generated by $\kappa^{(8)}$ and $\rho^{(8)}$, i.e.,

$$\mathbf{D}_4^{(8)} = \{\iota^{(8)}, \kappa^{(8)}, \rho^{(8)}, \kappa^{(8)}\rho^{(8)}, \kappa^{(8)}{\rho^{(8)}}^2, \kappa^{(8)}{\rho^{(8)}}^3, {\rho^{(8)}}^2, {\rho^{(8)}}^3\},$$

where $\iota^{(8)} = I_8$.

The group $\mathbf{D}_4^{(8)}$ is isomorphic to the symmetry group $\mathbf{D}_4$ of the square. To be precise, let $\iota = I_2$ be the identity matrix, κ the reflection matrix

$$\kappa = \begin{pmatrix} 1 & 0 \\ 0 & -1 \end{pmatrix},$$

and ρ the rotation matrix

$$\rho = \begin{pmatrix} 0 & -1 \\ 1 & 0 \end{pmatrix}.$$

Then $\mathbf{D}_4$ is the eight-element group

$$D_4 = \{\iota, \kappa, \rho, \kappa\rho, \kappa\rho^2, \kappa\rho^3, \rho^2, \rho^3\}.$$

Multiplication in $\mathbf{D}_4$ is matrix multiplication; the symbolic rules are obvious if we take into account that $\kappa^2 = \iota$, $\rho^4 = \iota$, $\rho\kappa = \kappa\rho^3$. The isomorphism of $\mathbf{D}_4^{(8)}$ and $\mathbf{D}_4$ maps $\iota^{(8)}$ to ι, $\kappa^{(8)}$ to κ, $\rho^{(8)}$ to ρ.

Now suppose that we follow a branch of fully symmetric equilibrium solutions to (8.108); i.e., each solution vector has the form

$$x = \begin{pmatrix} A \\ A \\ A \\ A \end{pmatrix}$$

with $A \in \mathbb{R}^2$. We first consider the one-dimensional absolutely irreducible representations of $\mathbf{D}_4$. Since κ and ρ generate $\mathbf{D}_4$ and are to be mapped onto ± 1, there are at most four such representations; it is easy to check that there are indeed four, characterized as follows:

$$\begin{array}{lll} \text{(i)} & \kappa \to 1, \rho \to 1, & \Sigma = \mathbf{D}_4. \\ \text{(ii)} & \kappa \to -1, \rho \to -1, & \Sigma = \{\iota, \kappa\rho, \kappa\rho^3, \rho^2\}. \\ \text{(iii)} & \kappa \to 1, \rho \to -1, & \Sigma = \{\iota, \kappa, \kappa\rho^2, \rho^2\}. \\ \text{(iv)} & \kappa \to -1, \rho \to 1, & \Sigma = \{\iota, \kappa\rho, \rho^2, \rho^3\}. \end{array} \tag{8.110}$$

Now consider the case where Ker G_x is one-dimensional; let it be spanned by

$$\mathbf{d} = \begin{pmatrix} A \\ B \\ C \\ D \end{pmatrix},$$

where $A, B, C, D \in \mathbb{R}^2$ are not all zero. Now Ker G_x has to be invariant under the action of the representation in $\mathbb{R}^{(8)}$ of one of the four groups described in (8.110). It is easily seen that this is possible only in the first two cases, so that either

$$\mathbf{d} = \begin{pmatrix} A \\ A \\ A \\ A \end{pmatrix}$$

or

$$\mathbf{d} = \begin{pmatrix} A \\ -A \\ A \\ -A \end{pmatrix}$$

($A \in \mathbb{R}^2$), respectively. In the first case, Ker G_x is in the fully symmetric space, so no symmetry breaking occurs. In the second, more interesting, case the isotropy subgroup of $\mathbf{d}$ is $\Sigma = \{S \in \mathbf{D}_4^{(8)} : S\mathbf{d} = \mathbf{d}\} = \{\iota^{(8)}, \kappa^{(8)}\rho^{(8)}, {\rho^{(8)}}^2, \kappa^{(8)}{\rho^{(8)}}^3\}$. This four-element group is isomorphic to $\mathbb{Z}_2 \times \mathbb{Z}_2$ when its elements are mapped onto $(0,0), (0,1), (1,0), (1,1)$ in that order. Furthermore, Fix(Σ) is the set of all vectors of the form

$$\begin{pmatrix} A \\ B \\ A \\ B \end{pmatrix},$$

where $A, B \in \mathbb{R}^2$. So the nonsymmetric branch in such a symmetry-breaking point will still have the symmetry of the group $\mathbb{Z}_2 \times \mathbb{Z}_2$. In the setting of Figure 8.2 opposite boxes have the same states. Furthermore, the action of $\mathbf{D}_4$ on Ker G_x is clearly a $\mathbb{Z}_2$-action, so the bifurcation is generically a pitchfork bifurcation.

Group theory tells us that all two-dimensional absolutely irreducible representations of $\mathbf{D}_4$ are isomorphic to the canonical one, i.e., the one determined by our definition of κ and ρ as 2×2 matrices (isomorphic means induced in the obvious way by an isomorphism of vector spaces). Furthermore, there are no absolutely irreducible representations in spaces with dimension higher than 2.

Now consider the case where Ker G_x is two dimensional. The action of $\mathbf{D}_4^{(8)}$ on Ker G_x must be the canonical action of $\mathbf{D}_4$ modulo the selection of a (not a priori known) vector base of this kernel. The isotropy vectors in the canonical representation of $\mathbf{D}_4$ are

$$\mathbf{d}_1 = \begin{pmatrix} 1 \\ 0 \end{pmatrix}, \quad \mathbf{d}_2 = \begin{pmatrix} \frac{\sqrt{2}}{2} \\ \frac{\sqrt{2}}{2} \end{pmatrix}, \quad \mathbf{d}_3 = \begin{pmatrix} 0 \\ 1 \end{pmatrix}, \quad \mathbf{d}_4 = \begin{pmatrix} -\frac{\sqrt{2}}{2} \\ \frac{\sqrt{2}}{2} \end{pmatrix}.$$

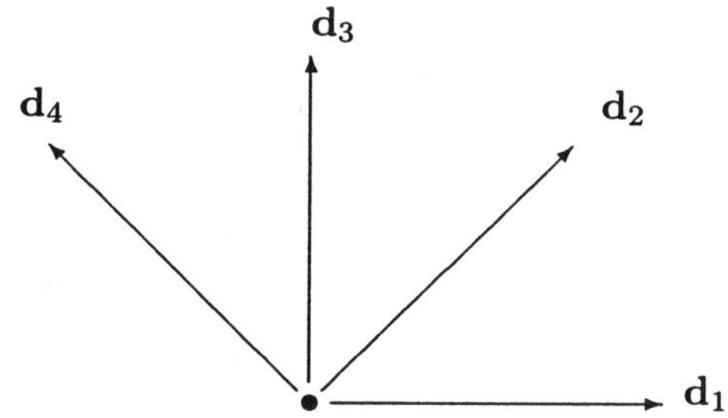

Figure 8.3: Isotropy vectors of $\mathbf{D}_4$.

See, e.g., [110, Ch. XIII, §5] and see Figure 8.3 for a picture. We note that $\mathbf{d}_1$ and $\mathbf{d}_3$ are conjugate isotropy vectors since $\rho\mathbf{d}_1 = \mathbf{d}_3$; similarly $\mathbf{d}_2$ and $\mathbf{d}_4$ are conjugate; on the other hand $\mathbf{d}_1$ and $\mathbf{d}_2$ are not conjugate since there is no element in $\mathbf{D}_4$ that maps $\mathbf{d}_1$ onto $\mathbf{d}_2$.

The isotropy subgroups of $\mathbf{d}_i (i = 1, \dots, 4)$ are

$$\Sigma(\mathbf{d}_1) = \{\iota, \kappa\}, \Sigma(\mathbf{d}_2) = \{\iota, \kappa\rho^3\}, \Sigma(\mathbf{d}_3) = \{\iota, \kappa\rho^2\}, \Sigma(\mathbf{d}_4) = \{\iota, \kappa\rho\}$$

with respective normalizers

$$\mathcal{N}_{\mathbf{D}_4}(\Sigma(\mathbf{d}_1)) = \{\iota, \kappa, \kappa\rho^2, \rho^2\}, \qquad \mathcal{N}_{\mathbf{D}_4}(\Sigma(\mathbf{d}_2)) = \{\iota, \kappa\rho, \kappa\rho^3, \rho^2\},$$
$$\mathcal{N}_{\mathbf{D}_4}(\Sigma(\mathbf{d}_3)) = \{\iota, \kappa, \kappa\rho^2, \rho^2\}, \qquad \mathcal{N}_{\mathbf{D}_4}(\Sigma(\mathbf{d}_4)) = \{\iota, \kappa\rho, \kappa\rho^3, \rho^2\}.$$

So we expect four simultaneous pitchfork bifurcations at the symmetry-breaking bifurcation point, each with its own symmetry type.

Let $\mathbf{d}_1^{(8)}, \mathbf{d}_2^{(8)}, \mathbf{d}_3^{(8)}, \mathbf{d}_4^{(8)}$ be corresponding isotropy vectors in $\mathbb{R}^8$. Let us denote

$$\mathbf{d}_1^{(8)} = \begin{pmatrix} A \\ B \\ C \\ D \end{pmatrix}$$

with $A, B, C, D \in \mathbb{R}^2$. The isotropy condition $\kappa^{(8)}\mathbf{d}_1^{(8)} = \mathbf{d}_1^{(8)}$ implies $A = D$, $B = C$. The normalizer action ${\rho^{(8)}}^2\mathbf{d}_1^{(8)} = -\mathbf{d}_1^{(8)}$ implies $A + C = 0$, $B + D = 0$. We infer that

$$\mathbf{d}_1^{(8)} = \begin{pmatrix} A \\ -A \\ -A \\ A \end{pmatrix}$$

with $A \in \mathbb{R}^2$. Since we deal with an irreducible representation, $\mathbf{d}_1$ actually determines Ker G_x; the other isotropy vectors are easily obtained from

$$\mathbf{d}_3^{(8)} = \rho^{(8)}\mathbf{d}_1^{(8)} = \begin{pmatrix} -A \\ -A \\ A \\ A \end{pmatrix}, \qquad \mathbf{d}_2^{(8)} = (\mathbf{d}_1^{(8)} + \mathbf{d}_3^{(8)})/2 = \begin{pmatrix} 0 \\ -A \\ 0 \\ A \end{pmatrix},$$

$$\mathbf{d}_4^{(8)} = \rho^{(8)}\mathbf{d}_2^{(8)} = \begin{pmatrix} -A \\ 0 \\ A \\ 0 \end{pmatrix}.$$

The remaining symmetry of the branch in the $\mathbf{d}_i^{(8)}$-direction ($i = 1, \ldots, 4$) is that of $\Sigma(\mathbf{d}_i)$. So points on the bifurcating branches have the respective forms

$$\begin{pmatrix} A \\ B \\ B \\ A \end{pmatrix}, \quad \begin{pmatrix} A \\ B \\ A \\ D \end{pmatrix}, \quad \begin{pmatrix} A \\ A \\ B \\ B \end{pmatrix}, \quad \begin{pmatrix} A \\ B \\ C \\ B \end{pmatrix}.$$

In Proposition 8.4.2(3) we claimed the existence of a symmetry-adapted base W. It can now be given in the present example as

$$W = \frac{1}{2}\begin{pmatrix} 1 & 0 & 1 & 0 & 1 & 1 & 0 & 0 \\ 0 & 1 & 0 & 1 & 0 & 0 & 1 & 1 \\ 1 & 0 & -1 & 0 & -1 & 1 & 0 & 0 \\ 0 & 1 & 0 & -1 & 0 & 0 & -1 & 1 \\ 1 & 0 & 1 & 0 & -1 & -1 & 0 & 0 \\ 0 & 1 & 0 & 1 & 0 & 0 & -1 & -1 \\ 1 & 0 & -1 & 0 & 1 & -1 & 0 & 0 \\ 0 & 1 & 0 & -1 & 0 & 0 & 1 & -1 \end{pmatrix}.$$

We note that further symmetry breaking is possible. This leads to the so-called secondary or tertiary bifurcations. The branches that arise from the corank-2 bifurcations have only a $\mathbb{Z}_2$-symmetry, so the next symmetry breaking will destroy all symmetry. On the other hand, the branch that arises from the corank-1 bifurcation can further bifurcate into branches with the symmetry of $\Sigma(\mathbf{d}_2)$ or $\Sigma(\mathbf{d}_4)$. So only a tertiary bifurcation will destroy all symmetry.

Remark. We saw that two of the four one-dimensional absolutely irreducible representations of $\mathbf{D}_4$ cannot lead to symmetry-breaking bifurcations in the present 4-box model. In other models they may quite well lead to symmetry-breaking bifurcations; see [69] for the case of a partial differential equation (PDE) on a square.

8.4.4 Numerical Implementation

Numerical applications and implementations of the equivariant branching lemma are discussed in many papers, e.g., [140], [189], [69], [5], [185], [100], [102], [99], [142], [239], [35]. Furthermore, the package SYMCON 2.0 [101] is available for continuation of symmetric states and detection and computation of symmetry-breaking bifurcation points.

An analysis of the underlying group, its irreducible representations, and isotropy directions is always necessary for a good understanding. Also, the cascade of possible further symmetry-breaking phenomena can be studied a priori; this cascade usually has several paths; cf. [100], [102]. The analysis can be a considerable task, but it has been performed for most groups that the reader is likely to encounter; see [214], [90], [110].

The further computational work on examples can exploit the same ideas that were explained in detail in §§8.1–8.3 in the case of $\mathbb{Z}_2$-symmetry. Basically, at least two approaches are possible. In the most robust and comprehensive setting a symmetry-adapted base of the state space (see §8.4.1) is constructed explicitly, and the diagonal blocks B_{kk} in (8.102) are computed in every fully symmetric point. We note that $B_{kk} = \tilde{B}_{kk} \otimes I_{n_k}$ in the sense of §4.4 where $\tilde{B}_{kk} \in \mathbb{R}^{m_k \times m_k}$ is defined by $(\tilde{B}_{kk})_{ij} = b_{ij}$, b_{ij} as in (8.102).

A symmetry-breaking bifurcation point with respect to a particular absolutely irreducible representation corresponds to a singular block $\tilde{B}_{kk}$; this can be monitored by any of the methods that we discussed for systems without symmetry. Implementation of this method may be quite a task for some groups but this is compensated for by a reduction in the computational work because the block diagonal structure simplifies the computations. See [227], [100].

In another approach (see [142], [239]), a Lyapunov–Schmidt reduction with symmetry-adapted borders is performed to monitor symmetry-breaking phenomena. This approach is more restricted since it focuses on one particular absolutely irreducible representation at a time (although it can be extended to take more into account). Let $\vartheta : \mathcal{S} \to \mathbf{GL}(k)$ be an absolutely irreducible representation of $\mathcal{S}$. We say that $B \in \mathbb{R}^{N\times k}, C \in \mathbb{R}^{N\times k}, D \in \mathbb{R}^{k\times k}$ form a *symmetry-adapted bordering* if

$$SB = B\vartheta(S), \quad S^T C = C\vartheta(S)^T, \quad \vartheta(S)D = D\vartheta(S) \tag{8.111}$$

for all $S \in \mathcal{S}$. If furthermore

$$M(x,\alpha) \equiv \begin{pmatrix} G_x & -B \\ C^T & -D \end{pmatrix} \tag{8.112}$$

is nonsingular in a neighborhood of the point where a symmetry-breaking bifurcation with representation ϑ is expected, then $g(x;y;\alpha;t) \in \mathbb{R}^k$, $v(x;y;\alpha;t) \in \mathbb{R}^N$ are defined for $y \in \mathbb{R}^k$, $t \in \mathbb{R}^m$, (y,t) in a neighborhood of $(0,0)$, by

$$G(x + v(x;y;\alpha;t);\alpha + t) - Bg(x;y;\alpha;t) = G(x;\alpha), \tag{8.113}$$

$$C^T v(x;y;\alpha;t) - Dg(x;y;\alpha;t) = y. \tag{8.114}$$

By multiplying (8.113) with S and (8.114) with $\vartheta(S)$ we obtain

$$g(x;\vartheta(S)y;\alpha;t) = \vartheta(S)g(x;y;\alpha;t), \tag{8.115}$$

$$v(x;\vartheta(S)y;\alpha;t) = Sv(x;\vartheta(y);\alpha;t). \tag{8.116}$$

By the assumption of absolute irreducibility of ϑ, D must be a multiple of the identity matrix and g_y has automatically the same form. Symmetry breaking with respect to ϑ occurs if $g_y = 0$. The choice of B, C that satisfy (8.111) usually does not cause problems because these can be generated from any vector in the isotypic component of the representation; cf. Proposition 8.4.2. We note that ideally C (respectively, B) should be an approximation to a base for the right (respectively, left) singular space of G_x in the suspected bifurcation point and $D = 0$; this guarantees the nonsingularity of $M(x, \alpha)$ at the critical point (cf. Proposition 3.2.2).

As in the $\mathbb{Z}_2$-case, the bifurcation point is a singular point of the system of equilibrium equations and therefore may cause numerical problems in the continuation of fully symmetric equilibria. The system may be regularized by introducing an artificial unknown vector $r \in \mathbb{R}^{n_k}$ (the dimension of the absolutely irreducible representation that characterizes the bifurcation) and replacing the equilibrium equations by the system

$$\begin{array}{rcrcl} G(x,\alpha) & - & Br & = & 0, \\ C^T x & - & Dr & = & 0, \end{array} \tag{8.117}$$

where B, C, D must be chosen exactly as in the Lyapunov–Schmidt reduction. The symmetry-breaking bifurcation point itself can be computed as the regular solution to a nonlinear system simply by adding the condition $(g_y)_{11} = 0$ to the system in (8.117). This also allows the continuation of symmetry-breaking bifurcation points if an additional parameter is freed.

The bifurcating branches in a branch point can be computed from the isotropy directions; the method in §8.1.6 applies. Degeneracies in the branching behavior caused by the nonlinear terms in the expansion of $G(x, \alpha)$ can be detected and computed by taking formal derivatives of (8.113) and (8.114) and evaluating at $y = t = 0$, similar to what we did in §§8.2.5–8.2.6. See [35] for details and an example in the case of $\mathbf{D}_4$-symmetry.

8.5 Notes and Further Reading

Numerical methods for Hopf bifurcation in systems with symmetry are discussed in [69], [19], [20], and also in the proceedings issue [7].

The relation between symmetry in a continuous problem and symmetry in its discretization is considered in several papers; we refer to [186], [34], [18], [20].

8.6 Exercises

1. Let $S \in \mathbb{R}^{n \times n}$ be an involution matrix, i.e., $S^2 = I$. Prove that the following are equivalent:
   (a) S is symmetric.
   (b) S is an orthogonal matrix.
   (c) The symmetric and antisymmetric spaces of S are orthogonal.
2. Give an example of an involution matrix that is not symmetric or orthogonal. Check that the symmetric and antisymmetric spaces are not orthogonal.
3. Describe all symmetric orthogonal 2×2 matrices. Show that there is an infinite number of them but that there only three similarity classes.
4. Give an example of a 2×2 orthogonal involution matrix T different from $\pm I_2$ and a nondiagonal matrix D such that $DT = TD$.

5. In the corank-2 point analyzed in §8.2.5 no unfolding by natural parameters (i.e., which respect the equivariance) is versal in the sense of Chapter 6. Prove this.

6. Repeat the analysis of §8.4.3 for the case of a similar system with D_3-symmetry. Give a symmetry-adapted base.

7. Consider the example in Exercise 4.8.22(c). Now use the system (8.29) with b, c as in §8.1.5 to locate the L-values close to $2L_1$ and $2L_2$ where the Jacobian matrix is singular (L_1, L_2 as in Exercise 4.8.21(c)).

8. Use the methods developed in Chapter 7 (applied in particular in §7.7) to classify the singular points computed in the previous exercise (pitchfork bifurcations).

Chapter 9

Bifurcations with Degeneracies in the Nonlinear Terms

We consider once more the dynamical system

$$\dot{x} = G(x, \alpha) \qquad (x \in \mathbb{R}^N, \alpha \in \mathbb{R}^m, G(x, \alpha) \in \mathbb{R}^N) \tag{9.1}$$

and its equilibrium solutions for which

$$G(x, \alpha) = 0. \tag{9.2}$$

Typically bifurcation points are characterized by the coefficients of the Taylor expansion of G. Let A be the Jacobian matrix of G and $B, C, D, \ldots$ the tensors of second-, third-, fourth-order and so on derivatives of G. In particular, for vectors $p, q, r \in \mathbb{R}^N$, $B(p, q)$, and $C(p, q, r)$ are in $\mathbb{R}^N$ with components

$$B_i(p, q) = \sum_{j,k=1}^{N} \frac{\partial^2 G_i(x, \alpha)}{\partial x_j \partial x_k} p_j q_k, \tag{9.3}$$

$$C_i(p, q, r) = \sum_{j,k,l=1}^{N} \frac{\partial^3 G_i(x, \alpha)}{\partial x_j \partial x_k \partial x_l} p_j q_k r_l \tag{9.4}$$

for $i = 1, 2, \ldots, N$.

Many important bifurcations are characterized by A; the nonlinear terms appear only in the nondegeneracy conditions. Chapter 4 dealt with the codimension-1 case, i.e., fold points (a single zero eigenvalue) and Hopf points (a single conjugate pair of pure imaginary eigenvalues). Chapter 5 dealt with the codimension-2 case, i.e., BT, ZH, and DH and also with some higher codimension cases.

Singularity theory deals only with the geometric properties of the solution set to (9.2). In Chapter 6 we discussed the numerical aspects of singularity theory. In Chapter 7 this study was expanded to include the case where there is a distinguished parameter, which leads to a refinement of the classification.

In this chapter we consider the remaining cases. The Lyapunov–Schmidt reduction, which was the essential numerical tool in Chapters 6 and 7, can be generalized to the case of nonlinear degeneracies in Hopf bifurcation [143], [244], but it does not lead to a global approach. It will be replaced by the center manifold reduction, which we briefly describe in §9.1. Also, we return to the fundamental setting with no distinguished bifurcation parameter. The inclusion of a distinguished bifurcation parameter is possible (see [107], [143], [244] for GH but leads to complicated refinements, which in our opinion are not really worthwhile.

9.1 Principles of Center Manifold Theory

Let (x^0, α^0) be a solution to (9.2). In §1.5 we saw that the dynamic behavior of (9.1) is fairly simple near (x^0, α^0) if A does not have eigenvalues on the imaginary axis. Now let n_c be the number of eigenvalues on this axis. For example, $n_c = 1, 2$ in the fold and Hopf cases, respectively (Chapter 4), $n_c = 2, 3, 4$ for BT, ZH, and DH, respectively (Chapter 5). It turns out that the interesting dynamics of (9.1) can be reduced to a space with n_c dimensions. To be precise, the state space contains a *center manifold*, i.e., an n_c-dimensional manifold that is tangential to the generalized eigenspace of A for the critical eigenvalues that is invariant under the flow of (9.1). Moreover, this manifold can be smoothly continued to nearby parameter values. If all noncritical eigenvalues of A have a negative real part, then the center manifold is also attractive. In all cases, the dynamic behavior of (9.1) is largely determined by the behavior on the center manifold and this explains why so much of the theory of dynamical systems concentrates on low-dimensional problems.

For reduced situations (i.e., $N = n_c$) the theory of bifurcations of equilibria is well understood, at least to codimension 2. We collect some of the basic results, referring to [13], [127], and [164] for details. Under generic conditions (some resonances being excluded like the one-to-one resonance that we met for DH in Chapter 5), the reduced system is equivalent (we omit the precise definition) to a *normal form* plus some higher-order terms. The theory tells which terms should be kept in the normal form. Unfortunately, it is not always possible to truncate the higher-order terms without changing the local topology of the flow, but the normal form nevertheless captures the most important features. The normal form coefficients are expressed in terms of the Taylor coefficients of the reduced system at the equilibrium point. If the reduced system is explicitly given (as usually assumed in such studies), then an important part of this task can be done by symbolic computation software; see [211].

However, if a codimension-2 bifurcation point is found in a model, then the center manifold and the reduced dynamical system are only implicitly known. To apply the classification results directly it would be necessary to compute the Taylor coefficients of the reduced system numerically. This is essentially a numerical task to which symbolic methods can contribute modestly. The reduction formulas are complicated and the numerical work is hard. See [32] for the present state of the art.

On the other hand, there is a more direct method, introduced in [63], [87] and further developed in [167] that avoids the computation of the Taylor coefficients of the reduced system and rather computes the normal form coefficients by imposing the normal form on the implicitly defined reduced equations. We consider this method in some more detail.

9.1.1 The Homological Equation for Dynamics in the Center Manifold

Consider system (9.1) and assume that (x^0, α^0) is a solution to (9.2) with n_c critical eigenvalues. The center manifold can be parametrized by $w \in \mathbb{R}^{n_c}$:

$$x = H(w), \qquad H : \mathbb{R}^{n_c} \to \mathbb{R}^N. \tag{9.5}$$

The reduced equations can be written as

$$\dot{w} = F(w), \qquad F : \mathbb{R}^{n_c} \to \mathbb{R}^{n_c}. \tag{9.6}$$

By substitution of (9.5) and (9.6) in (9.1) one gets the so-called homological equation

$$H_w F(w) = G(H(w)). \tag{9.7}$$

Now we formally expand the (unknown) functions H, F into Taylor series

$$F(w) = \sum_{|\nu|\geq 1} \frac{1}{\nu!} f_\nu w^\nu, \qquad H(w) = x^0 + \sum_{|\nu|\geq 1} \frac{1}{\nu!} h_\nu w^\nu. \tag{9.8}$$

Here $\nu = (\nu_1, \ldots, \nu_{n_c})$ with every ν_i an integer ≥ 1, $w^\nu = (w_1, \ldots, w_{n_c})^\nu = w_1^{\nu_1} w_2^{\nu_2} \ldots w_{n_c}^{\nu_{n_c}}$, and $|\nu| = \max(|\nu_1|, \ldots, |\nu_{n_c}|)$. The functions H, F are not defined uniquely by the homological equation (it is clear that a change of base in $\mathbb{R}^{n_c}$ would change H, F), but we can impose that F be in normal form. Also, the vectors $h_\nu, |\nu| = 1$, must span the generalized eigenspace of the critical eigenvalues of A.

The simplest example is, of course, the fold bifurcation. We have $n_c = 1$ and

$$F(w) = f_1 w + \frac{1}{2} f_2 w^2 + \cdots, \tag{9.9}$$

$$H(w) = x^0 + h_1 w + \frac{1}{2} h_2 w^2 + \cdots. \tag{9.10}$$

Here $f_1 = 0$ and h_1 must be a right singular vector of A. So by expanding the homological equation we obtain

$$\begin{aligned} &(h_1 + h_2 w + \tfrac{1}{2} h_3 w^2 + \cdots)(\tfrac{1}{2} f_2 w^2 + \cdots) \\ &= A(h_1 w + \tfrac{1}{2} h_2 w^2 + \cdots) + \tfrac{1}{2} B(h_1 w + \tfrac{1}{2} h_2 w^2 + \cdots, h_1 w + \tfrac{1}{2} h_2 w^2 + \cdots), \end{aligned} \tag{9.11}$$

from which we infer

$$\frac{1}{2} f_2 = \frac{1}{2} A h_2 + \frac{1}{2} B(h_1, h_1). \tag{9.12}$$

Multiplying with a left singular vector p of A we obtain

$$f_2 = \frac{p^T B(h_1, h_1)}{p^T h_1} \tag{9.13}$$

(since $n_c = 1$ we must have $p^T h_1 \neq 0$). From the analytic investigations it is known that for a fold bifurcation the reduced system (9.6) is equivalent to the truncated system

$$\dot{w} = aw^2 \tag{9.14}$$

whenever $a = \frac{1}{2}f_2 \neq 0$. So a is the normal form coefficient of (9.6). By (9.13) we have

$$a = \frac{1}{2}p^T B(q,q) \tag{9.15}$$

if p, q are left and right singular vectors of A normalized so that $p^T q = 1$. We note that this determines a only up to a nonzero factor. The differential equation (9.14) can easily be solved, and it is clear that the qualitative behavior of the solutions is independent of the sign of a; in other words, the relevant information is whether a is zero or not (this situation is not typical for the normal form coefficients in other bifurcations). The universal unfolding of (9.14) is

$$\dot{w} = \beta_1 + aw^2, \tag{9.16}$$

where β_1 is the unfolding parameter.

In other cases the details and results of the computations are much more complicated although the underlying idea is the same. Actually, the tedious work is best done by symbolic algebra techniques. We will sketch the main results in §9.1.2, discuss numerical details for cusp and GH in §9.2 and §9.3, respectively, and describe some applications in §9.4, leaving further examples to the Exercises in §9.6.

9.1.2 Normal Form Results

All codimension-1 bifurcations of (9.2) are determined by A. So are all except two codimension-2 bifurcations. The exceptions are the cusp points (CPs, characterized by A, B) and GH points (characterized by A, B, C). However, the coefficients of the normal forms require even higher-order derivatives, in fact, up to order 5.

We now give the normal forms for Hopf and for the codimension-2 cases (partially, with references for the most complicated cases). For proofs and more information on the dynamical significance of these points we refer to [127] and [164]. We follow the custom to assume for simplicity of notation that $x_0 = 0, \alpha_0 = 0$.

Hopf. Let the dynamical system (9.1) exhibit a Hopf bifurcation at $(0,0)$; i.e., A has eigenvalues $\pm i\omega_0$, $\omega_0 > 0$. Introduce the complex vectors p, q such that

$$Aq = i\omega_0 q, \qquad p^H A = i\omega_0 p^H$$

normalized so that $p^H q = 1$ (cf. Proposition 4.3.1). Then the restriction at $\alpha = 0$ to the two-dimensional center manifold is locally smoothly orbitally equivalent to the complex normal form

$$\dot{w} = i\omega_0 w + \ell_1 w|w|^2 + O(|w|^4), \qquad w \in \mathbb{C}, \tag{9.17}$$

where the normal form coefficient ℓ_1, known as the *first Lyapunov coefficient*, is given by

$$\ell_1 = \frac{1}{2}\mathrm{Re}\, \langle p, C(q,q,\overline{q}) - 2B(q, A^{-1}B(q,\overline{q})) + B(\overline{q}, (2i\omega_0 I_N - A)^{-1} B(q,q))\rangle.$$

If $\ell_1 \neq 0$, then the $O(|w|^4)$-term in (9.17) can be deleted without destroying the equivalence. Also, the restriction of (9.1) to the α_1-dependent center manifold then behaves like

$$\dot{w} = (\beta_1 + i\omega_0)w + \ell_1 w|w|^2, \tag{9.18}$$

where β_1 is an unfolding parameter.

The first Lyapunov value largely determines the dynamic behavior of (9.1) in the neighborhood of a Hopf point. If $\ell_1 < 0$, then a family of periodic orbits, stable in the center manifold, exists nearby and reduces to a fixed point in the Hopf point. This family of orbits can be parameterized by its amplitude, which becomes zero in the Hopf point. If $\ell_1 > 0$ then a similar result holds with periodic orbits that are unstable in the center manifold.

Cusp. This is the case that generically arises if the normal form coefficient a of the fold bifurcation vanishes. We have $n_c = 1$ and p, q can be defined as in the fold case. The reduced system is

$$\dot{w} = cw^3 + O(w^4), \qquad w \in \mathbb{R}. \tag{9.19}$$

Here c is a new normal form coefficient given by

$$c = \frac{1}{6} p^T (C(q,q,q) + 3B(q,h_2)), \tag{9.20}$$

where $h_2 \in \mathbb{R}^N$ is uniquely determined by the equations

$$Ah_2 = -B(q,q), \qquad p^T h_2 = 0.$$

This system is solvable since $p^T B(q,q) = 0$; the solution may be found by solving the bordered system

$$\begin{pmatrix} A & q \\ p^T & 0 \end{pmatrix} \begin{pmatrix} h_2 \\ s \end{pmatrix} = \begin{pmatrix} -B(q,q) \\ 0 \end{pmatrix}. \tag{9.21}$$

The square matrix in (9.21) is nonsingular by (a special case of) Proposition 3.2.1 since $p^T q \neq 0$.

If $c \neq 0$ then the perturbed reduced system is equivalent to

$$\dot{w} = \beta_1 + \beta_2 w + cw^3, \tag{9.22}$$

with two unfolding parameters β_1, β_2.

We note that 0 is a stable equilibrium if $c < 0$ and an unstable equilibrium if $c > 0$. The case $c < 0$ was considered in some detail in §1.2.3.

BT. At a BT bifurcation point $n_c = 2$ and A has a double zero eigenvalue with generalized right eigenvectors q_0, q_1, i.e.,

$$Aq_0 = 0, \qquad Aq_1 = q_0, \tag{9.23}$$

and generalized left eigenvectors p_0, p_1, i.e.,

$$p_0^T A = 0, \qquad p_1^T A = p_0^T. \tag{9.24}$$

Then $p_0^T q_0 = 0$ and q_1, p_1 can be chosen so that $p_0^T q_1 = p_1^T q_0 = 1, p_1^T q_1 = 0$. (This is proved most easily by a reduction to the Jordan normal form of A.)

The reduced system at the BT point is equivalent to the normal form

$$\begin{cases} \dot{w}_0 &= w_1, \\ \dot{w}_1 &= a w_0^2 + b w_0 w_1 + O(\|\|w\|^3). \end{cases} \tag{9.25}$$

The nonlinear nondegeneracy conditions are now $a \neq 0$, $b \neq 0$. If these are satisfied, then the reduced system on the perturbed center manifold is equivalent to the normal form

$$\begin{cases} \dot{w}_0 &= w_1, \\ \dot{w}_1 &= \beta_1 + \beta_2 w_0 + a w_0^2 + b w_0 w_1. \end{cases} \tag{9.26}$$

The most striking feature of this unfolding is the presence in the parameter space of a curve of points that originates at the BT point along which (9.1) has a homoclinic connection.

The normal form coefficients a, b are given by

$$a = \frac{1}{2} p_0^T B(q_0, q_0), \tag{9.27}$$

$$b = p_1^T B(q_0, q_0) + p_0^T B(q_0, q_1). \tag{9.28}$$

GH (= Bautin). A GH point is a Hopf point where ℓ_1 vanishes. Let p, q be as in the Hopf case. A normal form for the Bautin bifurcation in the reduced coordinates is

$$\dot{w} = i\omega_0 w + \ell_2 w|w|^4 + O(|w|^6), \tag{9.29}$$

where the *second Lyapunov coefficient* ℓ_2 is real. If $\ell_2 \neq 0$ then the reduced system on the perturbed center manifold is equivalent to

$$\dot{w} = (\beta_1 + i\omega_0)w + \beta_2 w|w|^2 + \ell_2 w|w|^4 \tag{9.30}$$

with two unfolding parameters β_1, β_2. The most striking feature in a generic unfolding is the presence of a curve in the parameter space that originates at the GH point along which (9.1) has a turning point of periodic orbits; i.e., a pair of stable and unstable periodic orbits collides and disappears (see §9.4 for a numerical example).

Formulas for the computation of ℓ_2 in terms of A, B, C, D, E and q_0, q_1, p_0, p_1 are provided in [167]. They are long and complicated and should best be handled by symbolic software.

ZH and DH. These two cases are also discussed in [167]. The dynamics of their unfoldings are even more complicated than in the previous cases; there are more normal coefficients and the higher-order terms cannot be truncated without destroying the equivalence. In both cases, invariant tori, Neimark–Sacker bifurcations of periodic orbits, Shilnikov homoclinic bifurcations, and chaotic motions are generic.

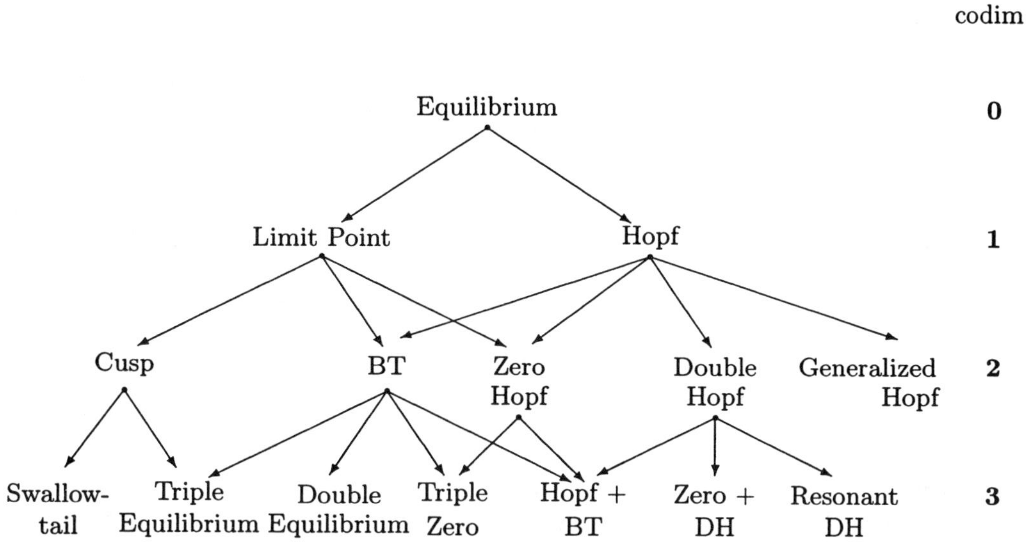

Figure 9.1: Interactions between bifurcations of codimension 0,1,2,3.

9.1.3 General Remarks on the Computation

The computation of fold and Hopf points was discussed in Chapter 4, computation of BT, ZH, and DH in Chapter 5. We did not discuss the computation of the normal form coefficients there but adding this is now a routine task. So for the codimension-2 cases only cusp and GH remain. We consider them in §9.2 and §9.3, respectively.

We will not discuss codimension-3 cases in any systematic way. However, it is clear that some can be computed easily. For example, the swallowtail singularity (ST) is found when the normal form coefficient c, defined in (9.20) for the cusp singularity, vanishes. The triple equilibrium bifurcation (ZA) is found on a curve of BT points if the coefficient b in (9.27) vanishes; generically it is the intersection of a BT curve and a cusp curve. The double equilibrium bifurcation (ZB) is found on a curve of BT points if the coefficient a in (9.27) vanishes.

In Figure 9.1 we present the generic interactions between the bifurcations of codimension 0,1,2, and some of codimension 3.

9.2 Computation of CPs

A CP is characterized by the fact that A has one zero eigenvalue, no other eigenvalues on the imaginary axis, and $p^T B(q,q) = 0$, where p, q are the left and right singular vectors of A.

9.2.1 The Manifold

It is convenient to consider A, B, C as objects in the spaces $\mathcal{M}_N$, $\mathcal{T}_N^2$, and $\mathcal{T}_N^3$ of $N \times N$ matrices, $N \times (N \times N)$ tensors, and $N \times (N \times N \times N)$ tensors, respectively. We note that the property of being a CP is fully determined by A, B. We will say that a pair $(A, B) \in \mathcal{M}_N \times \mathcal{T}_N^2$ is a CP pair if A, B satisfy the requested properties. The following result holds.

Proposition 9.2.1. The set of CP pairs is a smooth submanifold of $\mathcal{M}_N \times \mathcal{T}_N^2$ with codimension 2.

Proof. Let (A_0, B_0) be a CP pair. Let $b, c \in \mathbb{R}^N$, b not in the column range of A_0 and c not in the row range of A_0. For any $(A, B) \in \mathcal{M}_N \times \mathcal{T}_N^2$ we define

$$M(A,B) = \begin{pmatrix} A & b \\ c^T & 0 \end{pmatrix}. \tag{9.31}$$

Obviously there exists a neighborhood $\mathcal{U}$ of (A_0, B_0) in which $M(A, B)$ is nonsingular. In $\mathcal{U}$ we define the smooth functions $w(A, B), v(A, B) \in \mathbb{R}^N, g(A, B) \in \mathbb{R}$ by requiring

$$M(A,B) \begin{pmatrix} v(A,B) \\ g(A,B) \end{pmatrix} = \begin{pmatrix} 0_n \\ 1 \end{pmatrix}, \tag{9.32}$$

$$\begin{pmatrix} w^T(A,B) & g(A,B) \end{pmatrix} M(A,B) = \begin{pmatrix} 0_N^T & 1 \end{pmatrix}. \tag{9.33}$$

Now in $\mathcal{U}$ the CP pairs (A, B) are completely characterized by the two equations

$$\begin{aligned} g(A,B) &= 0, \\ w^T(A,B)B(v(A,B), v(A,B)) &= 0. \end{aligned} \tag{9.34}$$

In (9.34) all entries of A and B are independent variables; let us denote them for simplicity by the formal symbols $A(i,j)$ and $B(i,j,k)$, respectively. Then $g_{A(i,j)} = -w_i v_j$, $g_{B(i,j,k)} = 0$ and $(w^T B(v,v))_{B(i,j,k)} = w_i v_j v_k$ for all i, j, k. Since there exists at least one pair i, j for which v_i, w_j are nonzero at $(A, B) = (A_0, B_0)$, it follows that

$$\frac{\partial(g, w^T B(v,v))}{\partial(A,B)}$$

has full rank 2 at $(A, B) = (A_0, B_0)$. □

Now assume $m \geq 2$ and that at (x_0, α_0) the full Jacobian

$$\frac{\partial G}{\partial(x,\alpha)} \tag{9.35}$$

of (9.2) has full rank N so that (9.2) locally near (x_0, α_0) represents an m-dimensional manifold. This assumption will be called the *manifold condition*. Let T be a full rank $(N+m) \times m$ matrix whose columns span the kernel of (9.35). We note that T spans the tangent space to the equilibrium manifold and $(\frac{\partial G}{\partial(x,\alpha)})^T$ spans the space orthogonal to the equilibrium manifold.

Let $g_1(A,B), g_2(A,B)$ be any two functions that locally define the CP manifold regularly. We say that the *transversality condition* holds if the system

$$\begin{array}{rcl} G(x,\alpha) & = & 0, \\ g_1(A(x,\alpha),B(x,\alpha)) & = & 0, \\ g_2(A(x,\alpha),B(x,\alpha)) & = & 0 \end{array} \tag{9.36}$$

has full rank. Equivalently,

$$\frac{\partial(g_1,g_2)}{\partial(A,B)}\frac{\partial(A,B)}{\partial(x,\alpha)}T \tag{9.37}$$

has full rank. Also equivalently, the kernels of $\frac{\partial G}{\partial(x,\alpha)}$ and $\frac{\partial(g_1,g_2)}{\partial(x,\alpha)}$ intersect in a space with the minimal dimension $m-2$.

Obviously the transversality condition is independent of the choice of the two functions g_1, g_2, and of the choice of a particular method to compute CPs. On the other hand, a good computational method should lead to a regular defining system if the transversality condition holds.

9.2.2 A Minimally Extended Defining System

One method implemented in [165] follows Proposition 9.2.1 closely and therefore can be justified easily. It is essentially the same as the one proposed in §6.6.2. It uses vectors $b, c \in \mathbb{R}^N$ such that

$$M = \begin{pmatrix} A & b \\ c^T & 0 \end{pmatrix} \tag{9.38}$$

is nonsingular. We may identify b, c with the vectors used in the proof of Proposition 9.2.1. The defining system of the cusp is

$$\left\{\begin{array}{rcl} G(x,\alpha) & = & 0, \\ g(x,\alpha) & = & 0, \\ g^1(x,\alpha) & = & 0, \end{array}\right. \tag{9.39}$$

where g is obtained by solving the single bordered $(N+1)$-dimensional system

$$M\begin{pmatrix} v(x,\alpha) \\ g(x,\alpha) \end{pmatrix} = \begin{pmatrix} 0_N \\ 1 \end{pmatrix} \tag{9.40}$$

and g^1 is obtained by solving

$$M\begin{pmatrix} v^1(x,\alpha) \\ g^1(x,\alpha) \end{pmatrix} = \begin{pmatrix} -B(v,v) \\ 0 \end{pmatrix}. \tag{9.41}$$

Proposition 9.2.2. Suppose that the manifold condition is satisfied at (x_0, α_0). Then the system in (9.39) has maximal rank if and only if the transversality condition is satisfied.

Proof. Clearly the functions g, g^1 have the form $g(x,\alpha) = g_*(A(x,\alpha),\ B(x,\alpha))$, $g^1(x,\alpha) = g^1_*(A(x,\alpha),\ B(x,\alpha))$, where g_*, g^1_* are defined for all matrices in a neighborhood of the Jacobian at the CP, using formally the same equations (9.40) and (9.41). So the result is immediate from the definition of the transversality condition if g_*, g_*^1 form a regular defining system of the CP manifold. But obviously g_* is identical to the function g used in Proposition 9.2.1 and g_*^1 is identical to $-w^T B(v,v)$. □

9.2.3 A Large Defining System

In [205] a method was introduced which uses vectors $b, c \in \mathbb{R}^N$ such that

$$M = \begin{pmatrix} A & b \\ c^T & 0 \end{pmatrix} \tag{9.42}$$

is nonsingular. We may identify b, c with the vectors used in the proof of Proposition 9.2.1. The defining system of the cusp uses the unknowns x, α, q, p, where q, p are the right and left singular vectors of A in the CP:

$$\begin{cases} G(x,\alpha) & = & 0, \\ Aq & = & 0, \\ c^T q & = & 1, \\ A^T p & = & 0, \\ p^T b & = & 1, \\ p^T Bqq & = & 0. \end{cases} \tag{9.43}$$

Proposition 9.2.3. Suppose that the manifold condition is satisfied at (x_0, α_0). Then the system in (9.43) has maximal rank if and only if the transversality condition is satisfied.

Proof. Let $(z_1^T, z_2^T, V^T, W^T)^T$ be a singular vector of the Jacobian of (9.43), i.e.,

$$\begin{pmatrix} A & G_\alpha & 0 & 0 \\ Bq & G_\alpha q & A & 0 \\ 0 & 0 & c^T & 0 \\ p^T B & p^T G_{x\alpha} & 0 & A^T \\ 0 & 0 & 0 & b^T \\ p^T G_{xxx} qq & p^T G_{xx\alpha} qq & 2p^T Bq & (Bqq)^T \end{pmatrix} \begin{pmatrix} z_1 \\ z_2 \\ V \\ W \end{pmatrix} = 0. \tag{9.44}$$

The first block row in (9.44) is equivalent to the condition $z = (z_1^T, z_2^T)^T \in T$. The second and third block rows together are equivalent to the condition

$$\begin{pmatrix} A & b \\ c^T & 0 \end{pmatrix} \begin{pmatrix} V \\ 0 \end{pmatrix} = -\begin{pmatrix} Bqz_1 + G_{x\alpha} q z_2 \\ 0 \end{pmatrix}. \tag{9.45}$$

The fourth and fifth block rows of (9.44) together are equivalent to the condition

$$\begin{pmatrix} A & b \\ c^T & 0 \end{pmatrix}^T \begin{pmatrix} W \\ 0 \end{pmatrix} = -\begin{pmatrix} (p^T B z_1 + p^T G_{x\alpha} z_2)^T \\ 0 \end{pmatrix}. \tag{9.46}$$

If z denotes any variable in (x, α), then from (9.32) we derive

$$\begin{pmatrix} A & b \\ c^T & 0 \end{pmatrix} \begin{pmatrix} q_z \\ g_z \end{pmatrix} = -\begin{pmatrix} A_z q \\ 0 \end{pmatrix}. \tag{9.47}$$

Hence

$$\begin{pmatrix} A & b \\ c^T & 0 \end{pmatrix} \begin{pmatrix} q_x z_1 + q_\alpha z_2 \\ g_x z_1 + g_\alpha z_2 \end{pmatrix} = - \begin{pmatrix} Bqz_1 + G_{x\alpha} q z_2 \\ 0 \end{pmatrix}. \tag{9.48}$$

From (9.46) and (9.48) it follows that

$$V = q_x z_1 + q_\alpha z_2, \qquad 0 = g_x z_1 + g_\alpha z_2. \tag{9.49}$$

If z denotes any variable in (x, α) then from (9.33) we derive

$$\begin{pmatrix} A & b \\ c^T & 0 \end{pmatrix}^T \begin{pmatrix} p_z \\ g_z \end{pmatrix} = - \begin{pmatrix} A_z^T p \\ 0 \end{pmatrix}. \tag{9.50}$$

Hence

$$\begin{pmatrix} A & b \\ c^T & 0 \end{pmatrix}^T \begin{pmatrix} p_x z_1 + p_\alpha z_2 \\ g_x z_1 + g_\alpha z_2 \end{pmatrix} = - \begin{pmatrix} (p^T B^T z_1 + p^T G_{x\alpha} z_2)^T \\ 0 \end{pmatrix}. \tag{9.51}$$

From (9.46) and (9.51) it follows that

$$W = p_x z_1 + p_\alpha z_2, \qquad 0 = g_x z_1 + g_\alpha z_2. \tag{9.52}$$

Both (9.49) and (9.52) imply that

$$\frac{\partial g}{\partial(x, \alpha)} z = 0. \tag{9.53}$$

Combining the equality expressed in the last block row of (9.44) with (9.49) and (9.52) we obtain

$$p^T \frac{\partial B}{\partial(x, \alpha)} zqq + 2p^T Bq \frac{\partial q}{\partial(x, \alpha)} z + \left(\frac{\partial p}{\partial(x, \alpha)} z \right)^T Bqq = 0, \tag{9.54}$$

i.e.,

$$\frac{\partial(p^T Bqq)}{\partial(x, \alpha)} z = 0. \tag{9.55}$$

Now the linearized system of (9.43) has maximal rank if and only if the null spaces of

$$\frac{\partial G}{\partial(x, \alpha)}$$

and

$$\frac{\partial(g, p^T Bqq)}{\partial(x, \alpha)}$$

have an intersection with minimal dimension $m - 2$. This is precisely the transversality condition. □

9.3 Computation of GH Points

9.3.1 The Manifold

The property of being a GH point is fully determined by A, B, C. We say that a triplet $(A, B, C) \in \mathcal{M}_N \times \mathcal{T}_N^2 \times \mathcal{T}_N^3$ is a GH pair if A, B, C satisfy the requested properties. The following result holds.

Proposition 9.3.1. The set of GH triplets is a smooth submanifold of $\mathcal{M}_N \times \mathcal{T}_N^2 \times \mathcal{T}_N^3$ with codimension 2.

Proof. Let (A_0, B_0, C_0) be a GH triplet. Let $b, c \in \mathbb{R}^{b(N)}$, $b(N) = N(N-1)/2$, b not in the column range of $2A_0 \odot I_N$ and c not in the row range of $2A_0 \odot I_N$. For any $(A, B, C) \in \mathcal{M}_N \times \mathcal{T}_N^2 \times \mathcal{T}_N^3$ we define

$$M(A, B, C) = \begin{pmatrix} 2A \odot I_N & b \\ c^T & 0 \end{pmatrix}. \tag{9.56}$$

There exists a neighborhood $\mathcal{U}$ of (A_0, B_0, C_0) in which $M(A, B, C)$ is nonsingular. In $\mathcal{U}$ we define the smooth functions $w(A, B, C), v(A, B, C) \in \mathbb{R}^{b(N)}, g(A, B, C) \in \mathbb{R}$ by requiring

$$M(A, B, C) \begin{pmatrix} v(A, B, C) \\ g(A, B, C) \end{pmatrix} = \begin{pmatrix} 0_{N_b} \\ 1 \end{pmatrix}, \tag{9.57}$$

$$\begin{pmatrix} w^T(A, B, C) & g(A, B, C) \end{pmatrix} M(A, B, C) = \begin{pmatrix} 0_{N_b}^T & 1 \end{pmatrix}. \tag{9.58}$$

Let $q_0 = q_{01} + iq_{02}$ be a right eigenvector of A_0 for the eigenvalue $i\omega_0$. Since $i\omega_0$ is algebraically simple there exist smooth functions $\eta(A) + i\omega(A) \in \mathbb{C}$, $q_1(A) + iq_2(A) \in \mathbb{C}^N$, $p_1(A) + ip_2(A) \in \mathbb{C}^N$, such that $\eta(A_0) = 0, \omega(A_0) = \omega_0, q(A_0) = q_0$, and

$$\begin{array}{cc} A\, q(A) = (\eta(A) + i\,\omega(A))\, q(A), & A^T p(A) = (\eta(A) - i\,\omega(A))\, p(A), \\ \langle p(A), q(A) \rangle = 1, & \langle q_0, q(A) \rangle = 1. \end{array} \tag{9.59}$$

Now in $\mathcal{U}$ the GH triplets (A, B, C) are characterized by the two equations

$$\begin{array}{rcl} g(A, B, C) & = & 0, \\ \ell_1^*(A, B, C) & = & 0, \end{array} \tag{9.60}$$

where

$$\ell_1^*(A, B, C) = \mathrm{Re}\, \langle p, C(q, q, \overline{q}) - 2B(q, A^{-1}B(q, \overline{q})) + B(\overline{q}, (2i\omega I_N - A)^{-1} B(q, q)) \rangle \tag{9.61}$$

(ω, p, q being functions of A). By (9.18) we have $\ell_1^* = 2\ell_1$ in the Hopf point; in other points ℓ_1^* need not be related to the cubic normal form coefficient in (9.18).

In (9.60) all entries of A, B, C are independent variables; we denote them by formal symbols like $A(i, j), B(i, j, k), C(i, j, k, l)$, respectively.

For $1 \leq i, j \leq N$ we have $g_{A(i,j)} = -w^T(2A_{ij} \odot I_N)v$. We note that $\Sigma_{i=1}^N g_{A(i,i)} = -w^T(2I_N \odot I_N)v = -w^T v$. Since 0 is an algebraically simple eigenvalue of $2A \odot I_N$ it follows that $w^T v \neq 0$ and hence there exists an i such that $g_{A(i,i)} \neq 0$. Clearly $g_{B(i,j,k)} = 0$ and $g_{C(i,j,k,l)} = 0$ in all cases.

Next we note that

$$\frac{\partial \ell_1^*}{\partial C(i,j,k,l)} = \text{Re } (p_i q_j q_k \overline{q}_l).$$

Choosing $k = l$ so that $q_k \neq 0$ we find

$$\frac{\partial \ell_1^*}{\partial C(i,j,k,k)} = |q_k|^2 (p_{i1} q_{j1} - p_{i2} q_{j2}).$$

Since q cannot be a real vector we can choose a j such that $q_{j2} \neq 0$. Then

$$\frac{\partial \ell_1^*}{\partial C(i,j,k,k)} = |q_k|^2 q_{j2} (p_{i1}(q_{j1}/q_{j2}) - p_{i2}).$$

Since the real and imaginary components of p cannot be proportional, it follows that there exists an i with $\frac{\partial \ell_1^*}{\partial C(i,j,k,k)} \neq 0$. Hence

$$\frac{\partial(g, \ell_1^*)}{\partial(A,B,C)}$$

has full rank 2 at $(A,B,C) = (A_0, B_0, C_0)$. □

Now assume that $m \geq 2$ and that the manifold condition holds as in §9.2.2.

Let $g_1(A,B,C), g_2(A,B,C)$ be any two functions that locally define the GH manifold regularly. We say that the *transversality condition* holds if the system

$$\begin{aligned} G(x,\alpha) &= 0, \\ g_1(A(x,\alpha), B(x,\alpha), C(x,\alpha)) &= 0, \\ g_2(A(x,\alpha), B(x,\alpha), C(x,\alpha)) &= 0 \end{aligned} \tag{9.62}$$

has full rank. Equivalently,

$$\frac{\partial(g_1,g_2)}{\partial(A,B,C)} \frac{\partial(A,B,C)}{\partial(x,\alpha)}^T \tag{9.63}$$

has full rank. Also equivalently, the kernels of $\frac{\partial G}{\partial(x,\alpha)}$ and $\frac{\partial(g_1,g_2)}{\partial(x,\alpha)}$ intersect in a space with the minimal dimension $m-2$.

Obviously the transversality condtition is independent of the choice of the two functions g_1, g_2 and of the choice of a particular method to compute GH points. On the other hand, a good computational method should lead to a regular defining system if the transversality condition holds.

9.3.2 A Minimally Extended Defining System

We discuss two methods for GH points; both are implemented in CONTENT [165]. One of these uses only the minimal number $N+2$ of defining equations (the number of state variables plus the codimension of the bifurcation). At a GH point the bialternate product matrix $2A \odot I_N$ is singular. The vectors $v_{1b}, v_{2b}, w_{1b}, w_{2b} \in \mathbb{R}^{b(N)}$ and scalars d_{12}, d_{21} are chosen so that the matrix

$$M_b = \begin{pmatrix} 2A \odot I_N & w_{1b} & w_{2b} \\ v_{1b}^T & 0 & d_{12} \\ v_{2b}^T & d_{21} & 0 \end{pmatrix} \tag{9.64}$$

is nonsingular.

The defining equations for GH points are

$$\left\{ \begin{array}{rcl} G(x,\alpha) & = & 0, \\ \det G & = & 0, \\ \ell_1^* & = & 0, \end{array} \right. \tag{9.65}$$

where the matrix

$$G = \begin{pmatrix} g_{11} & g_{12} \\ g_{21} & g_{22} \end{pmatrix} \tag{9.66}$$

is obtained by solving the system

$$M_b \begin{pmatrix} V \\ G \end{pmatrix} = \begin{pmatrix} 0_{b(N)} \\ I_2 \end{pmatrix}. \tag{9.67}$$

We note that in (9.64) a doubly bordered extension of $2A \odot I_N$ is used instead of the singly bordered extension in the proof of Proposition 9.3.1. Indeed, on curves of GH points DH points (two pairs of pure imaginary eigenvalues) can be expected generically; in such points $2A \odot I_N$ has rank defect 2 and a singly bordered extension is necessarily singular, causing a breakdown in the defining system for GH. With two borders (wisely adapted during the continuation procedure) this can be avoided.

In (9.65) ℓ_1^* is defined by (9.61). The scalar $\omega \in \mathbb{C}$ and vectors $p, q \in \mathbb{C}^N$ are obtained by linear operations based on the solution of linear systems with (9.64) and its transpose as in §4.5.

Proposition 9.3.2. Suppose that the manifold condition is satisfied at (x_0, α_0). Then the system in (9.65) has maximal rank if and only if the transversality condition is satisfied.

Proof. The numerical method is close to the proof of Proposition 9.3.1, as is best seen from the similarity of (9.60) and (9.65), which both contain the condition $\ell_1^* = 0$. To relate the function $g(A, B, C)$ in (9.60) to the function $\det G$ in (9.65) we note that by (9.57) and (9.67)

$$\det(2A \odot I_N) = (\det M)g = (\det M_b)(\det G).$$

Taking derivatives with respect to any variable z and evaluating at a Hopf point ($g = \det G = 0$) we find that

$$(\det(2A \odot I_N))_z = (\det M)g_z = (\det M_b)(\det G)_z.$$

Since $\det M$ and $\det M_b$ are nonzero, it follows that g vanishes if and only if $\det G$ vanishes and that g_z vanishes if and only if $(\det G)_z$ vanishes. Hence the two functions $\det(G)$ and ℓ_1^* form a regular defining system for the GH manifold. □

9.3.3 A Large Defining System

The minimally extended system for GH points is based directly on the mathematical theory and can be implemented fairly easily. However, implementation of symbolic or automatic differentiation of ℓ_1^* is hard. We therefore present another method that can use symbolic derivatives of order up to 4. The price paid for that is the size of the system. The number of variables is $8N+5$.

The idea is to express explicitly that A has an imaginary eigenvalue $i\omega$ with right eigenvector $q \in \mathbb{C}^N$ and left eigenvector $p \in \mathbb{C}^N$ and then to add the condition that ℓ_1^* vanishes. To fix the right and left eigenvectors we add the normalization conditions $\langle q_0, q\rangle = \langle p, q\rangle = 1$, where $q_0 \in \mathbb{C}^N$ is the normalized right eigenvector q at a previously computed point on the curve. To simplify formally the expression for ℓ_1^*, we introduce $v \in \mathbb{R}^N, w \in \mathbb{C}^N$ as additional unknowns, where

$$v = A^{-1}B(q,\overline{q}), \qquad w = (2\,i\,\omega\,I_N - A)^{-1}\,B(q,q).$$

Thus, the unknowns of the system are the components of

$$(x, \alpha, q, \omega, p, \lambda, v, w).$$

The defining equations for GH are given by the complex system:

$$\left\{\begin{array}{rcl} G(x,\alpha) &=& 0,\\ Aq - i\omega q &=& 0,\\ A^T p + \lambda p &=& 0,\\ \langle q_0, q\rangle - 1 &=& 0,\\ \langle p, q\rangle - 1 &=& 0,\\ Av - B(q,\overline{q}) &=& 0,\\ (2i\omega I_n - A)w - B(q,q) &=& 0,\\ \mathrm{Re}\,\langle p, C(q,q,\overline{q}) - 2B(q,v) + B(\overline{q},w)\rangle &=& 0. \end{array}\right. \tag{9.68}$$

The complex variable λ is introduced artificially to regularize the system; formally, along the GH curve, $\lambda = i\omega$.

In the real form we decompose $q = q_1 + iq_2$, $p = p_1 + ip_2$, $\lambda = \lambda_1 + i\lambda_2$, $w = w_1 + iw_2$. Also the second, third, fourth, fifth, and seventh equations in (9.68) are decomposed into separate equations for the real and imaginary parts. In this way we obtain a real system with $8n+5$ equations and $8N+3+m$ variables.

Proposition 9.3.3. Suppose that the manifold condition is satisfied at (x_0, α_0). Then the real form of the system in (9.68) has maximal rank if and only if the transversality condition is satisfied.

Proof. In the real form the variables of the defining system for GH are

$$(x, \alpha, q_1, q_2, \omega, p_1, p_2, \lambda_1, \lambda_2, v, w_1, w_2)$$

and the corresponding Jacobian $J_{\max}$ has the form

$$\begin{pmatrix}
A & G_\alpha & 0 & 0 & 0 & 0 & 0 & 0 & 0 & 0 & 0 & 0 \\
Bq_1 & G_{x\alpha}q_1 & A & \omega I_n & q_2 & 0 & 0 & 0 & 0 & 0 & 0 & 0 \\
Bq_2 & G_{x\alpha}q_2 & -\omega I_N & A & -q_1 & 0 & 0 & 0 & 0 & 0 & 0 & 0 \\
p_1^T B & p_1^T G_{x\alpha} & 0 & 0 & 0 & J_{46} & -\lambda_2 I_N & p_1 & -p_2 & 0 & 0 & 0 \\
p_2^T B & p_2^T G_{x\alpha} & 0 & 0 & 0 & \lambda_2 I_N & J_{57} & p_2 & p_1 & 0 & 0 & 0 \\
0 & 0 & q_{01}^T & q_{02}^T & 0 & 0 & 0 & 0 & 0 & 0 & 0 & 0 \\
0 & 0 & -q_{02}^T & q_{01}^T & 0 & 0 & 0 & 0 & 0 & 0 & 0 & 0 \\
0 & 0 & p_1^T & p_2^T & 0 & q_1^T & q_2^T & 0 & 0 & 0 & 0 & 0 \\
0 & 0 & -p_2^T & p_1^T & 0 & q_2^T & -q_1^T & 0 & 0 & 0 & 0 & 0 \\
J_{10,1} & J_{10,2} & -2Bq_1 & -2Bq_2 & 0 & 0 & 0 & 0 & 0 & A & 0 & 0 \\
J_{11,1} & J_{11,2} & -2Bq_1 & 2Bq_2 & -2w_2 & 0 & 0 & 0 & 0 & 0 & -A & -2\omega I_N \\
J_{12,1} & J_{12,2} & -2Bq_2 & -2Bq_1 & -2w_1 & 0 & 0 & 0 & 0 & 0 & 2\omega I_N & -A \\
J_{13,1} & J_{13,2} & J_{13,3} & J_{13,4} & 0 & \mathcal{R} & \mathcal{I} & 0 & 0 & J_{13,10} & J_{13,11} & J_{13,12}
\end{pmatrix}, \qquad (9.69)$$

where $J_{46} = J_{57} = A^T + \lambda_1 I_N$, $J_{10,1} = Bv - G_{xxx}(q_1, q_1) - G_{xxx}(q_2, q_2)$, $J_{10,2} = G_{x\alpha}v - G_{xx\alpha}(q_1, q_1) - G_{xx\alpha}(q_2, q_2)$, $J_{11,1} = -Bw_1 - G_{xxx}(q_1, q_1) + G_{xxx}(q_2, q_2)$, $J_{11,2} = -G_{x\alpha}w_1 - G_{xx\alpha}(q_1, q_1) + G_{xx\alpha}(q_2, q_2)$, $J_{12,1} = -Bw_2 - 2G_{xxx}(q_1, q_2)$, $J_{12,2} = -G_{x\alpha}w_2 - 2G_{xx\alpha}(q_1, q_2)$, $\mathcal{R} = C(q_1, q_1, q_1) + C(q_1, q_2, q_2) - 2B(q_1, v) + B(q_1, w_1) + B(q_2, w_2)$, $\mathcal{I} = C(q_2, q_2, q_2) + C(q_1, q_1, q_2) - 2B(q_2, v) + B(q_1, w_2) - B(q_2, w_1)$, $J_{13,1} = p_1\mathcal{R}_x + p_2\mathcal{I}_x$, $J_{13,2} = p_1\mathcal{R}_\alpha + p_2\mathcal{I}_\alpha$, $J_{13,3} = p_1\mathcal{R}_{q_1} + p_2\mathcal{I}_{q_1}$, $J_{13,4} = p_1\mathcal{R}_{q_2} + p_2\mathcal{I}_{q_2}$, $J_{13,10} = p_1\mathcal{R}_v + p_2\mathcal{I}_v$, $J_{13,11} = p_1\mathcal{R}_{w_1} + p_2\mathcal{I}_{w_1}$, $J_{13,12} = p_1\mathcal{R}_{w_2} + p_2\mathcal{I}_{w_2}$.

Let $Z = (z_1^T, z_2^T, Q_1^T, Q_2^T, \Omega, P_1^T, P_2^T, \Lambda_1, \Lambda_2, V^T, W_1^T, W_2^T)^T$ be a singular vector of the Jacobian of (9.68), i.e.,

$$J_{\max} Z = 0. \qquad (9.70)$$

The first block row in (9.70) is equivalent to the condition $z = (z_1^T, z_2^T)^T \in T$. The second, third, sixth, and seventh block rows together are equivalent to

$$\begin{pmatrix} A & \omega I_N & q_2 \\ -\omega I_N & A & -q_1 \\ q_{01}^T & q_{02}^T & 0 \\ -q_{02}^T & q_{01}^T & 0 \end{pmatrix} \begin{pmatrix} Q_1 \\ Q_2 \\ \Omega \end{pmatrix} = \begin{pmatrix} B(q_1, z_1) + G_{x\alpha}q_1 z_2 \\ B(q_2, z_1) + G_{x\alpha}q_2 z_2 \\ 0 \\ 0 \end{pmatrix}. \qquad (9.71)$$

This system has full rank $2N+1$ at the GH point with left singular vector $(p_1^T, p_2^T, 0, 0)$. It is solvable if and only if

$$p_1^T(B(q_1, z_1) + G_{x\alpha}q_1 z_2) + p_2^T(B(q_2, z_1) + G_{x\alpha}q_2 z_2) = 0 \qquad (9.72)$$

or, equivalently, if and only if

$$\left\langle \mathrm{Re}\left\langle p, \frac{\partial A}{\partial(x, \alpha)} q \right\rangle, z \right\rangle = 0. \qquad (9.73)$$

This condition is linear in z. If it is satisfied, then Q_1, Q_2, Ω are uniquely defined by (9.71).

Next, the fourth, fifth, eighth, and ninth block rows of (9.70) together form a nonsingular square linear system in $(P_1^T, P_2^T, \Lambda_1, \Lambda_2)$ if $z \in T$ satisfies (9.73) and Q_1, Q_2, Ω

satisfy (9.71). Similarly, the tenth block row uniquely defines V and the eleventh and twelfth block rows uniquely define W_1, W_2.

The last row of (9.70) is equivalent to the condition

$$\frac{\partial \ell_1^*}{\partial(x,\alpha)} z = 0. \tag{9.74}$$

It follows from Proposition 4.3.2 that $\frac{\partial \eta}{\partial(x,\alpha)} = \text{Re } \langle p, \frac{\partial A}{\partial(x,\alpha)} q\rangle$. Also, by Proposition 4.4.16 and the definition of $g(A,B,C)$ in (9.57) one has $\eta(A,B,C) = h(A,B,C)g(A,B,C)$, where $h(A,B,C)$ is a smooth function in a neighborhood of the GH point and nonzero in the GH point itself. Since the two functions $\eta(A,B,C), \ell_1^*(A,B,C)$ together form a regular defining system for the GH manifold, the result follows. □

9.4 Examples

9.4.1 A Turning Point of Periodic Orbits in the Hodgkin–Huxley Model

The generic two-parameter unfolding of a GH bifurcation point is well known; for descriptions and further references see [127, §7.1] and [164, §8.3.2]. We recall that a Hopf curve passes through the GH point with ℓ_1^* changing sign at the GH point. On the side with $\ell_1^* < 0$, stable periodic orbits are born in the direction where the real part η of the critical eigenvalues is positive. On the side with $\ell_1^* > 0$, unstable periodic orbits are born in the direction where η becomes negative. Further details of the bifurcation diagram depend on the sign of ℓ_2.

The case with $\ell_2 < 0$ is represented in Figure 9.2. There are stable equilibria in the regions where $\eta < 0$ and unstable equilibria where $\eta > 0$. The stable periodic orbits continue to live on the side where $\beta > 0$ and $\eta < 0$. So there is a region in parameter space where a stable periodic orbit, an unstable periodic orbit, and a stable equilibrium exist together. There is also a boundary curve in parameter space, (T in Figure 9.2) along which the stable and unstable periodic orbits coalesce and disappear, i.e., where a turning point bifurcation of periodic orbits occurs.

If $\ell_2 > 0$, then a similar (or rather dual) phenomenon occurs in the part where $\beta < 0$ and $\eta > 0$; there is a region where an unstable equilibrium exists inside a stable periodic orbit inside an unstable periodic orbit. Again there is a boundary curve in parameter space where the stable and unstable periodic orbits coalesce and disappear. See Figure 9.3.

We illustrate this by some experiments in the Hodgkin–Huxley model that was first discussed in §4.6.2. We consider the bifurcation diagram in Figure 4.2; state and parameter values in the GH point are given in Table 4.5. Starting again with the values in Table 4.4 and computing an equilibrium curve with free parameter I we detect the Hopf point in the lower left corner of Figure 4.2 and find the normal form coefficients $\omega = 0.586234$ and $\ell_1^* = 0.0295828$. We compute again the Hopf curve through this point with free parameters I, T. We detect the GH point in Figure 4.2 again and find that it has normal form

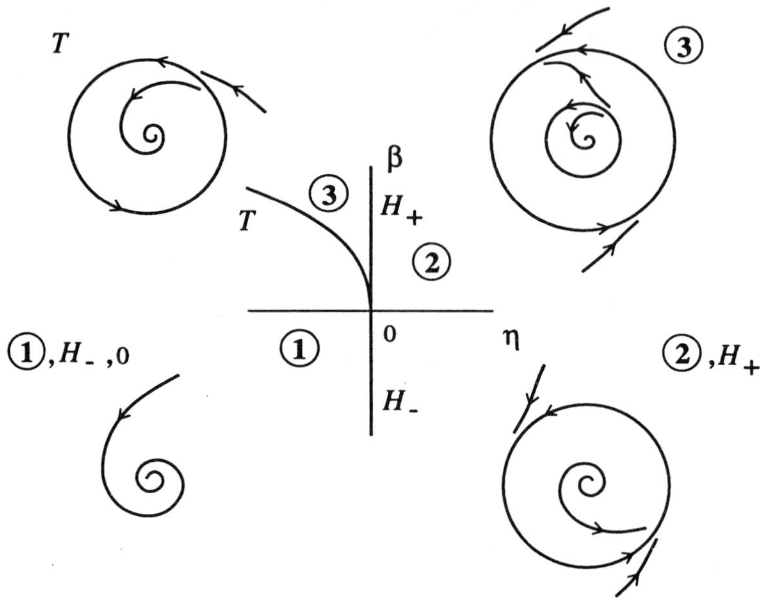

Figure 9.2: Stable and unstable equilibria and periodic orbits near a GH point with $\ell_2 < 0$.

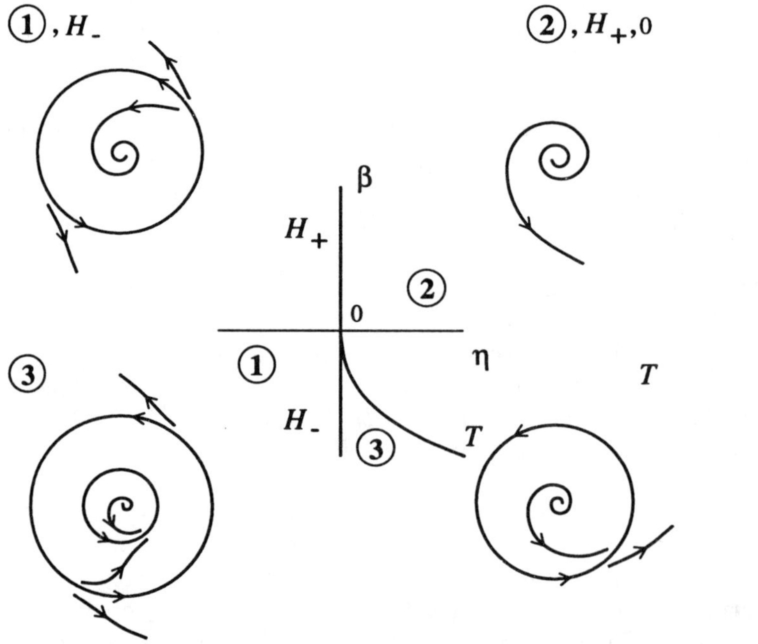

Figure 9.3: Stable and unstable equilibria and periodic orbits near a GH point with $\ell_2 > 0$.

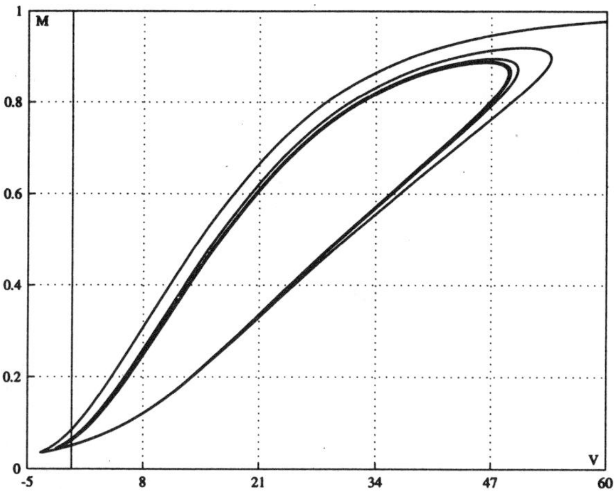

Figure 9.4: Convergence to a stable periodic orbit in the Hodgkin–Huxley model.

coefficients $\omega = 4.78029$ and $\ell_2 = -0.0077938876$. We note that only the sign of ℓ_2 matters and the exact value depends on the normalization of q. For the sake of completeness we note that the components of q were $-0.9992244609 - 2.1973762505i$, $-0.0304236265 - 0.0441721769i$, $-0.0171499218 - 0.0018090256i$, $0.0181868045 + 0.0017671412i$. The high-order derivatives in the computation of ℓ_2 were generated symbolically using MAPLE.

We then consider (fairly arbitrarily) a point P_1 on the Hopf curve prior to the GH point with $V = 11.1758$, $M = 0.177261$, $N = 0.493693$, $H = 0.232554$, $I = 33.6114$, $T = 25$. We have not yet passed the GH point, so the first Lyapunov coefficient is still positive and a hard loss of stability must occur at this Hopf point. To test eigenvalues, we then compute an equilibrium curve through the last computed Hopf point with T as a free parameter. We find a stable equilibrium for $V = 11.1758$, $M = 0.177261$, $N = 0.493693$, $H = 0.232554$, $I = 33.6114$, $T = 26.4763$ with eigenvalues -28.2322, $-0.116236 \pm 3.04341i$, and -1.65659. Starting with the perturbed values $V = 60, M = 1$ the orbit converges to a stable periodic orbit; see Figure 9.4; starting from the perturbed values $V = 13, M = 0.2$ the orbit converges to the stable equilibrium; see Figure 9.5 (note the difference in scale: the second orbit starts much closer to the equilibrium point).

From the unfolding analysis of GH points we now expect that for the same parameter values there also will be an unstable periodic orbit in the perturbed center manifold that forms the boundary between the domain of attraction of the stable equilibrium and the domain of attraction of the stable periodic orbit. The unstable orbit can be started from an Hopf bifurcation point but cannot be computed by simple orbit simulation. It can, however, be computed as the solution to a boundary value problem (cf. §4.7). In Figure 9.6 we show the result of such experiment done in CONTENT. We start from P_1 and compute the curve of periodic orbits with free parameter T. Figure 9.6 presents traces of points on the periodic orbits for increasing values of T. The parabolalike form of the projected manifold of periodic orbits near the Hopf point is clearly visible and so is the turning points of the curve of orbits for $T = 28.299$.

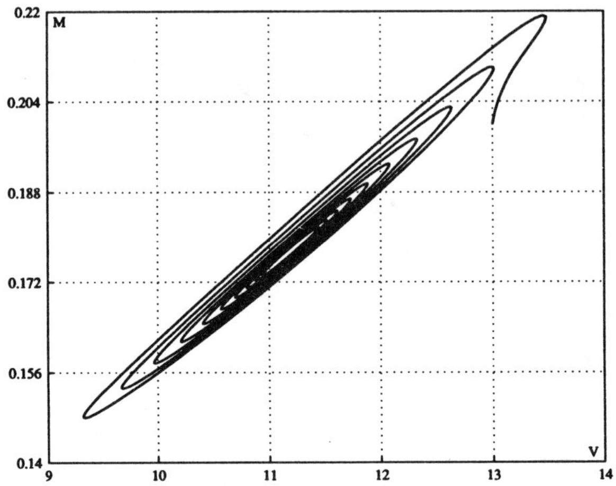

Figure 9.5: Convergence to a stable equilibrium in the Hodgkin–Huxley model.

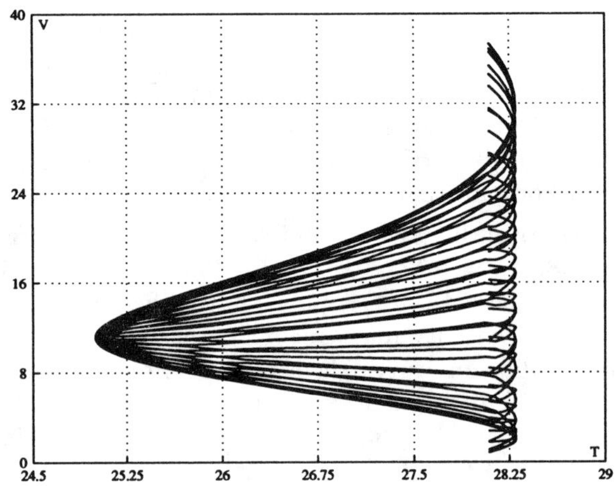

Figure 9.6: A turning point of periodic orbits in the Hodgkin–Huxley model.

9.4.2 Bifurcations with High Codimension in the LP-Neuron Model

We consider the LP-neuron model introduced in §5.4. By following paths of equilibria and fold points it is not hard to detect a CP. Coordinates of a CP are given in Table 9.1.

Starting from this point and with free parameters $I_{ext}, \bar{g}_{Af}, \bar{g}_{K(Ca)}$ a curve of CPs was computed and presented in Figure 9.7. This curve contains two swallowtail bifurcations (ST) and a triple equilibrium (ZA) (see §9.1.3). As expected, the projection of the curve of CPs on a two-parameter plane $(I_{ext}, \bar{g}_{Af})$ has the form of a cusp curve in the ST points.

Next we consider a curve of BT points; the coordinates of the starting point are given

Table 9.1: State variables and I_{ext}, $\bar{g}_{Af}$, $\bar{g}_{K(Ca)}$ at a CP in the LP-neuron model.

	State	Value	Param	Value
1	v	−23.13949	I_{ext}	0.1373037
2	h	0.01126695	$\bar{g}_{Af}$	−11.25685
3	Ca	0.2042418	$\bar{g}_{K(Ca)}$	5.260545
4	a_{Ca1}	0.1500495		
5	a_{Ca2}	0.001580395		
6	b_{Ca1}	0.03364835		
7	n	0.5273332		
8	$a_{K(Ca)}$	0.0176201		
9	$b_{K(Ca)}$	0.7460443		
10	a_A	0.7757871		
11	b_{Af}	0.001536437		
12	b_{As}	0.001536437		
13	a_h	0.0007179619		

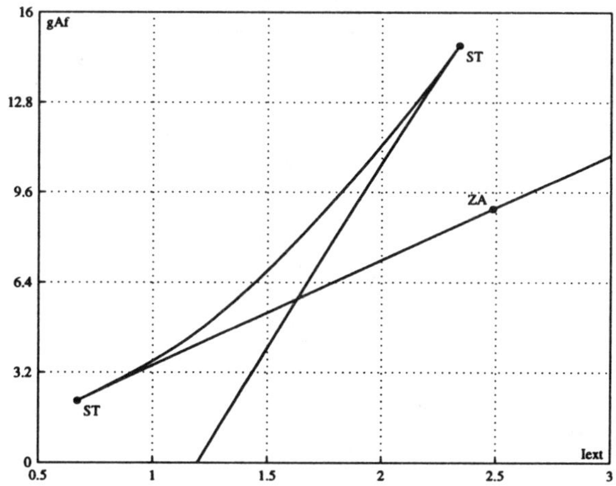

Figure 9.7: A curve of CPs in the LP-neuron model.

in Table 9.2. A global picture is presented in Figure 9.8. It contains a point with Hopf + BT bifurcation as well as a ZA and a ZB point. All these points are on the stability boundary; i.e., there are no eigenvalues with positive real part. Since the ZA and ZB points are very close we present a more detailed picture in Figure 9.9.

9.4.3 Dynamics of Corruption in Democratic Societies

The economics of supply and demand of bribes was pioneered in [208]; a survey of later work is given in [219]. We will discuss a dynamic model, introduced in [202] for the social phenomenon of political corruption in a democratic society. We discuss some numerical details of the study to show that most of the interesting dynamics can be explained qualitatively by the presence of a GH point in a typical two-parameter unfolding, apparently

Table 9.2: State variables and I_{ext}, $\bar{g}_{Af}$, $\bar{g}_{K(Ca)}$ at a BT point in the LP-neuron model.

	State	Value	Param	Value
1	v	−49.02293	I_{ext}	3.398637
2	h	0.6646423	$\bar{g}_{Af}$	12.96179
3	Ca	0.1179416	$\bar{g}_{K(Ca)}$	5
4	a_{Ca1}	0.004355952		
5	a_{Ca2}	3.922597E-5		
6	b_{Ca1}	0.4695044		
7	n	0.1957435		
8	$a_{K(Ca)}$	0.000130097		
9	$b_{K(Ca)}$	0.8357226		
10	a_A	0.4069875		
11	b_{Af}	0.1031369		
12	b_{As}	0.1031369		
13	a_h	0.006172727		

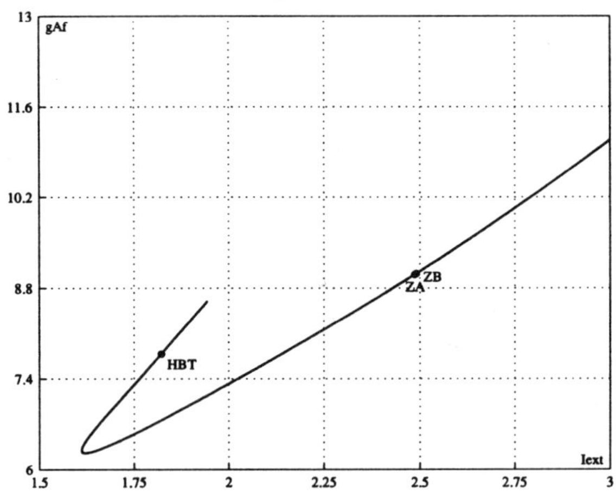

Figure 9.8: A curve of BT points in the LP-neuron model.

just a local phenomenon. For those interested in more background we refer to [91] and [202].

In the model the state variables are x, y, z with the following definitions.

- $x(t)$ = the public support (popularity, a proxy for power) of politicians at time t.
- $y(t)$ = the hidden assets that the corrupt politicians hold at time t.
- $z(t)$ = the investigation effort at time t, i.e., the sum of the activities by the police, courts, the press, and some individuals to reveal and punish corruption.

These variables could be measured in several ways and one has to make some assumptions

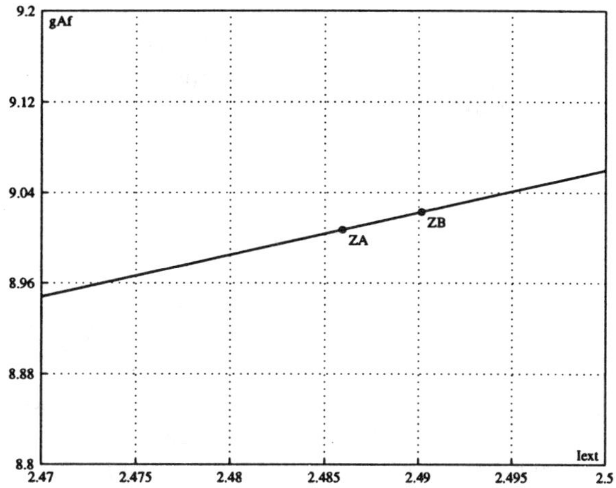

Figure 9.9: A zoom of Figure 9.8.

on the way that they influence each other. In [202] this leads to a dynamical system

$$\left\{\begin{array}{lcl} \dot{x} & = & x\mu^+(\frac{\alpha}{\beta+x} - k) - \mu^-\gamma xyz, \\ \dot{y} & = & y(\epsilon x - \gamma z - \rho), \\ \dot{z} & = & z(\sigma\gamma y - \delta), \end{array}\right. \tag{9.75}$$

where (x, y, z) has to be in the nonnegative octant (this is an invariant set). The 10 parameters have to be strictly positive, and a typical set of values is given by

$$\begin{array}{lllll} \mu^+ = 1, & \mu^- = 10, & k = 1, & \alpha = 1.5, & \beta = 0.5, \\ \rho = 0.1, & \epsilon = 0.4, & \delta = 2, & \gamma = 1, & \sigma = 2. \end{array} \tag{9.76}$$

We first note that the case where $\alpha - \beta k < 0$ is simple: (9.75) has a unique stable equilibrium at $(0, 0, 0)$ and all orbits converge to it. This case (anarchy) does not present mathematically interesting features, so we dismiss it. We further assume $\alpha - \beta k > 0$.

Now define $x^\star = \frac{\alpha-\beta k}{k} > 0$. We distinguish two further cases. If

$$\epsilon < \frac{\rho k}{\alpha - \beta k} \tag{9.77}$$

or, equivalently, $\epsilon x^\star < \rho$, then $(x^\star, 0, 0)$ is a stable equilibrium of (9.75) and all orbits that start in the positive octant converge to it. In [202] this is for obvious reasons called the uncorrupted situation.

The mathematically most interesting case is

$$\epsilon > \frac{\rho k}{\alpha - \beta k}. \tag{9.78}$$

In this situation the domain

$$0 \leq x \leq x^\star, \quad 0 \leq y, \quad 0 \leq z \tag{9.79}$$

is invariant under the dynamics of (9.75) and we can restrict to it. Furthermore, there is an unstable equilibrium at $(x^\star, 0, 0)$ and a heteroclinic loop consisting of three trajectories, namely, along the intervals

$$(0, 0, +\infty) \to (0, 0, 0) \to (x^\star, 0, 0) \to (x^\star, +\infty, 0) \tag{9.80}$$

and the numerical experiments all suggest strongly that this is an attracting heteroclinic loop. For example, for the parameter values in (9.76) all orbits starting in the interior of (9.79) converge to it. This means that they present a rather wild behavior, spending a long time in the "period of stagnation" near $(0, 0, 0)$, followed by an increase in popularity (closely following the x-axis) and a sudden increase in corruption when x is close to $x^\star$ (nearly parallel to the y-axis); then a collapse of both popularity and corruption is caused by an increased value of z and the system returns to a period of stagnation. This process repeates itself but is not periodic because the amplitude increases and the orbit moves toward the heteroclinic loop.

Under the assumption (9.78) there is always a unique equilibrium $E^+ = (x^+, y^+, z^+)$ of (9.75) in the interior of (9.79). Three regions with qualitatively different dynamic behavior of (9.75) were found in the parameter space:

- (I) The equilibrium is stable; the orbits that start outside its domain of attraction are attracted to the heteroclinic loop.
- (II) The equilibrium is unstable but surrounded by a stable periodic orbit. Orbits are either attracted by the stable periodic orbit or by the heteroclinic loop.
- (III) The equilibrium is unstable and all orbits starting from nonequilibrium points are attracted by the heteroclinic loop.

In [202] the first situation is described as the case of strongly controlled corruption, the second as weakly controlled corruption, and the third as uncontrolled corruption. Mathematically, the regions (I) and (II) meet in Hopf points with $\ell_1^* < 0$; (I) and (III) meet in Hopf points with $\ell_1^* > 0$; (II) and (III) meet in turning points of periodic orbits as we met already in §9.4.1. The three regions (I), (II), (III) meet in GH points.

We now describe some computations that were done with all parameters except δ, ϵ fixed at their values in (9.76). We first note that $x^\star = 1$ and the condition (9.77) becomes $\epsilon < 0.1$. So for $\epsilon < 0.1$ the point $(1, 0, 0)$ is a stable equilibrium and attracts all orbits that start in the positive octant. To study the dynamics in the region $\epsilon > 0.1$ it is natural to look first at the equilibria in the positive octant. We choose (fairly arbitrarily) the pair $\delta = 0.1, \epsilon = 0.4$ and computed the orbit that starts at $(0.5, 0.5, 0.5)$. A picture in the (x, y)-plane is shown in Figure 9.10; it is clear that the orbit converges to a stable equilibrium, which turns out to be at $(0.8413562, 0.05, 0.2365425)$. We note also that on several occasions it spends a lot of time near the unstable equilibrium at $(1, 0, 0)$.

In Figure 9.11 we present the equilibrium curves obtained by continuation of the found equilibrium with free parameters δ, ϵ, respectively. In particular, on the curve where δ is free, stability is lost at a Hopf point with coordinates $(0.542790, 0.374370, 0.117116)$ for the parameter values $\delta = 0.7487409, \epsilon = 0.4$; one has $\omega = 0.42765, \ell_1^* = 0.100832$. Hence the loss of stability is sharp; the Hopf point separates the regions (I) and (III).

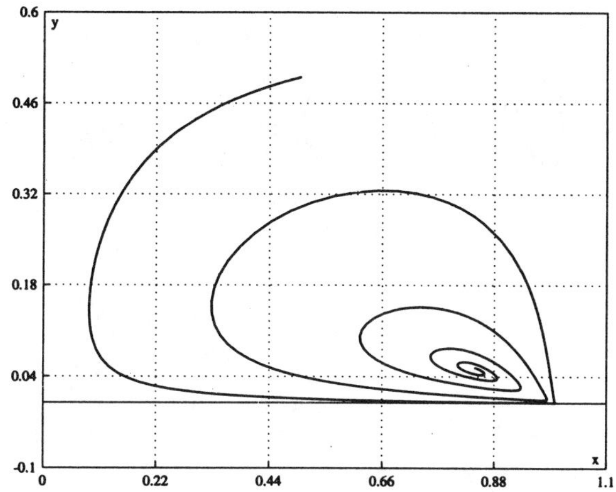

Figure 9.10: The corruption model: An orbit that converges to a stable equilibrium.

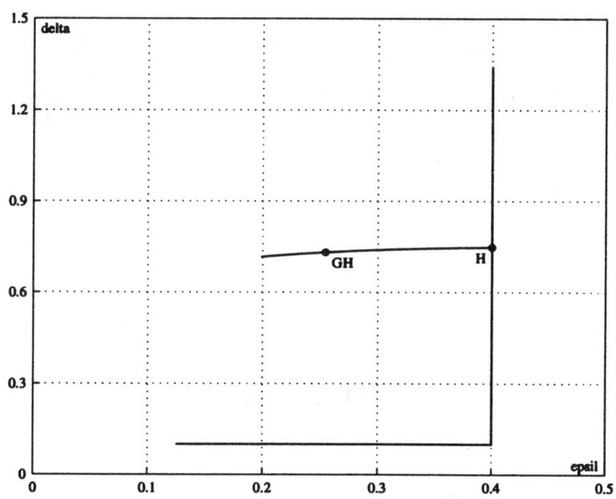

Figure 9.11: The corruption model: Two curves of equilibria, a Hopf curve, and a GH point.

We then computed the Hopf curve through the Hopf point and found a GH point at $(0.682074, 0.366103, 0.0734647)$ for the parameter values $\delta = 0.732206, \epsilon = 0.254320$. The normal form coefficients were $\omega = 0.316932$, $\ell_2 = 0.25955068$. The components of q were $0.8203260352 - 0.8203260352i$, $-0.5188507810 - 0.5188507810i$, $-0.2405391089 + 0.2405391089i$. The high-order derivatives in ℓ_2 were generated symbolically using MAPLE.

Starting from the lower left equilibrium point in Figure 9.11 we computed another curve of equilibria with free parameter δ. It intersects the Hopf curve in Figure 9.11 at the point (0.9231957, 0.3418442, 0.01578674) with parameter values $\delta = 0.6836885$, $\epsilon = 0.1254195$ with $\omega = 0.130544$, $\ell_1^* = -0.12632$. As expected there is now a soft loss

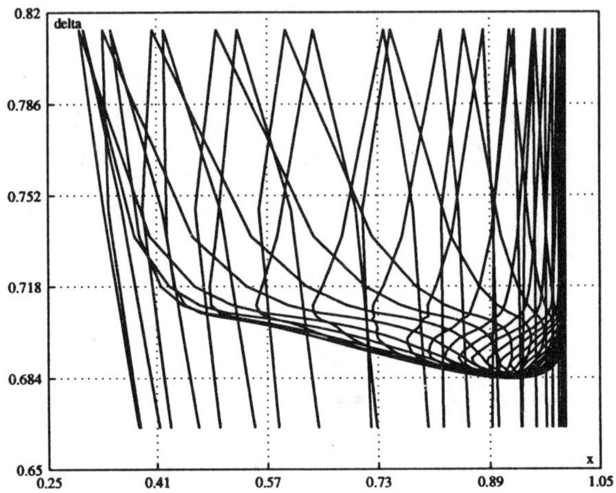

Figure 9.12: The corruption model: A turning point of periodic orbits.

of stability; i.e., a curve of stable periodic orbits continues the curve of stable equilibria. In other words, the Hopf point separates the regions (I) and (II). We continue these periodic orbits with free parameter δ and represent traces of the periodic orbits for increasing values of δ in Figure 9.12. For a value of δ near 0.81 the periodic orbits turn and become unstable; i.e., we reach the boundary between the regions (II) and (III). We note that the unstable periodic orbits in Figure 9.12 have their own interest: They separate the domains of attraction of the stable periodic orbit and the stable heteroclinic loop.

9.5 Notes and Further Reading

1. Classical references for the computation of CPs are [223], [124], [205], [125], [195], and [126]. The minimally extended system in §9.2.2 is essentially the same as the one obtained in §6.6.2 in the context of singularity theory.

2. The numerical continuation of GH points was also implemented in LOCBIF (for a recent version see [156]). In this software the Hopf condition is expressed using the Routh–Hurwitz determinant (cf. §4.7) and ℓ_1 is obtained via an intermediate reduction to the center manifold; cf. [132].

3. The numerical computation of invariant manifolds of dynamical systems has attracted a lot of recent work. We refer in particular to the Ph.D. theses [192] and [159], which also contain many further references.

9.6 Exercises

1. Consider the Hopf bifurcation example in §1.2.4. Compute the first Lyapunov coefficient and check that it predicts the correct bifurcation and stability behavior.

2. Consider the catalytic oscillator model in §4.6.1. Near which points on the Hopf curve in Figure 4.1 do you expect stable periodic orbits?

3. Consider system (9.26) with four parameters a, b, β_1, β_2. Check that $(0,0)$ is a Hopf bifurcation point whenever $\beta_1 = 0, \beta_2 < 0$. Then prove that in all such points ℓ_1^* is zero if and only if $ab = 0$. Prove also that ℓ_1^* has the sign of ab if $ab \neq 0$. Check it for some values of β_2 using dynamical systems software.

4. Consider (9.26) with $a = 1, b = -1, \beta_2 = -0.1$. Compute by orbit simulation stable periodic orbits for $\beta_1 = -0.1, -0.2, -0.21$. Look at the shapes of the orbits in (η_1, η_2)-space and the time spent near the rightmost intersection with the η_1-axis. Then simulate nearby orbits for $\beta_1 = -0.22$. You should see that they diverge to infinity; the periodic orbit has disappeared through a homoclinic bifurcation.

5. Consider the catalytic oscillator model in §4.6.1. Start with the equilibrium point in Table 4.2. Compute a curve of equilibrium points with free parameter q_2, detect a fold point, continue it with free parameters q_1, q_2, detect a CP, and continue it with free parameters q_1, q_2, k. Detect a ZA point on the cusp curve and compare with Figure 5.3.

6. Consider the Bazykin ecological model in Exercise §1.6.9. Use CONTENT to compute a curve of equilibrium solutions with free parameter δ, starting from the stable equilibrium computed in Exercise 9, §1.6. Find a limit point on this curve and continue it with free parameters γ, δ. Find a CP.

7. Use CONTENT to compute the following:

   (a) A curve of GH points in the Hodgkin–Huxley model through the GH point computed in §9.4.1 with free parameters $I, T, \bar{g}_{Na}$.

   (b) A curve of GH points in the corruption model through the GH point computed in §9.4.3 with free parameters ϵ, δ, μ^+.

   (c) A curve of GH points in the catalytic oscillator model that connects the two GH points in Figure 5.3.

8. Consider the system (9.75). Assume that $\epsilon < \frac{\rho k}{\alpha - \beta k}$ and set $x^\star = \frac{\alpha - \beta k}{k}$. Prove that the following intervals are heteroclinic connections:

   (a) $(+\infty, 0, 0) \to (x^\star, 0, 0)$,

   (b) $(0, 0, 0) \to (x^\star, 0, 0)$,

   (c) $(0, +\infty, 0) \to (0, 0, 0)$,

   (d) $(0, 0, +\infty) \to (0, 0, 0)$,

   (e) $(x^\star, +\infty, 0) \to (x^\star, 0, 0)$,

   (f) $(x^\star, 0, +\infty) \to (x^\star, 0, 0)$.

9. Consider system (9.75). Assume that $\epsilon > \frac{\rho k}{\alpha - \beta k} > 0$ and set $x^\star = \frac{\alpha - \beta k}{k}$. Prove that the following intervals are heteroclinic connections:

(a) $(+\infty, 0, 0) \rightarrow (x^\star, 0, 0)$,

(b) $(0, 0, 0) \rightarrow (x^\star, 0, 0)$,

(c) $(0, +\infty, 0) \rightarrow (0, 0, 0)$,

(d) $(0, 0, +\infty) \rightarrow (0, 0, 0)$,

(e) $(x^\star, 0, 0) \rightarrow (x^\star, +\infty, 0)$,

(f) $(x^\star, 0, 0) \rightarrow (x^\star, 0, +\infty)$.

10. Consider system (9.75) with parameter values (9.76). Compute several orbits starting with values in the interior of the region defined by (9.79). Do you find that they are attracted by the heteroclinic loop $(0, 0, +\infty) \rightarrow (0, 0, 0) \rightarrow (x^\star, 0, 0) \rightarrow (x^\star, +\infty, 0)$ where $x^\star = \frac{\alpha - \beta k}{k}$?

Chapter 10

An Introduction to Large Dynamical Systems

A time-evolution PDE is an equation of the form

$$u_t = F(u, u_x, u_{xx}, \ldots, x, \alpha), \tag{10.1}$$

where $u(x, t, \alpha)$ is a function defined for x in a domain of R^k ($k = 1, 2$, or 3) and subject to some initial and boundary conditions, $t \in [0, \infty[$ and $\alpha \in \mathbb{R}^m$.

The numerical computation of the solutions to (10.1) and even of its steady states is an active area of research. For a general introduction to computational methods for PDEs we recommend [231]. Computing singularities and bifurcations presents additional difficulties but is essential in the understanding of many problems that arise in fluid dynamics, aeronautics, temperature-driven convection, oceanography, atmospheric models, etc. Such applications are usually published in the literature of the field instead of numerical analysis or dynamical systems journals.

The precise relation between the undiscretized problem (10.1) and its discretized versions is the subject of approximation theory. It is not feasible to discuss this here; fortunately, a recent extensive discussion is available in [46].

In some simple cases (10.1) can be solved by a discretization with fixed mesh points and implicitly incorporated boundary conditions. This is precisely what we did in the example of the continuous Brusselator (§§4.2, 4.6.3, 7.7, and 8.1–8.3). In this case one is left with a large system of ODEs; from a numerical point of view the important issue is the efficient handling of the sparsity in the Jacobian matrix and higher-order derivatives.

In more difficult cases this approach often does not work reliably; the computations do not converge or lead to spurious solutions. However, there are many papers on the solution of problems of the type (10.1) that use better discretization methods. In particular, we mention the software package PLTMG [24] that allows us to solve a whole class of boundary value problems on regions in the plane, to continue the solution with respect to a parameter, and even to compute limit and branching points. This software combines a sophisticated finite element discretization with advanced linear algebra techniques.

Since a complete discussion is out of the question we will restrict ourselves to a few relevant topics that can serve as an introduction to the subject. In §10.3 we give some references to the literature that we feel to be of particular interest.

10.1 A Class of One-Dimensional PDEs

Many methods can be used to discretize (10.1). Obviously different classes of problems will require different discretization methods and therefore different linear algebra libraries for the storage and processing of matrices. By way of an example we will restrict to one particular class, which is at present implemented in CONTENT [165]. A feature that makes CONTENT particularly attractive is its built-in ability to associate specific linear algebra routines to each system and solution type. It seems likely that a similar approach can be used for other classes of PDEs, although each will involve considerable implementation work.

The class that we consider is that of nonlinear evolution problems with partial derivatives

$$\begin{array}{ll} u_t = F(u_{xx}, u_x, u, x, \alpha), & 0 < x < 1, \\ f^0(u_x, u, \alpha) = 0, & x = 0, \\ f^1(u_x, u, \alpha) = 0, & x = 1, \end{array} \tag{10.2}$$

where $F : \mathbb{R}^n \times \mathbb{R}^n \times \mathbb{R}^n \times \mathbb{R}^1 \times \mathbb{R}^m \to \mathbb{R}^n$ and $f^{0,1} : \mathbb{R}^n \times \mathbb{R}^n \times \mathbb{R}^m \to \mathbb{R}^n$ are sufficiently smooth nonlinear functions.

Such problems appear in numerous applications. For example, *reaction-diffusion systems* with one spatial variable $x \in [0, 1]$,

$$u_t = D(\alpha)u_{xx} + f(u, \alpha), \quad 0 < x < 1, \quad t \geq 0, \tag{10.3}$$

where $f : \mathbb{R}^n \times \mathbb{R}^m \to \mathbb{R}^n$ is a smooth function and $D(\alpha)$ is a positive diagonal $n \times n$ matrix smoothly dependent on $\alpha \in \mathbb{R}^m$, belong to the class (10.2). This includes the combustion model in §1.3 and the Brusselator in §4.2. The population dynamics models in §1.2 can also be generalized to reaction-diffusion models if the spatial distribution of the species is taken into account. Distribution of heat (the *heat equation*) is another standard example. See [231] for an introduction and [174], [221] for a more comprehensive treatment.

There are two simple but important types of solutions to (10.2):

(1) *Orbits*, which describe the evolution of an initial distribution $u(x, 0) = u_0(x)$ in time. In particular, asymptotic states (which are approached at $t \to \infty$) are important.

(2) *Stationary solutions*, which are independent of time and, therefore, satisfy the nonlinear boundary value problem

$$\begin{array}{ll} F(u_{xx}, u_x, u, x, \alpha) = 0, & 0 < x < 1, \\ f^0(u_x, u, \alpha) = 0, & x = 0, \\ f^1(u_x, u, \alpha) = 0, & x = 1. \end{array} \tag{10.4}$$

Both orbit integration and continuation of stationary states involve a *spatial discretization* of the problem (10.2), i.e., reducing it to a finite-dimensional system of differential algebraic equations (DAEs):

$$\mathcal{M}\dot{Y} = \mathcal{G}(Y, \alpha), \qquad \mathcal{G} : \mathbb{R}^K \times \mathbb{R}^m \to \mathbb{R}^K, \tag{10.5}$$

where $\mathcal{G}$ is smooth and $\mathcal{M}$ is a singular $K \times K$ matrix. Here $Y \in \mathbb{R}^K$ represents all the discretization data corresponding to problem (10.2).

We describe a technique to discretize (10.2), based on a finite-difference method for which the structure of linear systems to be solved in both integration and continuation is similar, since it is determined mainly by the selected discretization method and the implicit nature of the algorithms used.

In §10.1.1 we describe a second-order finite-difference approximation of solutions to problem (10.2) using a nonuniform mesh. An algorithm to adapt mesh distribution while computing a solution curve is also presented. It is based on the requirement for the approximation error of the x-derivatives to be uniformly distributed over the space interval. Sections 10.1.2, 10.1.3, and 10.1.4 describe specific algorithms implemented for integration and continuation. It happens that the appearing linear systems have similar structure and can be solved by the same linear algebra software.

10.1.1 Space Discretization

It is well known that the accuracy of a finite-difference approximation of a differential equation on any mesh strongly depends on the spatial behavior of the solution. It means that the mesh Θ_N should be selected according to this behavior. Since the shape of the solution usually varies with time and parameters, the mesh has to be adaptive.

To approximate a smooth function $u(x)$ by a finite-dimensional vector, we introduce a nonuniform mesh

$$\Theta_N = \{0 = x_0 < x_1 < \cdots < x_{N-1} < x_N = 1\} \tag{10.6}$$

and set $h_i = x_i - x_{i-1}$ for $i = 1, \ldots, N$. Denote the corresponding mesh values $u(x_i)$ by u_i, $i = 0, 1, \ldots, N$.

For the left end point we take

$$u_x(0) \approx \tilde{u}_x(0) = \hat{\varphi}_0 u_0 + \hat{\varphi}_1 u_1 + \hat{\varphi}_2 u_2, \tag{10.7}$$

where

$$\hat{\varphi}_0 = -\frac{2h_1 + h_2}{h_1(h_1 + h_2)}, \quad \hat{\varphi}_1 = \frac{h_1 + h_2}{h_1 h_2}, \quad \hat{\varphi}_2 = -\frac{h_1}{h_2(h_1 + h_2)}.$$

For each internal mesh point with $i = 1, 2, \ldots, N-1$, introduce

$$\tilde{x}_i = x_i + \frac{1}{3}(h_{i+1} - h_i),$$

and take

$$u(\tilde{x}_i) \approx \tilde{u}(\tilde{x}_i) = \eta_{i-} u_{i-1} + \eta_i u_i + \eta_{i+} u_{i+1}, \tag{10.8}$$

$$u_x(\tilde{x}_i) \approx \tilde{u}_x(\tilde{x}_i) = \varphi_{i-} u_{i-1} + \varphi_i u_i + \varphi_{i+} u_{i+1}, \tag{10.9}$$

$$u_{xx}(\tilde{x}_i) \approx \tilde{u}_{xx}(\tilde{x}_i) = \psi_{i-}u_{i-1} + \psi_i u_i + \psi_{i+}u_{i+1}, \tag{10.10}$$

where

$$\eta_{i-} = -\frac{(h_{i+1} - h_i)(2h_{i+1} + h_i)}{9h_i(h_i + h_{i+1})}, \qquad \eta_i = \frac{(2h_i + h_{i+1})(2h_{i+1} + h_i)}{9h_i h_{i+1}},$$

$$\eta_{i+} = \frac{(h_{i+1} - h_i)(2h_i + h_{i+1})}{9h_{i+1}(h_i + h_{i+1})},$$

$$\varphi_{i-} = -\frac{2h_i + h_{i+1}}{3h_i(h_i + h_{i+1})}, \quad \varphi_i = \frac{h_{i+1} - h_i}{3h_i h_{i+1}}, \quad \varphi_{i+} = \frac{h_i + 2h_{i+1}}{3h_{i+1}(h_i + h_{i+1})},$$

and

$$\psi_{i-} = \frac{2}{h_i(h_i + h_{i+1})}, \quad \psi_i = -\frac{2}{h_i h_{i+1}}, \quad \psi_{i+} = \frac{2}{h_{i+1}(h_i + h_{i+1})}.$$

Finally, for the right end point take

$$u_x(1) \approx \tilde{u}_x(1) = \hat{\varphi}_{N-2}u_{N-2} + \hat{\varphi}_{N-1}u_{N-1} + \hat{\varphi}_N u_N, \tag{10.11}$$

where

$$\hat{\varphi}_{N-2} = \frac{h_N}{h_{N-1}(h_{N-1} + h_N)}, \quad \hat{\varphi}_{N-1} = -\frac{h_{N-1} + h_N}{h_{N-1}h_N}, \quad \hat{\varphi}_N = \frac{2h_N + h_{N-1}}{h_N(h_{N-1} + h_N)}.$$

The above approximations have third-order accuracy for the function $u(x)$ and second-order accuracy for its first and second derivatives; see [78], [79]. More precisely, if $u(x)$ is a smooth vector function, then

$$\|u(\tilde{x}_i) - \tilde{u}(\tilde{x}_i)\| = O(h^3), \qquad i = 1, 2, \ldots, N-1,$$

while

$$\|u_x(0) - \tilde{u}_x(0)\| = O(h^2), \qquad \|u_x(1) - \tilde{u}_x(1)\| = O(h^2),$$

and

$$\|u_x(\tilde{x}_i) - \tilde{u}_x(\tilde{x}_i)\| = O(h^2), \qquad \|u_{xx}(\tilde{x}_i) - \tilde{u}_{xx}(\tilde{x}_i)\| = O(h^2)$$

for $i = 1, 2, \ldots, N-1$. Here

$$h = \max_{i=1,\ldots,N} h_i.$$

One of the possible criteria for mesh selection is the requirement for the approximation error for the first derivative to be uniformly distributed over the space interval (see, for example, [170], [209]).

The approximation (10.9) of the gradients on the nonuniform mesh (10.6) leads to the following expression

$$u_x(\tilde{x}_i) = \tilde{u}_x(\tilde{x}_i) + R(\tilde{x}_i), \qquad i = 1, 2, \ldots, N-1,$$

with the local error

$$R(\tilde{x}_i) \approx \frac{1}{18}u_{xxx}(\tilde{x}_i)(h_i^2 + h_i h_{i+1} + h_{i+1}^2),$$

where $h_i = x_i - x_{i-1} > 0$, $i = 1, \ldots, N$. To estimate the spatial dependence of the error, assume that N is big enough and consider a smooth function $h = h(x)$, $h(x_i) = h_i$. Then

$$h_{i+1} \approx h_i[1 + h_x(x_i)], \qquad i = 1, 2, \ldots, N-1.$$

Therefore,

$$R(\tilde{x}_i) \approx \frac{1}{6} u_{xxx}(x_i) h_i^2.$$

Clearly, if we take

$$h_i = C \|u_{xxx}(x_i)\|^{-\frac{1}{2}}$$

with a constant $C > 0$, then the norm of the leading term of the local error R will be uniformly distributed over the internal points having the value $\frac{1}{6}C^2$. Note that in this case we have $h_x = -\frac{1}{2}[\ln \|u_{xxx}\|]_x h$. Therefore, for sufficiently smooth functions $u(x)$, $|h_x| \ll 1$.

Finally, we can write our mesh point selection criterion in the form

$$S(x_i)(x_i - x_{i-1}) = C, \tag{10.12}$$

where $S(x) = \|u_{xxx}(x)\|^{\frac{1}{2}}$. In the implementation, we compute $u_{xxx}(x_i)$ using a five-point finite-difference approximation on the previous mesh and take

$$S(x) = \begin{cases} \|u_{xxx}(x)\|^{\frac{1}{2}}, & \|u_{xxx}(x)\| > \varepsilon^2, \\ \varepsilon, & \|u_{xxx}(x)\| \le \varepsilon^2, \end{cases}$$

with some $\varepsilon \ll 1$. This modification is natural, since for small $\|u_{xxx}\|$ equation (10.12) with modified $S(x)$ is satisfied on the uniform mesh, where the approximation formula (10.9) is exact for polynomial functions of degree 2.

To solve (10.12) for x_i, one can use different approaches. The simplest one is to consider (10.12) as the first-order analogue of

$$\int_{x_{i-1}}^{x_i} S(x)dx \;=\; C \;=\; \frac{1}{N}\int_0^1 S(x)dx \tag{10.13}$$

(see [209]). Solving equation (10.13) using the linear interpolation of $S(x)$ between the mesh points amounts to solving a quadratic equation for each x_i, $i = 1, 2, \ldots, N-1$. The linear interpolation guarantees the monotonicity of the mesh, i.e., $x_i > x_{i-1}$.

The obtained solution Θ_N to (10.12) may be improved by Newton iterations applied to the nonlinear system

$$\begin{cases} S(x_1)x_1 - S(x_2)(x_2 - x_1) = 0, \\ S(x_i)(x_i - x_{i-1}) - S(x_{i+1})(x_{i+1} - x_i) = 0, \quad i = 2, \ldots, N-2, \\ S(x_{N-1})(x_{N-1} - x_{N-2}) - S(1)(1 - x_{N-1}) = 0, \end{cases}$$

which is another form of the presentation of (10.12). The Jacobian matrix of this system is tridiagonal; therefore, we can solve the linear systems appearing at the Newton iterations by a standard elimination method.

When a new mesh is computed the approximate solution has to be recomputed by an interpolation method. At present CONTENT offers the choice between the global cubic spline or a linear interpolation. The global cubic spline is usually more accurate, but in the case of a wildly varying solution function linear interpolation is safer.

10.1.2 Integration by Crank–Nicolson

The aim of the integration is to approximate a time-dependent solution $u = u(x,t)$ to (10.2), given $u(x,0) = u_0(x)$. Let a mesh (10.6) be introduced, denote the corresponding mesh values of the solution $u(x_i,t)$ by $u_i(t)$, and consider u_i as time dependent in formulas (10.7)–(10.11). Substituting the approximations (10.7)–(10.11) into the problem (10.2), we obtain its Θ_N-discretization in the form

$$\mathcal{M}\dot{Y} = \mathcal{G}(Y,\alpha), \qquad \mathcal{G} : \mathbb{R}^K \times \mathbb{R}^m \to \mathbb{R}^K, \tag{10.14}$$

with $K = n(N+1)$,

$$Y(t) = \begin{pmatrix} u_0(t) \\ u_1(t) \\ \cdots \\ u_{N-1}(t) \\ u_N(t) \end{pmatrix},$$

$$\mathcal{M} = \begin{pmatrix} 0_n & 0_n & 0_n & & & & & \\ \eta_{1-}I_n & \eta_1 I_n & \eta_{1+}I_n & & & & & \\ & \eta_{2-}I_n & \eta_2 I_n & \eta_{2+}I_n & & & & \\ & & & \ddots & & & & \\ & & & & \eta_{(N-2)-}I_n & \eta_{N-2}I_n & \eta_{(N-2)+}I_n & \\ & & & & & \eta_{(N-1)-}I_n & \eta_{N-1}I_n & \eta_{(N-1)+}I_n \\ & & & & & 0_n & 0_n & 0_n \end{pmatrix} \tag{10.15}$$

(all undisplayed elements are $n \times n$ zero blocks) and

$$\mathcal{G}(Y(t),\alpha) = \begin{pmatrix} f^0(\tilde{u}_x(0,t), u_0(t), \alpha) \\ F(\tilde{u}_{xx}(\tilde{x}_1,t), \tilde{u}_x(\tilde{x}_1,t), \tilde{u}(\tilde{x}_1,t), \tilde{x}_1, \alpha) \\ F(\tilde{u}_{xx}(\tilde{x}_2,t), \tilde{u}_x(\tilde{x}_2,t), \tilde{u}(\tilde{x}_2,t), \tilde{x}_2, \alpha) \\ \cdots \\ F(\tilde{u}_{xx}(\tilde{x}_{N-2},t), \tilde{u}_x(\tilde{x}_{N-2},t), \tilde{u}(\tilde{x}_{N-2},t), \tilde{x}_{N-2}, \alpha) \\ F(\tilde{u}_{xx}(\tilde{x}_{N-1},t), \tilde{u}_x(\tilde{x}_{N-1},t), \tilde{u}(\tilde{x}_{N-1},t), \tilde{x}_{N-1}, \alpha) \\ f^1(\tilde{u}_x(1,t), u_N(t), \alpha) \end{pmatrix}, \tag{10.16}$$

where $\tilde{u}(\tilde{x}_i,t), \tilde{u}_x(\tilde{x}_i,t)$, and $\tilde{u}_{xx}(\tilde{x}_i,t)$ should be expressed in terms of $u_i(t)$ according to (10.7)–(10.11).

In principle, any standard method to solve DAEs can be used to integrate (10.14) numerically. We now describe the simplest second-order method, the *implicit trapezoid rule*, in the PDE context known as the Crank–Nicolson method. In §10.1.3 we will briefly discuss some stability concepts in ODE theory which suggest another second-order method, the *implicit midpoint rule*. Let Y^0 be the solution vector at time t. Then an $O((\Delta t)^2)$-approximate solution vector Y^1 at time $t + \Delta t$ can be found by solving

$$\mathcal{M}(Y^1 - Y^0) = \frac{\Delta t}{2}(\mathcal{G}(Y^0,\alpha) + \mathcal{G}(Y^1,\alpha)) \tag{10.17}$$

or, equivalently,

$$\mathcal{F}(Y^1) \equiv \mathcal{G}(Y^1,\alpha) - \frac{2}{\Delta t}\mathcal{M}Y^1 + \left[\mathcal{G}(Y^0,\alpha) + \frac{2}{\Delta t}\mathcal{M}Y^0\right] = 0. \tag{10.18}$$

We note that (10.17) is a special case of a family of methods with the general form

$$\mathcal{M}(Y^1 - Y^0) = (\Delta t)(\theta \mathcal{G}(Y^0, \alpha) + (1-\theta)\mathcal{G}(Y^1, \alpha))$$

with $\theta \in [0,1]$. For $\theta = 0.5$ we obtain (10.17); for $\theta = 0$ we obtain the method that is called *implicit Euler*. For $\theta \in [0,1[$ these methods can be implemented in the same way; we restrict ourselves to $\theta = 0.5$.

Now (10.18) is a system of nonlinear equations for Y^1 that can be solved by Newton's method. The Jacobian matrix of (10.18) has the form

$$\mathcal{F}_Y = \mathcal{G}_Y - \frac{2}{\Delta t}\mathcal{M},$$

where $\mathcal{M}$ is defined by (10.15), while

$$\mathcal{G}_Y = \begin{pmatrix} a_0 & b_0 & c_0 & & & & \\ a_1 & b_1 & c_1 & & & & \\ & a_2 & b_2 & c_2 & & & \\ & & & & & & \\ & & & a_{N-2} & b_{N-2} & c_{N-2} & \\ & & & & a_{N-1} & b_{N-1} & c_{N-1} \\ & & & & a_N & b_N & c_N \end{pmatrix}. \tag{10.19}$$

The undisplayed elements of $\mathcal{G}_Y$ are $n \times n$ zero blocks and $a_k, b_k, c_k,\ k = 0,1,\ldots,N$ are $n \times n$ matrices given by the following expressions:

$$a_0 = \frac{\partial f^0}{\partial u_0} = \frac{\partial f^0}{\partial u_x}\hat{\varphi}_0 + \frac{\partial f^0}{\partial u}, \quad b_0 = \frac{\partial f^0}{\partial u_1} = \frac{\partial f^0}{\partial u_x}\hat{\varphi}_1, \quad c_0 = \frac{\partial f^0}{\partial u_2} = \frac{\partial f^0}{\partial u_x}\hat{\varphi}_2,$$

$$\begin{aligned} a_i &= \frac{\partial F}{\partial u_{i-1}} = \frac{\partial F}{\partial u_{xx}}\psi_i^{-1} + \frac{\partial F}{\partial u_x}\varphi_i^{-1} + \frac{\partial F}{\partial u}\eta_i^{-1}, \\ b_i &= \frac{\partial F}{\partial u_i} = \frac{\partial F}{\partial u_{xx}}\psi_i + \frac{\partial F}{\partial u_x}\varphi_i + \frac{\partial F}{\partial u}\eta_i, \\ c_i &= \frac{\partial F}{\partial u_{i+1}} = \frac{\partial F}{\partial u_{xx}}\psi_i^{+1} + \frac{\partial F}{\partial u_x}\varphi_i^{+1} + \frac{\partial F}{\partial u}\eta_i^{+1}, \end{aligned}$$

where $i = 1, 2, \ldots, N-1$,

$$a_N = \frac{\partial f^1}{\partial u_{N-2}} = \frac{\partial f^1}{\partial u_x}\hat{\varphi}_{N-2}, \quad b_N = \frac{\partial f^1}{\partial u_{N-1}} = \frac{\partial f^1}{\partial u_x}\hat{\varphi}_{N-1}, \quad c_N = \frac{\partial f^1}{\partial u_N} = \frac{\partial f^1}{\partial u_x}\hat{\varphi}_N + \frac{\partial f^1}{\partial u}.$$

The time step Δt can be selected by using an estimate of the local $O((\Delta t)^3)$-error of (10.17). To make this clear in our DAE case we first return to the simpler case of ODEs. So consider a general dynamical system

$$\dot{x} = G(x), \qquad x(t_0) = x_0 \ (x, G(x) \in \mathbb{R}^n). \tag{10.20}$$

We can compute the higher-order derivatives of $x(t)$ by noting that

$$\ddot{x} = G_x\dot{x} = G_xG, \qquad \frac{d^3x}{dt^3} = (G_xG)_x\dot{x} = (G_xG)_xG.$$

We set $\Delta t = t - t_0$; up to fourth order of accuracy in Δt we have

$$x(t) = x_0 + \Delta t G(x_0) + \frac{1}{2}(\Delta t)^2(G_xG)(x_0) + \frac{1}{6}(\Delta t)^3((G_xG)_x)G)(x_0) + \mathcal{O}((\Delta t)^4). \quad (10.21)$$

Now the implicit trapezoid rule for solving (10.20) approximates $x(t)$ by the solution $\tilde{x}(t)$ to the system of equations

$$\tilde{x} - x_0 = \frac{\Delta t}{2}(G(x_0) + G(\tilde{x})). \quad (10.22)$$

This system usually has to be solved by an iterative method. By a Taylor expansion we obtain

$$\begin{aligned}\tilde{x}(t) - x_0 &= \tfrac{\Delta t}{2}(G(x_0) + G(x_0 + (\tilde{x} - x_0))) \\ &= \tfrac{\Delta t}{2}(G(x_0) + G(x_0) + G_x.(\tilde{x} - x_0) + \tfrac{1}{2}G_{xx}.(\tilde{x} - x_0).(\tilde{x} - x_0)) + \mathcal{O}((\Delta t)^4) \\ &= (\Delta t)G(x_0) + \tfrac{\Delta t}{2}G_x.(\tilde{x} - x_0) + \tfrac{\Delta t}{4}G_{xx}.(\tilde{x} - x_0).(\tilde{x} - x_0) + \mathcal{O}((\Delta t)^4).\end{aligned}$$

Inserting (10.22) several times into this expression we obtain

$$\begin{aligned}\tilde{x}(t) - x_0 &= (\Delta t)G(x_0) \\ &\quad + \tfrac{(\Delta t)^2}{2}G_x.G \\ &\quad + \tfrac{(\Delta t)^3}{4}(G_{xx}GG + G_xG_xG) \\ &\quad + \mathcal{O}((\Delta t)^4).\end{aligned} \quad (10.23)$$

Since $((G_xG)_x)G = G_{xx}GG + G_xG_xG$, it follows from (10.21) and (10.23) that

$$\tilde{x}(t) - x(t) = \frac{(\Delta t)^3}{12}(G_{xx}GG + G_xG_xG) + \mathcal{O}((\Delta t)^4). \quad (10.24)$$

This gives an error estimate for the implicit trapezoidal rule. If desired, we can get an estimate for $G_{xx}GG$ by finite differences of second order, i.e.,

$$G_{xx}GG \simeq \frac{G(x_0 + \delta G) - 2G(x_0) + G(x_0 - \delta G)}{\delta^2} \quad (10.25)$$

for a small number δ. An estimate for G_xG_xG may be found by twice taking finite differences of first order, i.e.,

$$H \simeq \frac{G(x_0 + \delta G) - G(x_0 - \delta G)}{2\delta}, \quad (10.26)$$

$$G_xG_xG \simeq \frac{G(x_0 + \delta H) - G(x_0 - \delta H)}{2\delta}. \quad (10.27)$$

From (10.25), (10.26), and (10.27) one obtains an estimate of $G_{xx}GG + G_xG_xG$ with five function evaluations of G.

We now return to our case of a discretized PDE. The Crank–Nicolson step (10.17) is equivalent to the implicit trapezoid rule applied to the $(N-1)$-dimensional system of ODEs

$$\dot{\tilde{u}} = G(\tilde{u}), \qquad \dot{\tilde{u}}_i = G_i(\tilde{u}_1, \ldots, \tilde{u}_{N-1}) \tag{10.28}$$

for $i = 1, \ldots, N-1$, where

$$G_i(\tilde{u}_1, \ldots, \tilde{u}_{N-1}) = \mathcal{G}_i(u_0, u_1, \ldots, u_N)$$

with $\mathcal{G}$ defined in (10.16). So $u = (u_0, \ldots, u_N)$ is implicitly considered a function of $\tilde{u} = (\tilde{u}_1, \ldots, \tilde{u}_{N-1})$. If F denotes the internal elements of $\mathcal{G}$ (the first and last n rows deleted), then

$$G_{\tilde{u}} = F_{u\cdot}\frac{\partial u}{\partial \tilde{u}}, \tag{10.29}$$

$$G_{\tilde{u}\tilde{u}} = F_{uu\cdot}\left(\frac{\partial u}{\partial \tilde{u}}\right)^2 + F_u\frac{\partial^2 u}{\partial \tilde{u}^2}. \tag{10.30}$$

By (10.8) we have

$$\tilde{u} = Mu, \tag{10.31}$$

where M is the constant $(N-1)n \times (N+1)n$ matrix obtained by removing the first and last n rows of the matrix $\mathcal{M}$ defined in (10.15). By differentiation we get

$$I_{(N-1)n} = M\frac{\partial u}{\partial \tilde{u}}. \tag{10.32}$$

Differentiating the boundary conditions $f^0(u) = 0, f^1(u) = 0$ and adding the resulting equations to (10.32) we get

$$I^e = M^e\frac{\partial u}{\partial \tilde{u}}, \tag{10.33}$$

where I^e is obtained by adding to $I_{(N-1)n}$ n zero rows at the top and n zero rows at the bottom. M^e is obtained by adding to M a first row consisting of the derivatives of f^0 with respect to u and a last row consisting of the derivatives of f^1 with respect to u, i.e., the same first and last block rows as in (10.19).

Taking further derivatives of (10.33) we obtain

$$0 = M^e_u\left(\frac{\partial u}{\partial \tilde{u}}\right)^2 + M^e\frac{\partial^2 u}{\partial \tilde{u}^2}. \tag{10.34}$$

From (10.33) and (10.34) the matrix $\frac{\partial u}{\partial \tilde{u}}$ and tensor $\frac{\partial^2 u}{\partial \tilde{u}^2}$ can be computed formally and substituted into (10.29) and (10.30). These in turn can be substituted formally into the expression

$$G_{\tilde{u}\tilde{u}}GG + G_{\tilde{u}}G_{\tilde{u}}G$$

to obtain an error bound for the time step by the Crank–Nicolson algorithm. Numerically this amounts to solving a few linear systems with the matrix M^e, which has the same sparsity as $\mathcal{G}_Y$ and matrix–vector and tensor–vector–vector products with F_u and F_{uu}, respectively. The sparsity of the matrix F_u and tensor F_{uu} can be exploited fairly easily.

An alternative is to use the finite difference approximations (10.25), (10.26), and (10.27) for the system (10.28). This requires at each of the five function evaluations of G the solution of a nonlinear system in u consisting of (10.31) and $f^0(u) = f^1(u) = 0$. The Jacobian of this system is precisely M^e.

10.1.3 B-stability and the Implicit Midpoint Rule

The notion of B-stability was introduced in [44]. Consider an autonomous system $\dot{x} = G(x)$, where $x \in \Omega \subset \mathbb{R}^N$. The phase flow $(\Phi^t)_{t\geq 0}$ of this system is defined by $\Phi^t(x_0) = \phi(t)$, where $\phi(t) = G(\phi(t)), \phi(0) = x_0$. The phase flow is called nonexpansive if $\|\Phi^t(x) - \Phi^t(y)\| \leq\leq \|x - y\|$ for all $x, y \in \Omega$. This is equivalent to the condition that G is *dissipative*, i.e., $\langle G(x) - G(y), x - y\rangle \leq 0$ for all $x, y \in \Omega$; see [76, §6.3.3].

A numerical integration method is called B-stable if the discrete phase flow is also nonexpansive whenever the system is dissipative. In other words, a very desirable property of the dynamical system is preserved by the integration method. Moreover, B-stability implies A-stability, i.e., the numerical method preserves (for all steplengths) the stability of linear dynamical systems; cf. [76].

The only method for which B-stbility is proved are the Gauss and the Radau methods. The simplest Gauss method is the implicit midpoint rule in which (10.17) is replaced by

$$\mathcal{M}(Y^1 - Y^0) = (\Delta t)\mathcal{G}\left(\frac{Y^0 = Y^1}{2}\right). \tag{10.35}$$

This method is second-order accurate. The simplest Radau method is implicit Euler, which is only first-order accurate. The next Gauss and Radau methods have order of accuracy higher than 2 but unfortunately their Butcher tableaus (again, see [76]) have at least dimension 2×2. This implies that in the implementation one has to solve linear systems with dimention $2N \times 2N$ where the nice block band structure of (10.15) and (10.19) is lost.

We conclude that there is much to recommend the implicit midpoint rule (10.35). The details of the method, including the discussion of sparsity in the matrices, are similar to those in §10.1.2.

Multistep methods might also be considered. Their obvious advantage is that they allow one to obtain a higher order accuracy while still preserving the sparsity structure of the matrices. However, they are not known to be B-stable. Also, they are not very suitable for interactive software because they need a starting procedure before full efficiency is obtained.

10.1.4 Numerical Continuation

In this case we reduce the problem (10.4) to a finite-dimensional continuation problem, i.e., computation of a curve defined by

$$\mathcal{F}(Z) = 0, \qquad \mathcal{F} : \mathbb{R}^{K+1} \to \mathbb{R}^K, \tag{10.36}$$

where $Z \in \mathbb{R}^K$ represents all the discretization data Y corresponding to problem (10.4) and the control parameters α.

Substituting the approximations (10.7)–(10.11) into the steady state problem (10.4), we obtain its Θ_N-discretization in the form $\mathcal{G}(Z) = 0$ with

$$Z = \begin{pmatrix} u_0 \\ u_1 \\ \cdots \\ u_{N-1} \\ u_N \\ \alpha \end{pmatrix}$$

and

$$\mathcal{G}(Z) = \begin{pmatrix} f^0(\tilde{u}_x(0), u_0, \alpha) \\ F(\tilde{u}_{xx}(\tilde{x}_1), \tilde{u}_x(\tilde{x}_1), \tilde{u}(\tilde{x}_1), \tilde{x}_1, \alpha) \\ F(\tilde{u}_{xx}(\tilde{x}_2), \tilde{u}_x(\tilde{x}_2), \tilde{u}(\tilde{x}_2), \tilde{x}_2, \alpha) \\ \cdots \\ F(\tilde{u}_{xx}(\tilde{x}_{N-2}), \tilde{u}_x(\tilde{x}_{N-2}), \tilde{u}(\tilde{x}_{N-2}), \tilde{x}_{N-2}, \alpha) \\ F(\tilde{u}_{xx}(\tilde{x}_{N-1}), \tilde{u}_x(\tilde{x}_{N-1}), \tilde{u}(\tilde{x}_{N-1}), \tilde{x}_{N-1}, \alpha) \\ f^1(\tilde{u}_x(1), u_N, \alpha) \end{pmatrix}.$$

Notice that $K = \dim Z = (N+1)n + m$, since Z now includes the parameters. To obtain a continuation problem of the general form (10.36), one has to append $(m-1)$ scalar equations

$$\tilde{g}^i(u_0, u_1, \ldots, u_{N-1}, u_N, \alpha) = 0, \qquad i = 1, 2, \ldots, m-1,$$

which represent the discretizations of some user-supplied conditions defined by $m-1$ equations

$$g^i[u] = 0, \qquad i = 1, 2, \ldots, m-1,$$

on the continued solution. In the case of one control parameter, no such extra conditions are required.

Therefore, the continuation problem (10.36) is defined by

$$\mathcal{F}(Z) = \begin{pmatrix} \mathcal{G}(Z) \\ \tilde{g}(Z) \end{pmatrix}, \tag{10.37}$$

where $\tilde{g} = (\tilde{g}^1, \ldots, \tilde{g}^{m-1})^T$. Thus, $\mathcal{F} : \mathbb{R}^K \to \mathbb{R}^{K-1}$.

Taking into account (10.7)–(10.11), the Jacobian matrix of (10.37) can be written in the form

$$\mathcal{F}_Z = \begin{pmatrix} a_0 & b_0 & c_0 & & & & & & \delta_0 \\ a_1 & b_1 & c_1 & & & & & & \delta_1 \\ & a_2 & b_2 & c_2 & & & & & \delta_2 \\ & & & & & & & & \\ & & & & a_{N-2} & b_{N-2} & c_{N-2} & & \delta_{N-2} \\ & & & & & a_{N-1} & b_{N-1} & c_{N-1} & \delta_{N-1} \\ & & & & & a_N & b_N & c_N & \delta_N \\ q_0 & q_1 & q_2 & & & q_{N-2} & q_{N-1} & q_N & \Psi \end{pmatrix},$$

where $a_k, b_k, c_k,\ k = 0, 1, \ldots, N$ are $n \times n$ matrices defined in §10.1.2, while $\delta_k,\ k = 0, 1, \ldots, N$ are $n \times m$ matrices, $q_0, q_1, \ldots, q_N$ are $(m-1) \times n$ matrices, and Ψ is an $(m-1) \times m$ matrix (all undisplayed elements of $\mathcal{F}_Z$ are zeros). These extra matrices are given by the following expressions:

$$\delta_0 = \frac{\partial f_0}{\partial \alpha}, \quad \delta_i = \frac{\partial F}{\partial \alpha}, \quad \delta_N = \frac{\partial f^1}{\partial \alpha}, \quad q_j = \frac{\partial \tilde{g}}{\partial u_j}, \quad \Psi = \frac{\partial \tilde{g}}{\partial \alpha},$$

where $j = 0, 1, \ldots, N$.

The continuation step can now be performed as in §2.3.

10.1.5 Solution of Linear Systems

In §§10.1.2–10.1.4 we saw that both integration and continuation amount to solving linear systems of the form

$$\begin{pmatrix} a_0 & b_0 & c_0 & & & & & & \delta_0 \\ a_1 & b_1 & c_1 & & & & & & \delta_1 \\ & a_2 & b_2 & c_2 & & & & & \delta_2 \\ & & & & & & & & \\ & & & & a_{N-2} & b_{N-2} & c_{N-2} & & \delta_{N-2} \\ & & & & & a_{N-1} & b_{N-1} & c_{N-1} & \delta_{N-1} \\ & & & & & a_N & b_N & c_N & \delta_N \\ p_0 & p_1 & p_2 & & & p_{N-2} & p_{N-1} & p_N & \Phi \end{pmatrix} \begin{pmatrix} Y_0 \\ Y_1 \\ Y_2 \\ \\ Y_{N-2} \\ Y_{N-1} \\ Y_N \\ \gamma \end{pmatrix} = \begin{pmatrix} Z_0 \\ Z_1 \\ Z_2 \\ \\ Z_{N-2} \\ Z_{N-1} \\ Z_N \\ \mu \end{pmatrix}. \tag{10.38}$$

When continuing steady states, the right-hand side of (10.38) is composed of the given vectors $Z_0, Z_1, \ldots, Z_N \in \mathbb{R}^n$ and $\mu \in \mathbb{R}^m$. The solution is $Y_0, Y_1, \ldots, Y_N \in \mathbb{R}^n$ and $\gamma \in \mathbb{R}^m$. The matrix rows $(a_0, b_0, c_0, \ldots, \delta_0)$ and $(a_N, b_N, c_N, \ldots, \delta_N)$ correspond to the discretization of the left and right boundary conditions, while the matrix rows $(a_i, b_i, c_i, \ldots, \delta_i)$ with $i = 1, 2, \ldots, N-1$ are due to the discretization of the differential equations along the space interval. Finally, the bottom matrix row $(p_0, p_1, \ldots, p_N, \Phi)$ corresponds to the user conditions and the appended vector v tangent to the continued curve. In §10.1.2 we saw that each correction iteration in the continuation requires the solution of two linear systems (10.38) with the same matrix but with different right-hand sides. Therefore, the solving procedure has been separated into a *decomposition* and a *back substitution*.

While computing orbits, the last row and the last column in the matrix of (10.38) are absent. However, we formally set $m = 1$ in (10.38) and take

$$p_i = \delta_i = 0, \qquad \Phi = \mu = 1$$

to end up exactly with the same structure of (10.38) as in the steady state case.

We recall that the matrix M^e in (10.33) also has the sparsity structure of the one in (10.38) without the last row and column.

To exploit the special structure of (10.38) it can be solved by a block elimination method; for the details in the implementation in CONTENT we refer to [212, Chapter 2.4] or [166].

10.1.6 Example: The Nonadiabatic Tubular Reactor

We consider the model

$$\begin{cases} \dfrac{\partial y}{\partial t} &= \dfrac{1}{P_{em}} \dfrac{\partial^2 y}{\partial x^2} - \dfrac{\partial y}{\partial x} - D\,y\,\exp\left[\gamma\left(1 - \frac{1}{\theta}\right)\right], \\ \dfrac{\partial \theta}{\partial t} &= \dfrac{1}{P_{eh}} \dfrac{\partial^2 \theta}{\partial x^2} - \dfrac{\partial \theta}{\partial x} - \beta(\theta - \theta_0) + B\,D\,y\,\exp\left[\gamma\left(1 - \frac{1}{\theta}\right)\right] \end{cases}$$

with boundary conditions

$$y_x(0,t) = P_{em}\,(y-1), \quad \theta_x(0,t) = P_{eh}\,(\theta - 1), \quad y_x(1,t) = 0, \quad \theta_x(1,t) = 0,$$

which describes the conservation of reactant and energy for a nonadiabatic tubular reactor with axial mixing in dimensionless form. Here $x \in [0,1]$ is the space variable, y, θ are the state functions; the model has seven parameters $D, P_{em}, P_{eh}, \beta, \theta_0, \gamma, B$. Like the CSTR in §6.5 this is one of the most extensively studied models of chemical reactors. It exhibits a complicated behavior, including multiple solutions and stable and unstable periodic orbits. We refer to [133] for the context of the problem and further references. We note that the tubular reactor model is not a reaction-diffusion problem, since it contains the first-order spatial derivatives that represent convection.

We will restrict ourselves to a few model computations in CONTENT.

To compute a curve of equilibria we start from the trivial equilibrium

$$y(x,0) \equiv y_{in}, \quad \theta(x,0) \equiv \theta_{in}, \quad x \in [0,1]$$

with $y_{in} = \theta_{in} = 1$ at the point with parameter values

$$D = 0, \quad P_{em} = 5, \quad P_{eh} = 5, \quad \beta = 2.35, \quad \theta_0 = 1, \quad \gamma = 25, \quad B = 0.5.$$

In the continuation the Damkohler number D is chosen as the free parameter so that nonconstant equilibria can be computed. The interesting region ($0.15 \leq D \leq 0.18$) is presented in Figure 10.1 in the (D, $\texttt{t40} \equiv \theta_{40}$)-plane, using 50 mesh points so that $x \approx 0.816$ in the fortieth mesh point. We note that there are two disjoint D-regions where the tubular reactor admits three different equilibrium solutions.

For an orbit computation we start from the initial distribution

$$y(x,0) \equiv y_{in}, \quad \theta(x,0) \equiv \theta_{in}, \quad x \in [0,1]$$

with $y_{in} = \theta_{in} = 1$ at the point with parameter values

$$D = 0.165, \quad P_{em} = 5, \quad P_{eh} = 5, \quad \beta = 2.35, \quad \theta_0 = 1, \quad \gamma = 25, \quad B = 0.5.$$

Using the Crank–Nicolson method we compute an orbit that visibly converges to a stable periodic orbit as shown in a zoom-in in Figure 10.2 in the ($\texttt{y40} \equiv y_{40}$, $\texttt{t40} \equiv \theta_{40}$)-plane. We used again 50 mesh points so that $x \approx 0.816$ in the fortieth mesh point.

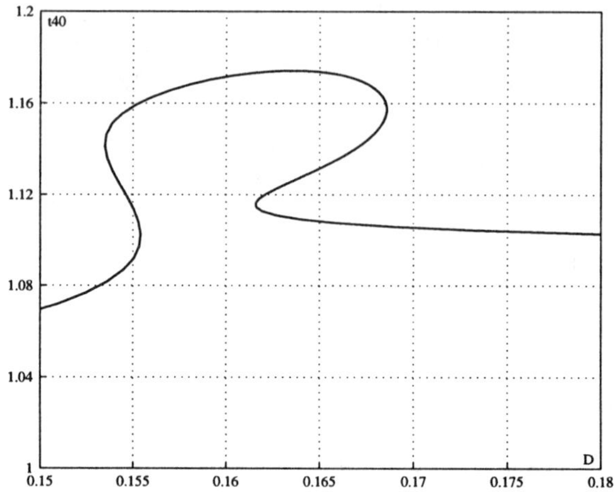

Figure 10.1: The nonadiabatic tubular reactor: A curve of equilibria.

10.2 Bifurcations: Reduction to a Low-Dimensional State Space

The obvious disadvantage of the bialternate product matrix (§4.4) is its dimension $b(N) = \frac{N(N-1)}{2}$. In the largest of our low-dimensional examples, the LP neuron in §5.4, $b(N) = \frac{13\times 12}{2} = 78$. This can be handled easily by small contemporary computers but already slows down the computations. For large problems the size of the bialternate product of the unreduced Jacobian is prohibitive so that good reduction techniques are essential. On the other hand, for large PDEs complete eigenvalue analyses of the Jacobian are also infeasible so that the issue of reduction has to be considered anyway.

The feasibility of a reduction of the Jacobian was shown in [119] by performing all bifurcation computations in the LP-neuron model after a reduction of the Jacobian to an (approximate) generalized eigenspace of the five eigenvalues with largest real part. Indeed, usually only a few eigenvalues (in this case four) with largest real part determine the equilibrium bifurcation behavior. The number 5 is chosen for safety reasons, namely, to monitor the possible interference of other eigenvalues. In this way we effectively deal with a matrix of order 5×5; we call this the reduced Jacobian A_r. Hence the bialternate product is reduced to a 10×10 matrix.

To be honest, reduction to a low-dimensional subspace is not essential in this model; we described the same computations in §5.4 without reduction. It cannot even be recommended since the iterative methods for computing eigenspaces are more time consuming than the use of the bialternate product of the full matrix would be. But it shows that subspace reduction can be combined with complicated bifurcation computations and thus should be tried for large problems. We now give more details.

In [119] a path of DH points is numerically followed and a one-to-one resonant point is detected and computed. Except for the reduction, the method is essentially the one described in §§5.3.2 and 5.3.4.

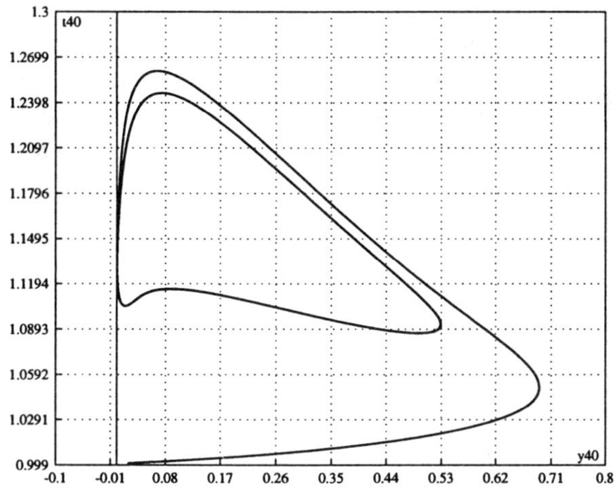

Figure 10.2: The nonadiabatic tubular reactor: An orbit that converges to a periodic orbit.

Let $G(x, \alpha) = 0$ describe the LP-neuron model, so $x, G(x, \alpha) \in R^{13}$, $\alpha \in R^{29}$. The reduction was done by subspace iteration, combined with a Cayley transformation as preconditioner. This method is studied in [56], [59], and [183], mainly with the aim of detecting Hopf bifurcations. For our purposes the accurate computation of invariant subspaces is more important. Therefore, we recall the basic ideas and some particular features of our case.

The (generalized) Cayley transform of a matrix A depends on two real parameters α_1, α_2 with $\alpha_1 > \alpha_2$ and is given by

$$C(A) = (A - \alpha_1 I)^{-1}(A - \alpha_2 I).$$

In practice, only matrix–vector products with $C(A)$ are computed; so it is sufficient to factor $(A - \alpha_1 I)$. The essential property of $C(A)$ is that if μ is an eigenvalue of A, then $\frac{\mu-\alpha_2}{\mu-\alpha_1}$ is an eigenvalue of $C(A)$. Hence all eigenvalues of A with real part greater than $\frac{\alpha_1+\alpha_2}{2}$ are mapped outside the unit circle; those with smaller real part are mapped inside the unit circle. A block power method based on left multiplication with $C(A)$ can then be used to compute the former without undue interference from the latter.

In the computations described in [119] there are always five eigenvalues with real part > -15 (some of them real, others complex). We therefore used $\alpha_1 = 0, \alpha_2 = -30$. The problem is not very sensitive to the choice of these parameters. Since a subspace iteration with a five-dimensional space would not converge in the last dimension if the fifth and sixth largest eigenvalues of $C(A)$ formed a conjugate pair of complex eigenvalues, the subspace iteration steps were done with six vectors.

To perform subspace iteration with the matrix $C(A)$ we start with an initial orthonormal set of vectors $Q_m^{(0)} = [q_1^{(o)}, \ldots, q_m^{(o)}]$ (in our case $m = 6$). Except for the very first point this initial approximation to an invariant space is obtained from a previous point, since we always work in a continuation context. Then the sequence $Q_m^{(1)}, Q_m^{(2)}, \ldots$ of orthonormal sets is generated by performing a block power step $V_m^l = C(A)Q_m^{(l-1)}$ followed by QR decomposition $V_m^{(l)} = Q_m^{(l)} R_m^{(l)}$. Convergence of this process to $Q_m^{(k)}$ is declared if in two consecutive sets of orthonormal vectors the first $m-1$ vectors essentially span the same space. To this end we require that each basis vector of the first space and its orthogonal projection on the second space differ by less than ϵ; the threshold $\epsilon = 1.0D-12$ was reached in all the computations involving the one-to-one resonant DH point where we wanted results as accurate as possible. For the continuation of DH points the threshold can be lowered to obtain a faster convergence; sometimes it even has to be lowered to obtain convergence at all.

Now $Q_{m-1}^{(k)T} A Q_{m-1}^{(k)}$ is the reduced Jacobian A_r whose eigenvalues are approximations to the $m-1$ rightmost eigenvalues of A. We choose $B, C \in \mathbb{R}^{10\times 2}$, $D \in \mathbb{R}^{2\times 2}$ such that

$$M(x,\alpha) = \begin{pmatrix} 2A_r \odot I_N & B \\ C^T & D \end{pmatrix} \tag{10.39}$$

is nonsingular in a neighborhood of the point that we are considering. We define the 10×2 matrix $Q(x,\alpha)$ and the 2×2 matrix $S(x,\alpha)$ by

$$M(x,\alpha)\begin{pmatrix} Q(x,\alpha) \\ S(x,\alpha) \end{pmatrix} = \begin{pmatrix} 0_{10,2} \\ I_{2,2} \end{pmatrix}. \tag{10.40}$$

By Corollary 3.3.4 the four entries of $S(x,\alpha)$ vanish together if and only if $2A_r \odot I_N$ has rank defect 2.

The derivatives of $S(x,\alpha)$ can be obtained in the familar way. Define the 10×2 matrix $W(x,\alpha)$ by

$$\begin{pmatrix} W^T(x,\alpha) & S(x,\alpha) \end{pmatrix} M(x,\alpha) = \begin{pmatrix} 0_{2\times 10} & I_{2,2} \end{pmatrix}. \tag{10.41}$$

Then S_z can be obtained from

$$S_z(x,\alpha) + W^T(2A_{rz}(x,\alpha) \odot I_n)Q = 0_{2,2}. \tag{10.42}$$

The defining system for the double Hopf curve is then $G(x,\alpha) = 0$, $S_{i_1,j_1}(x,\alpha) = 0$, $S_{i_2,j_2}(x,\alpha) = 0$, where (i_1,j_1), (i_2,j_2) are two index pairs chosen as in §5.3.

To compute the derivatives of A_r in a point (x,α) it is necessary to make a local choice for the bases of the invariant subspaces in points close to (x,α). (This is similar to choosing one of two possible directions for following a curve.) Essentially this is done by projecting a given orthonormal base of the invariant space at (x,α) onto the invariant space at a nearby point (x_1,α_1) and reorthonormalizing this base by a Gram–Schmidt procedure. With this choice we can compare the reduced Jacobian matrices in two nearby points and hence compute finite differences.

For the detection and computation of the one-to-one resonant point one proceeds as in §5.3.4 except for the subspace reduction. In fact, one computes the 10×2 matrix $Q_1(x,\alpha)$ and the 2×2 matrix $S_1(x,\alpha)$ by solving

$$M(x,\alpha)\begin{pmatrix} Q_1(x,\alpha) \\ S_1(x,\alpha) \end{pmatrix} = \begin{pmatrix} Q(x,\alpha) \\ S(x,\alpha) \end{pmatrix} \tag{10.43}$$

and uses $\det(S_1(x,\alpha))$ as the test function for one-to-one resonance. The resonant point is computed by using $\det(S_1(x,\alpha)) = 0$ in a root-finding routine.

10.3 Notes and Further Reading

1. Subspace iteration with the Cayley transformed Jacobian matrix as described in §10.2 is an expensive but robust procedure for the numerical detection, computation, and continuation of bifurcations. Other methods may be more efficient, but we do not expect them to be more robust. However, this is still an active area of research, so it is hard to draw conclusions.

2. There is a vast literature on the numerical continuation of equilibrium solutions to PDEs, including branching phenomena. An important part of the somewhat older literature deals with two- and three-dimensional PDEs using spectral methods or with a discretization on fixed grids. Structural mechanics and nonlinear eigenvalue problems are often considered. The continuation techniques are basically the same as for small systems, the new feature being the way in which the arising large sparse systems are solved. The stress therefore is on numerical linear algebra techniques. See [162], [163], [49] (multigrid), [3], [4] (conjugate gradients, GMRES), [6].

The more recent work is quite impressive, dealing with variable grids, finite element discretizations, and symbolic computation of the higher-order derivatives of the steady state equations. See in particular [60], [61], and the references therein. The conference proceedings [134] are also recommended to get an overview of current research on continuation and bifurcation methods in fluid mechanics.

3. There is an even more extensive literature on the computation of eigenvalues and eigenspaces of large, usually sparse matrices, partly motivated by bifurcation problems. In particular we draw the reader's attention to the freely available software package ARPACK [172], which is based on the implicitly restarted Arnoldi method [222]. Among other things, ARPACK allows us to compute a prescribed number of eigenvalues with largest real part for a large matrix, requiring only the possibility to compute matrix–vector products. This is made possible by a reverse communication interface with the driving program. In [171] ARPACK is used (with some modifications) to perform the linear stability analysis for the discretized Navier–Stokes equations (incompressible fluid flow).

4. It is clear from §10.2 that the numerical continuation of invariant linear subspaces of a given space and smooth decomposition of matrices are important topics in the study of bifurcations of large systems. See [159] for a discussion. This has also attracted attention in the study of the continuation of connecting orbits, i.e., homoclinic and heteroclinic orbits. In this case the stable and unstable invariant spaces at the end points of the orbits are important. See [71] and [77].

5. In recent years the recursive projection method (RPM) for the continuation of possibly nonstable equilibria (see [141], [220], and further references therein) has received much attention. The starting idea is the observation that time integration is sometimes easily available in code and will converge to a nearby stable equilibrium. So time stepping can be used as a corrector in a continuation procedure. Obviously this fails if the equilibrium is unstable. The idea of the RPM is to keep track of a base for a low-dimensional subspace that contains the eigenvectors corresponding to the critical eigenvalues near the imaginary axis or in the positive half-plane and perform Newton iterations in this subspace alone; time stepping leads to convergence in the other, stable directions. A fun-

damental advantage of this method is that some information on the bifurcation behavior (eigenvalues crossing the imaginary axis) can be obtained cheaply as a byproduct of the computation. We do not know how reliable this is in more complicated cases like BT, ZH, or DH.

6. Ideas from the RPM method have been extended to the computation and even numerical bifurcation analysis of periodic orbits of large systems; the resulting methods are called Newton–Picard methods. See [176] and [177].

10.4 Exercises

1. Prove by a direct computation that the approximation in (10.8) has third-order accuracy in $h = \max_{i=1,\dots,N} h_i$.

2. In the implicit Euler method for time integration (10.17) is replaced by

$$\mathcal{M}(Y^1 - Y^0) = (\Delta t)(\mathcal{G}(Y^1, \alpha)).$$

   The equivalent ODE method (cf. §10.1.2) is correct to first order with error

$$\frac{1}{2}(\Delta t)^2 (G_x G)$$

   in the second-order term. Give an explicit formula to compute this term numerically using only one solve with the matrix M^e in (10.33).

3. Consider the combustion model in §1.3. Use CONTENT to compute the equilibrium branches in Figure 1.13. Is there a value of μ for which the equilibrium branch has a hysteresis point in the sense of Chapter 6?

4. Consider the Bratu problem in §1.4 with boundary conditions $u(0) = u(1) = 0$. Compute a branch of equilibrium solutions starting with $\lambda = 0$, $u \equiv 0$. Compute the turning point on the curve for discretizations with 50, 100, 200, and 400 mesh points. Do you see a convergence in the λ-values for these turning points?

5. Repeat the computation of an equilibrium curve in §10.1.6 with $\beta = 2$ instead of $\beta = 2.35$. Find a value of D for which the tubular reactor admits five different equilibrium solutions.

6. Repeat the computation of an orbit in §10.1.6 with $D = 0.166$ instead of $D = 0.165$. Do you see a striking difference? What bifurcation behavior do you infer?

7. Consider the following reaction-diffusion equation:

$$\begin{array}{ll} u_t = u_{xx} + u(1-u), & 0 < x < 1, \\ u_x(0,t) = 0, & x = 0, \\ u_x(1,t) = 0, & x = 1, \end{array} \tag{10.44}$$

   where $u(x,t)$ is defined for $x \in [0,1]$, $t \in [0,\infty]$. The initial profile is $u(x,0) = f(x)$, where f is a given function, $f(x) \in [0,1]$ for $x \in [0,1]$. This is (a special case of)

Fisher's equation. It is a continuous analogue of the second population model in §1.2.1. Compute the time evolution of $u(x,t)$ for several initial profiles $f(x)$. What asymptotic behavior do you observe? Do you see a relation with §1.2.1?

Bibliography

[1] J. P. Abbott, *An efficient algorithm for the determination of certain bifurcation points,* J. Comput. Appl. Math. 4 (1978) 19–27.

[2] K. Alhumaizi and R. Aris, *Surveying a Dynamical System: A Study of the Gray-Scott Reaction in a Two-Phase Reactor,* Pitman Res. Notes in Math. 341, Addison–Wesley Longman Ltd., Harlow, UK, 1995.

[3] E. L. Allgower, C.-S. Chien, and K. Georg, *Large sparse continuation problems,* J. Comput. Appl. Math. 26 (1989) 3–21.

[4] E. L. Allgower, C.-S. Chien, K. Georg, and C.-F. Wang, *Conjugate gradient methods for continuation problems,* J. Comput. Appl. Math. 38 (1991) 1–16.

[5] E. L. Allgower, K. Böhmer, and Z. Mei, *A complete bifurcation scenario for the 2-d nonlinear Laplacian with Neumann boundary conditions on the unit square,* in [217], pp. 1–18.

[6] E. L. Allgower and K. Georg, *Numerical Continuation Methods: An Introduction,* Springer Ser. Comput. Math. 13, Springer, New York, 1990.

[7] E. Allgower, K. Böhmer, and M. Golubitsky (eds.), *Bifurcation and Symmetry,* ISNM 104, Birkhäuser Verlag, 1992.

[8] E. Allgower, K. Georg, and R. Miranda (eds.), *Exploiting Symmetry in Applied and Numerical Analysis,* Lectures in Appl. Math. 29, AMS, Providence, 1993.

[9] E. L. Allgower and K. Georg, *Numerical path following,* in [58], 3–208.

[10] J. M. Anderson, *Eutrophication of Lakes,* in Toward Global Equilibrium (eds. D. L. Maedows and H. Donella), Wright-Allen Press, Cambridge, MA, 1973.

[11] E. Anderson, Z. Bai, C. Bischof, J. Demmel, J. Dongarra, J. Du Croz, A. Greenbaum, S. Hammarling, A. Mckenney, S. Ostrouchov, and D. Sorensen, *LAPACK User's Guide,* SIAM, Philadelphia, 1992.

[12] V. I. Arnold, *On matrices depending on parameters,* Russian Math. Surveys 26 (1971) 29–43.

[13] V. I. Arnold, *Geometrical Methods in the Theory of Ordinary Differential Equations,* Springer-Verlag, New York, 1983.

[14] V. I. Arnold, A. Varchenko, and S. Goussein-Zadé, *Singularités des Applications Différentiables,* 1re partie, Editions Mir, Moscou, 1986.

[15] V. I. Arnold, *The Theory of Singularities and Its Applications,* The Press Syndicate of the University of Cambridge, Cambridge, 1991.

[16] V. I. Arnold, V. V. Goryunov, O. V. Lyashko, and V. A. Vasil'ev, *Singularity Theory* I, Springer-Verlag, Berlin, 1998.

[17] U. M. Ascher, J. Christiansen, and R. D. Russell, *A collocation solver for mixed order systems of boundary value problems,* Math. Comput. 33 (1979) 659–679.

[18] P. Ashwin, K. Böhmer, and Z. Mei, *A numerical Lyapunov-Schmidt method for finitely determined systems,* in [8], pp. 49–69.

[19] P. Ashwin and Z. Mei, *Normal form for Hopf bifurcation of partial differential equations on the square,* Nonlinearity 8 (1995) 715–734.

[20] P. Ashwin, K. Böhmer, and Z. Mei, *A numerical Lyapunov-Schmidt method with applications to Hopf bifurcation on a square,* Math. Comp. 64 (1995) 649.

[21] A. Back, J. Guckenheimer, M. Myers, F. Wicklin, and P. Worfolk, DSTOOL: *Computer assisted exploration of dynamical systems,* AMS Notices 39 (1992) 303–309.

[22] G. Bader and P. Kunkel, *Continuation and Gauss-Newton techniques,* in Computational Ordinary Differential Equations (eds. J. R. Cash and I. Gladwell), Clarendon Press, Oxford, 1992, pp. 225–236.

[23] C. Baker, J. Pryce, G. Vanden Berghe, and G. Monegato (eds.), *Ordinary differential and integral equations, Series "Numerical analysis in the 20th century," Vol.* 6, to appear as a special volume of J. Comput. Appl. Math., 2000.

[24] R. E. Bank, *PLTMG: A Software Package for Solving Elliptic Partial Differential Equations, Users' Guide* 8.0, SIAM, Philadelphia, 1998.

[25] N. W. Bazley and G. C. Wake, *The disappearance of criticality in the theory of thermal ignition,* Z. Angew. Math. Phys. 29 (1979) 971–976.

[26] A. D. Bazykin, *Mathematical biophysics of interacting populations,* Nauka, Moscow, 1985 [in Russian].

[27] A. D. Bazykin, *Nonlinear Dynamics of Interacting Populations,* World Scientific Series on Nonlinear Science, World Scientific, Singapore, 1998.

[28] W.-J. Beyn, *Defining equations for singular solutions and numerical applications,* in [162], pp. 42–46.

[29] W.-J. Beyn, *Global bifurcations and their numerical computation,* in [207], pp. 169–181.

[30] W.-J. Beyn, *The numerical computation of connecting orbits in dynamical systems,* IMA J. Numer. Anal. 9 (1990) 379–405.

[31] W.-J. Beyn, *Numerical methods for dynamical systems,* in Advances in Numerical Analysis, Vol. I: Nonlinear Partial Differential Equations and Dynamical Systems (ed. W. Light), Clarendon Press, Oxford, 1991, pp. 175–236.

[32] W.-J. Beyn and W. Kless, *Numerical Taylor expansions of invariant manifolds in large dynamical systems,* Schwerpunktprogramm der DFG: Dynamik: Analysis, effiziente Simulation und Ergodentheorie, Preprint 6/96.

[33] W.-J. Beyn, A. Champneys, E. Doedel, W. Govaerts, Yu. A. Kuznetsov, and B. Sandstede, *Numerical continuation and computation of normal forms,* in [38], Vol. III.

[34] K. Böhmer, *On a numerical Lyapunov-Schmidt method for operator equations,* Computing 53 (1993) 237–269.

[35] K. Böhmer, W. Govaerts, and V. Janovský, *Numerical detection of symmetry breaking bifurcation points with nonlinear degeneracies,* Math. Comput. 68 (1999), 1097–1108.

[36] R. I. Bogdanov, *Versal deformations of a singular point on the plane in the case of zero eigenvalues,* Funct. Anal. Appl. 9 (1975) 144–145.

[37] F. Brezzi, J. Rappaz, and P. A. Raviart, *Finite dimensional approximations of nonlinear problems, Part* III, *Simple bifurcation points,* Numer. Math. 38 (1981) 1–30.

[38] H. Broer and F. Takens (eds.), *Handbook of Dynamical Systems, Vol.* I (eds. B. Hasselblatt and A. Katok): *Ergodic Theory; Vol.* II (eds. H. Broer and F. Takens): *Bifurcation Theory; Vol.* III (eds. B. Fiedler, G. Iooss, and N. Kopell): *Towards Applications,* to be published by Elsevier.

[39] J. W. Bruce and P. J. Giblin, *Curves and Singularities,* Cambridge University Press, Cambridge, second edition, 1992.

[40] F. Buchholtz, J. Golowasch, I. R. Epstein, and E. Marder, *Mathematical Model of an Identified Stomatogastric Ganglion Neuron,* J. Neurophysiology 67 (1992) 332–340.

[41] J. D. Buckmaster and G. S. Ludford, *Theory of Laminar Flames,* Cambridge University Press, Cambridge, 1982.

[42] M. G. Bulmer, *Theoretical Evolutionary Ecology,* Oxford University Press, Oxford, 1994.

[43] J. R. Bunch, *The weak and strong stability of algorithms in numerical linear algebra,* Linear Algebra Appl. 88 (1987) 49–66.

[44] J. C. Butcher, *A stability property of implicit Runge–Kutta methods,* BIT 15 (1975) 358–361.

[45] V. I. Bykov, G. S. Yablonski, and V. F. Kim, *On the simple model of kinetic self-oscillations in catalytic reaction of CO oxidation,* Dokl. Akad. Nauk SSSR 242(3) (1978) 637–639 [in Russian].

[46] G. Caloz and J. Rappaz, *Numerical analysis for nonlinear and bifurcation problems,* in [58], 487–637.

[47] A. R. Champneys and Yu. A. Kuznetsov, *Numerical detection and continuation of codimension-two homoclinic bifurcations,* Internat. J. Bifur. Chaos 4 (1994) 785–822.

[48] A. Champneys, Yu. A. Kuznetsov, and B. Standstede, *A numerical toolbox for homoclinic bifurication analysis,* Internat. J. Bifur. Chaos 6 (1996) 867–887.

[49] T. F. Chan and H. B. Keller, *Arclength continuation and multigrid techniques for nonlinear eigenvalue problems,* SIAM J. Sci. Statist. Comput. 3 (1982) 173–194.

[50] T. F. Chan, *Deflation techniques and block elimination algorithms for solving bordered singular systems,* SIAM J. Sci. Statist. Comput. 5 (1984) 121–134.

[51] T. F. Chan, *Newton-like pseudo-arclength methods for computing simple turning points,* SIAM J. Sci. Statist. Comput. 5 (1984) 135–148.

[52] T. F. Chan, *Techniques for large sparse systems arising from continuation methods,* in [162], pp. 116–128.

[53] T. F. Chan and D. C. Resasco, *Generalized deflated block elimination,* SIAM J. Numer. Anal. 23 (1986) 913–924.

[54] C. S. Chien, C.-L. Shen, and Z. Mei, *Numerical continuation at double bifurcation points of a reaction-diffusion problem,* Internat. J. Bifur. Chaos 8 (1997) 117–139.

[55] S.-N. Chow and J. K. Hale, *Methods of Bifurcation Theory,* Grundlehren der mathematischen Wissenschaften 251, Springer-Verlag, Berlin, New York, 1982.

[56] K. N. Christodoulou and L. E. Scriven, *Finding leading modes of a viscous free surface flow: An asymmetric generalized eigenvalue problem,* J. Sci. Comput. 3 (1988) 355–406.

[57] K. Chu, W. Govaerts, and A. Spence, *Matrices with rank deficiency two in eigenvalue problems and dynamical systems,* SIAM J. Numer. Anal. 31 (1994) 524–539.

[58] P. G. Ciarlet and J. L. Lions (eds.), *Handbook of Numerical Analysis, Vol.* V, *Techniques of Scientific Computing* (Part 2), Elsevier, Amsterdam, 1997.

[59] K. A. Cliffe, T. J. Garratt, and A. Spence, *Eigenvalues of the discretized Navier-Stokes equation with application to the detection of Hopf bifurcations,* Adv. Comput. Math. 1 (1993) 337–356.

[60] K. A. Cliffe and S. J. Tavener, *Marangoni-Bénard convection with a deformable free surface,* J. Comp. Phys. 145 (1998) 193–227.

[61] K. A. Cliffe, S. J. Tavener, and H. Wilke, *Convectve effects on a propagating reaction front,* Phys. Fluids 10 (1998) 730–741.

[62] E. A. Coddington and N. Levinson, *Theory of Ordinary Differential Equations,* McGraw–Hill, New York, 1955.

[63] P. Coullet and E. Spiegel, *Amplitude equations for systems with competing instabilities,* SIAM J. Appl. Math. 43 (1983) 776–821.

[64] J. Damon, *Topological equivalence of bifurcation problems,* Nonlinearity 1 (1988) 311–331.

[65] G. Dangelmayr and D. Armbruster, *Classification of* $\mathbb{Z}$ *(2)-equivariant imperfect bifurcations with corank* 2, Proc. London Math. Soc 46 (1983) 517–546.

[66] C. de Boor and B. Swartz, *Collocation at Gaussian points,* SIAM J. Numer. Anal. 10 (1973) 582–606.

[67] B. De Dier, D. Roose, and P. Van Rompay, *Interaction between fold and Hopf curves leads to new bifurcation phenomena,* J. Comput. Appl. Math. 26 (1989) 171–186.

[68] D. W. Decker and H. B. Keller, *Multiple limit point bifurcation,* J. Math. Anal. Appl. 75 (1980) 417–430.

[69] M. Dellnitz and B. Werner, *Computational methods for bifurcation problems with symmetries—with special attention to steady state and Hopf bifurcation points,* J. Comput. Appl. Math. 26 (1989) 97–123.

[70] J. W. Demmel, *Applied Numerical Linear Algebra,* SIAM, Philadelphia, 1997.

[71] J. W. Demmel, L. Dieci, and M. J. Friedman, *Continuing connecting orbits via an improved algorithm for continuing invariant subspaces,* SIAM J. Sci. Comput., to appear.

[72] J. W. Demmel and A. Edelman, *The dimension of matrices (matrix pencils) with given Jordan (Kronecker) canonical form,* Linear Algebra Appl. 230 (1995) 61–87.

[73] C. Den Heyer and W. C. Rheinboldt, *On steplength algorithms for a class of continuation methods,* SIAM J. Numer. Anal. 18 (1981) 925–947.

[74] P. Deuflhard, B. Fiedler, and P. Kunkel, *Efficient numerical pathfollowing beyond critical points,* SIAM J. Numer. Anal. 24 (1987) 912–927.

[75] P. Deuflhard and A. Hohmann, *Numerische Mathematik,* W. de Gruyter & Co., Berlin, 1991.

[76] P. Deuflhard and F. Bornemann, *Numerische Mathematik* II, W. de Gruyter & Co., Berlin, 1994.

[77] L. Dieci and T. Eirola, *On smooth decomposition of matrices,* SIAM J. Matrix Anal. Appl. 20 (1999) 800–819.

[78] E. J. Doedel, *The construction of finite difference approximations to ordinary differential equations,* SIAM J. Numer. Anal. 15 (1978) 450–465.

[79] E. J. Doedel, *Finite difference collocation methods for nonlinear two point boundary value problems,* SIAM J. Numer. Anal. 16 (1979) 173–185.

[80] E. Doedel and J. Kernévez, AUTO: *Software for Continuation and Bifurcation Problems in Ordinary Differential Equations*, Report, Applied Mathematics, Caltech, Pasadena, CA, 1986.

[81] E. Doedel and M. Friedman, *Numerical computation of heteroclinic orbits,* J. Comput. Appl. Math. 26 (1989) 159–170.

[82] E. Doedel, H. Keller, and J. Kernévez, *Numerical analysis and control of bifurcation problems:* (I) *Bifurcation in finite dimensions,* Internat. J. Bifur. Chaos 1 (1991) 493–520.

[83] E. Doedel, H. Keller and J. Kernévez, *Numerical analysis and control of bifurcation problems:* (II) *Bifurcation in infinite dimensions,* Internat. J. Bifur. Chaos 1 (1991) 745–772.

[84] E. Doedel and X. Wang, AUTO94: *Software for Continuation and Bifurcation Problems in Ordinary Differential Equations,* Report, Applied Mathematics, Caltech, Pasadena, CA, 1994.

[85] E. J. Doedel, A. R. Champneys, T. F. Fairgrieve, Yu. A. Kuznetsov, B. Sandstede, and X. J. Wang, AUTO97: *Continuation and Bifurcation Software for Ordinary Differential Equations (with HomCont), User's Guide,* Computer Science Department, Concordia University, Montreal, Canada, 1997, ftp.cs.concordia/pub/doedel/auto.

[86] A. du Plessis and T. Wall, *The Geometry of Topological Stability,* Clarendon Press, Oxford, 1995.

[87] C. Elphick, E. Tirapegui, M. Brachet, P. Coullet, and G. Iooss, *A simple global characterization for normal forms of singular vector fields,* Phys. D 32 (1987) 95–127.

[88] J. P. Fink and W. C. Rheinboldt, *A geometric framework for the numerical study of singular points,* SIAM J. Numer. Anal. 24 (1987) 618–633.

[89] D. K. Fadeev and V. N. Fadeeva, *Computational Methods of Linear Algebra,* Freeman, San Francisco, 1963.

[90] A. Fässler and E. Stiefel, *Group Theoretical Methods and Their Applications,* Birkhäuser, Boston, 1992.

[91] G. Feichtinger and F. Wirl, *On the stability and potential cyclicity of corruption in governments subject to popularity constraints,* Math. Social Sci. 28 (1994) 113–131.

[92] M. Field, *Lectures on Bifurcations, Dynamics and Symmetry,* Pitman Res. Notes in Math. 356, Addison–Wesley Longman Ltd., Harlow, UK, 1996.

[93] L. J. Fradkin and G. C. Wake, *The critical explosion parameter in the theory of thermal ignition,* J. Inst. Math. Appl. 20 (1977) 471–484.

[94] M. Friedman and E. Doedel, *Numerical computation and continuation of invariant manifolds connecting fixed points,* SIAM J. Numer. Anal. 28 (1991) 789–808.

[95] M. Friedman and E. Doedel, *Computational methods for global analysis of homoclinic and heteroclinic orbits: A case study,* J. Dynam. Differential Equations 5 (1993) 37–57.

[96] A. T. Fuller, *Conditions for a matrix to have only characteristic roots with negative real parts,* J. Math. Anal. Appl. 23 (1968) 71–98.

[97] F. R. Gantmacher, *The Theory of Matrices, Vols.* I, II, Chelsea Publishing Company, New York, 1959 [translated from Russian].

[98] F. R. Gantmacher, *Applications of the Theory of Matrices,* John Wiley and Sons, New York, 1959 [translated from Russian].

[99] K. Gatermann, *Computation of bifurcation graphs,* AMS Lectures Appl. Math. 29 (1993) 187–201.

[100] K. Gatermann and A. Hohmann, *Symbolic exploitation of symmetry in numerical pathfollowing,* Impact Comput. Sci. Engrg. 3 (1991) 330–365.

[101] K. Gatermann and A. Hohmann, SYMCON 2.0, available by anonymous ftp from elib at ZIB.

[102] K. Gatermann and B. Werner, *Group theoretical mode interactions with different symmetries,* Internat. J. Bifur. Chaos 4 (1994) 177–191.

[103] K. Gatermann and B. Werner, *Secondary Hopf bifurcation caused by steady-state mode interaction,* Fields Inst. Commun. 5 (1996) 209–224.

[104] C. G. Gibson, *Singular Points of Smooth Mappings,* Res. Notes in Math. 25, Pitman, Boston, 1979.

[105] J. Golowasch and E. Marder, *Ionic currents of the lateral pyloric neuron of the stomatogastric ganglion of the crab,* J. Neurophysiology 67 (1992) 318–331.

[106] G. Golub and C. Van Loan, *Matrix Computations,* John Hopkins, Baltimore, third edition, 1996.

[107] M. Golubitsky and W. F. Langford, *Classification and unfoldings of degenerate Hopf bifurcations,* J. Differential Equations 41 (1981) 375–415.

[108] M. Golubitsky and V. Guillemin, *Stable Mappings and Their Singularities,* Grad. Texts in Math. 14, Springer, New York, 1973.

[109] M. Golubitsky and D. G. Schaeffer, *Singularities and Groups in Bifurcation Theory, Vol.* I, Springer-Verlag, Berlin, New York, 1985.

[110] M. Golubitsky, I. Stewart, and D. G. Schaeffer, *Singularities and Groups in Bifurcation Theory, Vol.* II, Springer-Verlag, Berlin, New York, 1988.

[111] W. Govaerts and J. D. Pryce, *A singular value inequality for block matrices,* Linear Algebra Appl. 125 (1989) 141–148.

[112] W. Govaerts, *Stable solvers and block elimination for bordered systems,* SIAM J. Matrix Anal. Appl. 12 (1991) 469–483.

[113] W. Govaerts and J. D. Pryce, *Mixed block elimination for linear systems with wider borders,* IMA J. Numer. Anal. 13 (1993) 161–180.

[114] W. Govaerts, *Computation of Takens–Bogdanov type bifurcations with arbitrary codimension,* SIAM J. Numer. Anal. 30 (1993) 1121–1133.

[115] W. Govaerts, *Solution of bordered singular systems in numerical continuation and bifurcation,* J. Comput. Appl. Math. 50 (1994) 339–347.

[116] W. Govaerts, *Bordered matrices and singularities of large nonlinear systems*, Internat. J. Bifur. Chaos 5 (1995) 243–250.

[117] W. Govaerts and A. Spence, *Detection of Hopf points by counting sectors in the complex plane,* Numer. Math. 75 (1996) 43–58.

[118] W. Govaerts, *Computation of singularities in large nonlinear systems*, SIAM J. Numer. Anal. 34 (1997) 867–880.

[119] W. Govaerts, J. Guckenheimer, and A. Khibnik, *Defining functions for multiple Hopf bifurcations*, SIAM J. Numer. Anal. 34 (1997) 1269–1288.

[120] W. Govaerts and B. Sijnave, *Rank-deficient matrices as a computational tool,* Numer. Linear. Algebra Appl. 4 (1997) 443–458.

[121] W. Govaerts, Yu. A. Kuznetsov, and B. Sijnave, *Implementation of Hopf and double Hopf continuation using bordering methods,* ACM Trans. Math. Software, 24 (1998) 418–436.

[122] P. Gray and S. K. Scott, *Chemical Oscillations and Instabilities: Nonlinear Chemical Kinetics,* Oxford University Press, Oxford, 1990.

[123] A. Griewank and G. W. Reddien, *The calculation of Hopf points by a direct method,* IMA J. Numer. Anal. 3 (1983) 295–303.

[124] A. Griewank and G. W. Reddien, *Characterization and computation of generalized turning points,* SIAM J. Numer. Anal. 21 (1984) 176–185.

[125] A. Griewank and G. W. Reddien, *The approximation of generalized turning points by projection methods with superconvergence to the critical parameter,* Numer. Math. 48 (1986) 591–606.

[126] A. Griewank and G. W. Reddien, *Computation of cusp singularities for operator equations and their discretizations,* J. Comput. Appl. Math. 26 (1989) 133–153.

[127] J. Guckenheimer and P. Holmes, *Nonlinear Oscillations, Dynamical Systems, and Bifurcations of Vector Fields,* Appl. Math. Sci. 42, Springer, New York, 1983.

[128] J. Guckenheimer, *Multiple bifurcation problems for chemical reactors,* Phys. D 20 (1986) 1–20.

[129] J. Guckenheimer, S. Gueron, and R. M. Harris-Warrick, *Mapping the dynamics of a bursting neuron,* Philos. Trans. Roy. Soc. London Ser. B. 341 (1993) 345–359.

[130] J. Guckenheimer, M. Myers, and B. Sturmfels, *Computing Hopf bifurcations* I, SIAM J. Numer. Anal. 34 (1997) 1–21.

[131] J. Guckenheimer and M. Myers, *Computing Hopf bifurcations* II: *Three examples from neurophysiology,* SIAM J. Sci. Statist. Comput. 17 (1996) 1275–1301.

[132] B. Hassard, N. D. Kazarinoff, and Y. H. Wan, *Theory and Applications of Hopf Bifurcation,* Cambridge University Press, London, 1981.

[133] R. Heinemann and A. Poore, *Multiplicity, stability and oscillatory dynamics of the tubular reactor,* Chemi. Engrg. Sci. 36 (1981) 1411–1419.

[134] D. Henry, H. Dijkstra, and L. Tuckermann (eds.), *Continuation Methods in Fluid Dynamics,* Proceedings of the meeting in Aussois, 6–9 September 1998, Vieweg series Notes on Numerical Fluid Dynamics, 1999.

[135] N. J. Higham, *Accuracy and Stability of Numerical Algorithms,* SIAM, Philadelphia, 1996.

[136] M. W. Hirsch and S. Smale, *Differential Equations, Dynamical Systems and Linear Algebra,* Academic Press, New York, 1974.

[137] A. Hodgkin and A. Huxley, *A qualitative description of membrane current and its application to conduction and excitation in nerves,* J. Physiol. 117 (1952) 500–544.

[138] M. Holodniok and M. Kubiček, *New algorithms for the evaluation of complex bifurcation points in ordinary differential equations. A comparative numerical study,* Appl. Math. Comput. 15 (1984) 261–274.

[139] E. Hopf, *Abzweigung einer periodischen Lösung von einer stationären Lösung eines Differentialsystems,* Ber. Math. Phys. Kl. Säcks. Akad. Wiss. Leipzig 94 (1942) 3–22.

[140] K. Ikeda and K. Murota, *Bifurcation analysis of symmetric structures using block-diagonalization,* Comput. Methods Appl. Mech. Engrg. 86 (1991) 215–243.

[141] H. Jarausch and W. Mackens, *Solving large nonlinear systems of equations by an adaptive condensation process,* Numer. Math. 50 (1987) 633–653.

[142] V. Janovský and P. Plecháč, *Numerical applications of equivariant reduction techniques,* in [7], 203–213.

[143] V. Janovský and P. Plecháč, *Local numerical analysis of Hopf bifurcation,* SIAM J. Numer. Anal. 33 (1996) 1150–1186.

[144] A. D. Jepson, *Numerical Hopf Bifurcation,* Thesis, Part II, California Institute of Technology, Pasadena, CA, 1981.

[145] A. Jepson and A. Spence, *Singular points and their computation,* in [162], 195–209.

[146] A. Jepson and A. Spence, *The numerical solution of nonlinear equations having several parameters* I: *Scalar equations,* SIAM J. Numer. Anal. 22 (1985), 736–759.

[147] A. Jepson and A. Spence, *Folds in solutions of two-parameter systems and their calculation. Part* I, SIAM J. Numer. Anal. 22 (1985) 347–368.

[148] A. Jepson and A. Spence, *On a reduction process for nonlinear equations,* SIAM J. Math. Anal. 20 (1989) 39–56.

[149] A. Jepson, A. Spence, and K. Cliffe, *The numerical solution of nonlinear equations having several parameters. Part* III: *Equations with* $\mathbb{Z}_2$*-symmetry,* SIAM J. Numer. Anal. 28 (1991) 809–832.

[150] H. B. Keller, *Constructive methods for bifurcation and nonlinear eigenvalue problems,* in Computing Methods in Applied Sciences and Engineering (eds. R. Glowinsky and J.-L. Lions), Springer-Verlag, New York, 1977, 350–384

[151] H. B. Keller, *Numerical solution of bifurcation and nonlinear eigenvalue problems,* in Applications of Bifurcation Theory (ed. P. Rabinowitz), Academic Press, New York, 1977, 359–384.

[152] H. B. Keller, *The bordering algorithm and path following near singular points of higher nullity,* SIAM J. Sci. Statist. Comput. 4 (1983) 573–582.

[153] H. B. Keller, *Numerical Methods in Bifurcation Problems,* Springer-Verlag, Berlin, New York, 1987.

[154] A. I. Khibnik, V. I. Bykov, and G. S. Yablonski, 23 *phase portraits of the simplest catalytic oscillator,* J. Fiz. Khim. 61 (1987) 1388–1390 [in Russian].

[155] A. I. Khibnik, *LINLBF: A program for continuation and bifurcation analysis of equilibria up to codimension three,* in [207], 283–296.

[156] A. Khibnik, Yu. Kuznetsov, V. Levitin, and E. Nikolaev, *LOCBIF, Version* 2: *Interactive LOCal BIFurcation Analyzer,* CAN Expertise Centre, Amsterdam, 1993.

[157] A. Khibnik, Yu. Kuznetsov, V. Levitin, and E. Nikolaev, *Continuation techniques and interactive software for bifurcation analysis of ODEs and iterated maps,* Phys. D 62 (1993) 360–371.

[158] A. I. Khibnik and A. S. Kondrashov, *Three mechanisms of Red Queen Dynamics,* Proc. Roy. Soc. London Ser. B 264 (1997) 1049–1056.

[159] W. Kless, *Numerische Approximation lokal invarianter Mannigfaltigkeiten in grossen dynamischen Systemen,* Ph.D. Thesis, University of Bielefeld, 1997; Hans Jacob Verlag, Paderborn, 1997.

[160] M. Kubiček and M. Holodniok, *Numerical determination of bifurcation points in steady state and periodic solutions—numerical algorithms and examples,* in [162] 247–270.

[161] P. Kunkel, *A tree-based analysis of a family of augmented systems for the computation of singular points,* IMA J. Numer. Anal. 16 (1996) 501–527.

[162] T. Küpper, H. Mittelmann, and H. Weber (eds.), *Numerical Methods for Bifurcation Problems,* ISNM 70, Birkhäuser Verlag, 1984.

[163] T. Küpper, R. Seydel, and H. Troger (eds.), *Bifurcation: Analysis, Algorithms and Applications: Proceedings of the Dortmund Conference, Aug.* '86, ISNM 79, Birkhäuser, Basel, 1987.

[164] Yu. A. Kuznetsov, *Elements of Applied Bifurcation Theory,* Appl. Math. Sci. 112, 2nd ed., Springer-Verlag, New York, 1998.

[165] Yu. A. Kuznetsov and V. V. Levitin, CONTENT: *A Multiplatform Environment for Analyzing Dynamical Systems,* Dynamical Systems Laboratory, CWI, Amsterdam, 1995–1997 (ftp.cwi.nl/pub/CONTENT).

[166] Yu. A. Kuznetsov, V. V. Levitin, and A. R. Skovoroda, *Continuation of Stationary Solutions to Evolution Problems in* CONTENT, Report AM-R9611, Centrum voor Wiskunde en Informatica, Amsterdam, 1996.

[167] Yu. A. Kuznetsov, *Numerical normalization techniques for all codim* 2 *bifurications of equilibria in ODEs,* SIAM J. Numer. Anal. 36 (1999) 1104–1124.

[168] I. S. Labouriau, *Degenerate Hopf bifurcation and nerve impulse,* SIAM J. Math. Anal. 16 (1985) 1121–1133.

[169] P. Lancaster, *Theory of Matrices,* Academic Press, New York, London, 1969.

[170] M. Lentini and V. Pereyra, *An adaptive finite difference solver for nonlinear two point boundary problems with mild boundary layers,* SIAM J. Numer. Anal. 14 (1977) 91–111.

[171] R. Lehoucq and J. Scott, *Implicitly Restarted Arnoldi Methods and Eigenvalues of the Discretized Navier-Stokes Equations,* Technical Report RAL-TR-97-058, Rutherford Appleton Laboratory, Didcot, England, 1997.

[172] R. B. Lehoucq, D. C. Sorensen, and C. Yang, *ARPACK Users' Guide: Solution of Large-Scale Eigenvalue Problems with Implicitly Restarted Arnoldi Methods,* SIAM, Philadelphia, 1998.

[173] V. V. Levitin, *Computation of Functions and Their Derivatives in* CONTENT, Report AM-R9512, Centrum voor Wiskunde en Informatica, Amsterdam, 1995.

[174] J. D. Logan, *An Introduction to Nonlinear Partial Differential Equations,* J. Wiley Interscience, New York, 1994.

[175] E. Lorenz, *Irregularity: A fundamental property of the atmosphere,* Tellus 36A (1984) 98–110.

[176] K. Lust, *Numerical Bifurcation Analysis of Periodic Solutions of Partial Differential Equations,* Ph.D. Thesis, Katholieke Universiteit Leuven, 1997.

[177] K. Lust and D. Roose, *Computation and Bifurcation Analysis of Periodic Solutions of Large-Scale Systems,* IMA Preprint Series # 1536, IMA, University of Minnesota, Minneapolis, February 1998.

[178] I. G. Macdonald, *Symmetric Functions and Hall Polynomials,* Clarendon Press, Oxford, 1995.

[179] J. E. Marsden and M. McCracken, *The Hopf Bifurcation and Its Applications,* Springer-Verlag, New York, 1976.

[180] J. Martinet, *Déploiements Versels des Applications Différentiables et Classification des Applications Stables,* Lecture Notes in Math. 535, Springer, New York, 1975.

[181] J. N. Mather, *Stability of C^∞ mappings,* VI. *The nice dimensions,* Proceedings of Liverpool Singularities Symposium, Vol. I (ed. C. T. C. Wall), Lecture Notes in Math. 192, Springer, New York, pp. 207–253.

[182] R. M. M. Mattheij and J. Molenaar, *Ordinary Differential Equations in Theory and Practice,* J. Wiley, New York, 1996.

[183] K. Meerbergen, A. Spence, and D. Roose, *Shift-invert and Cayley transforms for detection of rightmost eigenvalues of nonsymmetric matrices,* BIT 34 (1994) 409–423.

[184] Z. Mei, *A numerical approximation for the simple bifurcation points,* Numer. Funct. Anal. Optim. 10 (1989) 383–400.

[185] Z. Mei, *Solution branches of a semilinear elliptic problem at corank 2 bifurcation points,* Manuscripta Math. 75 (1992) 395–409.

[186] Z. Mei, *Utilization of scaling laws and symmetries in the path following of a semilinear elliptic problem,* in [7], pp. 263–273.

[187] G. Moore and A. Spence, *The calculation of turning points of nonlinear equations,* SIAM J. Numer. Anal. 17 (1980) 567–576.

[188] G. Moore, *Some remarks on the deflated block elimination method,* in [163], pp. 222–234.

[189] K. Murota and K. Ikeda, *Computational use of group theory in bifurcation analysis of symmetric structures,* SIAM J. Sci. Statist. Comput. 12 (1991) 273–297.

[190] J. D. Murray, *Mathematical Biology,* Biomathematics Texts 19, Springer-Verlag, Berlin, 1989.

[191] R. Neubert, *Predictor-corrector techniques for detecting Hopf points,* Internat. J. Bifur. Chaos 3 (1993) 1311–1318.

[192] H. Osinga, *Computing Invariant Manifolds,* Ph.D. Thesis, University of Groningen, 1996.

[193] L. Perko, *Differential Equations and Dynamical Systems*, Texts Appl. Math. 7, Springer, New York, 1991.

[194] G. Poenisch and H. Schwetlick, *Computing turning points of curves implicitly defined by nonlinear equations depending on a parameter,* Computing 26 (1981) 107–121.

[195] G. Poenisch, *Computing hysteresis points of nonlinear equations depending on two parameters,* Computing 39 (1987) 1–17.

[196] H. Poincaré, *Les Méthodes nouvelles de la Mécanique Céleste*, Gauthier-Villars, Paris, 1892.

[197] I. Prigogine and R. Lefever, *Stability and self-organization in open systems*, in Membranes, Dissipative Structures, and Evolution (eds. G. Nicolis and R. Lefever), J. Wiley, New York, 1974.

[198] W. C. Rheinboldt, *Solution fields of nonlinear equations and continuation methods,* SIAM J. Numer. Anal. 17 (1980) 221–237.

[199] W. C. Rheinboldt, *Numerical analysis of continuation methods for nonlinear structural problems,* Comput. Structures 13 (1981) 103–114.

[200] W. C. Rheinboldt and J. V. Burkardt, *Algorithm* 596, *A program for a locally parametrized continuation process*, ACM Trans. Math. Software 9(2) (1983) 236–241.

[201] W. C. Rheinboldt, *Numerical Analysis of Parametrized Nonlinear Equations,* J. Wiley, New York, 1986.

[202] S. Rinaldi, G. Feichtinger, and F. Wirl, *Corruption dynamics in democratic societies,* Complexity 3, (1998) 53–64.

[203] J. Rinzel, *Integration and propagation of neuroelectric signals,* in Studies in Mathematical Biology 15 (ed. S. Levin), Mathematical Association of America, Washington, DC, 1978, 1–66.

[204] D. Roose and V. Hlavaček, *A direct method for the computation of Hopf bifurcation points,* SIAM J. Appl. Math. 45 (1985) 879–894.

[205] D. Roose and R. Piessens, *Numerical computation of nonsimple turning points and cusps,* Numer. Math. 46 (1985) 189–211.

[206] D. Roose and B. De Dier, *Numerical determination of an emanating branch of Hopf bifurcation points in a two-parameter problem,* SIAM J. Sci. Statist. Comput. 10 (1989) 671–685.

[207] D. Roose, A. Spence, and B. De Dier (eds.), *Continuation and Bifurcations: Numerical Techniques and Applications,* Nato ASI Series C: Vol. 313, Kluwer Academic Series, Dordrecht, 1990.

[208] S. Rose and Ackerman, *Corruption, A Study in Political Economy,* Academic Press, New York, 1978.

[209] R. Russell and J. Christiansen, *Adaptive mesh selection strategies for solving boundary value problems,* SIAM J. Numer. Anal. 15 (1978) 59–80.

[210] Y. Saad, *Numerical Methods for Large Eigenvalue Problems,* Manchester University Press, Manchester, 1992.

[211] J. Sanders, *Versal normal form computation and representation theory,* in Computer Algebra and Differential Equations (ed. E. Tournier), Cambridge University Press, Cambridge, 1994, pp. 185–210.

[212] A. Samarskii and E. Nikolaev, *Numerical Methods for Grid Equations, Vol.* I *and* II, Birkhäuser-Verlag, Basel, Boston, Berlin, 1989.

[213] D. H. Sattinger, *Topics in Stability and Bifurcation Theory,* Lecture Notes in Math. 309, Springer, New York, 1973.

[214] J.-P. Serre, *Linear Representations of Finite Groups,* Springer-Verlag, New York, 1977.

[215] R. Seydel, *Numerical computation of branch points in nonlinear equations,* Numer. Math. 33 (1979) 339–352.

[216] R. Seydel, *From Equilibrium to Chaos. Practical Bifurcation and Stability Analysis,* Elsevier, New York, 1988 (second edition, Springer-Verlag, Berlin, New York, 1994).

[217] R. Seydel, F. W. Schneider, T. Küpper, and H. Troger (eds.), *Bifurcation and Chaos: Analysis, Algorithms and Applications,* ISNM 97, Birkhäuser-Verlag, Berlin, New York, 1991.

[218] A. L. Shilnikov, G. Nicolis, and C. Nicolis, *Bifurcation and predictability analysis of a low-order atmospheric circulation model,* Internat. J. Bifur. Chaos 5 (1995) 1701–1711.

[219] A. Shleifer and R. W. Vishny, *Corruption,* Quart. J. Econ. 108 (1993) 599–617.

[220] G. M. Shroff and H. B. Keller, *Stabilisation of unstable procedures: The recursive projection method,* SIAM J. Numer. Anal. 30 (1993) 1099–1120.

[221] J. Smoller, *Shock Waves and Reaction-Diffusion Equations,* Springer-Verlag, Berlin, New York, second edition, 1994.

[222] D. C. Sorensen, *Implicit application of polynomial filters in a k-step Arnoldi method,* SIAM J. Matrix Anal. Appl. 13 (1992) 357–385.

[223] A. Spence and B. Werner, *Non-simple turning points and cups,* IMA J. Numer. Anal. 2 (1982) 413–427.

[224] A. Spence, K. A. Cliffe, and A. D. Jepson, *A note on the calculation of paths of Hopf bifurcations,* J. Comput. Appl. Math. 26 (1989) 125–131.

[225] C. Stéphanos, *Sur une extension du calcul des substitutions linéaires,* J. Math. Pures Appl. 6 (1900) 73–128.

[226] G. W. Stewart, *Introduction to Matrix Computations,* Academic Press, New York, 1973.

[227] P. Stork and B. Werner, *Symmetry adapted block diagonalization in equivariant steady state bifurcation problems and its numerical applications,* Adv. Math. (China) 20(4) (1991) 455–487.

[228] A. M. Stuart and A. R. Humphries, *Dynamical Systems and Numerical Analysis,* Cambridge University Press, Cambridge, 1996.

[229] F. Takens, *Singularities of vector fields,* Publ. Math. IHES 43 (1974) 47–100.

[230] L. N. Trefethen and D. Bau, III, *Numerical Linear Algebra,* SIAM, Philadelphia, 1997.

[231] A. Tveito and R. Winther, *Introduction to Partial Differential Equations,* Texts in Appl. Math. 29, Springer, New York, 1998.

[232] A. Uppal, W. H. Ray, and A. Poore, *The classification of the dynamic behavior of continuous stirred tank reactors—Influence of reactor residence time,* Chem. Engrg. Sci. 31 (1976) 205–214.

[233] A. Vanderbauwhede, *Local Bifurcation and Symmetry,* Habilitation Thesis, University of Gent, Belgium, 1980.

[234] A. Vanderbauwhede, *Local Bifurcation and Symmetry,* Res. Notes Math. 75, Pitman, Boston, 1982.

[235] S. van Gils, M. Krupa, and W. Langford, *Hopf bifurcation with non-semisimple* 1:1 *resonance,* Nonlinearity 3 (1990) 825–850.

[236] W. C. Waterhouse, *The codimension of singular matrix pairs,* Linear Algebra Appl. 57 (1984) 227–245.

[237] B. Werner, *Computation of Hopf bifurcation with bordered matrices,* SIAM J. Numer. Anal. 33 (1996) 435–455.

[238] B. Werner and A. Spence, *The computation of symmetry-breaking bifurcation points,* SIAM J. Numer. Anal. 21 (1984) 388–399.

[239] B. Werner, *The numerical analysis of bifurcation problems with symmetries based on bordered Jacobians,* in Exploiting Symmetry in Applied and Numerical Analysis (eds. E. L. Allgower, K. Georg, and R. Miranda), Lectures Appl. Math. 29, AMS, Providence, 1993, pp. 443–457.

[240] S. Wiggins, *Introduction to Applied Nonlinear Dynamical Systems and Chaos,* Texts in Appl. Math. 2, Springer, New York, 1990.

[241] J. H. Wilkinson, *The Algebraic Eigenvalue Problem,* Clarendon Press, Oxford, 1965.

[242] F. A. Williams, *Combustion Theory,* Combustion Science and Engineering Series, Addison–Wesley, Reading, MA, second edition, 1985.

[243] R. W. Wittenberg and Ph. Holmes, *The limited effectiveness of normal forms: A critical review and extension of local bifurcation studies of the Brusselator PDE,* Phys. D 100 (1997) 1–40.

[244] H. Xu, V. Janovský, and B. Werner, *Numerical computation of degenerate Hopf bifurcation points,* Z. Angew. Math. Mech. 78 (1998) 12, 807–821.

[245] L. Yang and B. Werner, *Analysis of Takens-Bogdanov bifurcation by characteristic functions,* Nonlinear Anal. 26 (1996) 363–381.

[246] Z.-H. Yang and H. B. Keller, *A direct method for computing higher order folds,* SIAM J. Sci. Statist. Comput. 7 (1986) 351–361.

[247] Y. V. Zeldovich and U. A. Zisin, *On the theory of thermal stress. Flow in an exothermic stirred reactor,* II. *Study of heat loss in a flow reactor,* J. Tech. Phys. 11(6) (1941) 501–508 [in Russian].

Index